Springer Water

Series Editor

Andrey Kostianoy, Russian Academy of Sciences, P. P. Shirshov Institute of Oceanology, Moscow, Russia

The book series Springer Water comprises a broad portfolio of multi- and interdisciplinary scientific books, aiming at researchers, students, and everyone interested in water-related science. The series includes peer-reviewed monographs, edited volumes, textbooks, and conference proceedings. Its volumes combine all kinds of water-related research areas, such as: the movement, distribution and quality of freshwater; water resources; the quality and pollution of water and its influence on health; the water industry including drinking water, wastewater, and desalination services and technologies; water history; as well as water management and the governmental, political, developmental, and ethical aspects of water.

More information about this series at http://www.springer.com/series/13419

Abdelazim Negm · Ahmed Elkhouly
Editors

Groundwater in Egypt's Deserts

Springer

Editors
Abdelazim Negm
Water and Water Structures Engineering Department
Faculty of Engineering
Zagazig University
Zagazig, Egypt

Ahmed Elkhouly
Plant Ecology and Range Management Department
Desert Research Center
Egypt, Cairo, Egypt

ISSN 2364-6934 ISSN 2364-8198 (electronic)
Springer Water
ISBN 978-3-030-77621-3 ISBN 978-3-030-77622-0 (eBook)
https://doi.org/10.1007/978-3-030-77622-0

This Springer imprint is published by the registered company Springer Nature Switzerland AG
The registered company address is: Gewerbestrasse 11, 6330 Cham, Switzerland

Preface

No doubt that groundwater is an essential source of water resources in many regions in the world particularly in arid desert areas as in the case of Egypt. Egypt's deserts rely on groundwater where no other reliable source is available. The rapid growth of the population and the need for food at a reasonable price lead Egypt to extend its agricultural expansion to the deserts where 1.5 million feddans (1 Feddan = 4200 m^2) are entering the plan for cultivation. This book focuses on the groundwater in Egypt's deserts, its availability, quantity, quality, uses, and the future agricultural expansion. The book consists of 17 chapters in five parts.

Part "Introductory Section" is an introduction and contains two Chaps. "Introduction to "Groundwater in Egypt's Deserts"" and "An Overview of the Egyptian Deserts' Resources" as an introductory section to the book. Chapter "Introduction to "Groundwater in Egypt's Deserts"" introduces the book to the audience, where it presents a very brief description of the technical elements of the chapters. While in Chap. "An Overview of the Egyptian Deserts' Resources", an overview of the resources in Egypt's desert including the general features of the Egyptian deserts, the main geographical units, the main characteristics of the natural resources in Egyptian deserts.

Part "Groundwater Occurrence and Ecosystem Services" includes Chaps. "Groundwater Ocurrences in West Nile Delta, Egypt" and "Characterizing Ecosystem Services to Human Well-Being in Groundwater Dependent Desert Environments" that are devoted to groundwater occurrence and ecosystems associated with groundwater availability in the deserts. Chapter "Groundwater Ocurrences in West Nile Delta, Egypt" classifies and assesses the geological, geophysical, hydrological, and hydrochemical bases of the existed aquifers in the study area which are represented by the Oligocene and the Miocene aquifers. The chapter also aims to create baseline data to be used as a guide to monitor and detect the future changes in the groundwater level and quality clarifying the roles of lowering the groundwater levels, aquifer depletion, increase in water salinity and mixing between different aquifers. While Chap. "Characterizing Ecosystem Services to Human Well-Being in Groundwater Dependent Desert Environments" explores the merits of the ecosystem services framework as a means to characterize the services associated with groundwater availability and use in Egypt and to begin to capture their value with a focus on the Western Desert to illustrate the scope to apply the ecosystem service characterization and assessment approach and also highlights areas where Egyptian groundwater managers could invest in improving information systems for this purpose.

Part "Groundwater Exploration, Quantity, Quality, and Their Management" deals with Groundwater Exploration, Quantity, Quality and their Management in 8 Chapters from "Geophysical Groundwater Exploration in Arid Regions Using Integrated Land-Based Magnetic and DC Resistivity Measurements: A Case Study at Gilf Kebir Area, South Western Desert, Egypt" to "Assessment of Groundwater Resources in Egypt's Deserts". Chapter "Geophysical Groundwater Exploration in Arid Regions Using Integrated Land-Based Magnetic and DC Resistivity Measurements: A Case Study at Gilf Kebir Area, South Western Desert, Egypt" is titled (Geophysical groundwater exploration in arid regions using integrated DC resistivity and Magnetic modeling: A case study at Gilf Kebir area, Southwestern Desert, Egypt). It focuses mainly on evaluating the groundwater accumulations of the areas through the determination of the aquifer geometry as well as the depth to the water level that is done by applying surface geophysics techniques with remote sensing information. Additionally, Chap. "Ground Water Potentiality in Siwa and Baris Oases, Western Desert, Egypt" is about (Ground Water potentiality in Siwa and Baris Oases, Western Desert, Egypt). It deals with the employment of geological and geoelectrical and hydrogeological techniques for evaluation of the potentiality of groundwater in Siwa and Baris Oases that are located in the Western Desert. Chapter "Groundwater and Characteristics of the Tertiary-Quaternary Aquifer System West of Mallawi, Upper Egypt" is titled (Groundwater and Characteristics of the Tertiary-Quaternary Aquifer System West of Mallawi, Upper Egypt) and evaluates the characteristics of the existed aquifers and assessment of the groundwater potentialities and quality at the west of Mallawi area aiming to realize the sustainable development of such area.

On the other hand, Chap. "Groundwater Characterization and Quality Assessment in Nubian Sandstone Aquifer, Kharga Oasis, Egypt" which is titled (Groundwater Characterization and Quality Assessment in Nubian Sandstone Aquifer, Kharga Oasis, Egypt) assesses the hydro-geochemical characteristics in the Nubian Sandstone Aquifer (NSSA) in Kharga Oasis by identifying major variables affecting the quality of groundwater and evaluating the suitability of groundwater for domestic and agricultural purposes. GIS is used to create thematic maps for the most important parameters associated with groundwater quality. Moreover, Chap. "Groundwater Potential in the Bahariya Oasis, Western Desert, Egypt" under the title (Groundwater Potential in the Bahariya Oasis, Western Desert, Egypt) evaluates the hydrogeological conditions of the groundwater in Bahariya Oasis to determine its potentialities for optimum exploitation of groundwater, using the available data and field measurements and investigating the water exploitation effect on water quality of the aquifer. Additionally, Chap. "Groundwater Quality and Potentiality of Moghra Aquifer, Northwestern Desert, Egypt" with the title (Groundwater Quality and Potentiality of Moghra Aquifer, Northwestern Desert, Egypt) deals with the water quality and potentiality of the groundwater of the Lower Miocene aquifer in the Moghra area to delineate the subsurface geologic setting and the affecting structural elements (faulting and fractured zones). It shows the occurrence of groundwater resources in the Lower Miocene aquifer and investigates the hydrogeological characteristics of the lower Miocene aquifer by using geochemical and geophysical methods.

On the other hand, the topic (Transboundary groundwater management issues in the Nubian Sandstone Aquifer System (NSAS) is covered in Chap. "Transboundary Groundwater Management Issues in the Nubian Sandstone Aquifer System (NSAS)" which explores how Egypt can lead the way forward toward the sustainable utilization of the NSAS by sharing insights with the other riparian countries concerning available management technologies and innovations as it pushes the boundaries of science, engineering, and ecosystem management to respond to pressing shared water management challenges. Conserving the quality of groundwater and keeping it away from pollution by sewage system is discussed in Chap. "Groundwater Quality for Irrigation as an Aspect of Sustainable Development Approaches: A Case Study of Semi-Arid Area Around Ismailia Canal, Eastern Nile Delta, Egypt" which is titled (Groundwater quality for irrigation as an aspect of sustainable development approaches: a case study of semi-arid area around Ismailia Canal, Eastern Nile Delta, Egypt) with a focus on an area in the East of the Nile Delta where the land is semi-arid or arid. While Chap. "Assessment of Groundwater Resources in Egypt's Deserts" closes this section by making an assessment of the groundwater resources in Egypt's deserts and discusses its availability and sustainability.

In "Potential Use of Groundwater and Future Expansion", 3 chapters are presents to cover the theme (Potential Use of Groundwater and Future Expansion). This theme is devoted to the Potential Use of Groundwater and Future Expansion of agricultural projects and their needs for water. It is covered in three chapters. Chapter "Groundwater Exploitation in Mega Projects: Egypt's 1.5 Million Feddan Project" provides a brief description of the Mega agricultural project in Egypt. An overview of the general vision and strategy adopted by the MWRI highlights the MWRI's plans. On the other hand, Chap. "Optimum Economic Uses of Precious Costly Ground Water in Marginal and Desert Lands; Case Study in Egypt" explains the various possible ways to optimize the use of the scarce groundwater source which might be non-renewable. The last chapter in Part "Potential Use of Groundwater and Future Expansion" is dealing with the overall assessment of the water resources in Egypt including the share of the groundwater. The chapter presents the expected water resources demand in Egypt in the year 2025 and in the year 2050 to draw the attention of the concerning authorities to take care of this very critical file to the life of the Egyptians.

The last Part of the book is to conclude the book with a chapter titled (Update, Conclusions and Recommendations of the Groundwater in Egypt's Deserts". It closes the volume of the book with the main conclusions and recommendations for the benefits of the readers, policy planners, decision-makers, and stakeholders.

Last but not least, we want to thank all who contributed to this high-quality volume, which is a real source of knowledge and the latest research findings in the field of groundwater in Egypt's Deserts. We would love to thank all the authors for their invaluable contributions. Much appreciation and great thanks are also owed to the editors of the Earth and Environmental Sciences series at Springer for constructive comments, advice, and critical reviews particularly Andrey Kostianoy and Alexis Vizcaino. Acknowledgments are extended to include all members of the Springer team who have worked long and hard to produce this volume.

Additional Abdelazim Negm and Ahmed Elkhouly acknowledge the support of the Science, Technology, and Innovation Authority (STIFA) of Egypt in the framework of grant no. 30771 for the project titled "a novel standalone solar-driven agriculture greenhouse—desalination system: that grows its energy and irrigation water" via the Newton-Mosharafa funding scheme.

Comments and feedback from the current and audiences are most welcomed to improve the quality of the books in the next editions. New chapters are welcomed for the next edition from any potential qualified author.

Zagazig, Egypt, December 2019 Abdelazim Negm
Cairo, Egypt, September 2019 Ahmed Elkhouly

Contents

Introductory Section

Introduction to "Groundwater in Egypt's Deserts"

Abdelazim M. Negm, El-Sayed E. Omran, and Ahmed Elkhouly

Abstract In this chapter, a brief description of all the chapters of the book is presented. Therefore, the chapter contains the basic idea behind each chapter. The chapters are organized under five themes that are Introductory section (2 chapters), groundwater occurrence and ecosystem services (2 chapters), groundwater resources, quantity, quality and their management (8 chapters), and potential use of groundwater and future expansion (3 chapters) and the conclusion section (1 chapter). In addition to this, chapter "Overview of the Egyptian Desert's Resources" introduces briefly the natural resources in Egypt's deserts. Almost the information presented in this chapter does not have much about the results and the conclusion. The results and its analysis are provided in the body of the chapters. Each chapter ends with a set of conclusions and recommendations as well. A separate chapter to close the book is devoted to the update of the literature and to present the most important conclusions and recommendations from the book chapters.

Keywords Water logging · Assessment · Sustainability · Groundwater · Deserts · Environment · Egypt · Oasis · Agriculture

---

A. M. Negm (✉)
Faculty of Engineering, Water and Water Structures Engineering Department, Zagazig University, Zagazig 44519, Egypt
e-mail: amnegm@zu.edu.eg

E.-S. E. Omran
Faculty of Agriculture, Soil and Water Department, Suez Canal University, Ismailia 41522, Egypt

Institute of African Research and Studies and Nile Basin Countries, Aswan University, Aswan, Egypt

E.-S. E. Omran
e-mail: ee.omran@gmail.com

A. Elkhouly
Plant Ecology and Range Management Department, Desert Research Center, Cairo, Egypt
e-mail: elkhouly3000@gmail.com; elkhouly3000@hotmail.com

A. Negm and A. Elkhouly (eds.), *Groundwater in Egypt's Deserts*, Springer Water,
https://doi.org/10.1007/978-3-030-77622-0_1

Background

The Egyptian deserts are located in the hyper-arid regions of North Africa. The water resources in the desert are mainly groundwater. Most of the soil texture in the Egyptian desert is sand. The natural resources in the Egyptian deserts subjected to degradation and deterioration mainly by land salinization, water deficiency and wind erosion.

The Egyptian deserts comprise three main geographical units: the Western desert; the eastern desert; and the Sinai Peninsula. Most of the Egyptian desert is located in the hyper-arid regions with annual rainfall in most parts of less than 50 mm. Hyperarid provinces include all the area between Lat 22° and 30° N, except the coastal mountains along the Gulf of Suez (a representative figure is presented in chapter 2). The arid provinces include the northern section with winter rainfall, it extends along the Mediterranean coast and the Gulf of Suez. The main habitats in the Egyptian deserts are distinguished to wadies, sand formations, gravel formations, salt marshes and reed swamps. The desert's water resources are primarily groundwater. Egypt has two main groundwater aquifers the first comprises groundwater in the Nile Valley and Delta system and the second aquifer is the non-renewable type, which is located in the Western Desert of the Nubian Sandstone, where the groundwater exists and deep-seated. The main soil types occurring in the Egyptian deserts are: Calcic Yermosols, Haplic Yermosols, Orthic Solonchaks, Eutric Regosols, Calcaric Regosols, Haplic Xerosols, Lithosols and Shifting sand. Flora of Egypt comprises about 2121 species of vascular plants including 27 species cultivated. Egypt is the meeting point of floristic elements belonging to at least three phytogeographical regions: the African Sudano- Zambesian; the Asiatic Irano- Turanian; the Afro- Asiatic Saharo- sindian and the Euro- Afro- Asiatic Mediterranean. The desert vegetation is formed mainly of xerophytic shrubs and sub-shrubs. Domestic animals in Egyptian Deserts include camels, donkeys, sheep, and goats. The largest wild animal is the Aoudad (a type of bearded sheep), which survives in the southern fastness of the Western Desert. Other desert animals are the Dorcas gazelle, the Fennec (a small, desert-dwelling fox), the Nubian ibex, the Egyptian Hare, and two kinds of Jerboa (a mouse-like rodent with long hind legs for jumping). Interested reading can go for more detail in chapter 2 titled "**Overview of the Egyptian Desert's Resources**".

It is worth mentioning that this book comes follower to the book edited by Negm [1] and the one edited by Negm and El-Rawy [2] with the titles "**Groundwater in the Nile Delta**" and "**Groundwater in the Nile Valley Aquifer, Egypt**", respectively, to add additional insights to the picture of the groundwater availability, quality and quantities in Egypt particularly in Egypt's deserts.

Themes of the Book

Therefore, the book intends to address in more detail the following main themes:

- Groundwater Occurrence and Ecosystem Services.
- Groundwater Exploration, Quantity, Quality and Their Management.
- Potential Use of Groundwater and Future Expansion.

Chapter's Summary

The next subsections describe briefly the main technical elements of each chapter under its relevant theme.

Groundwater Occurrence and Ecosystem Services

This theme is covered in chapters 3 and 4. Chapter 3 is titled "**Groundwater Occurrences in Western Nile Delta, Egypt**". It presents an integration of the geoelectrical, hydrological, hydro-geochemistry and isotopic techniques that were applied to evaluate the groundwater potential in the West Nile Delta area. Since, the West Nile Delta suffered from the deterioration in groundwater due to the intensive exploitation which resulted in both groundwater depletion and salinity increase. For the potential management of groundwater supplies, monitoring and evaluating changes in groundwater level, performance, aquifer renewability and mixing between existing aquifers were therefore very necessary. The geoelectrical and hydrological results reveal that there are two aquifers separated by a basaltic sheet. The first one is represented by the Miocene aquifer (El-Moghra) and the second one is the Oligocene aquifer which has a lower resistivity than the upper one and this is mainly due to the increase of clay content and the water salinity. Also, these results reveal that the saturated thickness of Miocene aquifer decreased during the years (2003–2015) period which ranged between 1.3 and 1.7 m/year and the total dissolved solids (TDS) content of groundwater samples varied from 234 to 2458 mg/L implying significant deterioration and salinization problems. The $\delta18O$-$\delta2H$ relationship was suggesting that two nonrenewable groups. Results indicate that most wells operate at relatively high pumping rates followed by low efficiency and different capacities in addition to decreasing well yields and deteriorating water.

Moreover, chapter 4 with the title "**Characterizing Ecosystem Services to Human Well-Being in Groundwater Dependent Desert Environments**" is focusing on the exploration of the ecosystem services associated with groundwater reserves and extraction. These can be characterized, compared, and weighed in qualitative and quantitative terms by decision-makers. It will assist groundwater managers in assessing groundwater management decisions pros. and cons. Management decisions are often based on information that is very limited. When feasibility studies or economic assessments of groundwater decision-making have previously been made in Egypt, they have tended to focus only on the value of groundwater for agricultural production. By introducing a broader framework, ecosystem-service-based assessments should lead to better decision-making. Applying the framework for the evaluation of ecosystem services requires understanding, information and human capacity to effectively collect and use it. These can be mobilized through an iterative approach. The approach is illustrated through a case study located in the Nubian-Moghra Miocenene-Nile aquifer complex at Wadi Natrun in the Western Desert of Egypt. This provides insight into the potential for more ambitious use by decision-makers and scientists of the ecosystem service assessment method in areas where rapid groundwater extraction is already happening or may be accelerating in the future.

Groundwater Resources, Quantity, Quality and Their Management

This theme is covered in chapters from 5 to 13. Chapter 5 deals with the surface and subsurface mapping of the Nubian Sandstone Aquifer System (NSAS) and the delineation of subsurface structures predominant in the basement rocks, and their relationship with this aquifer in the area. The study area is the area of Gilf Kebir and its surroundings. They are considered as a case study for groundwater exploration using the conceptual geophysical program for mapping the groundwater potentialities in these hot climate and scarcity of rain desert areas. This work An integrated land-magnetic and DC resistivity methods were conducted with the employment of the geological and remote sensing data to delineate the shallow subsurface geological sections in terms of groundwater investigation. The study area is bounded by latitudes 220 00 and 230 00′ N, and longitudes 260 00 and 280 00′ E, covering an area of about 15,000 km^2. Four small locations were investigated namely; Eight Bills, Nosb El-Amir, Jabal Kamel and NE Jabal Kamel. Twenty two sounding points and 250 magnetic stations were carried out to illustrate aquifer geometry and basement relief, as well as prevailing hydrogeological conditions. The results will be helpful to guide the water resources management and environmental conservation in the Western Desert and other similar arid regions.

Moreover, chapter 6 is titled "**Groundwater Potentiality in Siwa and Baris Oases, Western Desert, Egypt**". This chapter deals with the employment of geological, geoelectrical and hydrogeological techniques to evaluate the potentiality of groundwater in Siwa and Baris Oases. The areas (Siwa and Baris oases) comprise two parts of the Western Desert, Siwa Oasis lies at 330 km southwest of Sallum and 65 km east of the Libyan borders, It is lying below the zero contour level elevation is usually considered as Siwa depression, while Baris Oasis lies at 90 km south of El-Kharga area. Hydrogeologically, the groundwater system that underlies Siwa Oasis consists of two productive aquifers. These are the Lower Cretaceous Nubian sandstone and the Middle Miocene fractured carbonates, while in Baris area consists of one productive aquifer that is the Nubian Sandstone aquifer.

However, chapter 7 is titled "**Groundwater and Characteristics of the Tertiary-Quaternary Aquifer System West of Mallawi, Upper Egypt**". The chapter explains the characteristics of groundwater in the west of Mallawi, where the groundwater exists I in three aquifers (Quaternary, Oligocene-Pleistocene and Middle Eocene Limestone Aquifer). The general trend of groundwater movement is from east to west. The water salinity of the three aquifers is less than 1000 ppm.

The dominated aquifers in the study area fall into two broad categories: The unconsolidated aquifers (granular) represented by Quaternary and Oligocene-Pleistocene. The consolidated fractured aquifer (fractured rock) represented by Eocene fractured limestone which considered as karst aquifer. The Quaternary aquifer in the flood plain is considered as highly productive where the aquifer has transmissivity values ranging between 5592 and 12,700 m^2/day and specific capacity ranging between 11.8 and 46.0 m^3/h/m. The water salinity of this aquifer ranges between 203.5 and 549.4 ppm increasing from east to west. The main source of recharge of this aquifer is the Nile water as indicated by the ion relationships. The Oligocene-Pleistocene aquifer is classified as moderately productive having transmissivity values ranging between 1256 and 6800 m^2/day and hydraulic conductivity ranges between 13.22 and 59.9 m/day with saturated thickness ranges between 91.5 and 113.5 m. The water salinity of this aquifer ranges between 719 and 801 ppm. For the Eocene fractured limestone aquifer, the aquifer is considered as highly productive aquifer having transmissivity values range between 1083 and 14,054 m^2/day. Its water salinity ranges between 462 and 845 ppm. The general flow direction of the granular aquifers (Quaternary and Oligocene-Pleistocene) is from east to the west as the Eocene fractured aquifer with some local change where there is a flow direction from southwest to the northeast direction which may be attributed to the over-exploitation in this locality or the effect of the fractures' orientation in this aquifer system.

Moreover, chapter 8 with the title "**Groundwater Characterization and Quality Assessment in Nubian Sandstone Aquifer, Kharga Oasis, Egypt**" provides baseline information on the groundwater quality and its suitability for different purposes of the Nubian sandstone aquifer in Kharga Oasis, Egypt. Hence, it could be useful for the management and sustainable development of the area. Also, it assesses the hydrogeochemical characteristics of groundwater in the Nubian Aquifer (NSSA) in Kharga Oasis and to evaluate its suitability for different purposes using field data and GIS techniques. Representative groundwater samples (144 samples) were collected and analyzed for various physiochemical attributes based on APHA [3]. The field surveys including wells location, well depth, ground elevation, and physicochemical properties were measured in situ. The results revealed that 93.8% of the samples are freshwater, whereas the rest 6.2% is slightly saline. Moreover, 35.4% of the samples are unsuitable for domestic purposes due to the high level of hardness. Salinity hazard, sodium percent, sodium adsorption ratio, residual sodium carbonate, magnesium hazard, Kelly's ratio, and soluble sodium percentage were used to evaluate groundwater quality for irrigation purposes, where most of the collected samples are unsuitable for agriculture under ordinary conditions.

Additionally, chapter 9 is titled "**Assessment of Groundwater Quality in Bahariya Oasis, Egypt**". It aims to evaluate the hydrogeological conditions of the groundwater aquifers (Nubian Sandstone Aquifer) in Bahariya Oasis. The hydrochemical analyses of 84 representative groundwater wells were carried out to determine the main hydrochemical characters, genesis and evaluation of the groundwater quality for different uses.

The physical and chemical characteristics were measured for the groundwater according to standard methods, such as temperature, turbidity, water electrical conductivity (EC), total suspended solids (TSS) and total dissolved solids (TDS), pH value, major anions and cations.

The average values of the EC, TDS, pH, major cations and anions did not violate the permissible limits of the drinking water set by the World Health Organization in 2011. Most of the three aquifer zones have freshwater averages of 286 mg/l. The average concentrations of heavy metals are higher than the permissible limit for manganese but lower for zinc.

Chapter 10 is titled "**Groundwater Quality and Potentiality of Moghra Aquifer, Northwestern Desert, Egypt**". It discusses the quality of groundwater in Moghra aquifer and its suitability for irrigation and drinking purposes. The Moghra area is a remnant of a larger paleolake including the mouth of a paleo-river. It is characterized by low relief and a mild topography with elevations −10 m below sea level to about 40 m above sea level. The geophysical study for 48 wells logs indicates great variations in lithology that consists of sand; sandstone and clay. The Moghra aquifer is under unconfined conditions. The formation of water resistivity of the aquifer generally increases towards the north. The decreasing of formation water resistivity in the south may be attributed to the decreasing in sand content and the existence of fine sand intercalated by shale and silt minerals. The formation factor increases towards the north. The effective porosity averages from 12 to 33%, and increases towards the west. The volume of shale is increased gradually towards the northwest. The origin of the Moghra aquifer is a mixture of water from different modern rainfalls, the water of post-Moghra aquifer, seawater and groundwater of Nubian Sandstone aquifer.

Chapter 11 is titled "**Transboundary Groundwater Management Issues in the Nubian Sandstone Aquifer System**". This chapter discusses Egypt's experiences in leading knowledge generation and sharing with other riparian countries to further the sustainable utilization of the major transboundary aquifer in Africa: the Nubian Sandstone Aquifer System (NSAS). It illustrates how the available management technologies and developments emerging in Egypt are pushing research, engineering, and ecosystem management boundaries to tackle the pressing challenges of shared water management.

The chapter first provides an overview of the growing sustainability challenges and innovations occurring in the NSAS. It then discusses the principles of sustainable utilization of groundwater in transboundary systems and provides a brief overview of the evolution of transboundary cooperation in the NSAS. The chapter observes that transboundary groundwater management initiatives have so far been state-led, and have focused on the generation of studies with the intent to share data between countries on the hydrological conditions of the aquifer.

A discussion of the challenges and way forward highlights the opportunity for increased engagement of local institutions to implement land and water management practices that conserve the health of the land that stores the groundwater reserves and enables their replenishment, to conduct or facilitate monitoring of groundwater conditions and to sustain the aquifer system. The extent to which this approach to groundwater management that is emerging in Egypt is innovative is underlined as one of the major conclusions.

Moreover, chapter 12 which is titled "**Groundwater Quality for Irrigation as an Aspect of Sustainable Development Approaches: A Case Study of Semi-arid Area Around Ismailia Canal, Eastern Nile Delta, Egypt**" discusses the need of conserving the quality of groundwater in the East of the Nile Delta where most of the land is semi-arid or arid. It is conducted to assessing the quality of the groundwater in the Quaternary aquifer of the semi-arid area Eastern Nile Delta region around Ismailia Canal, Egypt. The Suitability of groundwater for irrigation purposes is assessed too. This a basic evaluation investigation for understanding what management changes and practices that should be considered for conserving the soil from degradation, maintaining the soil productivity, restoring the maximum production capability and maintaining the full crop and long-term productivity under the given set of conditions.

Next to the above chapter, chapter 13 is titled "**Assessment of Groundwater Resources in Egypt's Deserts**" to provide an overall assessment of groundwater resources in Egypt. The chapter presents a general view on the Nile basin aquifers then the major six aquifers in Egypt. Different studies showing the groundwater abstraction rates, characteristics of the aquifers in the western, eastern and Sinai Peninsula.

The chapter begins by introducing the different Nile basin aquifers and their characteristics namely: Victoria artesian aquifer, Congo artesian aquifer, Upper Nile artesian aquifer, volcanic rock aquifers, Nubian sandstone aquifer, the Nile valley, and Nile Delta aquifers. Then, a brief introduction on the six major aquifers in Egypt is followed namely: Nile aquifer, Nubian sandstone aquifer, Moghra, Fissured carbonate, Coastal and the Hardrock aquifers. A brief history on the development and the exploitation of the groundwater and the consequences that latter followed it. The most recent studies and investigations are then outlined to show the groundwater extraction rates, salinity levels, the transmissivity of the soil aquifers, piezometric heads, etc. in different parts of Egypt's deserts. Models that simulate the groundwater flows n some areas are presented and simulation results for the future withdrawal rates are discussed. The recent advances in satellite imagery together with the integration of advanced techniques like GIS tools, Remote sensing, Watershed Modeling System, Enhanced Thematic Mapper Plus, and other surveying techniques are outlined to give the potential areas of groundwater across the country.

Potential Use of Groundwater and Future Expansion

This last theme in the book is devoted to discussing the economical use of the groundwater in Egypt's deserts, the agricultural expansion in Egypt's deserts and the overall assessment of water resources in Egypt to complete the picture of water resources and their distribution. In the chapter titled "Groundwater Exploitation in Mega Projects: Egypt's 1.5 million Feddan Project" the authors provide a brief description of the 1.5 Million Feddans project, including an overview of the general vision and strategy adopted by the MWRI to guarantee its sustainability and success. It highlights the MWRI's plans to avoid the mistakes that have been observed in past development projects that are dependent on groundwater management and to establish an effective system for predicting and monitoring economic and environmental impacts. The chapter presents the main points of both the proponents and the opponents of the project. The chapter stated that the establishment of the environmental monitoring system should be a high priority both for the groundwater resource users and for the government. This should be integrated with participatory planning for long-term sustainability, environmental risk assessment and disaster preparedness. While the chapter with the title "**Optimum Economic Uses of Precious Costly Groundwater in Marginal and Desert lands in Egypt**" explains the optimum economic using the precious, costly and nonrenewable desert groundwater in different sectors. In countries suffering from water shortage, particularly in arid and warm regions, the priority strategy of using groundwater should be planned as a result. The return of the water system in the industrial sector varies from 10 to 30 folds compared to the agricultural sector.

In countries that are suffering from water shortage especially in the arid and warm regions, the strategy of priority of using groundwater should by planned. The return back of the unit of water in the industrial sector is ranged between 10 and 30 folds than the agriculture sector. The industry sector uses a little of water which does not exceed 10% of the total water resources in the developing countries, but share in the GDP of the country 10 times more than the agriculture sectors which consume 85% of the total water resources. The labor income in the industrial sectors is higher as 2–10 times that of the labor income in the agriculture sector. Under specific conditions in some developing countries that are obliged to use the newly discovered groundwater in agriculture to reduce the food insecurity and the lack of foreign currency; scientific and logical policy should be followed. The priority of using groundwater in agriculture should be to the low temperature and seasonal rainfall area. Thus the new greening desert and agriculture extension projects in Egypt should be in the North and the Mediterranean coast of North Egypt which have a low temperature than south Egypt by 14 centigrade. The south, warm and low humidity region should have a priority for industry investments to reduce poverty and maximize the return back of water unite.

This theme ends with the chapter titled "**Assessment of Water Resources in Egypt: Current Status and Future Plan**" to introduce the current water resources supply and demands and the share of groundwater in the water resources budget in Egypt.

In this chapter, water resources in Egypt, both the conventional (Nile water, Deep groundwater, rainfall and flash floods and the desalinated water) and the non-conventional (shallow groundwater, reuse of the agricultural drainage water and the use of treated wastewater), have been presented. Data from different studies in the last 20 years have been collected and analyzed showing that the total water resources amount to approximately 82.8 billion cubic meters in the budget of 2017.

This is in addition to presenting the various sectors that consume water in Egypt like the agricultural, domestic, industrial, and other sectors (evaporation losses, environmental balance, fisheries, navigation, etc.). It is found that there is a gap that amounts to about 21 billion cubic meters in covering the water demands which pose pressure on the Egyptian government to accelerate the pace of an overall integrated water management policy.

By the end of the chapter, an overview of the different scenarios of the water budgets of the year 2025 and 2050 is presented. These scenarios take into account the accelerating population growth rate, climate change, the effect of the construction of projects in the upper Nile, the efficiency of the irrigation system and many other factors on the expected water consumption.

The book ends with the conclusions and recommendations chapter numbered 17. In the conclusions chapter, an update of the literature is made to cover some of the interesting topics which are related to the themes of the book. Some of these sources include [4–11].

Acknowledgements The writer of this chapter would like to acknowledge the authors of the chapters for their efforts during the different phases of the book including their inputs into this chapter.

Additionally, Abdelazim Negm and Ahmed Elkhouly acknowledge the support of the Science, Technology, and Innovation Authority (STIFA) of Egypt in the framework of the grant no. 30771 for the project titled "a novel standalone solar-driven agriculture greenhouse—desalination system: that grows its energy and irrigation water" via the Newton-Mosharafa funding scheme.

References

1. Negm (2019) Groundwater in the Nile delta. In: The handbook of environmental chemistry. Springer International Publishing AG, part of Springer Nature. https://doi.org/10.1007/978-3-319-94283-4
2. Negm AM, El-Rawy M (eds) (2021) Groundwater in the Nile Valley Aquifer, Egypt. ASTI series. Springer International Publishing AG, part of Springer Nature, (Under Production).
3. APHA (American Public Health Association) (2005) In: Eaton AD, Clesceri LS, Greenberg AE (eds) Standard methods for the examination of water and wastewater, 20th edn. APHA, AWWA, WPCF, New York
4. WWAP (2014) The United Nations world water development report 2014: water and energy. Paris, UNESCO. United Nations World Water Assessment Programme, UNESCO, Paris, p 230
5. UN (2014) System of environmental-economic accounting 2012 central framework UN, EU, FAO, IMF, OECD, WB, p 378

6. Tuinstra J, Jv W (2014) Ecosystem services in sustainable groundwater management. Sci Total Environ 485–486(485–486):798–803
7. MWRI, Ministry of Water Resources and Irrigation (Egypt) (2014) Water scarcity in Egypt: the urgent need for regional cooperation among the Nile Basin Countries. Report No. 6p
8. Kudoda AM, Abdalla OAE (2015) Hydrochemical characterization of the main aquifers in Khartoum, the capital city of Sudan. Environ Earth Sci 74:4771–4786. https://doi.org/10.1007/s12665-015-4464-x
9. Favretto N, Stringer LC, Dougill AJ, Perkins JS, Akanyang L, Dallimer M, Atlhopheng JR, Mulale K (2014) Assessing the socio-economic and environmental dimensions of land degradation: a case study of Botswana's Kalahari. Report for the economics of land degradation initiative ELD initiative, Leeds, UK, p 28
10. Ismail E, El-Rawy M (2018) Assessment of groundwater quality in West Sohag, Egypt. Desalin Water Treat 123:101–108. https://doi.org/10.5004/dwt.2018.22687
11. Fathi H, El-Rawy M (2018) GIS-based evaluation of water quality index for groundwater resources nearby wastewater treatment plants, Egypt. Poll Res 37:105–116

An Overview of the Egyptian Deserts' Resources

Ahmed A. Elkhouly, Abdelazim M. Negm, and El-Sayed E. Omran

Abstract Egypt total area found under arid and hyperarid climatic conditions, of which only a small portion (3% of total area) is agriculturally productive. Deserts in Egypt represent about 96% of Egypt's area. The Egyptian deserts comprise three main geographical units: the Western desert; the eastern desert; and the Sinai Peninsula. Egyptian deserts are located in the hyper-arid regions of North Africa and in Asia astride the Sahara and Arabian Desert with annual rainfall in most parts of less than 50 mm. The main habitats in the Egyptian deserts are distinguished to wadies, sand formations, gravel formations, salt marshes and reed swamps. The water resources in the desert are mainly groundwater. Most of the soil texture in the Egyptian desert are sand The Egyptian flora is a mixture of flora that characterizes three continents: Africa, Asia, Europe, and this related to the geographical location of Egypt between the three continents. The natural resources in the Egyptian deserts subjected to degradation and deterioration mainly by land salinization, water deficiency and wind erosion.

Keywords Deserts · Egypt · Degradation · Land salinization · Natural resources · Groundwater · Habitats · Water deficiency · Vegetation · Animals

---

A. A. Elkhouly (✉)
Plant Ecology and Range Management Department, Desert Research Center, Cairo, Egypt

A. M. Negm
Department of Water and Water Structures Engineering, Faculty of Engineering, Zagazig University, Zagazig 44519, Egypt
e-mail: Amnegm@zu.edu.eg

E.-S. E. Omran
Soil and Water Department, Faculty of Agriculture, Suez Canal University, Ismailia 41522, Egypt

Institute of African Research and Studies, Nile Basin Countries, Aswan University, Aswan, Egypt

A. Negm and A. Elkhouly (eds.), *Groundwater in Egypt's Deserts*, Springer Water,
https://doi.org/10.1007/978-3-030-77622-0_2

Introduction

Egypt can be considered an African/Mediterranean country, although Sinia is a part of Asia. It occupies an area of about one million Km^2. It lies, between Latitude 22° and 32°N and Long 25° and 35°E. It is bordered by the Mediterranean Sea on the North while on the South it is bordered by the Republic of Sudan, on the West by Libya, and on the East by Palestine, Gulf of Aqaba and the Red Sea [1]. Egypt's total area found under arid and hyperarid climatic conditions, of which only a small portion (3% of total area) is agriculturally productive. Deserts in Egypt represent about 96% of Egypt's area (Fig. 1).

Egypt is a part of the Sahara "North African Desert". Its area is divided geographically by the River Nile into two main parts: a western part (Western Desert) of about 681,000 km^2 and an eastern part comprising the Eastern "Arabian" Desert 223,000 km^2 and the Sinai Peninsula 61,000 km^2. Nile Valley including the Delta forms, a riparian Oasis 40,000 km^2 that's the densely inhabited farmlands of Egypt.

With developing concerns approximately climate change and conceivable affect on natural system interruption and advancement forms have had evident unfavorable fingerprints on keeping up a healthy environment in Egypt's Lakes [2] and depressions. Water shortage issue may be a part of the climate change, which could be a

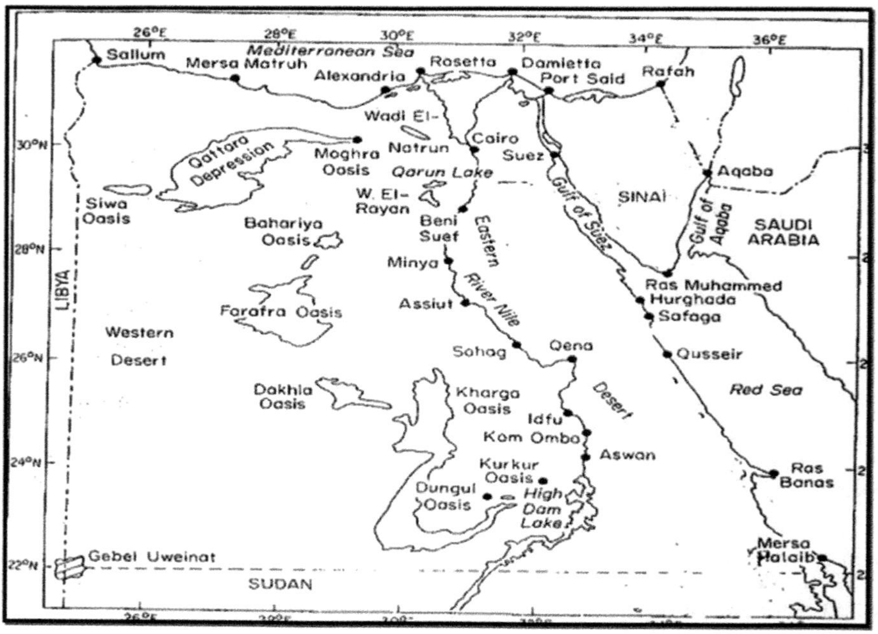

Fig. 1 Map of Egyptian deserts

developing issue around the world, with Egypt being hit especially difficult. Precipitation has relentlessly diminished, and temperatures have expanded. As temperatures take off and precipitation drops, crucial lakes are drying up [3].

Egypt is still attempting to reclaim the desert, to supply work and living space for its growing population and their growing settlements. Most fundamental nourishment items are imported, as generation is deficiently to meet population needs [4]. However, certain economists and environmentalists fear that the country's efforts to green the desert could be ill advised. This has prompted authorities in Egypt to develop plans—among other things—for large-scale resettlement. Egypt ought to utilize the desert to require care of the huge increment in population. population conditions in Egypt's desert are exceptionally distinctive from those within the valley and delta. Groundwater is the essential source of consumable water in a few nations, but there are challenges that influence supportability due to populace development, climate changeability, and land development [5]. The Nile River is the essential source of water in Egypt [6]. Egypt is confronting water shortage [7].

For sustainable socioeconomic developments in Egypt, resources management in desert is therefore important to protect limited resources. Egypt is divided into Agroecological zones. Each agroecological zone has critical varieties in environmental conditions. It is vital to focus on each area resources of these zones and monitoring of the resources and needs for development and capacity building and the development of appropriate legislation and recognize the beneficiaries of these areas of development [8].

The General Features of Egyptian Deserts

The Main Geographical Units

The Egyptian deserts comprise three main geographical units (Fig. 1):

i. The Western Desert;
ii. The Eastern Desert;
iii. The Sinai Peninsula

I. **Western Desert**

The Western Desert is occupying about 681,000 km^2. It extends over a vast area and composed of a large rocky surface with the highest portion in the southwestern corner where Gebel (mountain) Uweinat is found. North of Uweinat, the Gilf Eelkebir plateau (100 m amsl) occurred which forms from Nubian Sandstone [1]. This plateau is distinguished by slops decline sharply towards East and North to large depressions (Kharga and Dakhla depressions). To the North of this plateau, another plateau with arms extend in several directions. This plateau is composed of limestone and is lower in elevation than the Gilf Eel-kebir plateau, and constitutes the main landform feature West of the Nile Valley. In the heart of this plateau surface, two great depressions have occurred, Farafra and Bahariya depressions. The area of Farafra is more than 3000 km^2, and Bahariya has an area of about 1800 km^2. The Qattara—Siwa depression is considered to be part of a huge depression in the northern sector of the Western Desert.

II. **Eastern Desert**

The Eastern Desert occupied on the east border of Nile Valley and bordered by Suez Canal in the East with an area of about 223,000 km^2. This desert forms from a series of mountain chains (the Red Sea mountains), running parallel to the Red Sea and separated from it by a narrow coastal plain.

III. **Sinai Peninsula**

Sinai Peninsula occupies an area 61,000 km^2 in the northeastern portion of Egypt at South-West part of the Asian continent. The southern part of Sinai is formed of a complex of igneous and metamorphic rocks.

Climate

Climatic Regions

The Egyptian deserts is located in the hyper-arid regions of North Africa and lies in Asia astride the Sahara and the Arabian Desert with annual rainfall in most parts of less than 50 mm [1]. According to the aridity index P/ETP, the arid regions are classified to hyperarid ($P/ETP < 0.03$) and arid ($P/ETP = 0.03–0.20$). These classes are, in turn, subdivided according to the mean temperature of the coldest and the hottest month of the year. Consideration is also given to the time of the rainy period relative to the temperature regime. According to these bases, Egypt is distinguished into two climatic provinces, I. Hyper arid Province and Arid Province II as in (Figs. 2, 3 and 4) [1, 9].

I. Hyper arid Province and II. Arid Province. The hyperarid Province comprises three subprovinces: a. Hyperarid subprovince with mild winter and very hot summer: means of coldest and hottest months are 10–20 °C and >30 °C, respectively, which covers most of the Western Desert and the southern part of the Eastern Desert, b. Hyperarid subprovince with a mild winter and a hot summer (mean temperature of the hottest month = 20–30 °C) covering the Eastern Desert and the northeastern part of the Western Desert and Gebel Uweinat, and c. Hyperarid subprovince with very hot summer and very cold winter (<0 °C) prevailing in the montane country of the Sinai Peninsula [9].

The arid province comprises two subprovinces: the northern and southern: (a) The northern arid subprovince with winter rainfall which extends along the Mediterranean

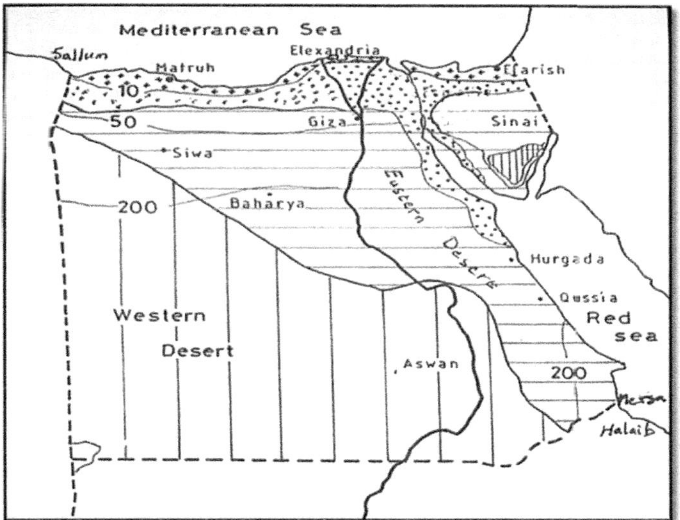

1. **Arid Province:** ++++ Arid (Attenuated); :::::: Arid (Accentuated)
2. **Hyperarid Province:** = Hyperarid: Hot summer and mild winter; | | | | Hyperarid: Very hot summer and mild winter; |||||| Hyperarid: Hot summer and very cold winter

Fig. 2 Climatic Provinces in Egypt [1]

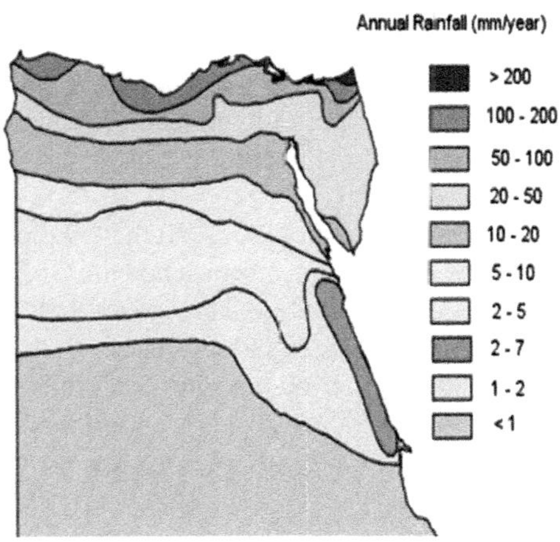

Fig. 3 Distribution of Rainfall in the Egyptian Deserts. (*Source* USGS GTOPOPO30; GADM global administrative areas; and UN Revision of World Urbanization Prospects.)

coast and Gulf of Suez. This subprovince is subdivided into 2 areas by UNESCO/FAO map of 1963: (i) the coastal area under the maritime influence of the Mediterranean sea with a shorter dry period (attenuated), annual rainfall may reach 200 mm and (ii) a more inland area with a longer dry period (accentuated), and an annual rainfall of 20–100 mm. Both areas are characterized by a mild winter and a hot summer. (b) The southern arid subprovince with a winter rainfall including Gebel Elba mountainous area of the Red Sea coast in the most southeastern corner of the Eastern Desert,

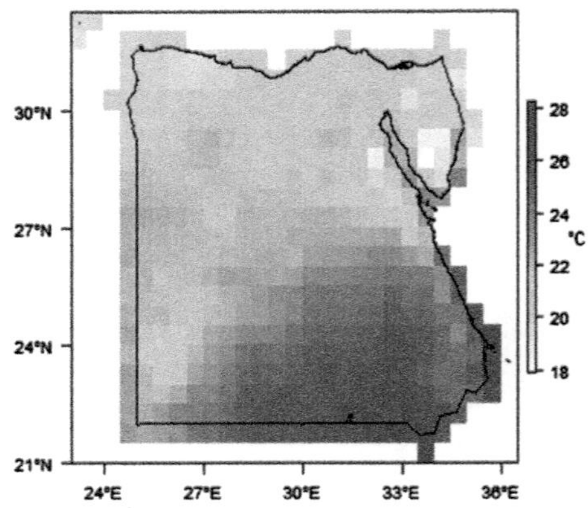

Fig. 4 Distribution of Temperature in the Egyptian Deserts. (*Source* USGS GTOPOPO30; GADM global administrative areas; and UN Revision of World Urbanization Prospects.)

here the annual orographic rainfall is significant where increase than 160 mm/year in Gebel Elba.

(i) Hyperarid provinces include all the area between Lat.22° and 30°N, except the coastal mountains along the Gulf of Suez. These are distinguished to:

- Hyperarid with a mild winter and a hot summer (mean temperature of the hottest month is 20–30 °C), this includes the Eastern Desert and the northeastern part of the Western Desert of Egypt, and Gebel Uweinat.
- Hyperarid with a cool winter (mean temperature of the coldest month is 0–10 °C), and hot summer, represented around the highlands of southern Sinai. Rain in hyperarid provinces is less than 30 mm/yr and is occasional and unpredictable.

(ii) The arid provinces include the northern section with winter rainfall, it extends along the Mediterranean coast and the Gulf of Suez. This section is distinguished into two provinces namely; –The coastal belt province under the maritime influence of the Mediterranean, with a shorter dry period (attenuated), –The more inland province with a long dry period (accentuated) and an annual rainfall from 20 to 100 mm. Both provinces are characterized by a mild winter and a hot summer. –The salient features of the main climatic variables in such provinces can be summarized in the following section.

Basic Climatic Elements

(a) Temperature Generally, summer is hot (mean of the very hottest month ranges between 20 and 30 °C), or very hot (mean of the hottest month is more than 30 °C). Winter is either warm (mean of the coldest month is 20–30 °C) or mild (mean minimum of the coldest month is 10–20 °C) expect on the highlands where the winter is cool with a mean minimum of the coldest month (between 0 and 10 °C).The temperature along the Red Sea coast varies between a mean minimum of the coldest month of about 10 °C towards the North and about 20 °C towards the South, and a mean maximum of the hottest month of about 33 °C towards the North and 40 °C towards the South. The range of variation becomes greater further inland (from about 4 to 38 °C in the oases of the Western Desert). In continental locations, temperature extremes of less than 4 °C in the coldest month (e.g. oases of the Western Desert) have been recorded. The coldest month is between December and February and the hottest month is between June and August in hyperarid and arid provinces, respectively (Fig. 4).

(b) Relative Humidity:

The relative humidity is influenced primarily by the relative nearness to the Mediterranean and the Red Sea. The most reduced records are those of inland areas of the parched and hyperarid provinces. The most noteworthy ones are those of areas closer to the Mediterranean coast. The most reduced records of relative humidity are by and large those of late spring, though the most elevated records are those of late autumn and early winter.

(c) Rainfall In general. Three precipitation belts may be recognized within the deserts of Egypt: (1) The Mediterranean coastal belt, (2) center Egypt with latitude 30°N as its southern boundary, and (3) upper Egypt. The first and second belts have a winter precipitation (Mediterranean regime). The stormy season expands from November to April, in spite of the fact that primarily concentrated in December and January. These belts compare generally to the attenuated and accentuated arid provinces of northern Egypt, where the normal annual rainfall ranges from 100 to 150 mm within the weakened parched territory, and from 20 to 100 mm within the accentuated arid province.

It expands south along the Gulf of Suez to Lat.26°N due to the orographic impact of the Red Sea coastal mountains. The third belt is nearly rainless; it compares generally to the hyperarid territories. Rain at this belt isn't an yearly repeating occurrence; 10 mm may happen once each ten a long time. The precipitation increments steadily to the North until comes to around 20 mm at the borders with the arid province (at Giza). One of the major highlights of precipitation in arid and semi-arid regions other than being meager, is its great temporal variability, the average deviation of annual precipitation from the mean, expressed as a percentage of the mean, is greatest in the hyperarid provinces (e. g. Siwa 83%). In the arid province the percentage variability is 65% at Giza which is close to the hyperarid provinces [1].

(d) Wind: In winter, the Sahara high-pressure system dominates the circulation and the northerlies bring cool dry air from the North Africa continental source region though occasionally the Arabian high brings warmer air to the eastern parts of Sudan. Both of these types are sometimes hindered by E-W depressions along the Mediterranean, and supplanted by cold dry air from the Eurasian landmass. In spring and autumn time, the Middle Arabian high is more overwhelming within the East, and the impact of Mediterranean depression is rarely felt, as air from both North Africa and the Arabian sources is considerably warmer than in winter. In Summer, the Saharan high is again dominant bringing hot dry air. Occasionally, very hot dust-laden winds blow (Khamaseen) which have numerous environmental consequences including possible effects on climate, soil formation, groundwater quality and crop growth. They may make issues counting considerable degrees of deflation and erosion [1].

Habitats

The environment can be broken down into a few essential units: biomes, ecosystems and habitats. A biome is like a city. An biological system is like a community in a city. There are numerous ecosystems in a biome. Each environment contains a number of diverse habitats. A habitat is like a home. There are many homes in an ecosystem. These habitats are the "homes" of the plant and animal species that live there. The hot desert ecosystem is the host of plants and animals which can be living in oppressive environments. This environment is represented by high solar energy, warm seasons throughout the year, little rainfall during winter, unpredictable rainfall, extreme variation in temperature between night and day as well as between winter and summer, and scattered vegetation. In this ecosystem, most plants are adapted to minimize the effects of too much solar energy. Plant annuals compress their life cycles and go dormant as conditions grow unfavorable [9]. According to [10], hot desert ecosystems cover all of Egypt and extend south to latitude 12 °N in Sudan. Such hot ecosystems are either arid or hyperarid and it is possible to distinguish between three main hyperarid and two arid provinces.

According to [11–19], the habitats in the Egyptian deserts are distinguished into the follow types:

I. **Gravel Formations**

It is a fluviatile deposit distinguished into two main classes:the gravel surfaces and the water runnels. The former comprises the flat gravel plains, the slops of gravel hills and the undulating surfaces of the gravel-covered mound and hillocks. The water runnels include the finer branches dissecting the gravel surface, the affluent branches receiving the water from the minor runnels, and the main channels collecting water from extensive catchment areas.

II. **Sand Formations**

The aeolian sand accumulation occupies an area of about 16% of the total surface area of Egypt. About 95% of this ratio is located in the Western Desert [20]. Sand formations represent a morphological feature of the coastal and inland deserts in both arid and hyper-arid provinces of Egypt [21, 22]. They vary in origin, size, height, structure, texture, water content and salinity. They may be categorized into: dunes, hillocks, hummocks, mounds and bars [22]. Psammophytes are plants that inhabit the sand formation, sand bars, sand hillocks and sand dunes which are usually associated with the lakes of the Oases and depressions [23, 24].

III. The Drainage Systems (wadis)

Each wadi comprises a main channel and a number of tributaries that receive drainage (run-off water) from branches and affluents. Viewed as a whole a wadi is a dried course of river. It has features of a natural drainage system covered under less arid conditions. The cutting of a water channel may follow either or both of two procedures; the first may be called deeper, the second cutting backward. In the former process the channel starts as a long shallow runnel extending at the surface of the plateau, guided by regional and local conditions of topography. According to the depth, deep and local conditions of topography, wadi habitat distinguishes to the following habitats:

(a) Mature Wadis

Wadi has a main channel which is wide, deep and well defined.It receives, along its long course, tributaries that are also well evolved. The mature wadi is a "fossil river".

(b) Shallow Wadis

These are a long runnels that are traverse erosion surface at higher levels. The course is shallow, the boundary banks do not exceed 2 m and are often formed of rubble material.

(c) Wadis Cutting Backward

These are among the results of the active erosional processes that are operative under the present climatic conditions. One may visualize the stages of its development by comparing various gullies and by observing the change in the same gully in a sequence of years.

IV. The Erosion Surfaces

The massive limestone body of the middle Eocene plateau includes bands of hard dolomitic limestone. These bands differ in their thickness. Each hard band marks an erosion surface or a potential erosion surface. There are in the form of erosion surfaces, at different levels, each floored by a dolomitic or silicified limestone band. three types of erosion surface are distinguished according to the level and component of rock fragments to: (a) plateau surface (is highest and the rock surface devoided of a veneer of detritus); (b) hammada (the rock surface covered with a veneer of rock fragments and boulders); (c) erosion pavement (the surface covered with a mantle of rock detritus including soft materials).

V. Mountainous Country

It is a rocky hills are characterizing by high elevations reaches to more than 1500 m above the sea level. its climate is describing as the cold desert and the humidity is very low. This habitat is characterizing by the aerographic rains. The soil in these habitats resulted from fragmentation rocks or movable soil by wind or by water erosion. The plants grow in the rock cracks.The highest mountain ranges in Egypt surround with many smaller valleys leading from the basin to the mountains in all directions. It is distributing in the Red Sea Coast or in the south of the Sinai Peninsula.

VI. **The Rocky Ridge**

An irregular higher in topography in the form of successive undulations running parallel to the coast. These undulations are in the form of calcareous rocky ridges alternating with depressions.

VII. **Salt Marshes**

Salt marshes are distinguished into two types: coastal salt marshes and inland salt marshes. Coastal salt marches form as a result of increasing the water table in soil due to the leaching of seawater to the soil adjacent to the seashore. Inland salt marshes are representing mainly in the oases and depressions in areas adjacent to lakes where water comes from the lateral seepage of lakes water and the underground water and in inland areas around springs where the water table is very shallow (or exposed). Under the prevalent climatic aridity, there is high evaporation of soil water and the accumulation of salts in the surface layers of soil. Also, the salt marshes have occurred in the land adjacent to the drains.These habitats are characterized by high mean values of EC and the water table.

VIII. **The Reed Swamps**

Reed swamps habitat is well represented in the shallow water or in the terrestrial borders of the lakes are distributed in the oases and depressions. Also, reed swamps habitat occurs in the reservoirs of the springs and the artesian wells, where the water is fresh to slightly brackish. These habitats are characterized by low values of soil salinity.

Natural Resources

Natural resources are divided into: non-renewable resources and renewable resources. Non-renewable resources include resources in the environment on the body of a consistent credit and taken it irreplaceable. Moreover,therefore these resources are in danger of depletion. Such as coal, oil, natural gas and radioactive metals and most of the groundwater.

Renewable resources include resources that have self renewed, these resources are not subject to depletion if exploited moderately away from the wasteful, Such as plant and animal resources (farms—natural pastures—Forestry—Fisheries).

Water Resources

(a) Ground Water

Egypt has two main groundwater aquifers the first comprises groundwater in the Nile Valley and Delta system. The total storage capacity of the Nile Valley aquifer system is about 200 billion m^3, with an average salinity of 800 ppm. The second aquifer is the non-renewable type, which is located in the Western Desert of the Nubian Sandstone, where the groundwater exists and deep-seated. Recent studies had indicated that this is not a renewable resource. Use of this fossil water depends on the cost of pumping and potential economic return over a fixed time period. With respect to the rainfall (1.5 BCM / yr) it occurs only in winter in the form of scattered showers along the coastal area and can not be considered a reliable source due to its locality and temporal variability.

The hydrogeological framework of Egypt comprises eight hydrogeological units (Fig. 5) [25]:

1. Nile Valley and Delta aquifers,
2. Coastal aquifers,
3. Nubian Sandstone aquifer,
4. Moghra aquifer,
5. Tertiary aquifer,

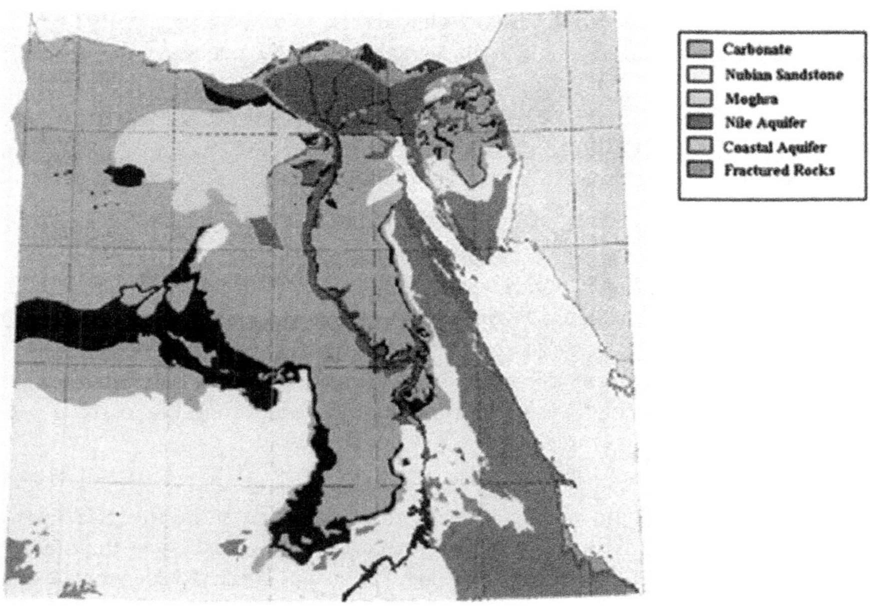

Fig. 5 Aquifer systems in Egypt [26]

6. Carbonate rocks complex aquifers,
7. Fissured basement complex aquifers and
8. Aquiclude rocks

The water-bearing rocks in Egypt are classified into the following two groups:

(a) Granular water-bearing rocks and
(b) Fissured and karstified water-bearing rocks.

Table 1 presents a summary the characteristics of the main aquifer systems of Egypt. Data are adopted from different sources, mainly, [25, 27–35].

Factors that threaten groundwater are depletion, salinity, salt-water intrusion, and pollution. Inefficient groundwater management can severely damage fragile ecosystems and act as a constraint for the national development plan.

(b) **Flash Flood**

Sometimes, Egypt received flash flood especially in the upper and middle Egypt, Red Sea Coast and Sinai.The US National Weather Service Weather Forecast Office defines a flash flood as "A flood caused by overwhelming or intemperate precipitation in a short period of time, for the most part less than 6 h". Flash floods are as a rule characterized by raging torrents after overwhelming downpours that tear through stream beds, urban streets, or mountain canyons clearing everything some time recently them. They can happen inside minutes or a couple of hours of intemperate precipitation. They can moreover happen indeed in case no rain has fallen, for the occasion after a levee or dam has fizzled, or after a sudden discharge of water by a debris or ice jam [36].

(c) **Soil Resources**

According to FAO/UNESCO [37] the main soils happening within the Egyptian deserts are:

I. **Calcic Yermosols**

These soils happen within the rocky desert east and west of the Nile as well as within the central portion of the Sinai Peninsula. They are brownish or yellowish-brown in color, emphatically calcareous and underlain by rock at shallow depth. Only small areas of wadis have moderately deep or deep soils but they are also strongly calcareous. Within the southwestern portion these soils are stony(Fig. 6).

II. **Haplic Yermosols**

These soils happen inside the narrow strips along the coast of the Red Sea, in two small areas. One area is in the southeastern part of the country and the other is opposite to the southern one-third of the Sinai Peninsula. These are deep loamy soils with weak structure. In parts they are saline (Fig. 6).

Table 1 Characteristics of the main aquifer systems of Egypt

Name of aquifer	Type locality	Depth of top aquifer (m)	Saturated thickness (m)	Depth to water table (m)	Hydraulic conductivity (m/day)	Porosity (%)	Salinity (ppm)
Nile Valley and Delta aquifer	Nile Valley	0–20	10–200	0–5	50–70	25–30	<1,500
	Nile Delta (south)	0–20	100–500	0–5	50–100	25–30	<1,500
	Nile Delta (north)	20–100	500–1,000	0–2	<50	< 30	>5,000
Coastal aquifers	Mediterranean	0	>5	±15	15–25	<30	1,000–6,000
	El-Qaa plain	50–100	60–80	50–70	5–10		600–2,500
	El-Arish aquifer	15–30	40–50	0–30	5–20		1,500–6,000
Nubian sandstone	Western desert						
	Kharga	50–200	500–700	0–30	2–4	20	<1,000
	Dakhla	200	500–1,000	0–20	6–7	20–25	<1,000
	Bahariya	150–300	1000–1,500	0–20	5–10	20–40	<1,000
	Farafra	200–500	1000–2,000	Flow	2–5	20–30	<1,000
	East oweinat	10–20	100–300	20–30	10–20	20–30	<1,000
	Nile basin fringes						
	Qena	100–250	500	Flow	Area 1–2		2,000–2,500
	Laqeita area	100–500	200–400	Flow	1–3		1500–2,000
	Eastern desert						
	Esh El Mallaha	0–30	<200	Flow			3,000–4,000
Nubian Sandstone	Sinai						
	Nakhla	1000	2,000	200			1,500–2,000
	Oyun Moussa	100–500	1,500	Flow			1,000–4,000
Mohgra aquifer	Natron/Qattara	0–200	500–900	100		20	1,000–1,2000
Fissured rocks							
Carbonates	Wadi Araba	0–100	500	Flow			1,000–1,2000
Hard rocks	South Sinai	0–50		+50			1.000–2,000

Depths are measured from the ground surface

III. Orthic Solonchaks

These are very strongly saline soils. They occupy a strip of delta area along the coast of the Mediterranean and a large area of the Qattara Depression about 200 km west of Cairo, These soils have little agricultural value. In the delta area however, they are being reclaimed and put under cultivation at a high cost which is justified because of

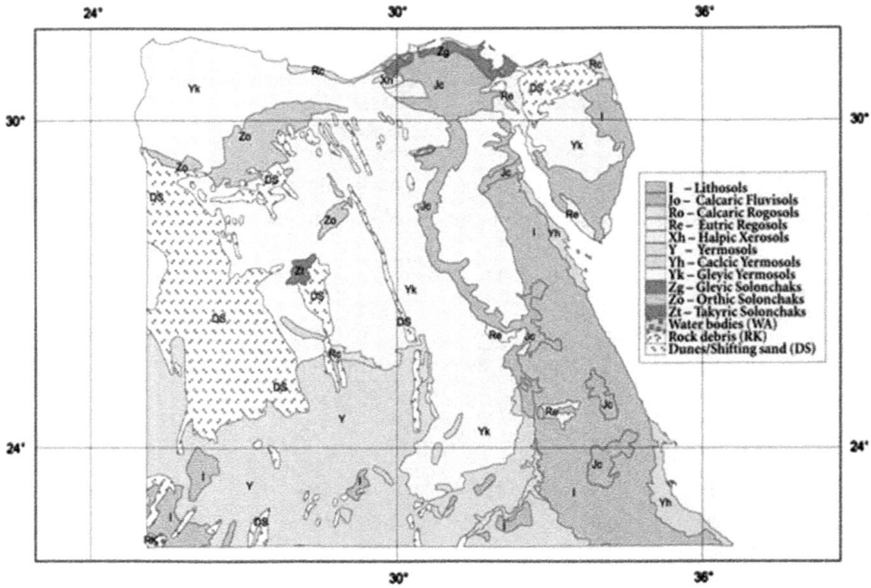

Fig. 6 Dominant soil map of Egypt (*Source* DSMW-FAO-UNESCO)

the good standard of farming and abundance of irrigation water after the construction of the Aswan dam (Fig. 6).

These soils are rough and gravelly and happen on slope slants and piedmont plains of the mountain locale east of the Nile Stream. They are non-calcareous. They have esteemed as it were as destitute brushing arrive (Fig. 6).

IV. Calcaric Regosols

These are deep clayey soils happening in a little plain range inside the desert west of the Nile Stream and a little zone within the northeastern corner of the country. These are shaped within the piedmont fields of limestone. These are good soils for irrigated agriculture (Fig. 6).

V. Haplic Xerosols

These are deep clayey soils of a piedmont plain near the northwestern tip of the Nile delta. They are soils of the semi-arid Mediterranean climate and support a poor crop of barley without irrigation; for profitable agriculture, supplemental irrigation is needed (Fig. 6).

VI. **Lithosols**

These are exceptionally shallow soils of the mountains within the range east of the Nile, along the coast of the Red Sea as well as the mountains within the southern portion of the Sinai peninsula. These are futile butfor poor grazing land (Fig. 6).

VII. **Shifting sand**

This is not soil in the real sense, but is soil material, occupying a large area in the western part of the desert west of the Nile River and the northern, one-third of the Sinai peninsula. It has little use (Fig. 6).

(d) **Plant Resources**

(i) **Flora**

In Egypt, desert vegetation is by far the most important and characteristic type of natural plant life. It covers vast areas and is formed mainly of xerophytic shrubs and sub-shrubs. Reference [38] recognized two types of desert vegetation, namely contracted and diffuse. Both types refer to permanent vegetation that can be accompanied by ephemeral (or annual) plant growth depending on the amount of precipitation in a given year. References [39, 40] added a third type termed 'accidental vegetation', where precipitation is so low and falls so irregularly that no permanent vegetation exists. It occurs mainly as contracted patches in runnels, shallow depressions, hollows, wadis and on old dunes with coarse sand.

According to [1], the Egyptian flora is a mixture of flora that characterizes three continents: Africa, Asia, Europe, and this related to the geographical location of Egypt between the three continents. Flora Egypt is rich in the Medicinal plants, which have been used since ancient times in the treatment of a large number of diseases such as species of genus (*Ammi, Artemisia, Hyoscyamus…* etc.).

Flora of Egypt comprises about 2121 species of vascular plants including 27 species cultivated. This means that the number of wild species about 2094 native and naturalized species under 712 genera and 121 families. Its divided as follows: 16 species represent pteridophytes, six species Gymnosperms, 1637 species Dicots, 435 species monocots. From these species 61 species endemic only in Egypt and this number represents about 2.9% of the total Egyptian flora—this means that the flora of Egypt richly diversity in spite of its growth in the country of the dry zones. This plant diversity may be due to the diversity of habitats in Egypt. Gramineae (Poaceae) is the largest family which contains 277 species (including 27 cultivated species) followed by Compositae (233), Cruciferae (230), Caryophyllaceae (102), Chenopodiaceae (85), Scrophulariaceae (77), Euphorbiaceae (62), Boraginaceae (57), and Labiatae (55) and Umbelliferae (51) species. The largest genera are *Euphorbia* (42), *Astragalus* (32), *silene* (24), *Convolvulus* (23), *Allium* and *Plantago* (22each), *Fagonia and Lotus* (20each), *Trifolium* (19), *Cyperus and Medicago* (18 each), *Heliotropuim* (17), *Atriplex, Bromus* and *Centaurea* (16 each), *Vicia* (15). *Anthemis, Indigofera,*

Salsola and *Stipagrostis (* 14each*)*, *Eragrostis, Erodium* and *Veronica (* 13each*), Bellevalia, Helianthemum and Solanum (* 12each*), Amaranthus, Kickxia, Launaea, Reseda and Trigonella (* 11each*), Acacia, Juncus, Muscari, Ranunculus, Tribulus* and *Zygophyllum (* 10each*)* and the rest of genus contain less than ten species and many of these genus represented by only one species. A striking feature in Egypt's flora is a large number of genera in proportion to that of the species amounting to about 3 species per genus. This is a very low figure compared to the average global proportion which amounts to about 14. The generic index, i.e., the number of genera per 100 species is relatively high which points to the marginal conditions of Egypt with respect to many genera, and also indicates the lack of accumulation and differentiation centers in Egypt.

Sinai Peninsula characterized by the largest number of endemic species for up to 37 species, including four species grow in certain other Egyptian geographic areas and the rest (33 species) grows only in the Sinai. Five species grow in the Eastern Desert only, and4 species distribute in the Eastern Desert and also in the other areas of Egypt. One species restricted in the Western Desert. Six species grow in the River Nile only, and 3 species is distributed in the Nile River and also in the other areas of Egypt. Three species grow in the oases only, and three species are distributed in the oases and also in the other areas of Egypt. Four species grow in the Mediterranean coast and other areas of Egypt. one species in Elba Mountains only.

Egypt is the meeting point of floristic elements belonging to at least three phytogeographical regions: the African Sudano- Zambesian; the Asiatic Irano- Turanian; the Afro- Asiatic Saharo- Indian and the Euro- Afro- Asiatic Mediterranean.

(ii) Vegetation

The vegetation in the Egyptian deserts distributes in three main geographical units:

iv. The Western Desert;
v. The Eastern Desert;
vi. The Sinai Peninsula.

i The Western Desert

Ecologically, Zahran and Willis [17] divided the area into three main regions: (1) the Western Mediterranean coastal belt; (2) the inland oases and depressions; and (3) the Gebel Uweinat.

(a) The Western Mediterranean Coastal Belt

It extends from Alexandria westward to El Sallum for about 500 km, and from the seashore inland for about 40-50 km. It is the richest part of Egypt in flowering plants, (50 % of the total number of species of the Egyptian flora). Out of which 154 species are confined in their distribution to this belt. Most of these species are annual weeds that flourish during the rainy season, giving the area a temporary gray grassland aspect. During the long dry periods the characteristic woody shrubs and perennial herbs, constitute the scrub vegetation of the area, scattered sparsely in parts and grouped in denser distinct patches in favored habitats. Five main types of ecosystems recognized in the coastal land:

1. Sand Dunes
2. Rocky Ridges
3. Saline depressions
4. Non-Saline Depression
5. Wadis.

– **Sand Dunes**

The following are the most common plant communities inhabiting sand dunes:

a. *Ammophila arenaria—Euphorbia paralias community*: this inhabits the mobile sand dunes bordering the sea.
b. *Ononis vaginalis- Crucianella maritima community* which occupies the older, advanced and higher dunes, where the sand may be consolidated impart.
c. The stabilized sand dunes dominated by Crucianella maritime, Elymus farctus, Pancratium maritimum, Thymelaea hirsuta, Euphorbia paralias and Echinops spinosissimus.

– **Rocky Ridges**

The vegetation of rocky ridges is an association of *Gymnocarpos decander* and *Thymelaea hirsuta.* The rocky sites with low moisture availability are dominated by communities of *Globularia Arabica* and *Thymus capitatus* while sites with fairly deep soils and high moisture availability are dominated by communities of *Asphodelus microcarpos, Herniaria hemistemon, Plantago albicans* and *Thymelaea hirsuta.*

– **Saline Depressions**

The saline depressions are dominated by halophytic communities. These communities are: *Juncus rigidus* community, *Halocnemum strobilaceum*community, *Zygophyllum album* community, *Sporobolus pungens* community,*Limoniastrum monopetalum* community, *Salsola tetrandra* community, *Arthrocnemum macrostachyum—Limoniastrum monopetalum* community, *Lygum spartum* community, *Salicornia fruticose—Limonium pruinosum* community, *Limoniastrum monopetalum—Lycium shawii* community and *Atriplex halimus—Picris radicata* community.

– **Non—Saline Depressions**

The non- saline depressions (barely fields) are favorable for cultivation. The vegetation of non- saline depressions belongs to the *Plantagineto-Asphodeletum microcarpae* associations, that dominated by *Anabasis articulta* community, *Zygophyllum album* community, *Plantago albicans* community and *Asphodelus microcarpos- Thymelaea hirsuta.*

– **Wadis**

The communities have been recognized in this habitat dominated by: Thymelaea hirsute, Gymnocarpos decander, Asphodelus microcarpos, Plantago albicans,Hammada scoparia, Anabasis articulate, Echiochilon fruticosum, and Artemisia inculta.

(b) **TheInland Oases and Depressions**

The plant life of these oases and depressions comprises four main types of vegetation namely: reed swamps vegetation, salt marsh vegetation, sand dune vegetation and desert plain vegetation. The reed swamps dominated by *Phragmites austoralis, Typha domingensis, Typha elephentina* and *Ceratophyllum demersum.*The salt marsh vegetation dominated by *Juncus rigidus, Juncus acutus, Cyperus laevigatus, Suaeda aegyptiaca, Suaeda vermiculata, Alhagi graecorum, Tamarix nilotica, Cressa cretica, Cyperus lavigatus, Cladium mariscus, Aeluropus lagopoides, Sporobolus pungens, Arthrocnemum macrostachyum, Suaeda aegyptiaca, Cynodon dactylon, Sarcocornia fruticosa, Salsola tetrandra.* The vegetation of sand dunes dominated by*Zygophyllum album, Imperata clyndrica, Alhagi graecorum, Nitraria retusa,Cornulaca monocantha* and *Desmostachya bipinnata.* The vegetation of desert plain dominated by *Zygophyllum coccineum, Fagonia arabica, Acacia raddiana,Alhagi graecorum, Nitraria retusa, Suaeda monica,Salsola imbricata* ssp. *Imbricate,Atriplex leucoclada, Phoenix dactylifera, Citrullus colocynthis, Tamareix amplexiculis, Calotropis procera, Chrozophora obliqua, Hyoscyamus muticus, Tamarix nilotica and Stipagrostis scoparia.*

The hydrophytic vegetation of the Oases dominated by Najas armata,Utricularia gibba, Lemna gibba, Najas graminea, Ceratophyllum domersum and Ruppia maritime.

(c) **Gebel Uweinat**

Gebel Uweinat is a mountain of little springs.The vegetation in Gebel Uweinat dominated by *Panicum turgidum, Zilla spinosa and Stipagrostis plumose* in the wadis of Gebel Uweinat. Near the wells, the vegetation dominatedby *Phragmites austoralis, Typha domingenses, Juncus rigidus and Imperata cylindrical.*The vegetation on the different altitudes of GebelUweinat dominated by *Aerva javanica, Acacia ehrenbergiana, A. raddiana and Ochradenus baccatus.*

(ii) **The Eastern Desert**

The Eastern Desert of Egypt occupies about 223,000 Km^2. i.e. 21% of the total area of Egypt. This area is extending from the Nile Valley eastward to the Gulf of Suez and the Red Sea which distinguished by high mountains, the peaks of many of them more than 1500 m.

(a) **The Red Sea Coastal land**

Within the Red Sea Coastal land, two halophytic types of vegetation have been recorded: mangrove and salt marsh. The Mangrove vegetation dominated by *Avicennia marina*, sometimes associated with *Rhizophora mucronata*. The salt marsh vegetationcomprises a number ofseveral community types include *Halocnemum strobilaceum, Arthrocnemum macrostachyum, Limoniastrum axillare, L. pruinosum, Sporobolus pungens, Suaeda monoica, Halopeplis perfoliata, Halopyrum mucronatum, Tamarix nilotica,T. passerinoides, Zygophyllum album, Nitraria retusa, Juncus rigidus, Cressa cretica, Salicornia fruticosa and Aleluropus massauensis.*

(b) **The Coastal Desert Wadis**

The vegetation classified into two main types: ephemerals and perennials. Ephemeral vegetation is dominated by *Zygophyllum simplex, Trianthema crystalline* and *Trribulus pentandrus*. Prennial vegetation dominated by *Zilla spinosa, Panicum turgidum, Salsola baryosma, Pennisetum dichotomum, Haloxylon salicornicum, Launaea spinosa, Acacia tortilis, Acacia ehrenbergiana, Acacia raddiana, Cleome droserfolia, Sphaerocoma hookeri, Lasiurus hirstus, Lycium shawii, Tmarix aphylla, Leptadenia pyrotechnica, Balanites aegyptiaca, Salvador persica, Tamarix aphylla and Capparis deciduas.*

(c) **The Coastal Mountains**

The Red Sea coastal mountains of Egypt may be categorized intoartesianmountain facing the Gulf of Suez and mountain facing the Red Sea proper.

(d) **Mountain facing the Gulf of Suez**

The vegetation dominated by two communities: *Zilla spinosa* and *Moringa peregrine.*

(e) **Mountain facing the Red Sea proper**

The wadis of these mountains dominated by Leptadenia pyrotechnica, Balanites aegyptiaca, Salvador persica, Acacia ehrenbergiana, Acacia raddiana, Moringa peregrine, Aerva javanica, Dracaena ombet, Euphorbia cuneata, Dodonaea viscosa, Abutilon pannosum, Euphorbia schimperi, Acacia etbaica, Commiphora opobalsamum, Panicum turgidum and Acacia tortilis.

(f) **The Inland Desert**

The inland part of the Eastern Desert lies between the range of the Red Sea coastal mountains in the east and the Nile Vally in the west. It divides into four geomorphological and ecological regions.

1. **The Cairo—Suez desert**

The vegetation comprises these communities: Panicum turgidum, Haloxylon salicornicum, Retama reatam, Anabasis articulate, Zygophyllum coccineum, Launaea spinosa, Artemisia monosperma, Ephedra alata, Zygophyllumdecumbens, Acacia tortilis andLasiurus hirstus.

2. **The Limestone desert**

This section includes the wadis in Helwan desert Beni-Suef—Minia desert, Assiut—Qena desert. The vegetation dominated by *Stachys aegyptiaca, Zygophyllum coccineum, Zygophyllum decumbens, Anabasis setifera, Zygophyllum album, Anabasis articulata, Zilla spinosa, Pennisetum divisum, Lycium shawii, Netraria retusa, Tamarix nilotica, Atriplex halimus Echium rawolfii , Echinops spinosissimus, Farsetia aegyptia, Francoeuria crispa , Aerva javanica,Calligonum comosum,Crotalaria aegyptiaca, Acacia ehrenbergiana* and *Leptadenia pyrotechnica.*

3. **The sandstone desert of Idfo-comombo**

The main communities in this section include Fagonia bruguieri, Schouwia thebaica, Zilla spinosa, Crotalaria aegyptiaca, Cassia senna, Francoeuria crispa, Citrullus colocynthis and Acacia ehrenbergiana.

4. **The Nubian desert**

The Wadis of Nubian desert that flow into the Nile include the following communities: Acacia tortilis,Morttia philaena , Fagonia indica, Cassia senna, Aerva javanica, Indigofera spinosa, Acacia ehrenbergiana, Salsola baryosma, Acacia raddiana, Tamarix nilotica, Salvador persica, Balanites aegyptiaca and Leptadenia pyrotechnica.

iii **The Sinai Peninsula**

The Sinai Peninsula is a triangular plateau in the northeast of Egypt with its apex. In the south, at Ras Muhammed, where the eastern coast of the Gulf of Suez meets the western coast of the Gulf of Aqaba. Its base, in the north, is along the Mediterranean Sea. The eastern section of the Egyptian Mediterranean coast that extends for about 240Km between Port Said and Rafah. The area of the Sinai Peninsula (61000 Km^2) is about 6% of that of Egypt [17].

(a) **The Mediterranean Coastal area**

The natural vegetation of the coastal area is very sparse: three main habitats can be recognized: sabkhas, sand dunes and open sand plains. The dominant species in these habitats are: *Halocnemum strobilaceum, Anabasis articulate, Tamarix nilotica, Artemisia monosperma, Zygophyllum album, Nitraria retusa, Panicum turgidum,*

Stipagrostis scoparia, Halocnemum strobilaceum, Retama raetam, Calligonum polygonoides, Deverra tortuosa, Anabsis articulata, Thymelaea hirsuta, Lycium shawi, Noaea mucronata, Asparagus stipularis, Stipagrostis scoparia, S. plumosa, Centropodia forsskaolii, Panicum turgidum, Artemisia monosperma, Moltkiopsis ciliate,Zygophyllum dumosum Suaeda aegyptiaca and Pharagimtes australis . Juncus arabicus, Retama reatam—Achillea fragrantissima and *Sarcocornia fruticosa.*

(b) **The Gulf of Suez coast**

Four types of habitat have been recognized in the western coast of Sinai: mangrove swamps, littoral salt marshes, oases and coastal area of the desert.

- **Mangrove swamps**: Pure community dominated by *Avicennia marina.*
- **Littoral salt marshes**: The dominant communities in this habitat are:*Halocnemum strobilaceum, Zygophyllum album, Nitraria retusa, Limonium pruinosum, Limonium pruinosum, Haplophyllum tuberculatum, Tamarix nilotica, Cressa cretica,Juncus rigidus and Aeluropus massauensis.*
- **The oases**: These oases like Ayon Mousa and Hammam Mousa. Its dominated by *Tamarix nilotica, Juncus rigidus, Alhagi graecorum* and *Desmostachya bipinnata.*
- **Coastal desert**: Vegetation in this habitat dominated by: *Retama reatam, Haloxylon salicornicum, Panicum turgidum, Varthemia montatna, Artemisia judaica, Phragmites austoralis, Typha domingensis, Alhagi graecorum, Desmostachya bipinnata, Haloxylon salicornicum, Zilla spinosa, Achillea fragrantissima, Zygophyllum coccineum. Capparis cartilaginea, Zygophyllum deccumbens* and *Tamarix nilotica.*

(c) **The Gulf of Aqaba coast**

The vegetation of of the western coast of the Gulf of Aqaba divided into: mangal, littoral salt marshes and coastal desert types. The mangal vegetation dominated by *Avicennia marina.* The littoral salt marshes dominated by *Limonium axillaris,* while the vegetation of the coastal desert dominated by *Artemisia judaica, Capparis cartilaginea, Panicum turgidum, Moringa pregrina, Acacia tortilis,Acacia raddiana, Zilla spinosa, Retama reatam, Calotropis procera Leptadenia pyrotechnica* and *Launaea spinose.*

(d) **The montane country**

The ranges of mountains that form the montane country of Sinai are in the southern and central subregions of the peninsula and comprise Gebel St. Catherine, Gebel Mosa, Gebel El-Tih, Gebel El-Maghara, Gebel El-Halal and others. The habitats of the montane country of Sinai include rocky wadis, upstream parts of large wadis that flow into the Gulf of Suez and Gulf of Aqaba, gullies, terraces, rock cervices and slopes of mountains. These habitats dominated by *Zilla spinosa,Gymnocarpos decadrum, Anabasis setifere, Fagonia mollis, Alkanna orientalis, Capparis spinosa, Launaea nudicaulis, Artemisia judaica, Acacia tortilis subsp. raddiana, Leptadenia pyrotechnica, Artemisia herba-alba, Fagonia arabica,*

Stachys aegyptiaca, Tanacetum sinaicum, Teucrium polium, Peganum harmala, Artemisia herba-alba, Matthiola arabica, Phlomis aurea, Achillea fragrantissima, Pulicaria undulate, Retama raetam, Haloxylon salicornicum, Heliotropium digynum, and Neurada procumbent, Ochradenus baccatus, Bufonia multiceps, Galium sinaica, Globularia arabica, Ephedra alata, Ballota undulata, Reseda pruinosa, Caylusea hexagyna, Bufonia multiceps, Globularia arabica, Ephedra alata, Crotalaria aegyptiaca and Chrozophora oblongifolia, Forsskaolea tenacissima, Lycium shawii, Seriphidium herba-alba, Hyoscyamus boveanus, Tamarix aphylla, Teucrium polium, Micromeria sinaica, Zygophyllum album, Reaumuria hirtella, Phoenix dactylifera, and Ficus pseudosycomorus, Pulicaria undulata, Artemisia judaica, Mentha lavandulacea; Hyoscyamus boveanus, Capparis spinos, Echinops spinossissimum, Typha domingensis, Phragmites australis, Pulicaria incise, Solanum nigrum, Fagonia tristis, Piturathos tortuosa, Cleome Africana, Cleome droserifolia, Zygophyllum coccineum, Phlomis aurea, Tanacetum santolinoides, Mentha longifolia, Varthemia montana and Nepeta septemcrenata.

(e) **Animal Resources**

Domestic animals in Egyptian Deserts include camels, donkeys, sheep, and goats, the final of which are especially discernible within the Egyptian farmland. The animals that figure so noticeably on the antiquated Egyptian friezes—hippopotamuses, giraffes, and ostriches—no longer exist in Egypt; crocodiles are found only south of the Aswan High Dam. The largest wild animal is the Aoudad (a type of bearded sheep), which survives in the southern fastnesses of the Western Desert. Other desert animals are the Dorcas gazelle, the Fennec (a small, desert-dwelling fox), the Nubian ibex, the Egyptian Hare, and two kinds of Jerboa (a mouselike rodent with long hind legs for jumping). The Egyptian Jackal (*Canis lupaster*) still exists, and the Hyrax is found in the Sinai mountains. There are two carnivorous mammals: the Caffre cat, a small Feline predator, and the Ichneumon, or Egyptian mongoose. Several varieties of lizards are found, including the large monitor. Harmful snakes incorporate more than one species of snake, the speckled snake is found all through the Nile valley and the Egyptian cobra (*Naje haje*) in agricultural areas. Scorpions are common in desert regions.

There are various species of rodents. Numerous assortments of insects are to be found, counting the grasshopper.

Egypt is wealthy in birdlife. Numerous birds pass through in large numbers on their spring and autumn time movements; in all, there are more than 200 moving sorts to be seen, as well as more than 150 inhabitant winged creatures. The hooded crow is a familiar resident, and the black kite is characteristic along the Nile valley and in Al-Fayyūm. Among the birds of prey are the lanner falcon and the kestrel. Lammergeiers and Golden Eagles live in the Eastern Desert and the Sinai Peninsula. The sacred ibis (a long-billed wading bird associated with ancient Egypt) is no longer found, but the great white egret and cattle egret appear in the Nile valley and Al-Fayyūm, as does the Hoopoe (a bird with an erectile fanlike crest). Resident desert birds are a distinct category, numbering about 24 kinds [41].

Discussion

Egypt is a desert country. Deserts in Egypt represent about 96% of Egypt area. About 98% of the Egyptians live in this area. The western desert of Egypt occupies about half of the country (44%). Most of these deserts are located in the hyper-arid region, except the north coast at the Mediterranean sea, which located in the arid region. Any strategies for development in Egypt should depend on the natural resources in these deserts. The main natural resources depend on the quantity and quality of available freshwater. Groundwater is the main source of freshwater in the Egyptian deserts, where the Nubian Sandstone and Carbonate rocks complex aquifers form the largest area in these deserts. Most of this area occupy the western desert of Egypt. So, the projects of sustainable development should focus on this area.About 420 thousand ha in the Egyptian soil is sandy and calcareous. The average results of physical and chemical analyses of soils, sampled at various locations to represent the various types of soils. The normal results about of physical and chemical investigations of soils, inspected at different areas to speak to the different sorts of soils. The organic matter content is low and so, accordingly, is the concentration of total nitrogen. As regards the alluvial soils (clayey and loamy clay), available phosphorous determined by Olsen's method is generally moderate. The available (soluble and exchangeable) potassium extracted with a neutral solution of ammonium acetate is high, and this is characteristic of most Egyptian alluvial soils [42].

In Egypt, desert vegetation is mostly patches distributed mainly in four main habitats, sand formation, gravels formations, salt marshes and reed swamps and is formed mainly of xerophytic shrubs and sub-shrubs. Monod [38], recognized two types of desert vegetation, namely contracted and diffuse. Both types refer to permanent vegetation that can be accompanied by ephemeral (or annual) plant growth depending on the amount of precipitation in a given year.The flora of Egypt includes 2121 species of vascular plants out of them 27 species cultivated and 61 species endemic.This flora is rich in the economic plants (medicinal, fodder, fiber, oil, wood, bioenergy, … etc.).The natural resources in the Egyptian deserts subjected to degradation and deterioration mainly by land salinization, water deficiency and wind erosion. These important resources in the Egyptian deserts need to more deep studies to use their potentialities for sustainable development in Egypt.

Conclusions

Desert in Egypt had a wide area represents about 96% of the total area of Egypt. This area comprises of three main geographical units: the Western desert, the eastern desert, and the Sinai Peninsula, where the Eastern desert is the biggest one (more than 50% o Egypt area). Most of the Egyptian deserts are located in the hyper-arid region except the Mediterranean coast.The coastal zone of the Mediterranean is rich with water resources (ground and surface water) comparing with the other deserts of

Egypt, while the Eastern desert is the poorest one in the water resources. On the other hand, the flora in this zone is more diverse than in the other deserts. Sinai Penninsula distinguished by the highest number of endemic plant species and the highest amount of rainfall at the far Eastern coast of the Mediterranean. The best soil properties for agriculture found in the edges of the Nile Valley and Nile Delta, also at the coast of the Mediterranean in the Western desert and Sinai Penninsula.

Recommendations

The future of development in Egypt started in the Western Desert which is promising especially in agriculture development, followed by Sinai Peninsula. To mitigate the population density in Nile Valley and Nile Delta, it is recommended to encourage the development projects and investments in the Western Desert and Sinai Peninsula which lead to the resettlement of Bedouins and reduce migration from these areas to the cities.

Acknowledgements Abdelazim M. Negm and Ahmed Elkhouly want to acknowledge the support of the Science and Technology Development Fund (STIDF) of Egypt in the framework of the grant no. 30771 for the project titled "A Novel Standalone Solar-Driven Agriculture Greenhouse—Desalination System: That Grows Its Energy and Irrigation Water" via the Newton-Mosharafa funding scheme, Call 4.

References

1. ENAPCD, Egyptian national action program to combat desertification (2005) Desert research center. Egypt, Cairo, p 118
2. Ahmed ISA (2015) Desertification in Egypt: current status and trends. In: The Proceedings of the 5th KIDF: conference proceedings kubuqi international desert forum 2015
3. Collins J (2017) Vital lakes disappearing around the world. http://www.dw.com/en/vital-lakes-disappearing-around-the-world/a-37401645
4. CAPMAS (2018) Egypt in figures, central agency for public mobilization and statistics, Egypt. https://www.capmas.gov.eg/Pages/StaticPages.aspx?page_id=5035. Accessed 25 Nov 2019; Gleick PH (2002) The world's water 2002–2003: the biennial report on freshwater resources. Isl Press, Washington, DC, USA
5. Lutz A, Thomas JM, Keita M (2011) Effects of population growth and climate variability on sustainable groundwater in mali West Africa. Sustainability 3:21–34
6. El-Din MMN (2013) Proposed climate change adaptation strategy for the ministry of water resources & irrigation in Egypt. UNESCO, Cairo, Egypt. http://www.eeaa.gov.eg/portals/0/eeaaReports/
7. CAPMAS (2017) Annual bulletin of the movement of production, foreign trade & available for consumption of agricultural commodities, cent. Agency Public Mobilization Stat. Cairo, Egypt. https://www.capmas.gov.eg/Pages/Publications.aspx?page_id=5104&Year=23426. Accessed 10 Dec 2019
8. Abayazid H, Al-Shinnawy I (2012) Coastal lake sustainability: threats and opportunities with climate change. J Mech Civ Eng Int Organ Sci Res (JMCE/IOSR) 1(5):33–41

9. El-Ghani MA, Salama F, Salem B, El-Hadidy A, Abdel-Aleem M (2013) Biogeographical relations of a hyperarid desert flora in eastern Egypt. Afr. J. Ecol. 52, 173–191. (John Wiley & Sons Ltd)
10. Ayyad MA, Ghabbour SI (1986) Hot deserts of Egypt and Sudan. In: Evenari M, Noy-Meir I, Goodall DW (eds) Ecosystems of the world, 12B, Hot desert and arid shrublands, B. Elsevier, Amsterdam, pp 149–202
11. Kassas M, Imam M (1959) Habitat and plant communities in the Egyptian desert. IV: the gravel desert. J Ecol 47:289–310
12. Kassas M, Girgis WA (1964) Habitats and plant communities in the Egyptian deserts: the limestone plateau. J Ecol 52:107–119
13. Kassas M, Girgis WA (1965) Habitat and plant communities in the Egyptian desert. VI: the units of desert ecosystem. J Ecol 53:715–728
14. Kassas M, Girgis WA (1970) Plant life in the nubian desert east of the nile Egypt. Extrait du Bulletin de l'Institute d'Egypte 51:47–71
15. Kassas M, Girgis WA (1972) Studies on the ecology of the eastern desert of Egypt: the region between lat. 27°30′N and lat. 25°30′N. Bulletin de la Société de Géogaphie d'Egypte 42:42–72
16. Elkhouly AA (1996) Ecological and phytochemical studies on Mesembryanthemmum species in Egyptian desert. PhD thesis, faculty of science, Menoufia University
17. Zahran MA, Willis AJ (1992) The vegetation of Egypt. Chapman & Hall, London, 424 p
18. El-Khouly AA, Khedr AA (2000) Species diversity and phenology of the wetland vegetation in Siwa Oasis, Western Desert, Egypt. Desert Inst Bull, Egypt 50(2):325–343
19. EL-Khouly AA, Zahran MA (2002) On the ecology of the halophytic vegetation of the Oases in Egypt. In: International symposium on optimum resources utilization in salt- affected ecosystems in arid and semi arid regions. Cairo, Egypt, pp 277–286
20. Gad MRM (1999) Environmental conditions affecting the growth of plants used for the stabilization of sand dunes: Siwa Oasis, Case study. In: Msc thesis institute of environmental studies and research. Ain Shams University, 141 p
21. Batanouny KH, Batanouny MH (1969) Formation of phytogenic hillocks. I: plant forming phytogenic hillocks. Acta Bot Acad Sci Hung. 14:243–252
22. Zahran MA, Mashaly IA (2000) Climate and vegetation of Egypt. In: Proceedings, 5th conference: metrology & sustainable development, Cairo, pp 196–210, 22–24 Feb 2000
23. Danin A (1983) Desert vegetation of Israel and Sinai. Cana PublishingHouse, Jerusalem
24. Zahran MA (1989) Principles of plant ecology and flora of Egypt. Dar El Nashr for Egyptian Universities-El Wafa Library 388 p
25. RIGW/IWACO Research institute for groundwater (1988, 1993) Hydrogeological map of Egypt, scale 1:2,000,000. Cairo
26. Elbeih SF (2015) An overview of integrated remote sensing and GIS for groundwater mapping in Egypt. Ain Shams Eng J 6:1–15
27. Abu Zeid M, Rady MA (1992) Water resources management and policies in Egypt. In: Country experiences with water resources management: economic, technical sand environmental issues. World Bank, Technical paper no. 175, Washington DC
28. Hefny K (1992) Groundwater assessment in Egypt. International Water Resources Association Magazine, Oxford
29. RIGW/IWACO (1999) Contributions to environmental management of Egypt's groundwater resources. Final report, EMGR Project, Cairo
30. Shata A (1982) Hydrogeology of the great nubian sandstone basin. Q J Eng Geol 15:127–133
31. Shata A (2001) Contribution to the hydrogeology of the nile delta region. In: Symposium on deltas, modern and ancient, Cairo
32. Allam AR, Saaf E, Dawoud MA (2002) Desalination of brackish groundwater in Egypt. Desalination 152, Elsevier, pp 19–26
33. El Tahlawi MR (2004) Geology of Egypt and its groundwater. In: Annual meeting, Egypt, commission on mineral and thermal waters, handouts, Cairo, pp 1–10. 25 Sept–04 Oct
34. El Tahlawi MR (2006) Mining geology. Assiut University Press, Assiut, p 246

35. El Tahlawi MR, Farrag AA, Ahmed SS (2008) Environ Geol 55:639. https://doi.org/10.1007/s00254-007-1014-1
36. National Weather Service Weather Forecast Office web site www.srh.noaa.gov/mrx/hydro/flooddef.php. Accessed 4 Feb 2010
37. FAO/UNESCO Soil map of the world, Map sheet VI. http://www.fao.org/docrep/008/y5863e/y5863e05.htm. Accessed on 15 Oct 2019
38. Monod T (1954) Mode contracté et diffus de la vegetation Saharienne. In: Cloudsley-Thompson JL (ed) Biology of desert (ed. by), Institute of Biology, London, pp 35–44
39. Kassas M (1966) Plant life in deserts. In: Hills ES (ed) Arid lands. Methuen/Paris, UNESCO, London, pp 145–180
40. Kassas M (1971) Die Pflanzenwelt. Teil, Pflanzenleben in der östlichen Sahara. In: Schiffers H (ed) Die Sahara und Ihre Randgebiete. I: Physiogeographie. Weltforum-Verlag, München, pp 477–497
41. Encyclopedia Britannica. Plant and animal life https://www.britannica.com/place/Egypt/Plant-and-animal-lifem. Accessed on 10 Oct 2019
42. FAO (2005) Fertilizer use by crop in Egypt. First version, published by FAO, Rome

Groundwater Occurrence and Ecosystem Services

Groundwater Ocurrences in West Nile Delta, Egypt

Ahmed Mohamed Ali Youssef, Sawsan Moselhy Mohamed Ibrahem, and Rasha Abdallah Hussien

Abstract The evaluation of the groundwater occurrences needs integration of more than one technique to give an accurate estimation of the groundwater potential. Since, the West Nile Delta suffered from the deterioration in groundwater due to intensive of exploitation which resulted in both groundwater depletion and salinity increase. So, the integrated geophysical, hydrological and isotopic investigations were carried out in Wadi El Farigh, western Nile delta to monitor and assessment of the changes in the groundwater level, quality, aquifer renewability and mixing between existing aquifers. Geophysical results reveal that there are two aquifer separated by basaltic sheet. The first one is represented by the Miocene aquifer (El-Moghra) and the second one is the Oligocene aquifer which has lower resistivities than the upper one and this is mainly due to increase of clay content and the water salinity. Hydrological results reveal that the saturated thickness of Miocene aquifer decreased during (2003–2015) period and this depletion was approximately ranged between 1.3 and 1.7 m/year. Total dissolved solids (TDS) content of groundwater samples varies from 234 to 2458 mg/L implying significant deterioration and salinization problems. Application of Drinking Water Quality Index (DWQI) reveals that water samples are ranked between excellent category (~33%) to poor category (~67%) which is almost always threatened or impaired. The $\delta^{18}O$–$\delta^{2}H$ relationship is suggesting that two distinct sample groups were classified according to their isotopic compositions. These two groups are nonrenewable and have isotopic signals varied completely according to the time of recharge. The relation between $\delta^{18}O$‰ and Cl concentration in (epm) suggested that both of two groups are exposed to dissolution as a result of an over-pumping process or mixing with more depleted strata (Oligocene aquifer) through the faults as in group 2 that exhibits more saline water and more isotopic depletion. Results indicate that most of the wells are operated at relatively high pumping rates accompanied with low efficiency and specific capacities besides decline of well yields and water deterioration.

A. M. A. Youssef (✉) · S. M. M. Ibrahem
Desert Research Center (DRC), El-Mataryia, Cairo, Egypt

R. A. Hussien
Egyptian Nuclear Radiological and Regulatory Authority (ENRRA), Cairo, Egypt

A. Negm and A. Elkhouly (eds.), *Groundwater in Egypt's Deserts*, Springer Water,
https://doi.org/10.1007/978-3-030-77622-0_3

Keywords Geophysical studies · Hydrological parameters · Isotopic techniques · Groundwater problems · Wadi El-Farigh · Western Nile Delta

Introduction

Groundwater is considered the major portion of the world's freshwater resources. One of the main challenges facing the sustainable development of Egypt is the need for better management of its limited fresh water resources. Groundwater exists in the Nile Valley, Nile Delta, Western Desert Oases and Sinai Peninsula. Some of the most important problems facing the groundwater management in Egypt are: Over-exploitation and lowering of its levels in some areas; waterlogging and/or salinization. West of the Nile Delta area is one of the most important areas in Egypt for agricultural, industrial and recreation investments due to its high groundwater potentials, accessibility and facilities provided by the government in Wadi El-Natrun, Sadat city and Wadi El-Farigh. The area of Wadi El-Farigh is one of the areas in the western Nile Delta fringes subjected to intensive land reclamation and other agricultural, industrial activities. Such activities are probably not controlled, have seriously affected the potentials of the groundwater resources in the area. This effect is quite noticed in the large farms where, groundwater over-pumping is very high to fulfill the crop requirements leading to serious problems in both the quality and quantity of withdrawing water. Where, gradual drawdown of water level yearly in these farms and deterioration of the groundwater quality are recorded in aquifers tapping this area. Western Nile Delta aquifers had been subjected to several studies since the mid of the twentieth century.

The investigated area lies between latitudes of 30°12′00″ & 30°18′00″ N and longitudes 30°28′48″ & 30°40′48″ E (Fig. 1). Its climate is characterized by a long hot summer and a short warm winter, low rainfall and high evaporation. The average monthly meteorological data provided by the Egyptian Meteorological Authority are recorded from Wadi El-Natrun station during the period from 1988 to 1993. The

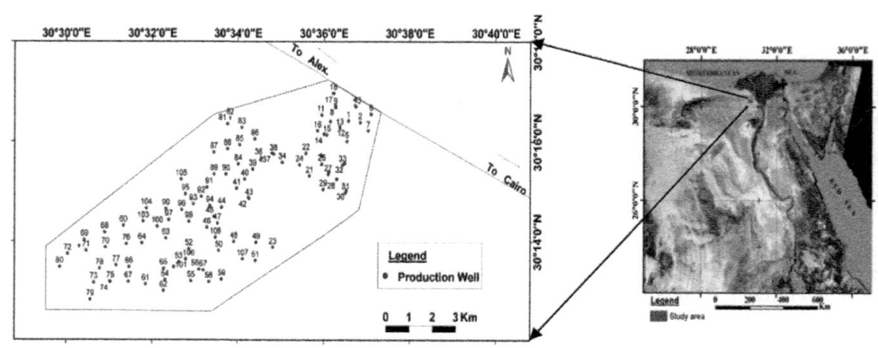

Fig. 1 A map showing the locations of the existed wells in the study area

climatic elements of the concerned area are summarized as follows: The recorded mean annual rainfall is 30.94 mm, the maximum air temperature value is 35.7 °C recorded in August, while the minimum recorded air temperature is 8 °C in January and the maximum relative humidity value is 66.5% in August and the minimum value is 44.4% in May.

The objectives of the present study are basically to classify and assess of geological, geophysical, hydrological and hydrochemical bases of the existed aquifers in the study area which are represented by the Oligocene and the Miocene aquifers. Also, to create baseline data that will be used as a guide to monitor and detect the future changes in the groundwater level and quality clarifying the roles of lowering of the groundwater levels, aquifer depletion, increase in water salinity and mixing between different aquifers.

Geomorphologic, Geologic and Hydrogeologic Settings

Geomorphologicaly, among the relevant studies of geomorphological aspects in the area, the followings are of particular interest: [1–8]. Based on these works, three geomorphological units are recognized as follows: the alluvial plains, the structural plains, and the tablelands. The structural plains are differentiated into two main units, namely; structural depressions and folded ridges and structural plateaus. The investigated area is a part of the structural depressions (Wadi El-Farigh depression. The elevation varies from less than 10 m (above mean sea level) at the northwest to more than 100 m at south.

Geologically, the west of the Nile Delta is covered by sediments and sedimentary rocks ranging in age from Late Cretaceous to Quaternary [1, 9, 10]. The study area as a part of the West of the Nile Delta comprised rock units varying from Oligocene to Quaternary. The stratigraphic succession from the base to the top is described as follows:

The tertiary deposits are related to Oligocene and Miocene. The Oligocene deposits are represented by Sand, ferruginous sandstone, sand clay, clayey sand and clay. The Oligocene deposits are exposed near Abu Roash domal structure to the southeast of the investigated area. It is unconformably underlain by Upper Cretaceous rocks [1, 5, 11, 12]. Oligocene deposits are overlain by dissected basaltic sheets of Abu Zaabal Formation. Through the drilled water wells by the Desert Research Center (DRC 1992–2013) and private companies along Cairo-Alex. Desert Highway, the basaltic sheets are detected in some localities at depth reaching down to more than 250 m with a thickness of about 20–30 m. The detection of the basaltic sheets at different levels reflects the effect of the structures on the geologic succession and the groundwater occurrence, flow and quality in consequence. The Miocene deposits cover wide parts of the area. These deposits are represented by El Moghra Formation which is mainly exposed on Wadi El Farigh area and its extension. El Moghra Formation is mainly composed of coarse sands, sandstone and clay interbeds with vertebrate remains, silicified woods and gravels at base [11]. The Quaternary deposits cover a large portion of the area under consideration and its adjacent areas. These deposits are composed of unconsolidated clastic deposits. It is not easy to differentiate between the Pleistocene and Recent deposits.

From the geologic structures point of view, the main regional structures affecting the study area and its vicinities have been discussed by many authors, [1, 11, 13]. The tectonic setting of the west Nile Delta is represented by folding and faulting. Twofold systems had affected the west Nile Delta area. According to their orientation, these fold systems are the Syrian arc system (NE–SW) as Abu Rwash anticline and the Clysmic system (NW–SE) as Wadi El Farigh and Wadi El Natroun anticline [14]. The faults are mainly arranged parallel to the surface folds. They belong to the Syrian arc system (NE–SW) and the Clysmic system (NW–SE), with low vertical displacements [15]. The southwestern portion of the Nile Delta had been subjected to intensive movements which resulted in dissecting the area with normal faults striking in NE–SW and NW–SE directions [13]. Some faults are major ones and define the regional structural pattern and other minor faults, however, had been initiated as direct effect of the major ones and form some local structural pattern such as horsts and grabens. These faults have a direct impact on the groundwater occurrence and flow direction. At some parts, these faults make a direct connection between the Miocene aquifer and the Oligocene aquifer with its saline water and consequently increase the salinity of the Miocene aquifer.

Hydrogeologically, the water-bearing formations in the west Nile Delta are represented by Pleistocene, Pliocene, Lower Miocene, Oligocene and Eocene aquifers. These aquifers are connected hydraulically. Lower Miocene and Oligocene aquifers are represented in the study area [16, 17].

Lower Miocene aquifer is represented by the Moghra Formation which covers the western part of the study area. This aquifer composed of sand, gravels, sandstone and clay interbeds with vertebrate remains and silicified wood [18–20]. This aquifer is present under semi-confined Condition, where it overlains by Quaternary deposits and underlains by the Oligocene basaltic sheet. The groundwater salinity is almost fresh.

Oligocene aquifer is composed of sand, sandy clay, clayey sand, sandy gravels, clay and sandstone interbeds with thin limestone band. The Oligocene aquifer is found to exist under confined to semi-confined conditions. The confined conditions prevail in the localities where the basaltic sheet and the underlying clay beds extent over a wide subsurface area. The semi-confined conditions are due mainly to the sandy clay and clay interbeds within the aquifer. The Oligocene aquifer is more clayey and/or higher in salinity or both than the Miocene aquifer [13]. The environmental isotopes indicate that the main recharging sources for the Oligocene are paleo-water [21].

Field Measurements

The field measurements include the geoelectrical measurements, hydrogeological data and collecting water samples.

Geoelectrical Measurements

The Schumberger four electrode configurations were used with current electrode spacing (AB) starting from 2 to 4000 m. This electrode separation is sufficient to reach the required depth that fulfills the aim of the study in view of the geologic and hydrogeologic information and is sufficient to detect the Oligocene aquifer which is underlying the basaltic sheet. The potentiometer (Terrameter SAS1000) was used for the geoelectrical measurements. This instrument directly measures the resistance (R) at each electrode separation with high accuracy.

Twelve Vertical Electrical Soundings (VES) were carried out (Fig. 2). Some of these soundings were carried out beside wells for verifying the geoelectrical interpretation. These geoelectrical soundings were carried out with the aims of detecting the Oligocene aquifer which is underlying the basaltic sheet to complete the subsurface pictures below the basaltic sheet because the all existed wells tapping only the Miocene aquifer. So, one of these soundings was measured beside well No.3 which having high salinity (1274 mg/L) and another sounding was carried out beside well No.20 which having the lowest salinity (293 mg/L) in the study area to be a reference

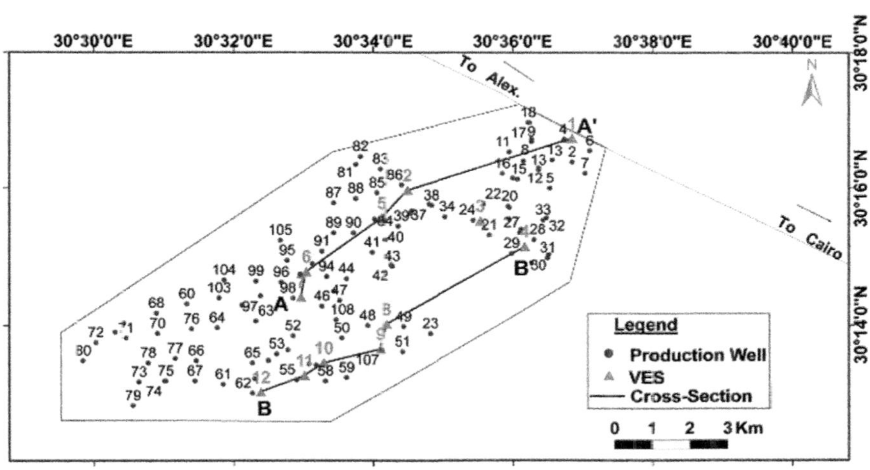

Fig. 2 A map showing the locations of the vertical electrical soundings, wells and the hydrogeoelectrical cross sections

salinity to find out the relative salinity of the Oligocene aquifer comparable with that of the Miocene aquifer.

Hydrological and Hydrochemical Measurements

In the present work, detailed study of the negative impact of the intensive aquifer exploitation was undertaken through the investigation of 108 production wells tapping the Miocene aquifer (Fig. 2). The study area is equipped with a good monitoring system which helps in the creation of base line data and will be used as a guide to monitor future deterioration of the Miocene aquifer, and also, the evaluation of the changes occurred by time.

Monitoring needs to focus on the depth to groundwater, groundwater level and saturated thickness of the aquifer and groundwater quality. The data of three different years were selected (2003, 2012 and 2015) and through these monitoring observations, effective management strategies can be put in place to control future occurrence of aquifer depletion.

For hydrochemical measurements about 23 groundwater samples were selected from the studied area.

Results and Discussion

The results and discussions of the geoelectrical, hydrogeological and hydrogeochemical studies will be discussed as the followings.

Geoelectrical Results

Prior to the detailed quantitative interpretation of the Vertical Electrical soundings data, a qualitative interpretation of the field curves was carried out to obtain preliminary information about the subsurface succession and the resistivity values. Generally, the field curves (Fig. 3) are terminated with K-type ($\rho1 < \rho2 > \rho3$), which is reflecting the presence of the Oligocene aquifer underlying the basaltic sheet with its high resistivity values ($\rho2$). Since the all existing wells taping only the Miocene aquifer, so the geoelectrical studies detected the Oligocene aquifer that is underlying the basaltic sheet.

The quantitative interpretation of the field curves was carried out by using computer programs [22, 23] for non-automatic iterative interpretation. The lithological data of the drilled wells were used for the construction of the initial model.

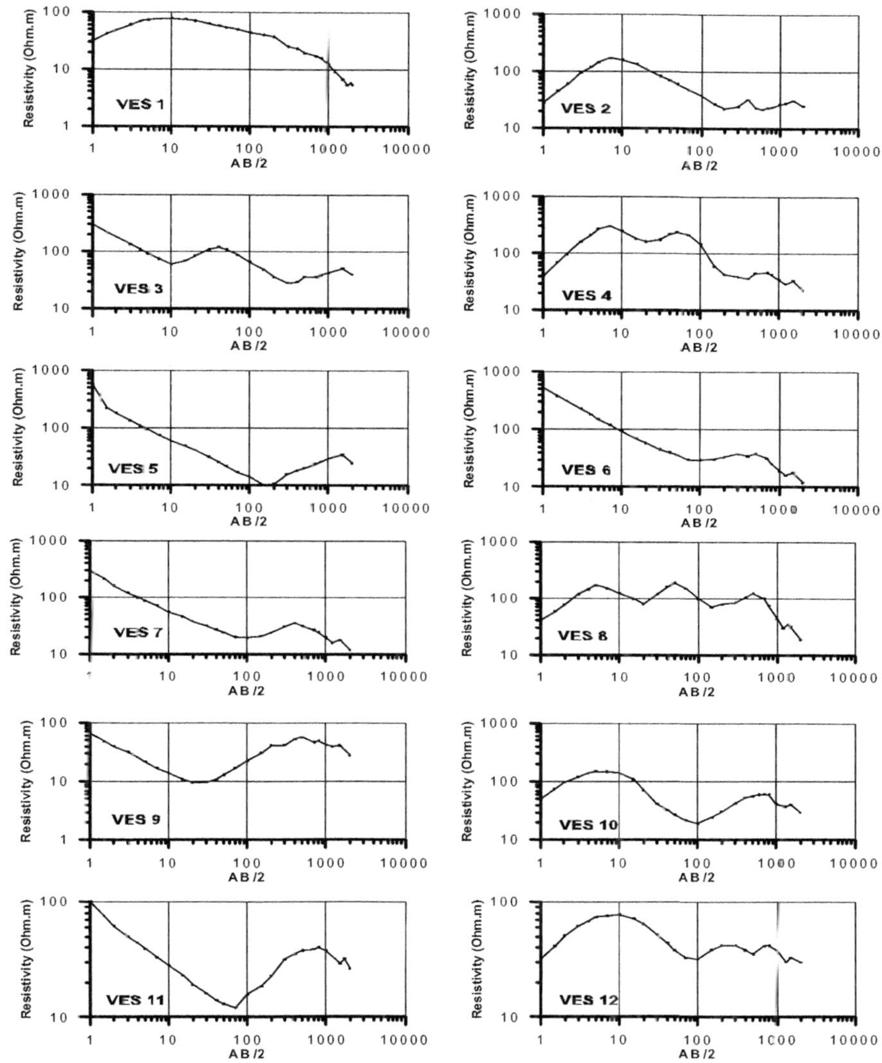

Fig. 3 A graph showing the different of the curve's type of the conducted vertical electrical sounding

The interpretation of the sounding curves and the hydrogeological data indicated that the subsurface succession area consist of 4 main geoelectrical units (A, B, C and D). The ranges of resistivities, thicknesses and corresponding lithology of these units are shown in Table 1.

Table 1 Ranges of resistivities, thicknesses and corresponding lithology of the geoelectrical units at the sounding stations

Geoelectrical unit	Resistivity (Ohm m)		Thickness (m)		Lithology
	Minimum	Maximum	Minimum	Maximum	
A	1 (VES 5)	1032 (VES 8)	80.8 (VES 6)	122.6 (VES 10)	Dry gravel, sand and clay
B	15 (VES 1)	110 (VES 10)	74 (VES 11)	210 (VES 1)	Water-bearing sand and clay of lower miocene aquifer (Moghra formation)
C	138 (VES 1)	1260 (VES 3)	28 (VES 2)	35 (VES 9)	Basaltic sheet
D1	8 (VES 1)	12 (VES 3)	29 (VES 3)	31 (VES 9)	Waterbearing clayey sand (top of Oligocene aquifer)
D2	4 (VES 1)	8 (VES 6)	–	–	Waterbearing sandy clay (bottom of Oligocene aquifer)

Geoelectrical Successions

To illustrate the horizontal and vertical distribution of the four geolectrical units (A, B, C and D) in the investigated area, 2 hydro-geoelectrical cross sections A–A′ (Fig. 4) and B–B′ (Fig. 5) were constructed in SW–NE direction. The description of these geoelectrical units can be summarized from top to bottom as follows.

Unit "A" consists of a group of dry thin layers of gravel, sand and clay. It has a wide resistivity range of 1–1032 Ω m (Table 1). The low resistivities reflect the presence of clay and silt, while, the high resistivity values are attributed to the presence of gravel and sand. This unit has a thickness varies from 80.8 m at VES No.6 to 122.6 m at VES No.10. This unit overlies unit "B" which is represented the water bearing formation and accordingly its thickness represents the depth to water in the study area.

Unit "B" represents the upper water bearing formation (Moghra aquifer) of Lower Miocene and consists of sand with clay intercalations. The resistivity of this unites ranges between 15 Ω m at VES No.1 and 110 Ω m at VES No.10. The low resistivity values reflect the presence of saline water (TDS = 1274 ppm) as recorded from well No.3 beside VES No.1, while the high resistivity values reflect the presence of fresh water (TDS = 293 ppm) as recorded from well No.20 beside VES No.12. Consequently, the resistivity values reflect to great extent the relatively salinity of

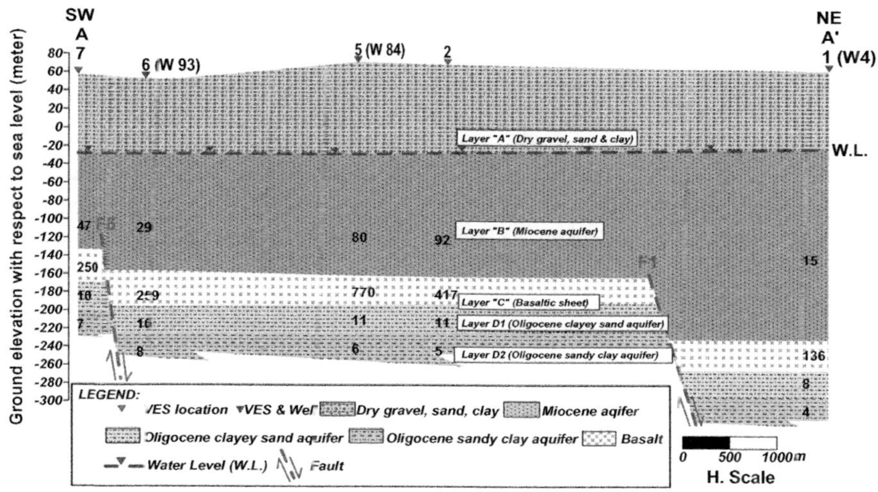

Fig. 4 Hydro-geoelectrical cross section A–A′ showing the vertical and horizontal distribution of the lithological succession, the water level and the inferred faults

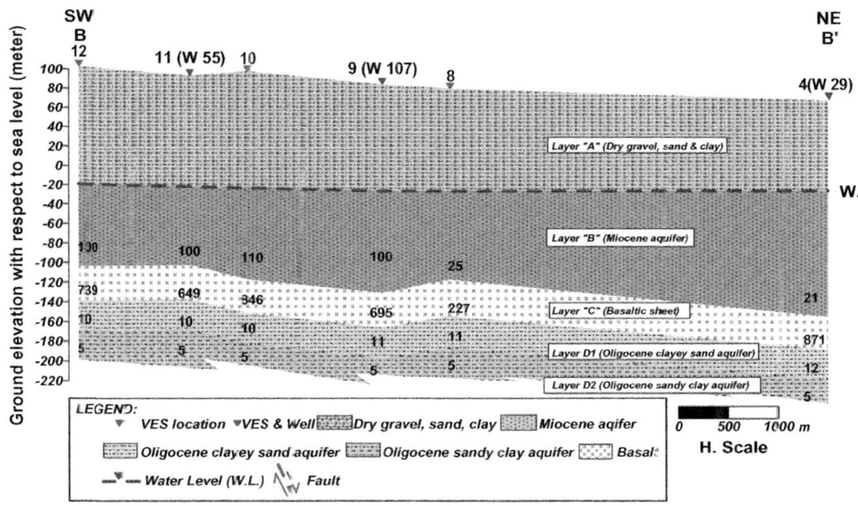

Fig. 5 Hydro-geoelectrical cross section B–B′ showing the vertical and horizontal distribution of the lithological succession, the water level and the inferred faults

the aquifer. This unit has a thickness ranging between 74 m at VES No.11 to 210 m at VES No.1 (Table 1).

Unit "C" represents the basaltic sheet that is separating the overlying Lower Miocene (Moghra aquifer) and the underlying Oligocene aquifer. The resistivity of this unite ranges between 138 Ω m at VES No.1 and 1260 Ω m at VES No.3. The thickness of this unit ranges between 28 m at VES No.2 to 35 m at VES No.9.

Unit "D" represents the lower water bearing formation of Oligocene aquifer which consists of sand and clay. According to the resistivity values of this unit, it can be differentiated it into two parts. The upper part (D1) has resistivity values ranging between 8 Ω m at VES No.1 and 12 Ω m at VES No.3. The thickness of this part ranges between 29 m at VES No.3 and 31 m at VES No.9 (Table 1). The lower part (D2) has resistivity values ranging between 4 Ω m at VES No.1 and 8 Ω m at VES No.6. It is clear that, the resistivities of this unit decrease downwards and generally are lower than that of the Miocene aquifer. This reflects the presence of more saline water in the Oligocene aquifer (TDS = 2990 ppm) than that of the Miocene aquifer at Well No.3 (TDS = 1274 ppm). This unit represents the base of the subsurface.

From the above mentioned discussion of the geoelectrical results the followings can be concluded:

- The resistivity of the water bearing layers reflects the salinity of the groundwater of the Miocene aquifer and the Oligocene aquifer where, the high resistivity value of the Miocene aquifer at VES No.12 reflects the presence of fresh water and the low resistivity values of the Miocene and the Oligocene aquifers reflect the presence of saline water.
- The detected faults F1 and F6 along the hydro-geoelectrical cross section A–A′ (Fig. 4) play an important role in the effect of the saline water of the Oligocene aquifer on the Miocene aquifer through the hydraulic connection between them as shown on the hydro-geoelectrical cross section.
- This is clear through the decreasing of the resistivity values of the Miocene aquifer at VES Nos.7, 6 and 1 which are adjacent to the faults. This is confirmed through the discussion of the hydrological and hydrogeochemical results in the next paragraphs.
- Due to the high salinity of the Oligocene aquifer under the basaltic sheet, the groundwater exploitation should be from the fresh water of the Miocene aquifer and the drilling should be down to the upper surface of the basaltic sheet only.

Hydrological Results

In this section, the hydrological setting and monitoring of the depth to water, water flow, saturated thickness and water salinity of the Miocene aquifer will be discussed and clarified.

The geologic setting of Tertiary basalt (geoelectrical unit C) is mostly used to differentiate between the overlying Moghra aquifer (geoelectrical unit B) and the underlying Oligocene aquifer (geoelectrical unit D). Also, the depth of the basaltic sheet and its level play an important role on the saturated thickness and the hydraulic connections between the two aquifers. The depth to the basaltic sheet (Fig. 6) ranges between 189 m in the west and 269 m in the east with a general increase northeastward resulting in an increase of the Miocene aquifer in the same direction. The basaltic

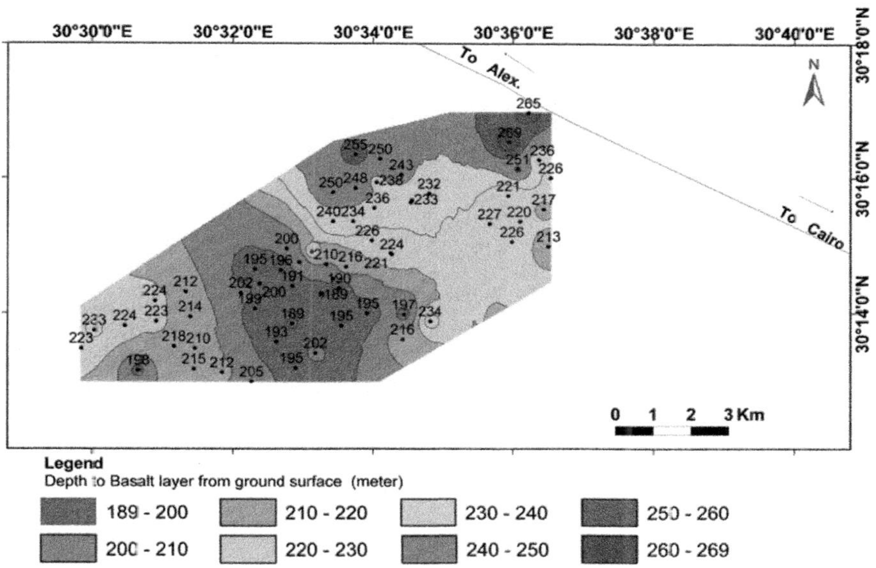

Fig. 6 A map showing the depth to the basaltic sheet (m) from the ground surface

sheet helps in determining the drilling depth of the Miocene aquifer to be fully penetrated and in calculating its saturated thickness by knowing the depth to the groundwater from the ground surface. On the other hand, the level of the top of the basaltic sheet (Fig. 7) shows that this level ranges between −99 m below mean sea

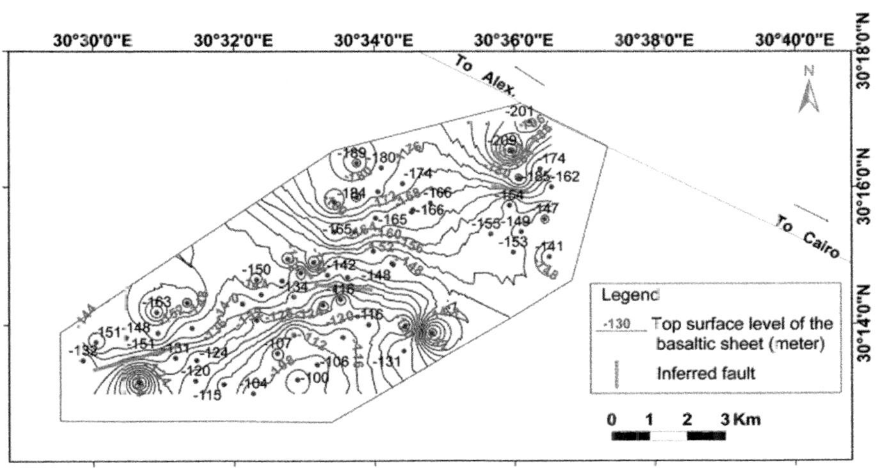

Fig. 7 A map showing the level of the top surface of the basaltic sheet (m) with respect to mean sea level showing the detected inferred faults

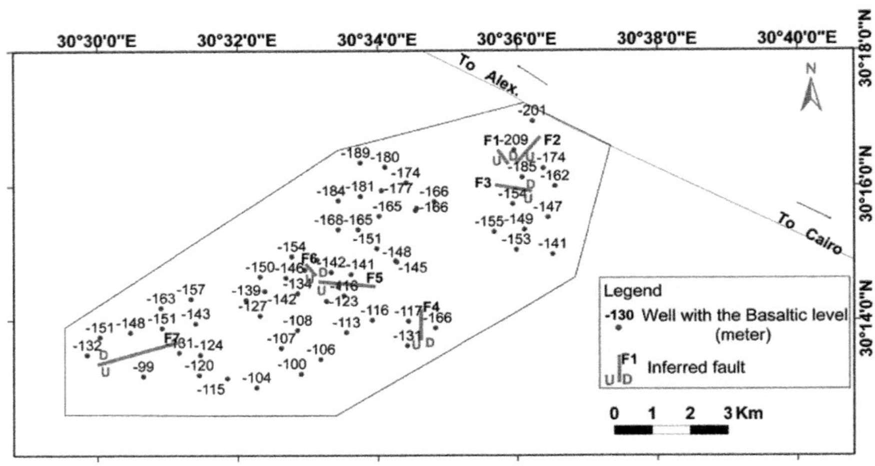

Fig. 8 A map showing the detected inferred faults in the study area

level in the southwestern part of the study area to −209 m below mean sea level in the northeastern part. This variation in the relief of the basaltic sheet helped in detecting a number of seven faults having NW–SE, NE–SW and E–W directions (Fig. 8) which are confirmed with the general geologic structures in the west of the Nile Delta area. These faults have a direct impact on the groundwater occurrences in the study area causing the convergence of Moghra aquifer and Oligocene aquifer as shown in the hydro-geoelectrical cross section A–A′ (Fig. 4). As the Moghra aquifer is fresher than the Oligocene aquifer, this convergence decreases the quality of Moghra aquifer due to the hydraulic connection between them.

Monitoring of Changes in Depth to Groundwater and Water Level in Miocene Aquifer

Three maps of the depth to groundwater were created for the 3 years (2003, 2012 and 2015). These maps show that depth to groundwater ranges between 61–116, 67–126 and 71–128 m respectively as shown in (Figs. 9, 10 and 11) with general increasing by time. To clarify the depletion in the Miocene aquifer with time between 2003 and 2015, a resultant depth to water map was created (Fig. 12). This map shows that, the depletion in the groundwater ranges from 6 m to more than 30 m during this period with a rate of about 1.3 and 1.7 m/year. Also, the groundwater movement in the study area is discussed through the construction of water level contour maps for the 3 years 2003, 2012 and 2015 (Figs. 13, 14 and 15). The water levels vary from −20.5 to −6.6, −34 to −16.8 and −47.1 to −20.9 m below mean sea level in 2003, 2013 and 2015, respectively. In general, the direction of the water flow in the 3 years has different directions due to the interference of the cone of depression at the different

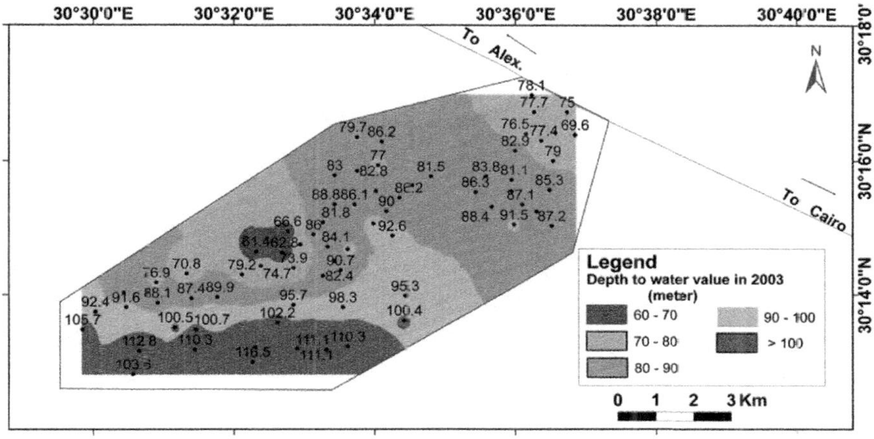

Fig. 9 A map showing the depth to water from ground surface (m) in year 2003

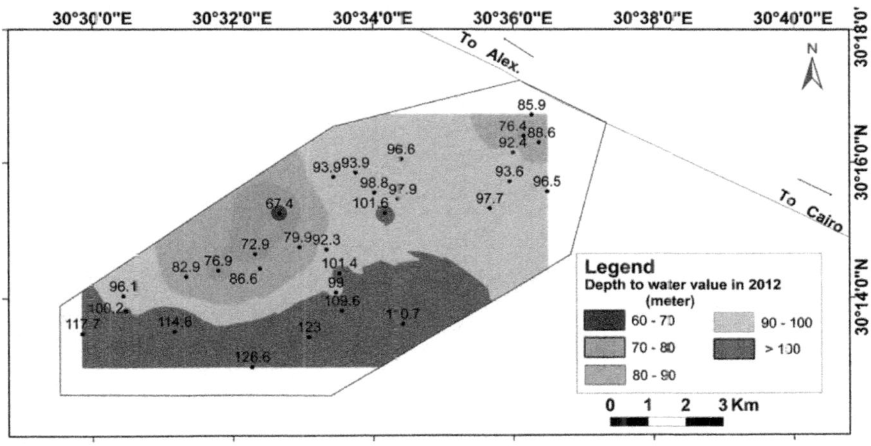

Fig. 10 A map showing the depth to water from ground surface (m) in year 2012

wells, where the distance between them was reached 200 m at some parts and the discharge is reached about 180 m^3/h with working hours of about 20–24 h/day. This makes closures (sinks) at different parts in the study area. On the other hand, a mound was made at well 93 (VES 6) due to the presence of faults (F5 and F6, see Figs. 2 and 8) which cause the hydraulic connection between the Miocene and the Oligocene aquifers. This caused depletion in the groundwater levels and increases of its salinity that reflecting mismanagement of the groundwater resources in the study area.

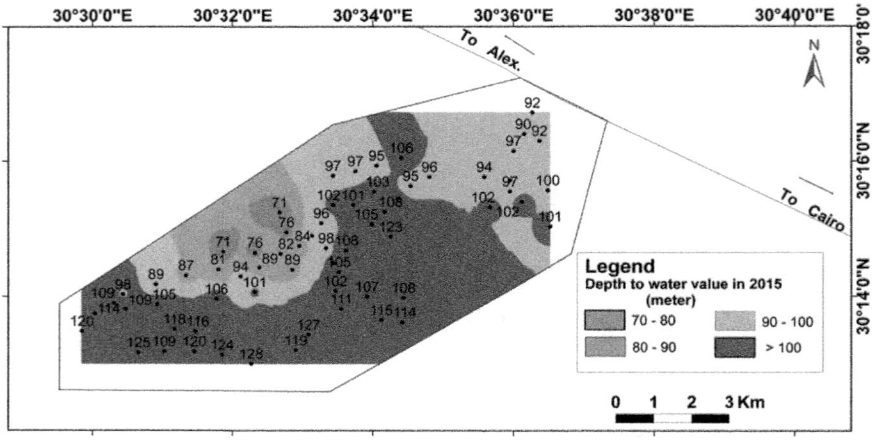

Fig. 11 A map showing the depth to water from ground surface (m) in year 2015

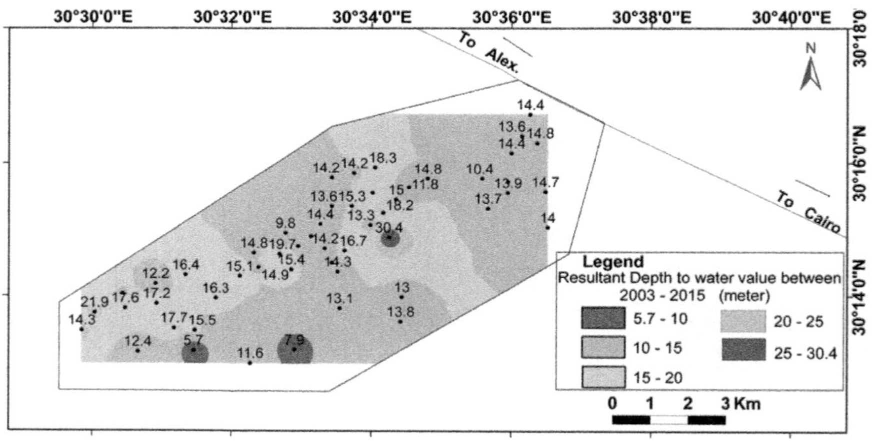

Fig. 12 A map showing the depletion of the groundwater surface (m) between the years 2003 and 2015

Monitoring the Changes in the Saturated Thickness of the Miocene Aquifer

The change in the depth to the groundwater by time results in change in the saturated thickness of the Miocene aquifer. Three layers of the saturated thickness of the aquifer were created for the three different years 2003, 2012 and 2015. The ranges of this saturated thickness are 83.9–175.3, 78.5–156.1 and 73–153 m, respectively with general decrease by time as shown in (Figs. 16, 17 and 18). From these figures, it is noticed that, in 2003 the maximum recorded saturated thickness of the Miocene

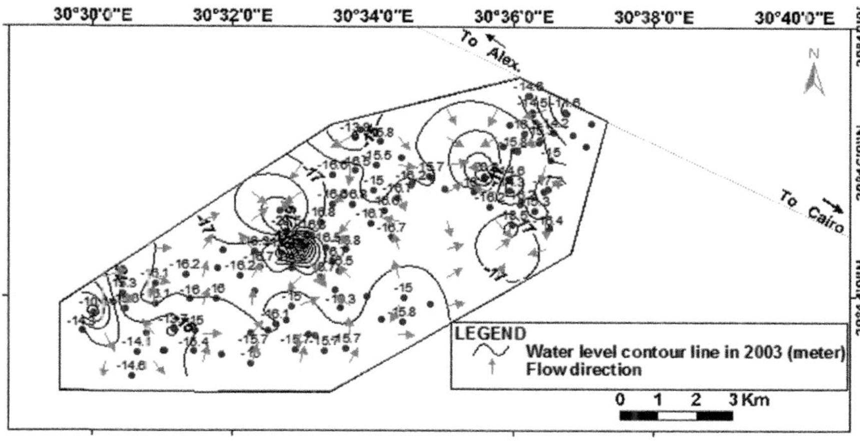

Fig. 13 A map showing the water level with respect to sea level (m) of the Miocene aquifer in year 2003

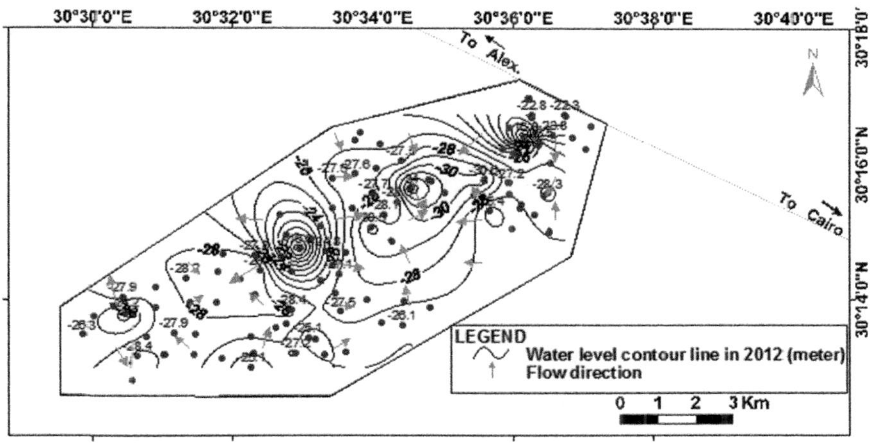

Fig. 14 A map showing the water level with respect to sea level (m) of the Miocene aquifer in year 2012

aquifer was 175 m while in 2015, the maximum thickness was 153 m and reaching in some areas to 73 m. This is clarifying to what extent the aquifer was deteriorated and needs good management.

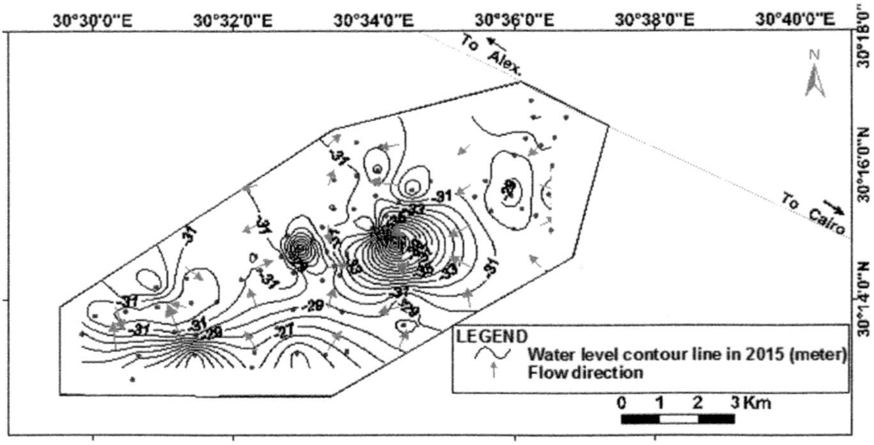

Fig. 15 A map showing the water level with respect to sea level (m) of the Miocene aquifer in year 2015

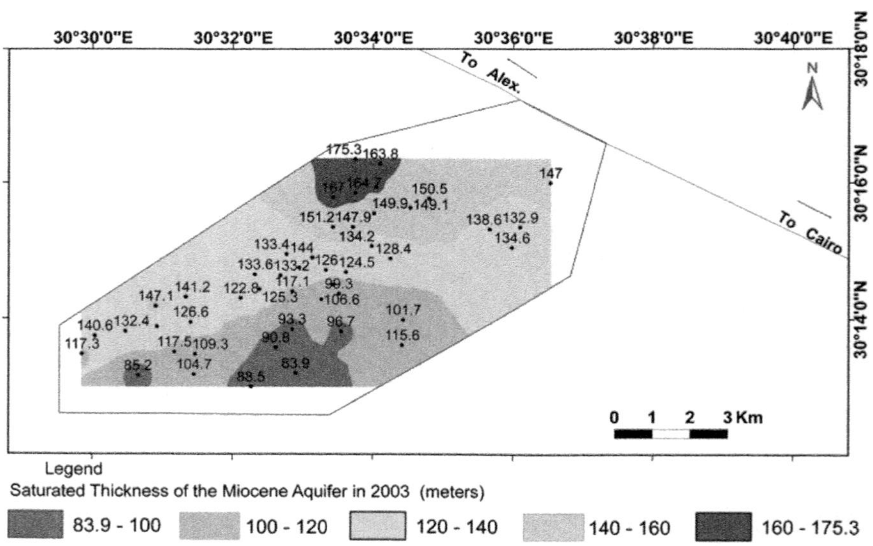

Fig. 16 A map showing the saturated thickness (m) of the Miocene aquifer in year 2003

Monitoring the Changes in the Water Quality of the Miocene Aquifer

In 2003, there are areas with salinity values less than 300 ppm and the dominant water salinity ranges between 300 and 500 ppm with exceptional areas having salinities up to 1000 ppm. In 2012, and 2015, the water salinities increased rapidly to reach in some

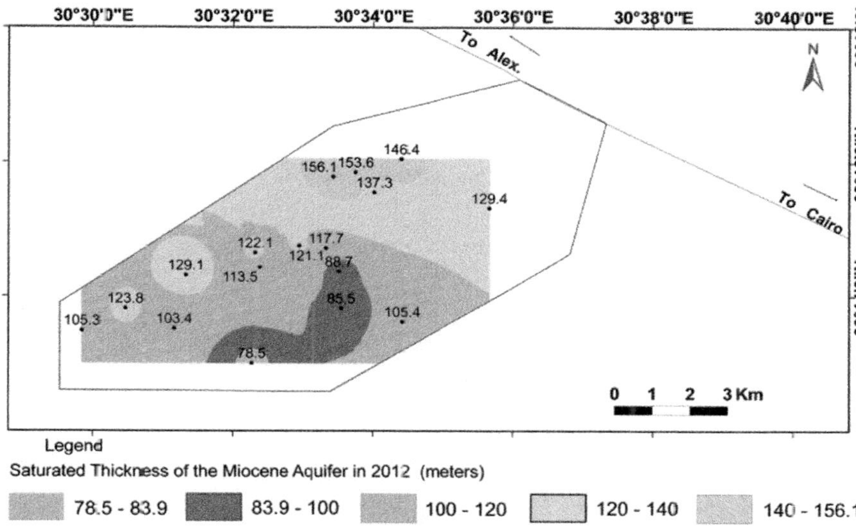

Fig. 17 A map showing the saturated thickness (m) of the Miocene aquifer in year 2012

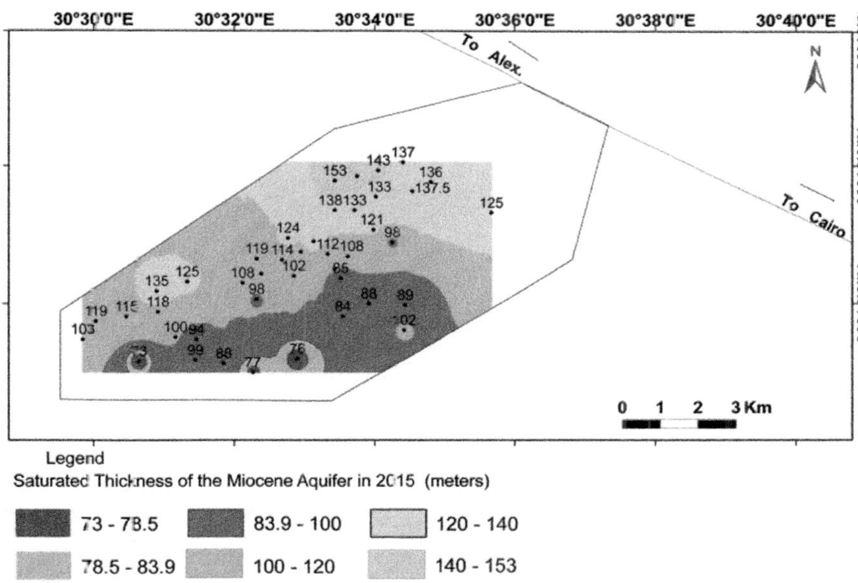

Fig. 18 A map showing the saturated thickness (m) of the Miocene aquifer in year 2015

localities around 2000 ppm, as shown in (Figs. 19, 20 and 21). The resultant salinity map between 2003 and 2015 (Fig. 22) show that the increase in the water salinity during this period had ranged between less than 200 ppm to more than 900 ppm in some localities in the study area. The increase in the water salinity of the Miocene aquifer may be attributed to the intensive exploitation which led to the interaction between the Miocene and the Oligocene aquifers via the existing faults.

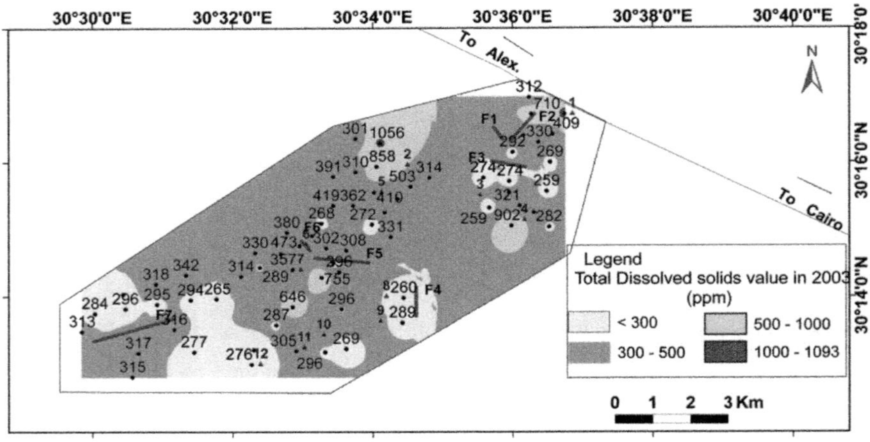

Fig. 19 A map showing the water salinity (ppm) of the Miocene aquifer in year 2003

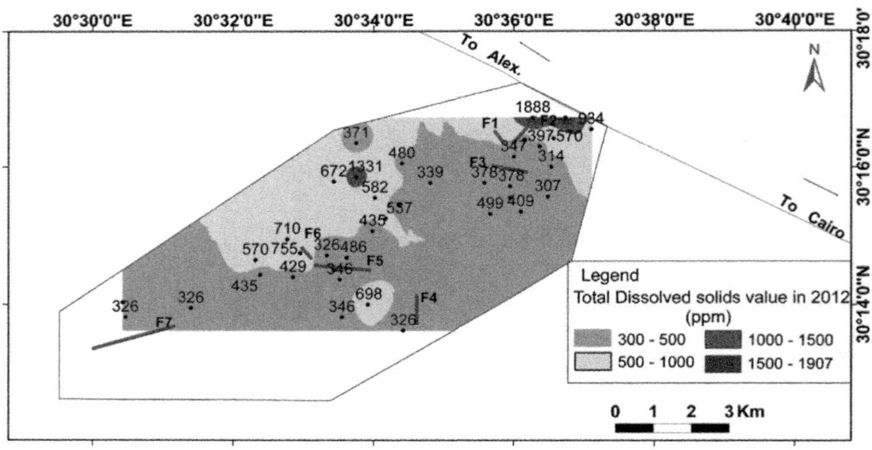

Fig. 20 A map showing the water salinity (ppm) of the Miocene aquifer in year 2012

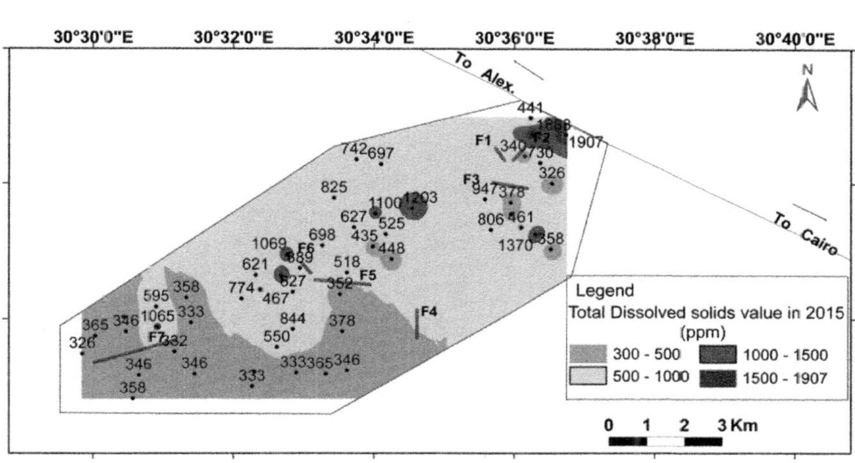

Fig. 21 A map showing the water salinity (ppm) of the Miocene aquifer in year 2015

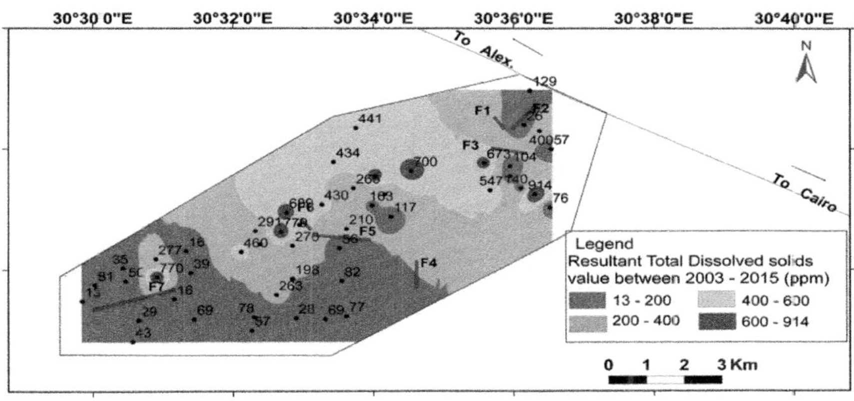

Fig. 22 A map showing the resultant salinity increase (ppm) between the years 2003 and 2015

Physical–chemical Parameters and Major Elements

The results of the physico-chemical analysis of groundwater samples from Wadi El-Farigh area expressed epm are shown in Table 2 as follow:

- PH values of the analyzed samples vary from 6.8 to 7.8 with an average value of 7.43. In general, PH of water samples was slightly acidic to slightly alkaline compared with [24] standards, all samples fall within the recommended limit (6.5–8.5) for human consumption.

Table 2 Physical and chemical characteristics of the collected groundwater samples in (meq/L) with some ionic ratios (2015)

S. No.	pH	EC (μs/cm)	TDS (mg/L)	Na^+ (meq/L)	K^+ (meq/L)	Ca^{2+} (meq/L)	Mg^{2+} (meq/L)	Total cations	HCO_3^- (meq/L)	Cl^- (meq/L)	$SO4^{2-}$ (meq/L)	Total anions	Error (%)	Na/Cl	SO_4/Cl	HCO_3/Cl	Ca/Mg	Ca/SO_4	Mg/Cl	Cl/HCO_3
3	7.57	1990	1274	13.43	0.92	5.30	2.50	22.16	3.20	11.83	6.35	21.38	1.79	1.14	0.54	0.27	2.12	0.83	0.21	3.70
9	7.53	1838	1176	13.26	1.28	5.12	0.20	19.86	4.16	14.69	1.11	19.96	0.25	0.90	0.08	0.28	25.59	4.61	0.01	3.53
20	7.65	458	293	2.17	0.33	1.60	1.00	5.11	1.92	1.98	0.93	4.83	2.80	1.10	0.47	0.97	1.60	1.71	0.51	1.03
36	7.60	580	371	2.78	0.10	2.01	0.99	5.89	2.10	2.32	1.09	5.51	3.30	1.20	0.47	0.91	2.03	1.84	0.43	1.10
43	7.10	2876	1841	16.52	0.21	6.47	4.38	27.57	1.95	18.68	7.71	28.34	1.37	0.88	0.41	0.10	1.48	0.84	0.23	9.58
45	7.07	845	541	4.00	0.85	2.91	0.75	8.50	2.46	3.80	1.75	8.01	2.90	1.05	0.46	0.65	3.88	1.66	0.20	1.55
47	7.77	584	374	3.07	1.36	1.60	0.50	6.52	1.89	3.55	0.52	5.96	4.40	0.86	0.15	0.53	3.20	3.07	0.14	1.88
48	7.52	1030	659	5.78	0.72	3.00	3.38	12.88	1.87	8.48	1.54	11.89	3.90	0.68	0.18	0.22	0.89	1.95	0.40	4.54
49	7.50	2660	1702	17.61	0.77	5.75	5.42	29.54	1.75	26.20	1.04	28.99	0.39	0.67	0.04	0.07	1.06	5.51	0.21	14.97
53	6.80	523	335	2.26	0.10	1.91	0.86	5.13	2.10	2.68	0.69	5.47	3.20	0.84	0.26	0.78	2.22	2.78	0.32	1.28
54	7.87	365	234	1.96	0.64	0.65	0.83	4.08	2.46	1.10	0.32	3.89	2.38	1.78	0.29	2.24	0.78	2.01	0.76	0.45
56	7.88	404	259	2.17	0.72	0.75	1.00	4.65	2.51	1.58	0.31	4.40	2.60	1.38	0.20	1.59	0.75	2.41	0.63	0.63
68	7.20	1199	767	5.00	0.15	4.86	3.03	13.04	2.1	8.69	2.54	13.33	1.09	0.58	0.29	0.24	1.60	1.91	0.35	4.14
72	7.50	558	357	2.43	0.10	2.06	0.97	5.56	1.95	2.75	0.78	5.48	0.54	0.88	0.28	0.71	2.13	2.65	0.35	1.41
74	7.50	563	360	3.30	0.10	1.51	0.72	5.63	2.1	2.53	0.83	5.47	1.16	1.30	0.33	0.83	2.09	1.81	0.28	1.21
87	7.20	3840	2458	20.87	0.21	11.18	5.13	37.38	2.16	19.77	17.71	39.65	2.80	1.06	0.90	0.11	2.18	0.63	0.26	9.14
88	7.76	3830	2451	22.30	1.23	10.75	5.42	39.70	1.83	33.88	4.19	39.90	0.25	0.66	0.12	0.05	1.98	2.57	0.16	18.49
93	7.70	976	625	5	0.77	3.40	1.25	10.20	2.96	5.66	1.21	9.83	1.80	0.85	0.21	0.52	2.72	2.80	0.22	1.91
94	7.10	377	241	2.00	0.08	1.27	0.80	4.14	2.05	1.11	0.75	3.91	2.60	1.81	0.68	1.85	1.59	1.68	0.72	0.54
103	7.76	2170	1389	17.83	1.21	3.75	2.08	24.86	3.44	17.24	2.50	23.18	3.40	1.03	0.15	0.20	1.80	1.50	0.12	5.01
104	7.50	977	625	5.00	0.13	2.80	1.72	9.65	2.05	5.57	1.70	9.32	1.70	0.90	0.30	0.37	1.63	1.65	0.31	2.72

(continued)

Table 2 (continued)

S. No.	pH	EC (µs/cm)	TDS (mg/L)	Na^+ (meq/L)	K^+ (meq/L)	Ca^{2+} (meq/L)	Mg^{2+} (meq/L)	Total cations	HCO_3^- (meq/L)	Cl^- (meq/L)	$SO4^{2-}$ (meq/L)	Total anions	Error (%)	Na/Cl	SO_4/Cl	HCO_3/Cl	Ca/Mg	Ca/SO_4	Mg/Cl	Cl/HCO_3
105	7.70	501	321	2.09	0.08	2.27	0.96	5.39	2.50	1.88	0.97	5.35	4.30	1.11	0.51	1.33	2.37	2.34	0.51	0.75
106	7.70	383	245	1.65	0.08	1.78	0.75	4.25	2.45	1.01	0.44	3.91	4.20	1.63	0.44	2.42	2.38	4.02	0.74	0.41
Min	6.80	365	234	1.65	0.08	0.65	0.20		1.75	1.01	0.31			0.58	0.04	0.05	0.75	0.63	0.01	0.41
Max	7.88	3840	2458	22.30	1.36	11.18	5.42		4.16	33.88	17.71			1.81	0.90	2.42	25.59	5.51	0.76	18.49
Average	7.49	1349.3	864	7.85	0.54	3.78	2.01		2.39	9.27	3.00			1.07	0.35	0.79	3.78	2.36	0.35	4.35
Sea water	8.20	59,700	39,350	0.516	11.6	23.4	0.12		2.7	0.61	0.066			0.84	0.108	4.432	0.2	4.49	0.196	0.225

TD Total dissolved solids, *EC* Electrical conductivity
Q = (Ca+Mg)/(SO_4+HCO_3)

- The EC values vary from 365–3840 μScm^{-1} at 25 °C with a mean of 1349 μScm^{-1}. EC values are above the permissible limits of [24] of 1500 μScm^{-1} in 29% of groundwater samples.
- Total dissolved solids (TDS) vary from 233–2457 mg/L (average 863 mg/L) classified as 46% fresh water except 29% classed as slightly saline groundwater and about 21% moderately saline groundwater according to [25, 26].
- Possible sources of saltwater enrichment in the lower Miocene aquifer are either the leakage of saline water from shale and clay interbeds or due to upward—leakage of old marine water along fault planes from the underlying Oligocene aquifer, which characterized by relative high salinity. It is worthy to mention that [18] recorded that the salinity of the Oligocene aquifer in the southwestern part of the Nile Delta reached 2566 mg/L. The cation chemistry is dominated by sodium followed by calcium and magnesium in 83% of the samples. Values of Na^+ and K^+ ions are 1.65–22.3 and 0.08–1.36 epm respectively compared with [24] about 29% of samples which exceeding the permissible limits for drinking water for (Na = 8.69 epm). Whereas, Ca^{2+} and Mg^{2+} ions vary from 0.65–11.18 and 0.2–5.42 epm respectively. The anion chemistry is dominated by chloride followed by bicarbonate and sulphate in about 58% of the samples. Values of Cl^- concentration vary from 0.31–17.71 epm with 37% of samples exceeding [24] limits for ($Cl^- = 7.04$ epm). Bicarbonate vary from 4.16 to 1.75 epm with average value of 2.36 epm. Sulphate concentration varies from 0.31 to 17.71 epm (average 2.39 epm). Total Hardness values for the groundwater samples are range from (74–811 mg/L) categorized as 37% moderately hard and 29% are very hard according to Table 3.

To assess the geochemical evolution in groundwater flow system, a graphical representation of Piper's diagram [27] is extensively used. The different water samples have been classified according to their chemical composition using the Piper diagram in the study area (Fig. 23). The water type undergoes an evolution process along the flow line towards more additions in Na^+ and Cl^- ions. This shift can be attributed to mixing effect of (Miocene aquifer) with more saline water (Oligocene aquifer) through the fault that act as a conduit permit upward movement of the groundwater (samples 3, 9 and 49) also, other factor was the over-exploitation which enhanced dissolution processes of halite, dolomite and gypsum. In other words, water with increased abstraction changed to more Ca, Mg, Cl and SO_4 type (samples 87, 88 and 43).

Water–Rock Interaction Processes

Interactions between groundwater and surrounding host rocks are believed to be the main processes responsible for the observed chemical characteristics of groundwater in Wadi El-Farigh area. Evaluation of such processes requires the description of the main mineral assemblage of the rocks in which water is found, and the identification of chemical reactions responsible for the geochemical evolution of groundwater [28].

Table 3 Total hardness, saturation index and groundwater quality index with stable isotopic composition results of groundwater samples (2015)

S. No.	TH (mg/L)	Hardness as $CaCO_3$	SI calcite	SI dolomite	SI gypsum	SI cibbsite	SI anhydrite	SI gypsum	SI halite	CCME WQI	δO-18 ‰	δD ‰	d-excess ‰
3	388	Very hard								Poor	−1.83	5.37	20.23
9	266	Hard	0.47	0.53	−1.79	2.71	−1.432	−1.143	−4.975	Poor	−1.46	6.76	20.75
20	129	Moderately hard	0.14	0.85	−1.91	2.61	−2.082	−1.793	−5.416	Poor	−1.98	4.14	20.35
36	149	Moderately hard								Poor	0.06	4.71	4.23
43	539	Very hard								Poor	−0.05	3.9	4.3
45	182	Hard	−0.40	−0.27	−1.69	3.73	−1.976	−1.686	−6.461	Excellent	−1.88	5.31	20.35
47	105	Moderately hard	0.26	1.05	−2.01	2.50	−2.295	−2.006	−6.346	Excellent	−1.93	5.59	21.03
48	316	Very hard	0.30	1.80	−1.86	2.72	−2.147	−1.858	−5.598	Poor	−1.86	6.12	19.93
49	554	Very hard	0.68	2.48	−1.61	2.72	−1.903	−1.614	−4.707	Poor	−1.89	5.11	18.93
53	137	Moderately hard			−2.25		−2.909	−1.909	−6.714	Excellent	−1.91	5.47	4.58
54	74	Soft	0.32	1.61	−2.49	2.40	−2.784	−2.494	−6.901	Excellent	−1.95	4.75	20.01
56	87	Moderately hard	0.29	1.44	−2.53	2.39	−2.822	−2.532	−6.935	Poor	−1.91	4.9	17.2
68	392	Very hard	−0.26	0.37	−1.51	3.11	−1.801	−1.512	−6.071	Poor	−0.2	1.36	2.96
72	151	Hard	−0.27	0.22	−2.20	3.59	−2.486	−2.196	−6.843	Poor	−0.09	3.72	4.44
74	111	Moderately hard	−0.37	0.04	−2.29	2.69	−2.577	−2.288	−6.746	Excellent	−0.02	3.47	3.63
87	811	Very hard	−0.09	0.57	−0.62	2.85	−0.913	−0.623	−5.166	Poor	0.21	7.3	5.62

(continued)

Table 3 (continued)

S. No.	TH (mg/L)	Hardness as $CaCO_3$	SI calcite	SI dolomite	SI gypsum	SI cibbsite	SI anhydrite	SI gypsum	SI halite	CCME WQI	δO-18 ‰	δD ‰	d-excess ‰
88	804	Very hard	1.22	2.93	−0.47	2.44	−0.755	−0.467	−4.448	Poor	−1.75	3.2	21
93	232	Hard					−2.01	−1.721	−5.798	Excellent	−2.01	3.85	18.44
94	102	Moderately hard	−0.84	−0.79	−2.37	4.11	−2.658	−2.368	−7.314	Poor	0.2	6.32	4.72
103	290	Hard	0.85	2.58	−1.18	2.62	−1.464	−1.175	−4.811	Poor	−2.04	2.61	19.98
104	225	Hard			−1.81		−2.102	−1.813	−6.242	Excellent	0.01	6.15	6.07
105	186	Hard	−0.03	0.58	−1.99	4.75	−2.276	−1.987	−7.077	Poor	−0.12	5.72	6.68
106	125	Moderately hard	−0.01	0.70	−2.46	2.69	−2.751	−2.461		Excellent	−0.12	2.51	3.47

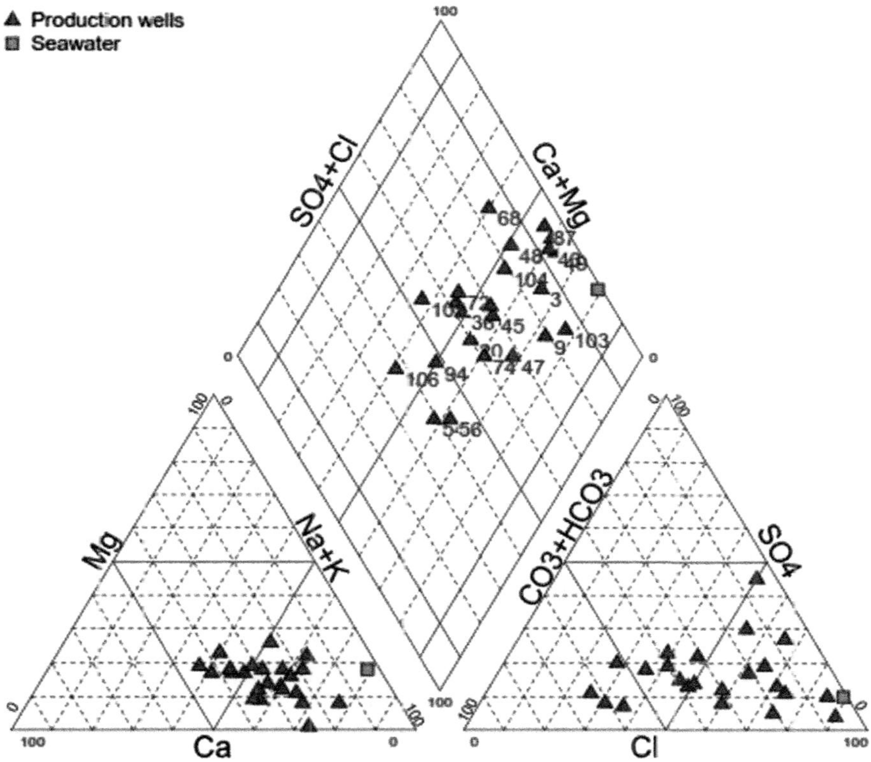

Fig. 23 Piper diagram of the collected groundwater samples

From the available studies in the literature, such reactions generally include chemical weathering of rock-forming minerals, dissolution–precipitation of secondary carbonates and ion exchange between clay minerals and water Ionic ratios expressed as epm Na^+/Cl^-, SO_4^{2-}/Cl^-, Ca^{2+}/Mg^{2+}, Ca^{2+}/SO_4^{2-}, Mg^{2+}/Cl^- are shown in Table 2. In general, sodium increases linearly with chloride (Fig. 24a) with a high correlation coefficient (R2 = 0.91); most groundwater samples fall on or below the 1:1 halite dissolution line, suggesting a Na^+ excess over Cl^-. Excess Na^+ can be extracted from aquifer bedrock sources such as plagioclase (albite) dissolution, as shown in equation:

$$2NaAlSi_3O_8 + 2CO_2 + 11H_2O \rightarrow Al_2Si_2O_5(OH)_4 + 2Na^+ + 4H_4SiO_4 + 2HCO_3$$

or cation exchange activity as shown by equation:

$$\tfrac{1}{2}\,Ca^{2+} + Na - clay \rightarrow \tfrac{1}{2}\,Ca - clay + Na^+$$

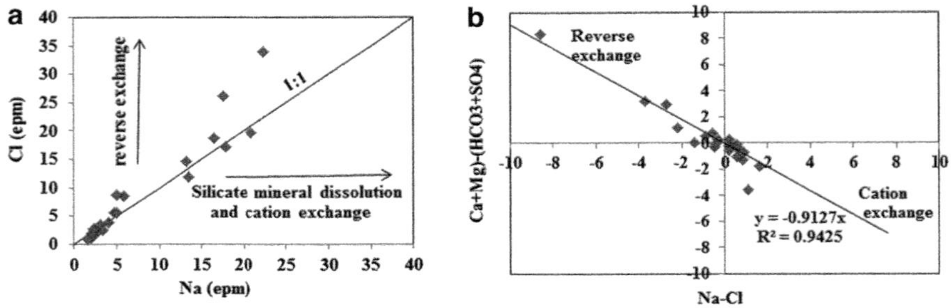

Fig. 24 **a** A plot of Cl^- against Na^+ (epm) **b** Bivariate plots of $(Ca^{2+}+Mg^{2+})-(HCO_3^-+SO_4^{2+})$ against (Na^+-Cl^-)

$(Ca^{2+} + Mg^{2+})$–$(HCO_3^- + SO_4^{2-})$ (epm) was plotted against (Na^+-Cl^-) (epm) to investigate the occurrence of cation exchange reaction in groundwater (Fig. 24b).

Since calcite and dolomite dissolution are likely additional sources of Ca^{2+} and Mg^{2+} in the groundwater other than cation exchange, potential contributions of Ca^{2+} and Mg^{2+} from calcite and dolomite dissolution to lithogenic Ca^{2+} and Mg^{2+} in the groundwater were accounted for by subtracting the equivalent concentrations of HCO_3^- and SO_4^{2-} in the groundwater [28].

To account for lithogenic Na^+ available for trade, it was believed that Na^+ contribution from meteoric origin would be balanced by equivalent Cl^- concentration, so the equivalent Cl^- concentration was subtracted from the Na^+ concentration. The samples in (Fig. 24b) are clustered along a line with a slope of −0.94 and a value of −1. In the $(Ca^{2+} + Mg^{2+}) - (HCO_3^- + SO_4^{2-})$ plot region, water undergoes ion exchange or reverse ion–exchange reactions along a line with a slope of -1. Ion exchange and/or reverse ion exchange are important processes that enable Na^+, Ca^{2+}, and Mg^{2+} to partially enter or exit the groundwater system.

Also this was seen from plot of Ca against Mg or SO_4^{2-} in epm (Fig. 25a, b), where increase in Ca^{2+} or Mg^{2+} The interaction between groundwater and the clay fraction of the aquifer content controls the water chemistry, according to concentrations. The presence of SO_4^{2-}in about 12% of the analyzed groundwater samples may be attributed to gypsum dissolution. Except for 24% of the fresh water samples, where $Mg^{2+}/Cl^- > 0.5$ [29], the Mg^{2+} versus Cl^- graph (Fig. 25c) shows an abundance of chloride. In the case of SO_4^{2-} versus Cl^- (Fig. 25d), Cl^- concentrations rise in the groundwater samples. Furthermore, as the samples become more Cl^- enriched, the concentration of SO_4^{2-} decreases until it reaches a SO_4^{2-}/Cl^- ratio of about 0.1. (Seawater ratio). This suggests that the Lower Miocene aquifer's groundwater is suffering from a salt-water crisis as a result of excessive withdrawals.

Calculating the chemical equilibrium of the water with the mineral phase may be used to determine the potential for a chemical reaction. Calculating a saturation index (SI) using analytical data [30, 31] can be used to determine the equilibrium state of water with respect to a mineral phase. The SI is the logarithm of the ratio of the ion activity product (IAP) to the mineral equilibrium constant at a given temperature,

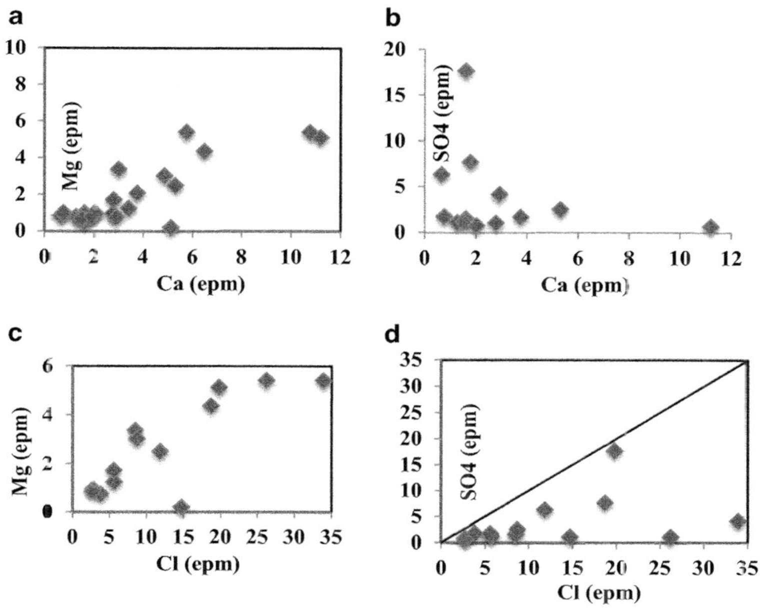

Fig. 25 **a** A plot of Ca against Mg (epm) **b** A plot of Ca against SO_4 (epm) **c** A plot of Cl against Mg (epm) **d** A plot of Cl against SO_4 (epm)

expressed as:

$$SI = \log\ (IAP/K_{SP}).$$

where; IAP = ion activity product, K_{SP} = solubility product at given temperature.

A positive SI indicates that the mineral phase in equilibrium can precipitate because the water is supersaturated with respect to the particular mineral phase and thus incapable of dissolving more of the mineral. A negative index (SI) implies under-saturation and mineral phase dissolution, while a neutral SI indicates that the mineral phase is in equilibrium. Results of saturation index is performed using SOLMINEQ.GW are summarized in Table 3. In terms of calcite and dolomite, the majority of water samples are supersaturated (37 and 63%, respectively), and SId values are higher than SIc values. Calcium precipitation as Ca and/or Ca–Mg carbonate is suggested by super-saturation. Calcite and dolomite are under-saturated in around 33 and 8% of the analysed water samples, respectively. This form of water would become a source of calcium and phosphorus.

Trace Metals Distribution

Human and anthropogenic forms of trace elements can be found in subsurface environments. Mineral weathering is one of the most important natural sources of feldspar, which can be found in the Miocene aquifer of the study region. Fertilizers, agricultural effluents, and leaks from service pipes are also anthropogenic causes. Comparing values of trace elements concentrations include: Al, Cd, Co, Cr, Cu, B, Mo, Ni, Pb, Sr, V, and Zn in the study area according to [32, 33] as shows in Table 4. All the results of trace elements are within the permissible limits for drinking water purposes except sample (No.94) have Al concentration (0.38 mg/L) and sample (No.20) have Fe concentration (0.93 mg/L) and these may within the measurement error or from weathering of some minerals as a natural sources e.g. gibbsite in the aquifer media.

Drinking Water Quality Index (DWQI)

Although there have been a number of attempts to establish such a water quality index, the index suggested in the Canadian Environmental Quality Guidelines [34] appears to be the most successful to date. Scope F1, frequency F2, and amplitude F3 are the three variables that make up this index. From the data collection of chemical analyses of groundwater samples, sixteen hydrochemical parameters were chosen as variables. These have been compared with the corresponding objectives values in Table 5 according to [33].

Following the definition of the variables and objectives, the three factors (F1, F2, and F3) that make up the CCME WQI index are determined using the method proposed by [33]. The following are the meanings of the various parameters used in the calculation of the index, as well as the equations used:

***F1* (Scope)** represents the proportion of variables that fail to achieve their goals at least once during the time span under consideration ("failed variables"), as a percentage of all variables measured:

$$F1 = (\text{Number of Failed Variables}/\text{Total Number of Variables}) \times 100.$$

***F2* (Frequency)** represents the percentage of individual assessments that fail to achieve their goals (also known as "failed tests"):

$$F2 = (\text{Number of Failed Variables}/\text{Total Number of Variables}) \times 100.$$

***F3* (Amplitude)** represents the amount by which failed test values do not meet their objectives. $F3$ is calculated in three steps.

(i) The number of times by which an individual concentration is greater than (or less than, when the objective is a minimum) the objective is termed an "excursion" and is expressed as follows. When the test value must not exceed

Table 4 Trace elements concentration in (mg/L) of the collected groundwater samples

S. No.	Al	B	Cd	Co	Cr	Cu	Fe	Mn	Mo	Ni	Pb	Sr	V	Zn
3	<0.04	0.5629	<0.0002	<0.0002	<0.01	<0.01	0.0451	<0.003	<0.0004	<0.001	0.0031	2.021	<0.01	0.017
9	<0.04	0.4472	<0.0002	<0.0002	<0.01	<0.01	0.0813	<0.003	<0.0004	<0.001	0.0041	1.72	<0.01	0.0061
20	<0.04	<0.01	<0.0002	<0.0002	<0.01	<0.01	0.9277	<0.003	<0.0004	0.0023	0.0078	0.3902	0.0214	0.0733
36	0.1156	0.0201	<0.0002	0.0009	<0.001	0.0131	0.2969	0.007	0.0023	0.0009	0.0045	0.381	0.007	0.1401
43	0.0429	0.0217	<0.0002	<0.0002	0.0095	<0.002	0.0792	<0.001	<0.001	0.0009	0.0041	1.646	<0.005	0.0131
45	0.1566	0.0222	<0.0002	<0.0002	<0.01	<0.01	0.1672	<0.003	0.0012	<0.001	0.0028	0.6223	0.0299	0.0175
47	<0.04	<0.01	<0.0002	<0.0002	<0.01	<0.01	0.0999	<0.003	0.0008	<0.001	<0.001	0.4262	<0.01	0.0068
48	<0.04	<0.01	<0.0002	<0.0002	<0.01	<0.01	0.548	<0.003	<0.0004	<0.001	0.0026	0.982	<0.01	0.0162
49	<0.04	<0.01	<0.0002	<0.0002	<0.01	<0.01	0.1725	<0.003	<0.0004	<0.001	0.002	3.028	0.023	0.0177
53	<0.01	<0.005	<0.0002	<0.0002	<0.0055	0.0035	0.0542	<0.001	0.0001	<0.0006	0.0021	0.3284	<0.005	0.0095
54	<0.04	<0.01	<0.0002	<0.0002	<0.01	<0.01	0.1866	<0.003	<0.0004	<0.001	0.0021	0.31	0.0232	0.0069
56	<0.04	<0.01	<0.0002	<0.0002	<0.01	<0.01	0.2144	<0.003	0.0009	<0.001	0.0038	0.3056	<0.01	0.0047
68	0.0484	<0.005	0.0007	<0.0002	0.0042	<0.002	0.0572	0.0023	<0.001	<0.0005	0.0038	1	<0.005	0.0345
72	0.2663	<0.005	0.0003	<0.0002	0.0047	0.0116	0.1695	0.0071	<0.001	<0.0005	0.0192	0.3589	<0.005	0.0753
74	0.0333	<0.005	<0.0002	<0.0002	<0.001	0.0058	0.0505	<0.001	0.0036	<0.0005	0.0039	0.2565	<0.005	0.017
87	0.028	0.8964	<0.0002	<0.0002	0.0124	0.0039	0.1476	0.0038	0.002	0.0006	0.0045	2.3	<0.005	0.0158
88	<0.04	0.594	<0.0002	<0.0002	<0.01	<0.01	0.069	<0.003	<0.0004	<0.001	<0.001	4.474	<0.01	0.0069
93	<0.04	<0.01	<0.0002	<0.0002	<0.01	<0.01	0.2915	<0.003	0.0005	<0.001	0.0051	1.328	<0.01	0.0067
94	0.38	0.0277	<0.0002	<0.0002	0.0028	0.0272	0.6716	0.0152	0.0043	0.0043	0.0079	0.2917	0.0145	0.1398
103	0.0584	0.1615	<0.0002	<0.0002	<0.01	<0.01	0.355	<0.003	0.0006	<0.001	0.0023	4.422	<0.01	0.0105
104	<0.01	<0.005	<0.0002	<0.0002	0.0052	<0.002	< 0.004	<0.001	0.0045	<0.0005	0.0045	0.6887	<0.005	<0.003

(continued)

Table 4 (continued)

S. No.	Al	B	Cd	Co	Cr	Cu	Fe	Mn	Mo	Ni	Pb	Sr	V	Zn
Min	0.028	0.0201	0.0003	0.0009	0.0028	0.0035	0.0451	0.0023	0.0005	0.0006	0.002	0.2565	0.007	0.0047
Max	0.38	0.8964	0.0007	0.0009	0.0124	0.0272	0.9277	0.0152	0.0045	0.0043	0.0192	4.474	0.0299	0.1401
Average	0.140	0.334	0.0005	0.0009	0.007	0.012	0.257	0.008	0.002	0.002	0.005	1.392	0.019	0.035

Table 5 Selected objective based on the Egyptian high committee of water [33]

Objective	Value	Objective (mg/L)	Value
pH	6.5–8.5	B	0.5
TDS (mg/L)	1000	Al	0.2
Hardness (mg/L)	500	Cr	0.05
Ca^{2+}(mg/L)	350	Cd	0.003
Mg^{2+}(mg/L)	150	Cu	2.0
Na^{+}(mg/L)	200	Mn	0.4
$SO_4{}^{2-}$(mg/L)	250	Ni	0.02
Cl^{-}(mg/L)	250	Pb	0.01
Mo (mg/L)	0.07	Zn	3.0
Fe (mg/L)	0.3		

the objective:

$$\text{excrusion}_i = \{(\text{Failed Test Value}_\text{i}/\text{Objective}_\text{j})\} - 1$$

(ii) The collective amount by which individual tests are out of compliance is calculated by summing the excursions of individual tests from their objectives and dividing by the total number of tests (both those meeting objectives and those not meeting objectives). This variable, referred to as the normalized sum of excursions, or *nse*, is calculated as:

$$nse = \sum_{i=1}^{n} \text{excrusion}_i / \text{no of Tests}$$

(iii) *F3* is then calculated by an asymptotic function that scales the normalized sum of the excursions from objectives (*nse*) to yield a range between 0 and 100.

$$F3 = \{\text{nse}/(0.01\text{nse} + 0.01)\}$$

Once the F1, F2 and F3 factors have been determined, the overall water quality index **CCMEWQI** can be calculated by summing the three factors as if they were vectors according to the following equation:

$$\text{CCMEWQI} = 100 - \{(\text{F}_1^2 + \text{F}_2^2 + \text{F}_3^2)\}/1.732$$

The overall value of the index helps to classify the water samples into different categories according to drinking purposes suitability as indicated in Table 5.

The data of 19 objectives were used to calculate CCME-WQI model [34]. Water quality index is ranked between excellent category (about 33% of water samples) which is protected with a virtual absence of threat to poor category (about 67% of water samples) which is almost always threatened or impaired (Table 6).

Table 6 Categorization of the CCMEWQI method [33]

Categorization	Index value	Water quality
Excellent	95–100	Virtual absence of threat
Good	80–94	Minor degree of threat
Fair	65–79	Occasionally threatened
Marginal	45–64	Frequently threatened
Poor	0–44	Almost always threatened

Isotopic Composition of Groundwater

Stable isotopes ($\delta^{18}O$ and H^2) of water samples were measured to provide basic information on the origin and sources of recharge for the Miocene aquifer in the studied area. The stable isotope composition of groundwater samples collected from the study area, together with those of the d-excess values are compiled in Table 3. Values of $\delta^{18}O$ are range between (−2.04 and +0.21‰) with average value of −1.015‰ and δD values are between (+1.36 and +7.3‰) with an average value of 4.66‰ with d-excess values range from 2.96 to 21.03‰. The plot of $\delta^{18}O$ and δD (Fig. 26) offers a complementary insight into the origin of groundwater in Wadi El-Farigh area. Two distinct sample groups were distinguished according to their isotopic compositions and to their position regarding GMWL (Global Meteoric Water Line) and MMWL (Mediterranean Meteoric Water Line).

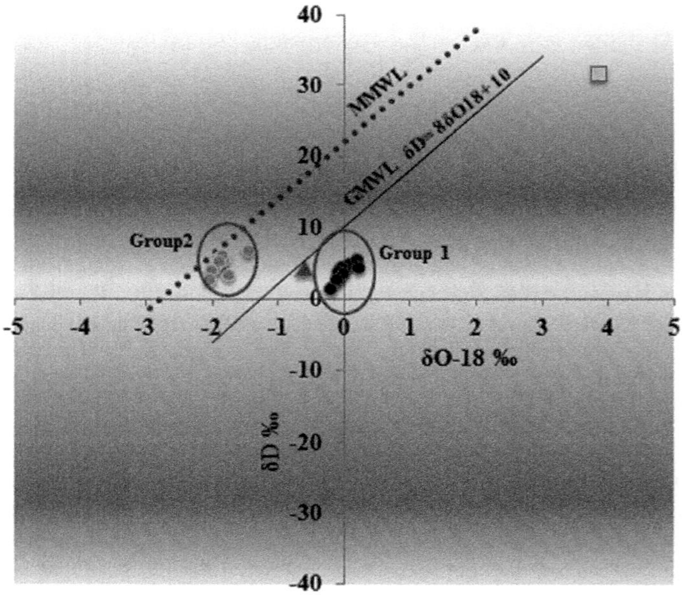

Fig. 26 Plot of δO^{18}‰ against δD‰ for the collected groundwater samples

The First Group1 comprises (11 sample) distributed in north and northwestern parts of Wadi El-Farigh area (Fig. 26) which is slightly enriched in stable isotopes ($\delta^{18}O$ between −0.20 and + 0.21‰) and (δD between +1.36 and +7.3‰) with Cl^- concentration ranging from 1.01 to 19.77 epm as appeared in the Miocene aquifer samples. This low-saline groundwater samples (group1) have slightly stable isotopic enrichment indicating that this group may undergo a tiny portion of evaporation infiltration. The Second Group2, comprises (12 sample) located in northeastern and southern parts of Wadi El-Farigh area (Fig. 27) is depleted in stable isotopes ($\delta^{18}O$ between −2.04 and −1.46 ‰) and (δD between +2.61 and +6.76‰) and exhibits more salinity where, Cl^- concentration ranging from 1.10 to 33.88 epm. As shown in (Figs. 28 and 29) most of this high –saline groundwater samples (group 2) are located on/beside these faults (F1, F2 and F4) that reflecting mixing of two different recharging sources Miocene and Oligocene aquifers (more depleted water), where these faults act as conduits that permit upward movement of the groundwater as indicated previously from the hydro-geoelectrical cross section A–A′ (Fig. 4). A combination of the chemical and isotopic content of groundwater samples is represented in Fig. 29 in which $\delta^{18}O$ is plotted against Cl concentration in epm. Both of group1 and group2 have the tendency of increasing salinities that reflected by water–rock interactions mechanisms where, dissolution/precipitation and mixing effects occur along the groundwater flowpath but for group2 mixing of Miocene with more saline water of Oligocene aquifer at the faults reflects more salinity than group1 also, more depletion of $\delta^{18}O$ isotopic composition. [35] analyzed tritium content for Miocene aquifer water samples. The results of tritium content indicated that a small number of the analyzed water samples located at eastern part adjacent to the Nile (El Rayah El-Nasery) have tritium content in the range from (1.77 to 5.55 TU); suggesting a contribution to this groundwater from recently infiltered Nile water.

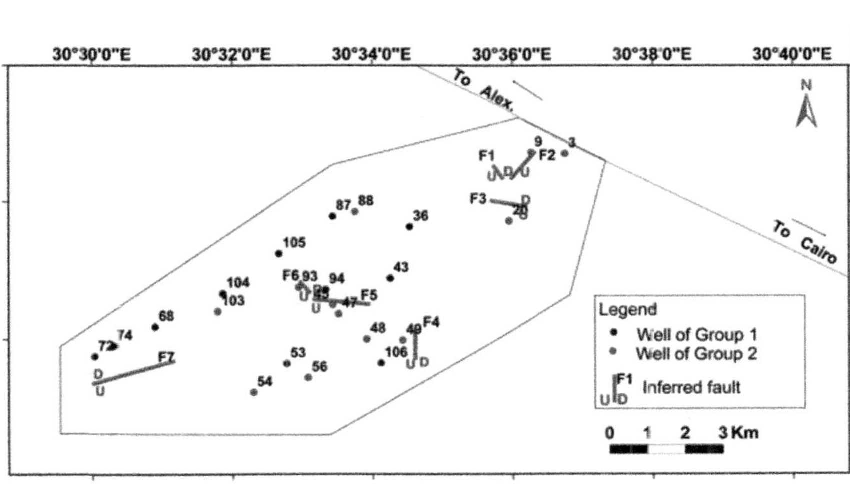

Fig. 27 Distribution of groundwater groups with faults

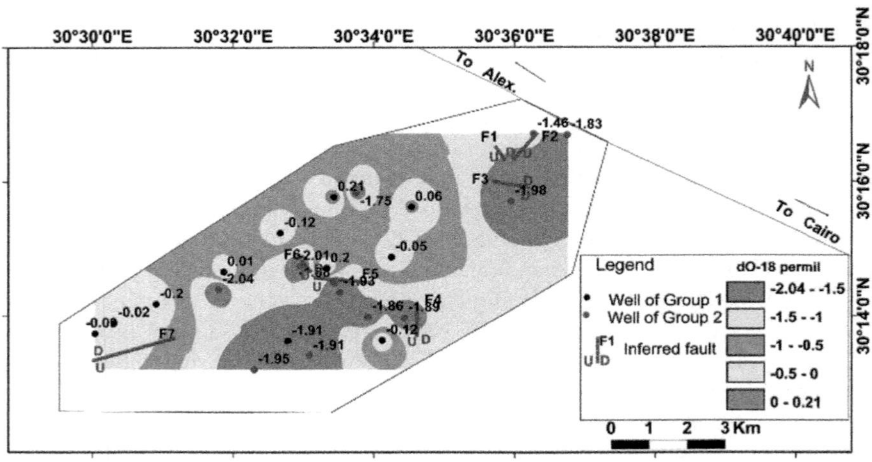

Fig. 28 Spatial distribution of δO^{18}‰ isotopic composition in the study area

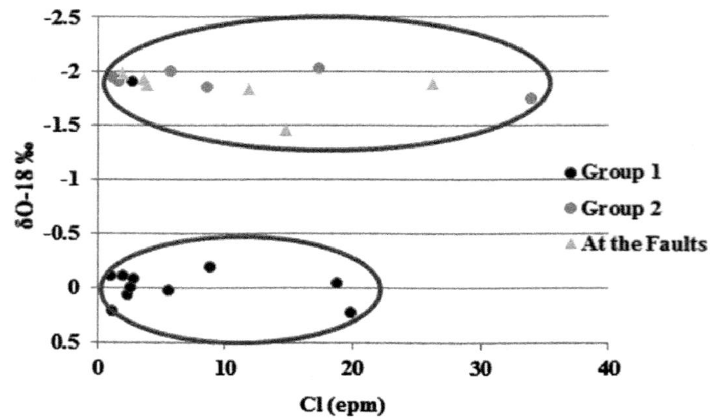

Fig. 29 Cl concentration in (epm) against $\delta^{18}O$‰

On the other hand, most of water samples at western part of Wadi El-Farigh have tritium content undetectable or varying in the range of the lower limit of detection confirming the presence of paleorecharge source of more than 45 years (may be Oligocene aquifer or deeper strata) confirming non-renewablity and absence of recent recharge of this aquifer. For C-14 radiocarbon measurement. [35] found that the radiocarbon age of Miocene aquifer ranges from 2570 to 9426 y.b.p. The nearly absence of tritium content in most of the studied groundwater samples in Miocene aquifer and the combination of depleted value of $\delta^{18}O$ and the high radiocarbon age are used collectively to put constrains on the Paleo-water recharge component. This conclusion indicated the possibility that remnants of Paleo-Nile water is entrapped for long times in the buried channels and was fed by paleo-meteoric water during Holocene period (8500–5000 y.b.p.). This source would have an isotopic composition strongly different than that of the present day Nile water.

Conclusions

From the above mentioned discussions the followings can be concluded:

1. The condition in Wadi El-Farigh is a serious indicator for the mismanagement of the water resources in the Miocene aquifer tapping this area. Actually, the continuous decrease in the saturated thickness of the aquifer will lead to a tragedic problem in the agricultural development in this area. Bearing in mind that the saturated thickness of the Miocene aquifer in Wadi El-Farigh was decreasing with time as clear from the three years 2003, 2012 and 2015, where the thickness is 83.9–175.3, 78.5–156.1 and 73–153 m respectively. It is noticed that, in 2003 the dominant value of the saturated thickness of the Miocene aquifer is more than 120 m while in 2015; the dominant value is less than 120 m reaching in some areas to 73 m. One should give a warning for the applied water exploitation rates.
2. The groundwater depletion ranges from 6 m to more than 30 m during this period of time and the dominant depletion is approximately ranges between 1.3 and 1.7 m/year.
3. Also, the quality of groundwater for the study area was deteriorate (salinization problems, low water quality for drinking only 33% are suitable for drinking purposes and high concentrations of some trace elements due to natural or anthropogenic sources). All these problems are considered a very good indicator for mismanagement of the groundwater resources in the study area.

Recommendations

From the above mentioned discussions the followings can be recommended:

1. The drilling should penetrate the full thickness of the Miocene aquifer to increase the saturated thickness and, hence, the aquifer transmissivity through the reaching of the drilling to the top surface of the basaltic sheet which is considered as the base of the Miocene aquifer.
2. The well design should be constructed I a proper way to isolate the salty layers from the productive zone of the aquifer.
3. The spacing between each two successive drilled wells should be not less than 500 m to avoid well interference, and hence, the water quality degradation.
4. The whole area of the west Nile Delta should be have a monitoring networks of observation wells.

5. Well drilling in the area, should be strongly controlled and licensed.
6. The modern irrigation systems should be applied
7. The crop pattern in the study area should be changed to avoid the crops of high consumptive use of water.

References

1. Shata AA (1962) Geology, hydrology and groundwater hydrology of Wadi El-Natrun and adjacent areas, Internal report, PT. 1, Cairo, U.A.R., Desert Institute
2. Pavlov M (1962) Hydrogeology, in preliminary report on the geology, hydrogeology and groundwater hydrology of Wadi El-Natrun and the adjacent areas, Part 2, Cairo, U.A.R., Desert Institute, The General Development Organization, p 63
3. El-Fayoumy IF (1964) Geology of groundwater supplies in Wadi El-Natrun area. MSc thesis, Faculty of Science, Cairo University, p 109
4. Sanad S (1973) Geology of the area between Wadi El-Natrun and the Moghra depression. PhD thesis, Faculty of Science, Assiut University, Assiut, Egypt, p 184
5. El-Shazly EM, Abdel Hady M, El-Ghawaby H, El-Kassas K, Khawasik SM, El-Shazly MM, Sanad S (1975) Geologic interpretation of Landsat satellite images for West Nile Delta area. Remote Sensing Center, Academy of Scientific Research and Technology, Cairo, Egypt, p 38
6. Attia SH (1975) Pedology and soil genesis of the Quaternary deposits in the region West of the Nile Delta, North and East of Wadi El-Natrun. PhD thesis, Faculty of Science, Ain Shams University, Cairo, Egypt, p 258
7. El-Ghazawi MM (1982) Geological studies of the Quaternary-Neogene aquifers in the area northwest Nile Delta. MSc thesis, Faculty of Science, Al-Azhar University, Cario, Egypt, p 170
8. Abdel Baki AMA (1983) Hydrogeological and hydro-geochemical studies in the area west of Rosetta branch and South El Nasr canal. PhD thesis, Faculty of Science, Ain Shams University, p 156
9. Shata AA (1959) Geological problems related to the groundwater supply of some desert areas of Egypt. Internal report, Desert Institute, Cairo, Egypt
10. Shata AA, El-Fayoumy IF (1967) Geomorphological and morphopedological aspects of the region west of the Nile Delta with special reference of Wadi El-Natrun area. Bull Inst Desert d'Egypte T XVIII(1):1–28
11. Said R (1962) The geology of Egypt. Elsevier Publishing Co., Amsterdam, New York, p 380
12. Shata AA (1955) An introductory note on the geology of the northern portion of the Western Desert of Egypt. Bull Inst Desert d'Egypte T 5(3):96–106
13. Abd El Rahman AA (1996) Geophysial study on the groundwater conditions in the area southwest of the Nile Delta between Abu Raoash and El Khatatba road. MSc thesis, Faculty of Science, Ain Shams University, p 116
14. Omara SM, Sanad S (1975) Rock stratigraphy and structural features of the Rea between Wadi El Natron and Moghra Depression. Western Desert, Egypt, pp 1–54
15. El Ghazawi MM, Atwa SM (1994) Contributions of some structural elements to the groundwater conditions in the southwestern portion of the Nile Delta. Geological Society of Egypt, Cairo, Egypt, 38(2), p 49
16. Sobeih MM, El-Arabi NE, Esam E-D, Helal BSA (2017) Management of water resources to control groundwater levels in the southern area of the western Nile delta, Egypt. Water Sci 31:137–150
17. Nofal ER, Fekry AM, Ahmed MH, El-Kharakany MM (2018) Groundwater; extraction versus recharge; vulnerability assessment. Water Sci 32:287–300

18. Mohamed Ibrahem SM (2020) Groundwater hydrology and characteristics of the tertiary aquifers, Northwest Cairo, Egypt. NRIAG J Astron Geophy 9(1):420–432
19. Ibrahim SMM (2005) Groundwater resources management in Wadi El-Farigh and its vicinities for sustainable agricultural development. PhD thesis, Faculty of Engineer, Ain Shams University, Cairo, p 152
20. Gad M, El Hattab M, Galal Sh (2016) Assessment of the Miocene aquifer in Wadi El-Farigh area by using GIS techniques. NRIAG J Astron Geophys 5(2016):463–473
21. Youssef AMA, Ibrahim SMM, Hussien RA (2017) Monitoring and assessment of groundwater in the area of Km80, Cairo-Alexandria highway using geophysical, hydrological and isotopic techniques—case study. Egypt J Geol 61(2017):41–67
22. Van Der Velpen BPA (1988) RESIST, version 1.0, a package for the processing of the resistivity sounding data. MSc Research Project, ITC, Deft, the Netherlands
23. Interpex Limited (1996) RESIX- PLUS, Reistivity data interpretation software, v. 2.39. Golden, Colorado, USA
24. WHO (2011) Guidelines for drinking water quality, 4th edn. World Health Organization Geneva, Recommendations, pp 1–4
25. Konikow L, Reilly T (1999) Seawater intrusion in the United States. Seawater intrusion in coastal Aquifers—concepts, methods and practices. Springer, pp 463–506
26. Rhoades JD, Kandiah A, Mashali A (1992) The use of saline waters for crop production. FAO
27. Piper AM (1944) A graphic procedure in the geochemical interpretation of water analysis. In: American geophysical union transactions, vol 25, pp 914–923
28. Appelo C Williemsen A (1987) Geochemical calculations and observations on salt water intrusion, I: a combined geochemical mixing cell model. J Hydrol 94:313–330
29. McLean W, Jankowski J (2000) Groundwater quality and sustainability in an alluvial aquifer, Australia. In: Sililo A (ed) XXX IAH congress on groundwater: past achievements and future challenges. A.A., Balkema, Rotterdam
30. Custodio E, Bruggeman GH (1987) Groundwater problem in coastal areas. UNESCO publication Bielt Frers, Fleurs, Belguim
31. Garrels RM, Mackenzie FT (1971) Gregor's denudation of the continents. Nature 231:382–383
32. Stumm W, Morgan JJ (1981) Aqautic chemistry. Wiley Interscience, New York
33. EHCW (2007) Egyptian higher committee for water and Egyptian standard for drinking water and domestic uses (in Arabic)
34. CCME (Canadian Council of Ministers of the Environment) (2001) Canadian water quality guidelines for the protection of aquatic life: CCME Water Quality Index 1.0, Technical Report. In: Canadian environment quality guidelines, 1999, Canadian Council of Ministers of the environment, Winnipeg
35. Hagagg KH (2009) Water resources evaluation and environmental assessment of the southwest area of the Nile Delta. MSc thesis, Faculty of Science, Cairo University

Characterizing Ecosystem Services to Human Well-Being in Groundwater Dependent Desert Environments

Caroline King-Okumu, Ahmed Abdelkhalek, and Boshra Salem

Abstract As groundwater use intensifies, the stakes in management decisions grow larger and more complex. Ecosystem service analysis connects groundwater management decision-making to a range of valuable outcomes for human well-being. This chapter explores conceptual developments enabling the ecosystem services associated with groundwater reserves and extraction options to be characterized, compared, and weighed in qualitative and quantitative terms by decision-makers. The application of these concepts and associated methods is illustrated through a case study located in the Nubian-Moghra Miocenene-Nile aquifer complex at Wadi Natrun in the Western Desert of Egypt. The chapter reveals a wide range of provisioning services provided by groundwater in the case study area of the Egyptian desert. It is unlikely that these could all be fully assessed. However, the case study demonstrated that groundwater managers can use the ecosystem services framework to identify and assess some of the most important benefits that groundwater provides to society, including its essential services as a buffer against water security threats. The iterative approach offers options for groundwater managers in Egypt and elsewhere to coordinate across sectors of government, engaging the public and the scientific community to overcome assessment challenges in a progressive manner.

Keywords Ecosystem services · Dryland ecosystems · Human well-being · Vulnerability

---

C. King-Okumu (✉)
The Borders Institute (TBI), Africa, Nairobi, Kenya
e-mail: caroking@yahoo.com; carkin@ceh.ac.uk

GeoData Institute, University of Southampton, Southampton, UK

A. Abdelkhalek
Technical Office of Groundwater Sector, Ministry of Water Resources and Irrigation, Giza, Egypt
e-mail: ahmedimail@yahoo.com

B. Salem
Pharos University (www.pua.edu.eg), Alexandria, Egypt
e-mail: boshra.salem@dr.com

A. Negm and A. Elkhouly (eds.), *Groundwater in Egypt's Deserts*, Springer Water,
https://doi.org/10.1007/978-3-030-77622-0_4

Introduction

In areas affected by drought and long-term desiccation trends, groundwater can secure human life, providing a continual source of water for drinking, food production, human settlement and other economic activities [1, 2]. Economic benefits that can be achieved from groundwater use in the short term are often seen to be more important than conserving resources for an uncertain future. However, under conditions of very low and erratic rainfall, such as those that occur in many dry and desert areas, reserves of groundwater may be either non-renewable or recharged over long timeframes from areas that are remote from the point of use. The presence of these reserves provides a buffer against periods of increased climatic stress and uncertain surface water supplies [3, 4]. Continuous depletion of this buffer and the use of groundwater reserves in excess of recharge rates undermines their availability for the future [5, 6].

Threats to the continued supply of groundwater for drinking and economic development are causing increasing concern for human security in many of the drier regions of the world, where dwindling supplies of groundwater are extracted at rates that exceed natural recharge [7]. Since regulatory measures often cannot stop over-extraction under conditions of water scarcity [2], there is increasing interest in economic approaches and the use of financial (dis)incentives to manage private-sector use of groundwater [8, 9]. There is also a growing realization that economic decisions concerning the expenditure of groundwater reserves need to be weighed more carefully in order to better understand the likely tradeoffs between short- and long-term benefits and losses to human well-being for society as a whole [10, 11].

Building systems to undertake the economic assessment, manage groundwater and improve decision-making requires investment to generate relevant information and human capacities to marshal and use it [12]. To be effective, a groundwater management institution needs to have the capacity to regulate extractions, recharge and threats to water quality [13]. The higher the stakes in groundwater management decisions [10], the greater the justification to invest more in understanding the present state of the system, including the environmental and social dynamics that operate within it, and the potential trade-offs that could be associated with different management options.

The challenge to use economic valuation to weight groundwater management options has received increasing attention internationally [14–17]. During the last millennium, economic assessments of groundwater management decision-making in Egypt and elsewhere often focused on the value of groundwater for agricultural production [18, 19]. However, during the present millennium, the broader value of ecosystem services has emerged as an important conceptual frame of reference [11, 20–23]. Ecosystem services are benefits provided by nature that enhance the welfare of humans and can, therefore, sometimes be valued in monetary units. Ecosystem services include food production but also a range of other valuable benefits that societies need from the environment. Decision-makers can use the ecosystem services framework to explore whether the total value of the mix of services obtained from the environment under one management regime is greater or less than the total economic value of the mix provided under an alternative management approach [24–27].

Human needs for groundwater cut across all sectors—including drinking water supply and health needs. Aquifers supply these safely with minimal energy and infrastructure for delivery or purification [28]. Conceptually, ecosystem service assessment would appear an ideal frame through which to generate a well-balanced assessment of the value of services provided by groundwater to secure human water needs [29, 30]. Management scenarios including different future configurations and valuation of ecosystem services may involve either qualitative or quantitative valuation [31]. This evolving approach to environmental economic decision-making has been the focus of a series of recent conceptual guides for decision-makers [32, 33] and also practical applications for integrated resource management in dryland contexts [34–37].

In this chapter, the merits of the ecosystem services framework are explored as a means to characterize the services associated with groundwater availability and use in Egypt and to begin to capture their value. A case study in the Western Desert illustrates the scope to apply the ecosystem service characterization and assessment approach and also highlights areas where Egyptian groundwater managers could invest in improving information systems for this purpose. This draws on a series of previous studies in the area [37–39]. The discussion reflects on the challenges of effective valuation of ecosystem services to inform national and local decision-makers. The conclusions focus on the justification for increased investment in groundwater management capacities to improve decision-making. Recommendations highlight the scope for an iterative inclusive and participatory approach to review groundwater management decision-making, increasing resource allocation, public participation and engagement of the private sector.

Characterizing and Valuing Ecosystem Services from Groundwater in Dryland Ecosystems

Ecosystem services to human well-being are generally categorized as provisioning services (e.g. drinking water, food, fuel), regulatory services (e.g., climate regulation, water purification), cultural services (e.g., education, recreation) and supporting services (e.g., water cycling, soil formation). Ecosystem services associated with groundwater have been identified in each of these categories [32] (Box 1).

Box 1: Groundwater—Related Ecosystem Services [28]
Providing services:

- Drinking water
- Water for food and beverage industry
- Water supply for agricultural activities, such as irrigation
- Strategic groundwater resources
- Process water for industry

- Cooling water for industry.

Regulating services:

- Groundwater as a storage medium for heat or coolness
- Groundwater as a supplier of coolness and warmth
- Maintain groundwater level and prevent subsidence
- Maintain groundwater level and stability of civil engineering constructions
- Water retention, drainage (water buffering)
- Water supply in groundwater-dependent surface water regimes
- Water supply in groundwater-dependent seepage areas
- Purifying and filtering effect of groundwater and soil.

Supporting services:

- Groundwater functions and processes (indirectly) related to ecosystem services
- Role of groundwater in the biogeochemical cycles.

Cultural services

- Preservation of cultural-historical and archaeological values
- Esthetical and ethical values of the groundwater ecosystem.

Ecosystem service assessments draw on environmental economics to identify valuation methods for each service [40]. Methods to value both direct uses and also indirect use values, including option value and non-use values incorporating existence and bequest values have been explored through an extensive literature on Total Economic Valuation (TEV) [41]. The remainder of this section explores how each ecosystem-service type may be valued.

Provisioning Services

Direct use values often apply to the provisioning services (i.e., groundwater that is consumed for drinking, agriculture, and other uses). Various controversies surround the selection of foods, fibres and other biotic resources that should be valued as groundwater products in both wetlands [42, 43], and dryland environments [44]. Few assessments succeed in capturing the full diversity and value of cultivated crops and livestock in any ecosystem, let alone the array of wild plants and animals that grow spontaneously due to soil water-availability caused by a capillary rise or the pooling of water at natural depressions in the landscape. This is partly due to the difficulties of quantifying and valuing a large range of products under smallscale or subsistence uses—or which are not marketed for other reasons, e.g. relating to lack of market access. Furthermore, while most assessments assign value according to local market prices, longer market chains may add to the value generated within the economy [45].

Regulating and Supporting Services

Many of the regulating, supporting, and cultural ecosystem services are considered to represent 'indirect' use values that contribute to the direct use values of the provisioning services. Often, it has been observed that assigning a separate and additional value to these services is not necessary and could lead to 'double-counting' [e.g. 46]. However, some regulating services can be valued as an option for future conditions of scarcity, or have non-use values associated with existence or bequest functions that would be accommodated separately from direct uses within the TEV framework [14, 41]. For this reason, some environmental economists have argued that their values should not be excluded from ecosystem service assessments [47].

Aquifers provide services by delivering water and maintaining pressure so that it can be available when and where it is needed. To calculate the replacement cost of the lost pressure where water tables have fallen, the additional unit cost to pump a fixed volume of water from an increased depth has been used [48, 49]. Additional energy use for pumping can enable water users to continue to use their wells as the water table falls as long as the water level remains within the reach of the well. However, if the water level falls below the reach of the well, or the level of salinity in the well becomes too high, the well will need to be replaced. The replacement cost for the loss of artesian pressure must then include the cost of drilling a new well, in addition to the increased requirement of energy for pumping.

Other in situ services provided by groundwater that may be taken into account in valuing groundwater systems include the support of ecosystems and phreatophytic agriculture, the maintenance of spring flows and base flows, the prevention of land subsidence and seawater intrusion, and the potential for exploiting geothermal energy or storing heat [1].

The timing and location of groundwater availability when and where it is needed—e.g., during droughts or in remote areas can create a value that is not captured in relation to productive uses of water under non-water-scarce conditions [3, 4, 50–52]. As long as the buffer is not used, it may be regarded as natural capital [20] rather than as a service. As a result of this distinction, the groundwater buffer value is not always valued in all recent dryland ecosystem service assessments [e.g. 34, 53, 54]. However, if decision-makers do not recognize the value of groundwater reserves and are only aware of the value of groundwater extraction, they may decide that extraction should be accelerated. This could threaten the sustainability of the resource.

If availability of the option to use groundwater could prevent the future loss of natural or physical capital (e.g. enabling timely irrigation of an orchard that would otherwise die during a drought—requiring years of investment to replace) the value of the buffer should relate to the averted loss, either at the full replacement cost, or in a form similar to an insurance premium [55]. It has been suggested that buffer value may be defined as the difference between the maximal value of a stock of groundwater under uncertainty and its maximal value under certainty where the supply of surface water is stabilized at its mean [52]. However, under uncertainty groundwater can also preserve human life, and valuation of this maximal use can be controversial. Even in less extreme cases, assessment of the maximal value is time-dependent because as scarcity increases, causing the buffer to be used, the market price is likely also to increase. Furthermore, in some cases higher value uses (e.g. for human drinking-water) may replace lower-valued agricultural uses [e.g. 56]. In a limited number of developed country contexts, hedonic pricing has been used to identify a difference in land value attributable to different levels of groundwater access [14].

Cultural Services

Cultural services associated with recreational uses or spiritual values of ecosystems—including those associated with spring-waters and watering points in desert landscapes—are notoriously difficult to place a value on. However, various previous studies conducted in dryland systems have used travel cost calculations to reveal hidden market prices paid to access and enjoy such services. For example, visitor numbers and prices paid to stay in local hotels have been used to assess touristic value [e.g. 35].

Case Study Materials and Methods

Approach to the Characterization of Ecosystem Services

An iterative consultative process enables system characterization, service identification and valuation [after 32, 57]. This includes building an understanding of the resource—its physical characteristics as well as the uses and pressures it faces—and determining how valuation might help inform decisions about the use of groundwater. Two major components explored in this review are:

- Characterization of ecosystem services
- Valuation of selected ecosystem services.

A 2-tiered approach to the valuation [57], allows an initial assessment to be made based on information available, including recommendations for further investigation needed to refine the first tier assessment. This accommodates a relatively quick and less data-intensive route (Tier 1) to provide an initial conservative assessment of the ecosystem service value which can then be used to indicate priorities for deepening and refining the evidence base through investments in research to pursue key knowledge gaps and enhance certainty (Tier 2). The rapid Tier 1 assessment identifies and provides a justification for strategic decisions to be taken and the investment of resources that might be needed to achieve the Tier 2 valuation. It is important to recognize that the valuation will not ever be fully exhaustive. What is essential is that it should be sufficient to provide decision-makers with an effective indication of the nature and magnitude of the stakes involved in the decisions that need to be made.

Case Study Area

Wadi Natrun is an administrative centre (*Markaz*), in the Western Desert of Egypt (Fig. 1). It includes a rapidly growing urban area, five surrounding villages, a number of small hamlets, and a large expanse of farm and desert land. Although the climate is arid, with an average annual rainfall below 50 mm, a natural depression in the desert landscape supplies the water table and a series of lakes with drainage water from the river and canal systems and cultivated lands at the western edge of the Nile delta. This water forms a shallow aquifer, known as the Pleistocene aquifer. The lakes have been estimated to contain an average annual water level of 63 $\times$ 106 m^3/yr [58]. Drainage-water and lake water are saline and vulnerable to contamination [59].

Located beneath the Pleistocene aquifer, the local Pliocene is the main aquifer in the depression and has been used for irrigation since the 1960s. The Pliocene aquifer has been estimated to receive around 60 mm^3 of recharge per year from both the Pleistocene aquifer, above, and the Miocene aquifer, below [60, 61]. The Miocene aquifer, containing non-renewable groundwater resources extends West and Southwards beneath the Western Desert. This aquifer is also increasingly exploited. In the multilayer system as a whole, water resources have been extracted in excess of recharge. The water table has been falling since the 1990s and salinization has been observed [62].

The total population of the Wadi Natrun administrative area is around 79,000 [63]. This has risen since the previous census in 2006, which recorded over 70,000 people [64]. More rapid growth took place during the previous decade, when the population almost tripled from around 25,000 in 1996. According to the Egyptian Agricultural Census [65, 66], the cultivated area around the villages and town approximately doubled in size over the period 2000–2010, to cover an extent close to 1000 ha. Administrative divisions for the surrounding area had changed between census dates, but by 2010, these outlying areas added a further 3000 ha, bringing the total cultivated area to around 4000 ha—all dependent on groundwater for irrigation, and still growing. If a new agricultural census is conducted in 2020, it may reveal the continued expansion of the cultivated area, even if the rate of agricultural expansion may be slowing down, and degradation is occurring in some areas.

According to the population census [64], irrigated agriculture accounted for 55% of local economic activity. Other economic activities were in industries servicing the agricultural sector, e.g., transportation and storage (8%), retail and servicing of vehicles (10%). Intensive agricultural use of each hectare of land in this region is estimated to provide four full-time jobs for agricultural workers, plus an equal number in the service sectors (i.e., a total of 8 jobs per hectare) [67]. However, not all of the farmland is cultivated intensively. Farming patterns present in the area

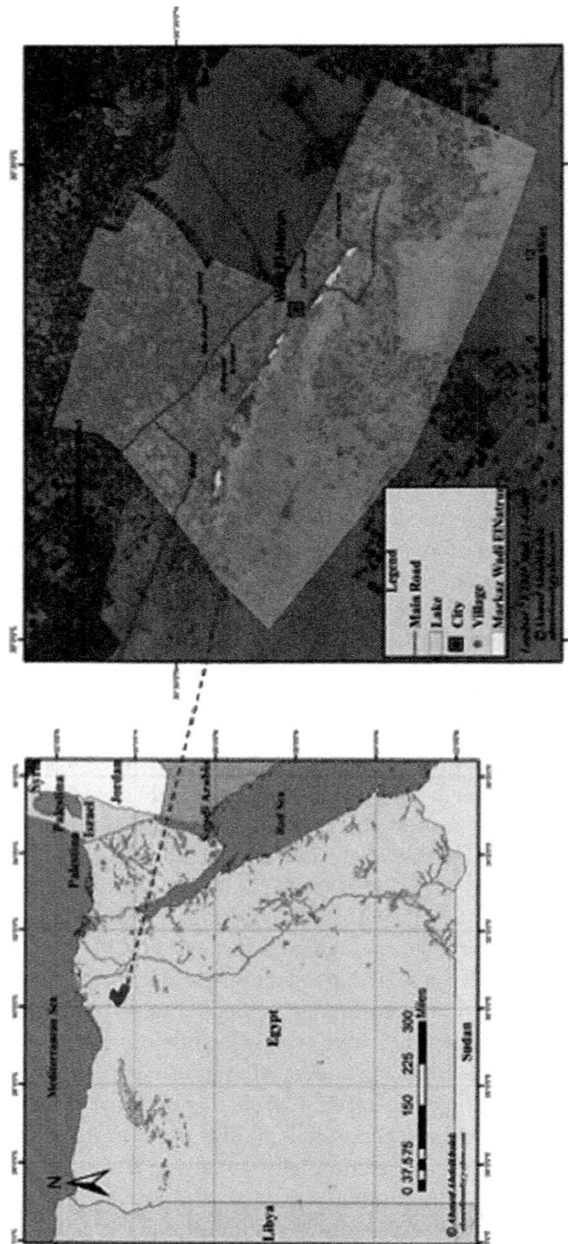

Fig. 1 Location of case study area (Modified by Ahmed Abdelkhalek from administrative maps and landsat images)

Table 1 Scenarios for Nile flow and groundwater use (BCM) [68]

	Drying scenario			Wetting scenario		
	Nile flows	Deep gwater	Shallow + reuse	Nile flows	Deep gwater	Shallow + reuse
2015	52.5	2	21.75	52.5	2	21.75
2025	47.47	2.3	19	55.55	2	22.3
2050	40.38	3.4	15.3	57.48	2.2	23.2
2075	36	4.5	13.5	55.8	2.5	23.35
2100	29.33	6	12.8	53.98	2.75	23.8

also include small-scale subsistence farming and sporadic speculative cultivation by absentee investors [38]. An industrial area has also been expanding since the 1990s.

Over the coming decades, the requirements for water to support development in the Nile Delta including all sectors is anticipated to increase, the flow of water arriving through the Nile system is expected to decrease, and increasingly groundwater is anticipated to make up some of the shortfall [68]. Additional water supplies from other sources—e.g., seawater and treated wastewater are also expected to be used. Groundwater management scenarios for the coming century include a local reduction in recharge from the Nile system, further reductions due to increasing temperatures and evapotranspiration, and also increased extractions under both wetting and drying conditions (Table 1). In addition to the increasing use of both deep and shallow groundwater, non-treated water reuse is projected to increase.

Characterization of Ecosystem Services

Previous valuation work in the study area was identified to have focused exclusively on the value of groundwater use for irrigated crop production [69–71]. However, previously published field survey investigations [38] indicated that in addition to field crops, livestock that grazed in the wetland and fed on irrigated fodder crops were also an essential component of smallholder livelihoods. Some of the larger farms in the area also operate intensive dairy production.

In ecosystem service assessments, ecosystem components are usually identified according to landcover classes visible in the area using a combination of field observation and remote sensing methodologies. Geographical Information systems for landcover mapping offer the option to subdivide agricultural land covers into a manageable range of different classes according to farm type, and to overlay available topographic, landcover and hydrogeological maps [61, 72]. Even though hydrogeological information has not been updated since 1991 in our case study area and long-term continuous groundwater monitoring data is not publicly available, this enables a general estimation of agricultural groundwater use patterns. It does not capture the use of ecosystem services in industry, construction and service sectors.

Tier 1 Assessment of Ecosystem Services

This step was based on a review of published literature, including published social and ecological investigations completed in the study area [e.g. 38, 73], and conceptual literature (as briefly reviewed earlier in this chapter). A tier 1 assessment [after 57], does not seek to present its findings as an exhaustive valuation of the ecosystem services, but rather presents the state of available knowledge, and highlights priorities requiring deeper exploration through a Tier 2 assessment. The ecosystem services were mapped into a TEV [after 74]. This enabled identification of services that could be valued, valuation methods, data sources and gaps remaining.

Characterization of Ecosystem Services

The ecosystem service characterization in each landcover class included the identification of agricultural produce, including both crops and livestock (Table 2). Wildlife, particularly birds and insects, were also observed around the drainage lakes. Other products were identified, including fibrous plants used to make fences and papyrus, salts with potential for use in industry, natron for glassmaking, sand and construction products. Artesian pressure and groundwater quality were identified to have economic importance as regulating services. The historic, cultural and amenity values of the area also provide economic opportunities for the attraction of tourists to local hotels and museums. These ecosystem services were classified into different types and a rapid assessment of vulnerability to groundwater degradation was made. Management decisions might be expected to affect these vulnerabilities in different ways.

Defining the Scope for Assessment of Ecosystem Service Values

A manageable number of key ecosystem services for weighing of management scenarios and tradeoffs at the site (Table 3) were selected from amongst those initially identified (Table 2).

Creation of management scenarios including different future configurations and valuation of ecosystem services may involve either qualitative or quantitative valuation [31]. The availability of either quantitative or qualitative data concerning the selected ecosystem services was assessed, and possible methods for valuation explored. Based on information available during this assessment, a minimum set of initial values for an initial Tier 1 assessment could include those for provisioning of water, crops, and livestock, as well as water regulating services and cultural services (Table 4).

Table 2 Characterization of ecosystem services in the study area and their exposure to groundwater threats [after 38, 72, 75]

Ecosystem component	Ecosystem services				Vulnerabilities to groundwater degradation
	Cultural	Provisioning	Supporting	Regulating	
Subsurface and water sources (springs and wells)	Natural wonder of springs artesian pressure and purification	Water for domestic use, bottling, irrigation, and industry	Water cycling, hydrogeological structure and function	Regulate artesian pressure Separate fresh water from saline water and contamination threats	Loss of artesian pressure and falling water table Loss of water quality due to mixing between aquifer layers and with surface infiltration
Agricultural land	Green environment	Food crops and income	Generation of nutrient rich soil and grazing fodder for animals	Carbon sequestration and cooling effect where date palm and other trees remain under cultivation	Altered quality of water limits cultivation options Falling groundwater table requires digging of deeper wells. Increasing salinity may affect diversity of fruit and vegetable species
Built environment (houses, factories, roads, water and energy infrastructure)	Architectural beauty, facilities for settlement, hospitality and economic activity	Building materials (sand, glass, tiles), energy supplies, piped water	Waste removal and drainage systems	Regulate recharge patterns, human activities, numbers, and movement/transportation	Increase water demand and extraction Altered runoff and recharge patterns introduce contamination threats
Sand and gravel desert land	Religious significance, space and amenity for Coptic temples mental health, prisons, military uses, hunting	Sand and gravel extraction, Limited grazing and hunting of wild animals and birds	Wild plants, insects, reptiles and mammals create a food chain to break down and remove wastes	Waste decomposition and removal	Alteration of water availability in soil, rocks and gravels affects the survival of sparse vegetation and wild species

(continued)

Table 2 (continued)

Ecosystem component	Ecosystem services				Vulnerabilities to groundwater degradation
	Cultural	Provisioning	Supporting	Regulating	
Salt-marsh and halfa grassland	Traditional common land 'curf', Ecotourism, biodiversity	Grazing for animals, plants for fiber, rare papyrus reeds (*Cyperus papyrus*), bird species	Filtration, sedimentation, soil formation	Drainage water purification, carbon sequestration and other elements in soils, silt and plant matter	Falling groundwater tables affect the extent of salt-marsh vegetation
Saline lakes	Landmark and tourist attraction	Limited fish-farming and production of other species, salts for domestic, cosmetic and industrial use	Habitat for birds and microorganisms, evaporation	Natural drainage area, ensuring the quality of the surrounding land for cultivation, humidity regulates climate	Alteration of the quantity and quality of groundwater inflow to lakes affects water levels and chemical conditions for salt production and species habitats

Table 3 Scope of assessment of groundwater dependent ecosystem value components [after 74]

Value type	Value subtype	Meaning	Service	Service type	Valuation approach	Units	Landcover class and subdivisions
Use values	Direct use	Consumptive use: provisioning services	Water for agricultural use	Provisioning	Market value derived from net value of additional unit of crop production (production function)	US$ profit from market value of crop less inputs per m^3 water per hectare per Year	Agricultural land: commercial horticulture (grape, citrus), or small farms
			Habitats for animal use (grazing, hunting, apiculture)		Market value—derived from net value of additional unit of livestock meat, milk, wool production	US$ avoided expenditure from market value of animal products per small farm per year	Salt marsh, desert
			Minerals for extraction and industrial uses (e.g. sand, salts, construction materials, agrochemicals)		Market value—derived from net value of additional unit of produce (production function)	– commodities (ceramics, fertilizers, etc.) US$/unit/$m^3$ of water/year	Desert
			Water for industrial and service uses (e.g. water bottling, vehicle services, agroprocessing, hospitality)		Market value—derived from net value of additional unit of produce (production function)	Market value of: – expenditure by service station visitors US$/visitor/year – bottled water US$/$m^3$/factory/year	Built environment: service centres, factories, bottling plants
			Water for domestic uses		Market value: price to purchase water delivery	Domestic water tariff US$/$m^3$/year	Built environment: homes

(continued)

Table 3 (continued)

Value type	Value subtype	Meaning	Service	Service type	Valuation approach	Units	Landcover class and subdivisions
	Indirect use	Regulating services	Drainage Aquifer recharge and storage Water purification Water pressure	Regulating	Cost based: Avoided or replacement cost: pumping and treatment costs for water	Additional m depth of pumping US$/m^3/year treatment cost for hotel: US$/m^3/year	Subsurface: aquifer layers and well depth
	Option	Future availability for private benefit	Agricultural production, settlement or industrial use of land	Cultural	Revealed preference: Hedonic price of land w water supply	Additional value of land with fresh water supply US$/ha	Agricultural land and sand and gravel desert
Non-use values	Bequest value	Knowledge that services will be available to future generations	Value of the land as capital				
	Altruist value	Knowledge that services are available to others	Biodiversity Green environment Natural springs Archaeological remains		Revealed preference: Travel cost or green brand preference	Travel cost to Coptic Monastery or ecotour	Salt marsh, desert
	Existence value	Knowledge that services continue to exist					

Note The valuation methods are not exclusive and may often be combined—e.g. contingent valuation can be used to assess willingness to pay replacement costs [adapted from 27, 48, 76]

Table 4 Summary of minimum initial assessment components and present market prices

Service type	Service	Commodity with market price in 2009
Provisioning	Water for drinking	Local water rates: 30LE × 10,000 households × 12months
	Irrigated crops	Crop value/m^3/ha/yr
	Grazing and forage	Animals/household (e.g. 1 cow, 6 goats or sheep and 10 birds per year per household for an estimated 5000 (50% of total) households engaged in small farming in the study area)
Regulating	Water quality and pressure	Cost of pumping water from increased depth each year and periodic replacement of wells
Cultural	Aesthetic and spiritual quality of springs, lakes and wetland	Estimated 400 tourist visits/year at cost of hotel ($50)

Provisioning

Approximate domestic water rates in 2009 were identified through household interviews (Table 4). The actual volume of water used by households and the cost paid by the water service provider to extract water and provide the service is not in the public domain. However, it could be possible to ascertain these values through a Tier 2 assessment involving the service provider. This would also be likely to reveal increases in domestic water, energy and fuel prices that have taken place over the past decade.[1]

Since an industrial estate is located in the study area, including various factories and also around six water bottling facilities, the omission of water provisioning services to them in this assessment is significant. It could be possible to overcome this challenge through a Tier 2 assessment involving the local electricity provider since electricity could be assumed to be the main source of power used in extracting the groundwater, and a unit ratio could be established. The value per unit of water extracted for industrial uses would depend on the products.

Groundwater provisioning to agriculture can be estimated on a per hectare basis according to the value of the crop that is produced. Based on previous farm budget studies in the study area, the gross margin per hectare and m^3 of groundwater for crops can be estimated from the market price per unit and the input costs. Additional by-products for animal consumption would also likely be produced from non-fodder crops [77]. For the future, increasing prices may be assumed due to the growing scarcity of water to maintain current cropping patterns. Although evapotranspiration is expected to increase due to anticipated increases in temperature, irrigation efficiency may also increase [68]. Hydrologic modeling [after 78] could be used to

[1] Note that these changes in prices will have been related to factors other than groundwater conditions. Some further analysis of the effects of the removal of subsidies might be available from the World Bank in light of their work on this issue.

refine estimates of agricultural water use and return flows to the near-surface for a Tier 2 assessment.

A rapid estimate of the present value of livestock production could be based on assumed average ownership of, e.g. 1 cow, 6 goats or sheep and 10 birds per year per household for an estimated 5000 (50% of total) households engaged in small farming in the study area. A Tier two assessment could employ household survey techniques to refine these estimates, incorporate consideration of seasonality and market chains affecting the value of livestock production and incorporate consideration of any uses and value made of wild plants and animals. It might also take larger-scale intensive livestock production into account.

Regulating

To value the regulation of water supply (pressure and quality), farms can be classified into two categories [after 38]: small farms (around 2.5 ha) with wells of an average 30 m depth in 2009, and larger farms (the average irrigated area around 25 ha) and wells of an average 100 m depth in 2009.

An estimate of an average annual increase of 1 m in depth to the water affecting larger farms in the study area was identified [62], and reports of well failure from farm employees in previously published surveys [38]. If it is assumed that this rate of decline will continue, and that the majority of pumps in use on farms in the area will continue to rely on the use of diesel fuel to pump water from an additional metre of depth each year, with the replacement of one well occurring every five years on the larger farms, these assumptions can enable an approximate calculation of the cost of the loss of groundwater pressure on the large farms. Based on this assessment, use of GIS [after 78, 79] could be pursued to refine this 'bathtub' representation of groundwater system characteristics and vulnerability. The effects of alternate fuel sources to supply the amounts of energy needed should also be explored as critical determinants of cost and value.

For the smaller farms, field surveys suggested a slower rate of well failure and replacement (e.g., every ten years), which was concluded to result from the lower elevation of most of the smaller farms [described in 80]. However, the expense to replace a well was more difficult for small farmers to afford and the additional social costs associated with the loss of small farming livelihoods are more difficult to capture. A Tier 2 assessment might be proposed to evaluate the cost of an unemployment benefit and pension scheme to replace the social safety net and retirement occupation that is provided by smallholder farming in the study area. Nevertheless, it is difficult to conceive of an assessment method that could effectively capture the full costs in terms of household diets, health and family dislocation associated with the loss of family farming traditions.

Cultural

Concerning the aesthetic, spiritual and recreational dimensions of value associated with the study area, the most convenient option for economic assessment consists in either visitors' willingness to pay (e.g. to overnight in local hotel accommodation) or consumers' willingness to pay for local branded produce. For the purposes of the initial assessment, the travel cost method is the most feasible option. Although visitor statistics have not been collected in any systematic way, a cost of around US$50/person might be assumed for an estimated 400 visitors/year. However, over the past decade, the numbers of visitors have been affected by security concerns. This would reduce the extent to which visitor numbers could indicate the cultural value placed on the resource in the study area. Therefore other assessment approaches might also be needed for refinement of these estimates in a Tier 2 assessment. Rapid participatory methods might offer a starting point for identification of relevant value types.

Discussion of Scope for Management Scenarios Using the Ecosystem Service Framework

This chapter has demonstrated that there are many valuable ecosystem services that could be affected by groundwater management decisions in the Egyptian desert. Management scenarios may need to consider uncertainties concerning the volumes of surface water that will be available in Egypt in the future, due to climate change and upstream land management decisions. Specific scenarios could also consider the effects of different levels of investment in water management technologies and regulatory systems as well as land-use changes.

It is unlikely that all of the benefits that groundwater in Egypt's deserts provides could be fully assessed. For practical reasons, assessments will need to be selective, and start by assessing a few key services. The case study demonstrated how groundwater managers can use the ecosystem services framework to identify and assess some of the most important benefits that groundwater provides to society, including its essential services as a buffer against water security threats.

The case study also demonstrated that there are information constraints affecting the scope for assessment of some of the key services provided by groundwater. For example, the volumes and value of water use by industries, such as water bottling plants, could not be assessed without the engagement of the companies involved. However, these challenges could be overcome through an iterative participatory approach by groundwater managers with water service providers, industrial users, and general public, as well as with agricultural water users and scientists.

When it comes to supporting and regulating services, the chapter demonstrated that there are methods available to assess the value of these. This can enable comparison between management options that involve accelerating groundwater extraction, versus more conservative strategies. However, comparing these values will be complicated and controversial. This is why a discursive and iterative approach that enables different stakeholders' perspectives to be discussed is important. Cultural values could also require further exploration and discussion.

In the selected case study, like most other dryland contexts, the time and costs required to overcome current limitations on data availability are potentially significant. However, this chapter has demonstrated that the initial phase of an iterative assessment of groundwater-related ecosystem services can be achieved fairly inexpensively. This should guide any additional investments in data collection and analysis. Where further investment in data collection could be justified, the first phase assessment would strengthen, order, and prioritize the case for a second-Tier assessment.

Improving groundwater management decision-making through the application of the ecosystem service assessment framework requires not only data, but also understanding and human capacities to prioritize, collect and make use of this information effectively. These capacities can be mobilized progressively through an iterative approach, according to the needs for decision-making and the scope of the groundwater management challenges to be faced.

Conclusions

The higher the stakes in groundwater management decisions, the greater the justification to invest more time and resources in understanding the present state of the system, including the environmental and social dynamics that operate within it and the potential trade-offs that could be associated with different management options.

The chapter revealed a wide range of provisioning services provided by groundwater in the case study area of the Egyptian desert. It is unlikely that these could all be fully assessed. However, the case study demonstrated that groundwater managers can use the ecosystem services framework to identify and assess some of the most important benefits that groundwater provides to society, including its essential services as a buffer against water security threats.

Improving groundwater management decision-making through application of the ecosystem service assessment framework requires understanding, information and human capacities to collect and make use of it effectively. These can be mobilized through an iterative approach by groundwater managers with water service providers, industrial users, and general public, as well as with agricultural water users and scientists.

Recommendations

The following recommendations can be made for decision-makers and stakeholders to explore through future studies:

- An iterative, inclusive and participatory approach is needed to review groundwater management decision-making, increasing resource allocation, public participation and the engagement of the private sector.
- A broad participatory assessment involving both local stakeholders and global change researchers should help to ensure that critical buffering or safety-net services provided by groundwater are not overlooked.
- In order not to encourage over-extraction of groundwater, and to incentivise improved use of opportunities for recharge there is a need to value the services associated with groundwater storage and buffering against water scarcity and risks.
- The assessment of ecosystem services provided by groundwater and their economic value must be compared to management costs, including increasing demands for energy to pump and treat water supplies
- Groundwater management scenarios should consider groundwater management decisions in the context of the wider availability of water from surface-water, recycling and treatment options
- The iterative approach offers options for groundwater managers in Egypt to coordinate across sectors of government and overcome these assessment challenges in a progressive manner.

Acknowledgements Many people have provided inputs into the compilation of this chapter. They bear no responsibility for the final chapter, or any errors that it may contain. We wish to express particular thanks to Dr. Abdelazim Negm for encouragement to complete and publish this chapter. We are grateful also to Fabrice Renaud, Jaroslav Vrba and colleagues taking part in the United Nations University (UNU)—UNESCO International Hydrological Programme (IHP) international research project on Groundwater and Human Security Case Studies (GWAHS-CS) for intellectual and financial support. Additional funding for field research was provided through an Award for Doctoral study funded by ESRC/NERC, UK. We also thank Harris Switzman for advice during the preparation of the manuscript.

References

1. Van der Gun J (2012) Groundwater and global change: trends, opportunities and challenge. UNESCO, Paris, p 44
2. WWAP (2014) The United Nations world water development report 2014: water and energy. UNESCO. United Nations World Water Assessment Programme, Paris, p 230
3. Foster T, Brozovic N, Speir C (2017) The buffer value of groundwater when well yield is limited. J Hydrol 547:638–649
4. Tsur Y, Issar A (1989) The Buffer role of groundwater when supply of surface water is uncertain. Dev Water Sci 39:373–379
5. Gleeson T, Wada Y, Bierkens MFP, van Beek LPH (2012) Water balance of global aquifers revealed by groundwater footprint. Nature 488(7410):197–200. https://doi.org/10.1038/nature11295
6. Mazzoni A, Heggy E, Scabbia G (2018) Forecasting water budget deficits and groundwater depletion in the main fossil aquifer systems in North Africa and the Arabian Peninsula. Global Environ Change 53:157–173. https://doi.org/10.1016/j.gloenvcha.2018.09.009

7. UNDP (2013) Water governance in the Arab Region—managing scarcity and securing the future. Regional Bureau for Arab States (RBAS), Republic of Lebanon, p 182
8. Roumasset J, Wada CA (2013) A dynamic approach to PES pricing and finance for interlinked ecosystem services: watershed conservation and groundwater management. Ecol Econ 87:24–33. https://doi.org/10.1016/j.ecolecon.2012.11.023
9. Burnett K, Pongkijvorasin S, Roumasset J, Wada CA (2015) Incentivizing interdependent resource management: watersheds, groundwater and coastal ecology. In: Handbook of water economics, pp 150–161. https://doi.org/10.4337/9781782549666.00015
10. OECD (2015) Drying wells, rising stakes. Towards sustainable agricultural groundwater use. OECD, Paris, France
11. Hassan R, Scholes R, Ash N (2005) Ecosystems and human well-being: current state and trends. Millennium Ecosystem Assessment, Island Press
12. Roumasset JA, Wada CA (2014) Integrated groundwater resource management. In: Sustainable economic development: resources, environment, and institutions, pp 77–89. https://doi.org/10.1016/B978-0-12-800347-3.00005-4
13. Vaux H (2011) Groundwater under stress: the importance of management. Environ Earth Sci 62:19–23. https://doi.org/10.1007/s12665-010-0490-x
14. Qureshi ME, Reeson A, Reinelt P, Brozović N, Whitten S (2012) Factors determining the economic value of groundwater. Hydrogeol J 20:821–829. https://doi.org/10.1007/s10040-012-0867-x
15. Gohar AA, Cashman A (2018) The economic value of groundwater irrigation for food security under climate change: implication of representative concentration pathway climate scenarios. Water Resour Manage 32:3903–3918. https://doi.org/10.1007/s11269-018-2026-1
16. Koundouri P, Dávila OG, Anastasiou Y, Antypas A, Mavrogiorgis T, Mousoulides A, Mousoulidou M, Vasiliou K (2014) An econometric analysis of agricultural production, focusing on the shadow price of groundwater: policies towards socio-economic sustainability. Global Issues in Water Policy 7:87–100
17. Koundouri P, Roseta-Palma C, Englezos N (2017) Out of sight, not out of mind: developments in economic models of groundwater management. Int Rev Environ Resour Econ 11:55–96. https://doi.org/10.1561/101.00000091
18. Attia FAR (1992) Economic evaluation of groundwater development plans in Upper Egypt. Int J Water Resour Dev 8:135–144. https://doi.org/10.1080/07900629208722544
19. Attia FAR (1989) Use of groundwater for irrigation in the Nile Valley. Water Res Develop 5:91–98. https://doi.org/10.1080/07900628908722419
20. Costanza R, D'Arge R, De Groot R, Farber S, Grasso M, Hannon B, Limburg K, Naeem S, O'Neill RV, Paruelo J et al (1997) The value of the world's ecosystem services and natural capital. Nature 387:253–260. https://doi.org/10.1038/387253a0
21. El Serafy S (1998) Pricing the invaluable: the value of the world's ecosystem services and natural capital. Ecol Econ 25:25–27. https://doi.org/10.1016/S0921-8009(98)00009-3
22. Toman M (1998) Why not to calculate the value of the world's ecosystem services and natural capital. Ecol Econ 25:57–60. https://doi.org/10.1016/S0921-8009(98)00017-2
23. Pearce D (1998) Auditing the earth: the value of the world's ecosystem services and natural capital. Environment 40:23–28. https://doi.org/10.1080/00139159809605092
24. Kovacs K, West G, Xu Y (2017) The use of efficiency frontiers to evaluate the optimal land cover and irrigation practices for economic returns and ecosystem services. J Hydrol 547:474–488. https://doi.org/10.1016/j.jhydrol.2017.01.059
25. Kovacs K, West G (2016) The influence of groundwater depletion from irrigated agriculture on the tradeoffs between ecosystem services and economic returns. PLoS ONE 11. https://doi.org/10.1371/journal.pone.0168681
26. Kovacs KF, Xu Y, West GH, Popp M (2016) The tradeoffs between market returns from agricultural crops and non-market ecosystem service benefits on an irrigated agricultural landscape in the presence of groundwater overdraft. Water (Switzerland) 8. https://doi.org/10.3390/w8110501

27. DeFries R et al (2005) Analytical approaches for assessing ecosystem condition and human wellbeing. In: Hassan R, Scholes R, Ash N (eds) Ecosystems and human well-being: current state and trends: findings of the Condition and Trends Working Group. Island Press, Washington, DC, pp 37–67
28. Tuinstra J, van Wensem J (2014) Ecosystem services in sustainable groundwater management. Sci Total Environ 485–486:798–803. https://doi.org/10.1016/j.scitotenv.2014.03.098
29. Griebler C, Avramov M (2015) Groundwater ecosystem services: a review. Freshwater Sci 34:355–367
30. Hérivaux C, Grémont M (2019) Valuing a diversity of ecosystem services: the way forward to protect strategic groundwater resources for the future? Ecosyst Serv 35:184–193. https://doi.org/10.1016/j.ecoser.2018.12.011
31. Henrichs T, Zurek M, Eickhout B, Kasper K, Raudsepp-Hearne C, Ribeiro T, Vuuren D, Volkery A (2010) Chapter 5 Scenario development and analysis for forward-looking ecosystem assessment. In: Ash N, Blanco H, Garcia K, Brown C et al (eds) Ecosystems and human well-being a manual for assessment practitioners, pp 151–220
32. Ash N, Blanco H, Brown C, Garcia K, Henrichs T, Lucas N, Raudsepp-Hearne C, Simpson R, Scholes R, Tomich T et al (2010) Ecosystems and human well-being a manual for assessment practitioners. Island Press, Washington/Covelo/London
33. Wittmer H, Berghofer A, Forster J, Almack K (2010) The value of nature for local development. In: Sukhdev P (ed) TEEB—the economics of ecosystems and biodiversity for local and regional policy makers, pp 12–27
34. Favretto N, Luedeling E, Stringer LC, Dougill AJ (2017) Valuing ecosystem services in semi-arid rangelands through stochastic simulation. Land Degrad Dev 28:65–73. https://doi.org/10.1002/ldr.2590
35. Ericksen P, De Leeuw J, Said M, Silvestri S, Zaibet L (2012) Mapping ecosystem services in the Ewaso Ng'iro catchment. Int J Biodiv Sci Ecosyst Serv Manage 8:122–134. https://doi.org/10.1080/21513732.2011.651487
36. Ahmed M (2013) Ecosystem services. In: Arab millennium ecosystem assessment synthesis report. UNEP, pp 58–84
37. King C, Salem B (2013) Assessing the cost of groundwater degradation in the urbanizing desert area of Wadi El Natrun. In: The economy of green cities. Springer, pp 295–311
38. King C, Salem B (2012) A socio-ecological investigation of options to manage groundwater degradation in the western desert, Egypt. Ambio 41:490–503. https://doi.org/10.1007/s13280-012-0255-8
39. Salem B, Hak MGE, Khalifa S, King-Okumu C (2017) Beyond drastic in Wadi Natrun. In: Renaud F, Schuster-Wallace C (eds) Groundwater and human security. UNU, Hamilton, Ontario, pp 25–96
40. Pascual U, Muradian R (2010) Chapter 5: The economics of valuing ecosystem services and biodiversity. In: The economics of ecosystems and biodiversity: the ecological and economic foundations
41. CVG (1997) Valuing ground water: economic concepts and approaches committee on valuing ground water. Water Science and Technology Board, Commission on Geosciences, Environment, and Resources, National Research Council
42. Barbier E, Acreman M, Knowler D (1997) Economic valuation of wetlands: a guide for policy makers and planners. Ramsar Convention Bureau, Gland, Switzerland, p 138
43. De Groot R, Stuip M, Finlayson M, Davidson N (2006) Valuing Wetlands: guidance for valuing the benefits derived from wetland ecosystem services. Ramsar Convention Secretariat, Gland p 66
44. Davies J (2007) The economics of mobile Pastoralism: a global summary. Nomadic Peoples 91–116
45. Davies J (2007) Total economic valuation of Kenyan Pastoralism. World Initiative for Sustainable Pastoralism (WISP), Nairobi
46. Landers D, Nahlik A (2013) Final Ecosystem Goods and Services Classification System (FEGSCS). United States Environmental Protection Agency

47. Haines-Young R, Potschin M (2011) Common International Classification of Ecosystem Services (CICES) Paper prepared for discussion at the expert meeting on ecosystem accounts organised by The UNSD, the EEA and the World Bank European Environment Agency. London, p 17
48. Bolt K, Ruta G, Sarraf M (2005) Estimating the cost of environmental degradation—a training manual in English, French and Arabic. World Bank, Washington, DC, pp E1-79
49. Croitoru L, Sarraf M (eds) (2010)The cost of environmental degradation: case studies from the Middle East and North Africa (English). World Bank, Washington, DC, p 192
50. GWP (2014) The links between land use and groundwater—governance provisions and management strategies to secure a 'sustainable harvest'. Global Water Partnership, p 20
51. Tsur Y (2015) Conjunctive management of water resources in agriculture. In: Dinar A, Schwabe K (eds) Handbook of water economics. Edward Elgar Publishing, Cheltenham, UK
52. Tsur Y (1990) The stabilization role of groundwater when surface water supplies are uncertain: the implications for groundwater development. Water Resour Res 26:811–818
53. Favretto N, Stringer LC, Dougill AJ, Dallimer M, Perkins JS, Reed MS, Atlhopheng JR, Mulale K (2016) Multi-criteria decision analysis to identify dryland ecosystem service trade-offs under different rangeland land uses. Ecosyst Serv 17:142–151. https://doi.org/10.1016/j.ecoser.2015.12.005
54. Reed MS, Stringer LC, Dougill AJ, Perkins JS, Atlhopheng JR, Mulale K, Favretto N (2015) Reorienting land degradation towards sustainable land management: linking sustainable livelihoods with ecosystem services in rangeland systems. J Environ Manage 151:472–485. https://doi.org/10.1016/j.jenvman.2014.11.010
55. UN (2012) System of environmental economic accounting for water. United Nations, New York, p 216
56. Jabarin A (2013) Environmental economic valuation of the HIMA system: the case of Zarqa River Basin—Jordan natural resource economic valuations. IUCN, Amman, Jordan, p 36
57. Bann C, Wood SC (2012) Valuing groundwater: a practical approach for integrating groundwater economic values into decision making—a case study in Namibia, Southern Africa. Water SA 38:461–466. https://doi.org/10.4314/wsa.v38i3.12
58. Idris H, Nour S (1990) Present groundwater status in Egypt and the environmental impacts. Environ Geol Water Sci 16:171–177. https://doi.org/10.1007/BF01706041
59. Ahmed MA, Samie SGA, El-Maghrabi HM (2011) Recharge and contamination sources of shallow and deep groundwater of pleistocene aquifer in El-Sadat industrial city: isotope and hydrochemical approaches. Environ Earth Sci 62:751–768. https://doi.org/10.1007/s12665-010-0563-x
60. Fekry A, Salam MA (2002) Evaluation of groundwater development scenarios in Wadi El Natrun depression. Water Sci 31–38
61. RIGW (1990) Hydrogeological map of Egypt scale 1:100 000 Wadi El Natrun Research Institute for Groundwater (RIGW), Ministry of Public Works and Water Resources, p 11
62. Ibrahim S (2005) Groundwater resources management in Wadi El-Farigh and its vicinities for sustainable agricultural development. Ain Shams University, Cairo
63. CAPMAS (2017) Population and housing census. Central Agency for Public Mobilisation and Statistics, Arab Republic of Egypt, Cairo
64. CAPMAS (2007) Population and housing census. Central Agency for Public Mobilisation and Statistics, Arab Republic of Egypt, Cairo
65. MALR (2000) Results of the agricultural census for the agricultural year 1999/2000: Behayra governorate. Ministry of Agriculture and Land Reclamation, Economic Affairs Section, Central Office for Agricultural Economy, Public Office for Agricultural Census, Arab Republic of Egypt, p 122
66. MALR (2010) Results of the agricultural census for the agricultural year 2009/2010: Behayra governorate. Ministry of Agriculture and Land Reclamation, Economic Affairs Section, Central Office for Agricultural Economy, Public Office for Agricultural Census, Arab Republic of Egypt, p 122

67. IBRD (2007) Project appraisal document on a proposed loan in the amount of US$145 Million to the Arab Republic of Egypt for West Delta Water Conservation and Irrigation Rehabilitation Project. 22 May 2007. World Bank, Washington, DC, p 104
68. Nour El-Din M (2013) Climate change risk management in Egypt proposed climate change adaptation strategy for the Ministry of Water Resources & Irrigation in Egypt Prepared for UNESCO-Cairo Office EEAA. Ministry of Water Resources and Irrigation, Cairo, p 116
69. Attia FAR, Fahmy HS, Eid MJH, Slootweg R (eds) (2005) The West Delta Water Conservation and Irrigation Rehabilitation Project (WDWCIRP) Environmental and social impacts and a framework management—plan part I: environmental and social impact assessment. Part II: environmental and social management plan—final draft. Arab Republic of Egypt Ministry of Water Resources and Irrigation and World Bank, p 86
70. Attia F, Fahmi H, Gambarelli J, Hoevenaars R, Slootweg R, AbdelDayem S (ed) (2007) The West Delta Water Conservation and Irrigation Rehabilitation Project (WDWCIRP) environmental and social impacts and a framework management—plan part I: environmental and social impact assessment, part II: environmental and social management plan—final draft. Arab Republic of Egypt Ministry of Water Resources and Irrigation and World Bank, p 119
71. Slootweg R, Beukering P (2008) Valuation of ecosystem services and strategic environmental assessment lessons from influential cases. Netherlands Commission for Environmental Assessment, Utrecht, p 40
72. Awad M (2002) Land use planning of Wadi El-Natrun depression towards sustainable development. University of Alexandria
73. Salem B, Gad-El-Hak M, Khalifa S, KingOkumu C (2017) Egypt case study: beyond drastic in Wadi Natrun. In: Renaud F, Schuster-Wallace C (eds) Groundwater and human security, pp 25–96
74. Kumar P (2010) The economics of ecosystems and biodiversity: ecological and economic foundations. London and Washington, DC, Earthscan
75. Zahran MA, Willis AJ (2009) The vegetation of Egypt. Springer. https://doi.org/10.1007/978-1-4020-8756-1
76. Hussain S, Gundimeda H (2010) Tools for valuation and appraisal of ecosystem services in policy making. In: Wittmer H (ed) TEEB—the economics of ecosystems and biodiversity for local and regional policy makers, pp 41–63
77. MALR (2009) Sustainable Agricultural Development Strategy towards 2030 (SADS). Agricultural Research & Development Council, Arab Republic of Egypt, Ministry of Agriculture and Land Reclamation, Cairo
78. Switzman H, Salem B, Gad M, Adeel Z, Coulibaly P (2018) Conservation planning as an adaptive strategy for climate change and groundwater depletion in Wadi El Natrun Egypt. Hydrogeol J 26:689–703. https://doi.org/10.1007/s10040-017-1669-y
79. Abdelkhalek A (2013) Groundwater vulnerability assessment using modified drastic GIS based index at the Western Nile Delta Region, Thesis submitted in parial fulfillment of the requirements for the Master's degree in Intergrated Water Resoures Management. Institute for Technology and Resources Management in the Tropics and Subtropics (ITT) Cologne University of Applied Science, Germany, and Faculty of Graduate Studies, The University of Jordan, 2013
80. King C, Thomas DSG (2014) Monitoring environmental change and degradation in the irrigated oases of the Northern Sahara. J Arid Environ 103:36–45. https://doi.org/10.1016/j.jaridenv.2013.12.009

Groundwater Exploration, Quantity, Quality, and Their Management

Geophysical Groundwater Exploration in Arid Regions Using Integrated Land-Based Magnetic and DC Resistivity Measurements: A Case Study at Gilf Kebir Area, South Western Desert, Egypt

Khaled Gemail, Amin Ibrahim, and Sayed Bedair

Abstract Groundwater in hyper-arid regions as the southern parts of Egyptian Western Desert is considered as a key component for providing drinking water resources and supporting new national projects. Therefore, exploration programs and sustainable management strategies should be established by decision-makers for assessment of the groundwater resources for such areas. In the present work, the area south to Gilf Kebir plateau was considered as a case study for groundwater exploration using the integrative approach between regional-scale geological information with land-based geophysical measurements. This exploration program deals with the mapping of the Nubian Sandstone Aquifer System (NSAS) and delineation of subsurface structures and uplifts predominant in the basement bedrocks that have subdivided the area into sub-basins and controlled the aquifer recharge and groundwater potentialities in the Southern Western Desert. In order to achieve that target, land-based magnetic survey was applied along oriented long profiles, followed by Direct Current (DC) resistivity sounding to detect promising sites of groundwater. To fill the knowledge gap due to the absence of water wells in the region, the pervious remote sensing (Radarsat-1) and geological information were employed for picking four spots with different conditions and orientation to collect the land geophysical data. Geophysical results delivered that the region NE Jabal Kamel is promising for drilling due to the absence of igneous intrusions that act as natural barriers in the direction of water flow from the south, in addition to the great thickness of the sedimentary cover. On the other hand, some places as Eight Bells is characterized by shallow basement and basaltic intrusions with a depth of 30 m, while increase towards NE Jabal Kamal to reach about 500 m depth. The constructed geoelectrical and magnetic models give a sufficiently well-founded picture of subsurface structures and uplifts and their role in the groundwater potentialities in this arid region. The expected depth to water ranges from 69 to 81 m. The thickness of the proposed

K. Gemail (✉) · A. Ibrahim
Environmental Geophysics Lab (ZEGL) at Geology Department, Faculty of Science, Zagazig University, Zagazig 44519, Egypt
e-mail: khaledgemail@zu.edu.eg

S. Bedair
The Research Institute for Groundwater (RIGW), Ministry of Water Resources & Irrigation, Kanater Khairia, Egypt

A. Negm and A. Elkhouly (eds.), *Groundwater in Egypt's Deserts*, Springer Water,
https://doi.org/10.1007/978-3-030-77622-0_5

aquifer varies from 62 to 82 m with generally decreasing westwards due to igneous intrusions that interfering with the sedimentary section. While the chance of groundwater occurrence is nihilistic at the other three small areas Eight Bells, Jabal Kamel and Nosb Al-Amir.

Keywords Western Desert · Gilf Kebir · Nubian Sandstone Aquifer System (NSAS) · DC resistivity · Land-based magnetic · Radarsat-1

Introduction

The shortage of water resource has become a major concern in the hyper-arid region as in the northern Africa, where scarcity of water is a defining characteristic [1–4]. Where, in hyper-arid regions as in the case of southern parts of Egyptian Western Desert, it needs something special to fill a knowledge gap in the water resources where the region is characterized by dry hot climate and shortage of rains. In such conditions, the groundwater is considered the only water resource based on the depth and thickness of the expected aquifers. In these regions, soil properties and water resources, can only be determined from regional-scale maps and Landsat images, remote sensing techniques, or by vague interpolation of records from stations several hundreds of kilometers away [5, 6]. However, the main question will be what are the suitable conditions for groundwater accumulations in those regions? In such a case, geophysical information with different scales from regional-scale i.e. satellite or area-born geophysics to local land scale geophysics is necessary to understand the links between subsurface structures and groundwater potentialities [6, 7]. Accordingly, the integrative relationship is essential between the geophysicist and the hydrogeologist to extract the maximum information from a geophysical survey; since there are very few instances when a unique model is applicable for one set of geophysical data [8]. In other words, the accuracy of the geophysical survey for the prediction of groundwater potentialities is a spatial decision problem that typically based on prior geological and hydrogeological information.

The southern and southwestern foreland of Gilf Kebir area is mostly made up of Precambrian basement of irregular and broken relief overlooked by the Jabal Uweinat intrusion situated at the corner between Egypt, Sudan and Libya in the west as shown in Fig. 1. The study area lies between long. 26° 00 and 28° 00 E and lat. 22° 00 and 23° 00 N. According to the available geological information, four sites in the area south to Gilf Kebir plateau were selected for geophysical field measurements. These sites are: Eight Bells, Nosb Al-Amir, Jabal Kamel and NE Jabal Kamel (Fig. 1). This area gives special importance to the present study in achieving the national project in constructing the promised localities at the southern borders of the Egyptian Western Desert.

Mapping and delineating the subsurface structures in the selected sites that are categorized by different types of geological structures (faults and local igneous intrusions) and controlled on the groundwater movement and lowering the water table is

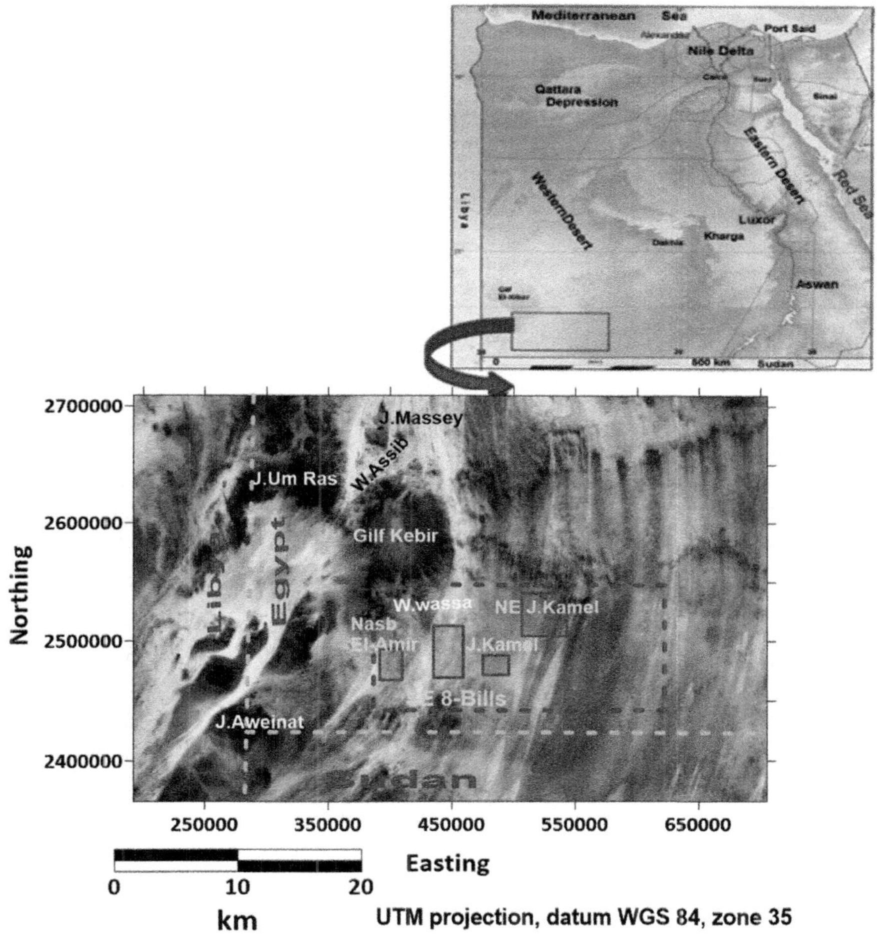

Fig. 1 Location map and Landsat image of southwestern corner of Egyptian Western Desert with the expanded map showing the details of the selected site for geophysical measurements

very important for groundwater exploration. In order to achieve that target, a land-based magnetic survey was applied along oriented long profiles, followed by direct current (DC) resistivity sounding using Schlumberger array to select promising sites of groundwater.

The applied geophysical regime in the present work was integrated with previous remote sensing and geological information and considered as a primarily program for groundwater exploration in the southwestern corner of Western Desert. In such hard conditions as in the considered area, the magnetic method was employed as indirect method for groundwater exploration to describe the subsurface geological structures and mapping the sub-basins in basement surface where the groundwater is most probably associated with the low reliefs and fractured basement bedrock. On the other hand, Schlumberger resistivity sounding technique was used as direct tool in groundwater investigations and integrated with the magnetic method with a view to providing information on the geological interpretation, basement topography, subsurface structural features and their hydrogeological significance [9], in order to determine possible regions for groundwater potential zones. Additionally, the resistivity method can give a clear idea about the water-bearing formations from the surface measurements.

Climate and Geomorphology

From the climate point of view, the area is located within the hyper- arid belt of North Africa. It is characterized by scarcity of rainfall, high temperature and low humidity with an average yearly humidity of 30% [10]. The aridity index exceeds 200, i.e. the evaporation rate exceeds 210 times the precipitation. The area has been subjected to some arid and wet period in its past geologic history, which exerts their influence on the present surface features. Temperature is typical as that of the desert; with high temperature during the day then drops sharply during the night. The regular temperature during the year is around 31 °C with a maximum of 51 °C in summer and a minimum as low as 5 to 12 °C in winter [11]. At a depth of 50 cm (below ground level), the temperature of the soil ranges from 16° on January to 35° on August. Wind regimes depend upon temperature and the presence of high and low pressures cells. Under normalized atmospheric pressure conditions, the temperature is the main factor where wind speed starts minimum at early morning then wind speed rises to reach its maximum daily hours. High wind speeds carrying sand dunes with 100 m/year for fine sand.

Geomorphologically, the area can be separated into two main features. An offshoot of the Great Sand Sea in the north and sand sheets to the east with low reliefs as seen in Fig. 2, while the western parts are characterized by high topographical features which separate the mountain from its highland neighbors, including the prominent Kamel El-Din Plateau to the north and Jabal Uweinat to the south. The highest elevation in the area 1100 m (a.m.s.l) that is observed in the northwestern corner of the studied areas, while the lowest elevation 300 in the eastern lowlands with an average elevation 600 m at the center (Fig. 2).

Four geomorphologic units are defined in the area. These are sand sheets in the east, table land of Jabal Kamel, sharp hilly ground, basement and sand dunes of crescent-shaped in longitudinal lines (Fig. 2). According to Issawi and Sallam [12] the ring complex structure in Gebel Uweinat was intruded into a high metamorphic basement terrain extending north along the Libyan–Egyptian border exposed in Jabal Mehashmat in Egypt as seen in Fig. 2. To the north of Jabal Uweinat uplift, the sandstone plateaus of Gilf Kebir and Abu Ras have been known as the Gilf Kebir.

Darius [5] used SRTM data and constructed the drainage system of the Gilf Kebir-Uweinat area as shown in Fig. 3. Accordingly, the boundaries of the uplands are

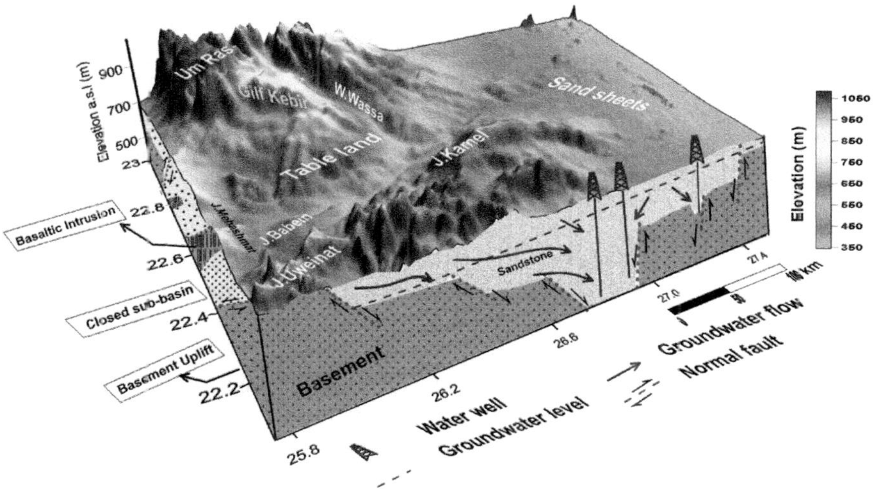

Fig. 2 3D topographic block of the southwestern corner of Egypt shows the geomorphologic units and landforms. The southern boundary illustrates a hydrogeological cross section from the nearby wells in eastern parts of the area

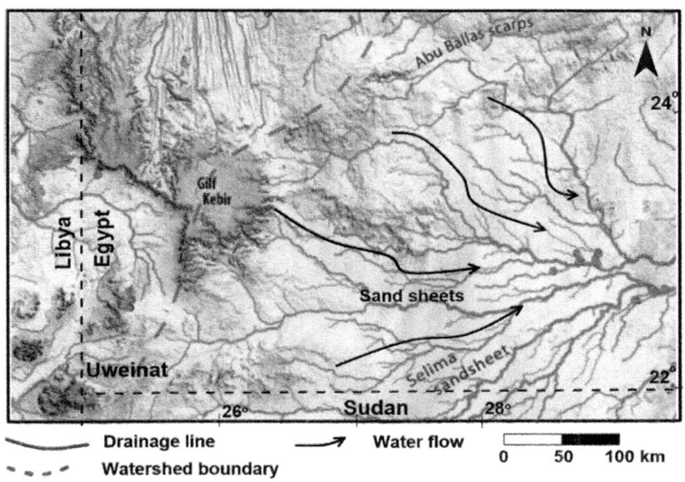

Fig. 3 Drainage networks from SRTM data of Gilf Kebir-Uweinat area shows the main watershed running from Gilf Kebir plateau to south-east boundary [5]

controlled by deep tectonically structures which are formed the main drainage lines in the area, These drainage lines are characterized by steep walls, considerable sediment insides, and cliffs that collect all run-off from the western and northern highlands to eastward into the plains (Fig. 3). The eastern drainage system is one of the major tributaries to the East of Jabal Uweinat and Gilf Kebir Plateau.

Geological and Hydrogeological Setting

From the surface geology point of view, the area involves two plateaus which rise in all, while the northern part abruptly upwards at Jabal Um Ras from the surrounding lowlands towards the East (Fig. 2). The southeastern and eastern regions are appeared as massive peneplaines and covered with immense sand sheets and dune belts as shown in the geological map of seeing Fig. 4.

Geologically, the area southwestern region is characterized by a rolling basement with swells, troughs and basin structures, which are filled with ferruginous sandstones of the Nubian system of partly considerable thickness (Fig. 2). These sandstone series of mainly Mesozoic origin are characterized by their reserves in utilizable fossil groundwater of Nubian Sandstone Aquifer System (NSAS). Figure 4 shows the geological map of and the stratigraphic sequence of the different rock units of the area. These geologic units that appear on the surface of the area are separated by the exposed basement rocks in the southwestern corner. Tertiary Basalt intrusions spread in some localities of the area (see Fig. 4). Gilf Kebir Formation located at the north and northwest is the oldest formation of the Nubian Sandstone [12, 13]. The longitudinal sand dunes take the NE-SW directions and the sand sheets spread on the eastern side of the region (see Fig. 4).

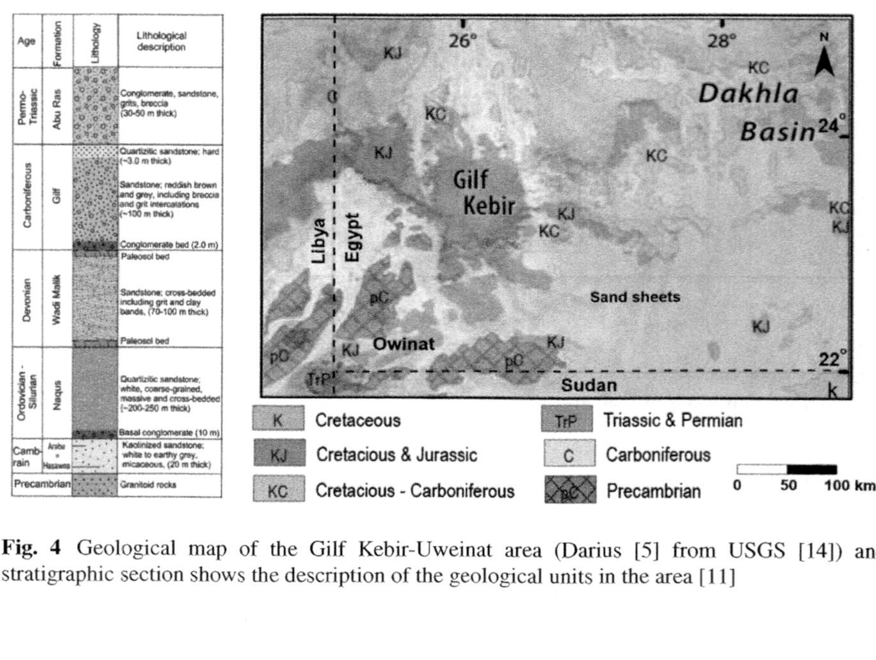

Fig. 4 Geological map of the Gilf Kebir-Uweinat area (Darius [5] from USGS [14]) and stratigraphic section shows the description of the geological units in the area [11]

Structurally, the area is affected by some joints and fractures in the earth's surface and there are no vertical or lateral throws on both sides [15]. It is considered as a dome shape in the basement rocks beneath the sediments. Faults, mostly E–W dissecting the country into blocks and ridges. To the east of the Gilf Kebir Plateau, a confused country of the irregular scarped plain is named Gilf Kebir Pediplain. Basaltic flows were observed to the west of this Pediplain, whereas iron bands are common in.

According to El Alfy [16], the E–W fault system is common in the area and can be recognized in cross-sections, however it is invisible on the land surface due to the aeolian sand sheets and dunes. These faults have subdivided the area into local basins (sub-basins) within the Nubian Sandstone controlling the groundwater movement in the area. Some disconnected local structural lows are restricted by no-flow boundary conditions, as exposed by long-duration pumping tests in the east of Uweinat regions [17].

In many places in the considered area, the sand sheets and dune belts are covered the near-surface structures that are controlled the groundwater flow and aquifer distribution as fractures, faults, and igneous intrusion. Reasonably, Laake [18] used the satellite radar data for mapping the ancient buried channels in the hard sandstone and the near-surface structures below the sand sheets in the considered area (Fig. 5). According to this investigation, the shape of the ridges and the drainage lines crossing the Gilf Kebir plateau disclose fairly straight fault lines of W–E and SW–NE trends (Fig. 5a). Therefore, the use of satellite radar which can penetrate dry soil up to 20 m to detect the buried paleo-river beds as seen in Fig. 5b, c.

The Nubian Sandstone Aquifer System extends over an enormous area to the south in Sudan and Chad and extends to the north in Egypt and Libya (Fig. 6). The groundwater in this aquifer system is considered the main source of fresh water for the development of the Egyptian Western Desert [19]. According to Foster and Loucks [20] no recent significant groundwater recharge is measurable, thus planning the use or pumping from NSAS should be based on the probability that sometime in the future the water in these aquifers will be fully down. Thus, the NSAS in the southern Western Desert needs an achievable pumping regime with continuously monitoring policy to manage and prevent the intensive use of this non-renewable resource.

Geophysical Investigations

To fill the knowledge gap of ambiguous geophysical methodology, a conceptual exploration program was offered in the considered area. This program involves collecting the regional geological and hydrogeological information from forgoing investigations as an initial concept for adjustment and modification of the obtained models to draw a clear picture of the groundwater potentialities in this extremely arid region. Thus, in the present work, the land geophysical modeling as local-scale was integrated with regional-scale information inferred from remote sensing, SAR and SRTM radar data [18, 21, 22], and GRACE satellite gravity data [23, 24].

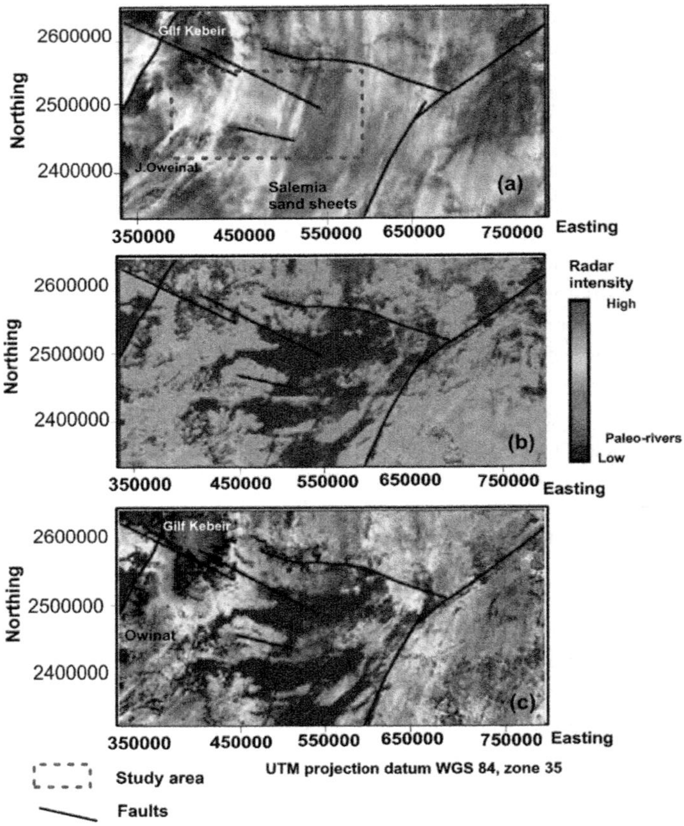

Fig. 5 Structural and paleo-rivers inferred from Radarsat data close to Gilf Kebir-Uweinat area [18], **a** multispectral image (Landsat 742 RGB), **b** low intensity radar data (Radarsat1) the blue color indicates the paleo-rivers below sand sheets, and **c** combine of radar data with Landsat image reflecting the boundaries of paleo-rivers in the area

Geophysical Data Acquisition

For mapping the groundwater potentiality in the area south to Gilf Kebir plateau, land-based magnetic survey was employed and integrated with DC resistivity survey along long profiles with different directions in four selected sites (Fig. 1). These field measurements were carried out by Research Institute for Groundwater during winter of 2009 [25] and reinterpreted here in the present work using an integrative approach with available regional-scale geological and hydrogeological information. The magnetic survey in the selected regions aims to achieve the depth to the basement as well as determine the thickness of sedimentary cover. Additionally, this survey was carried out for mapping the expected igneous intrusions, faults and uplifts which

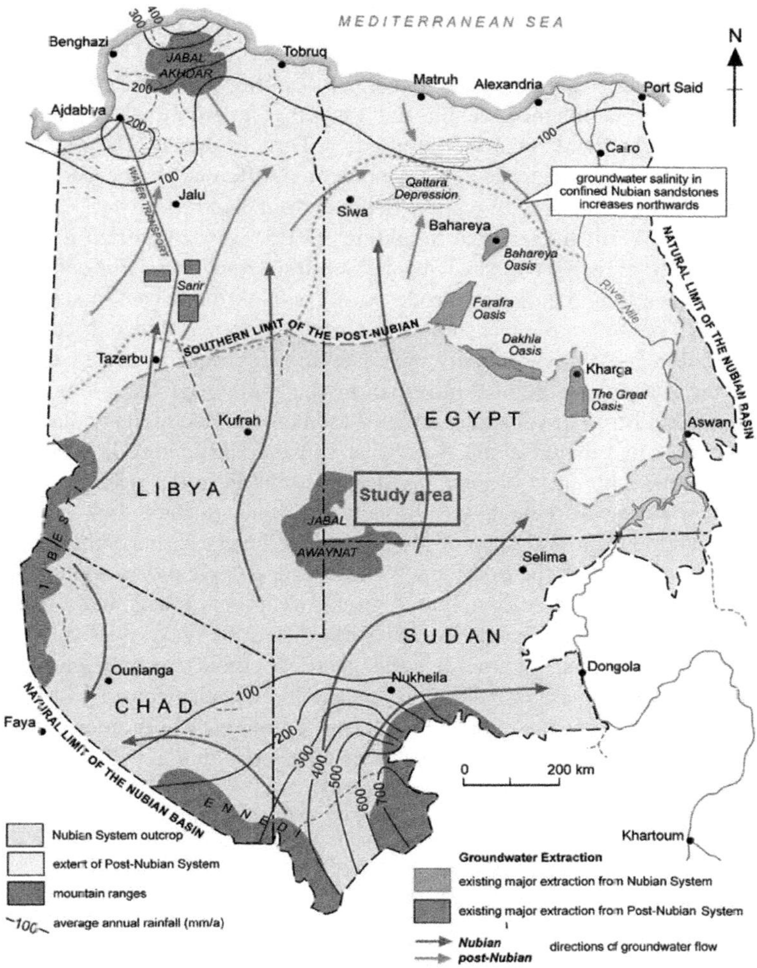

Fig. 6 Hydrogeological map shows the regional distribution of Nubian Sandstone Aquifer System (NSAS) and post Nubian system [20]

are common in the area and to be avoided in case of well drilling. In this survey, the total magnetic intensity field was measured along 5 profiles scattered in different directions to cover a considerable part of the investigated area. The distribution of magnetic profiles was governed by the accessibility of the ground conditions and georeferenced using the GPS system. A total of 200 magnetic readings were collected with sampling interval ranging between 300 and 1300 m depending on the field condition. Two magnetometers were used in which one was used as a base station (Proton precision type) and the other (Cesium Vapor type) was used for measuring the total field intensity. Practically, both instruments were employed for diurnal correction, where the diurnal variations of the geomagnetic field were subtracted from the measurements at each station [26].

After the magnetic survey was done in the 4 localities, DC resistivity survey was carried to confirm the magnetic results and mapping the groundwater potentiates in the area. In the electrical resistivity methods, a Direct Current (DC) is introduced into the ground by two electrodes, and the potential difference is measured between two points suitably chosen with respect to the current electrodes.

A total of 24 Vertical Electrical Soundings (VES) were measured using Syscal Junior R2 (IRIS) resistivity meter with the Schlumberger array. For each VES the distance between the potential electrodes was gradually increased in steps starting from 0.5 to 75 m to obtain a measurable potential difference. The half current electrode separation (AB/2) was usually increased in steps starting from 1 to 1000 m. Based on the available geological information, the applied spread is long enough to penetrate the sandstone cover and reach the basement surface in many places. These soundings were distributed along 4 profiles and marked using GPS instruments. To overcome the ambiguity problem, which is common in the resistivity interpretation [27], some measurements were conducted beside surface rock exposures to define the parametric resistivities of the subsurface layers by correlating the resistivity measurements with the lithological and hydrogeological data and also finding key relationships between resistivity and groundwater characteristics [28]. In some places as in the Jabal Kamel and Nosb El-Amir the surface layer was very resistive ferruginous sandstone and to provide good electrical contact with the ground, some precautions were considered, such as digging a hole for each electrode, adding water in the hole and increasing the number of current electrodes on both sides, particularly at large distances. Additionally, the used instrument is automatically estimated the standard deviation of errors from repeat cycles.

Data Processing and Interpretation

The measured geophysical data were interpreted qualitatively where the first appraising of the subsurface structures and depth to the bedrock can be achieved by looking at the shape of the field measurements and the ranges of the apparent resistivities and total magnetic field values. For example (Fig. 7), the field data and the smoothed curve of the VES 1 at Jabal Kamel site reveals shallow fresh/unweathered basement rocks where the last layer increases abruptly as appeared in Fig. 7a with acute angle at the end, while the apparent resistivity distribution at VES 3 along the same profile produce smooth curve to indicate relatively deep basement with increasing the probability of shallow groundwater aquifer. Accordingly, the geophysical measurements were processed using different techniques to be more significant.

In land-based magnetic survey, because the earth's magnetic field naturally varies from the poles to the equator, the International Geomagnetic Reference Field (IGRF) correction was done [29]. Additionally, the Reduction-To-the-Pole (RTP) technique

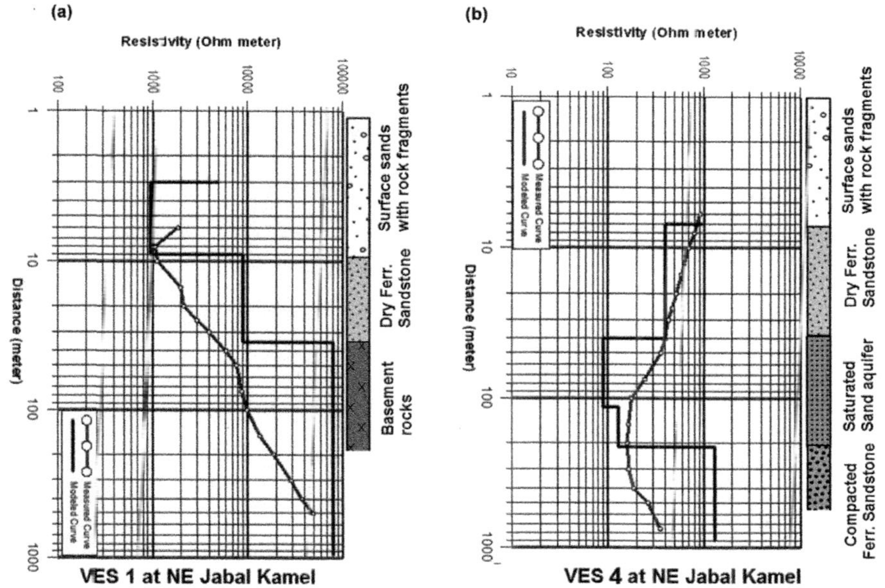

Fig. 7 Examples of resistivity sounding curves and geological interpretation at NE Jabal Kamel site. **a** The last layer is basement rock in VES1. **b** Saturated sands in VES4

of Mendonca and Silva [30] was applied to correct the shifting in the magnetic anomalies positions due to the variation of inclination and declination angles of the magnetic vector. After RTP technique, the values of total magnetic intensity express the true earth's magnetic intensity of the rocks and structures.

Magnetic data was applied to describe the subsurface fault trends or related structures responsible for recharging the aquifer in the area. As mentioned earlier, this kind of structure plays an important role in controlling the sand aquifer depths and most probably water movement from the southern parts of NSAS. Further enhancement analysis on the total magnetic intensity data such as Vertical Derivative (VD), Horizontal Derivative (HD) and Analytical Signal (AS) that are yielded information on the basement surface and igneous intrusions as well as the thickness of the sedimentary overburden along the studied profiles [31]. For geologic interpretation, the shape of the (AS) anomaly is a bell-shaped symmetric positioned above the intrusion body or magnetic contacts and its maxima occurs directly over the edges of magnetic sources.

The depths to the basement surface along the measured profiles were calculated using spectral analysis and 2D Euler deconvolution (ED) which is confirmed with DC resistivity results in some places. Figure 8 shows the interpreted power spectrum curves along the magnetic profiles carried out at the investigated sites. Conversely, the power spectrum curves along Jabal Kamel and Northeast Jabal Kamel show a

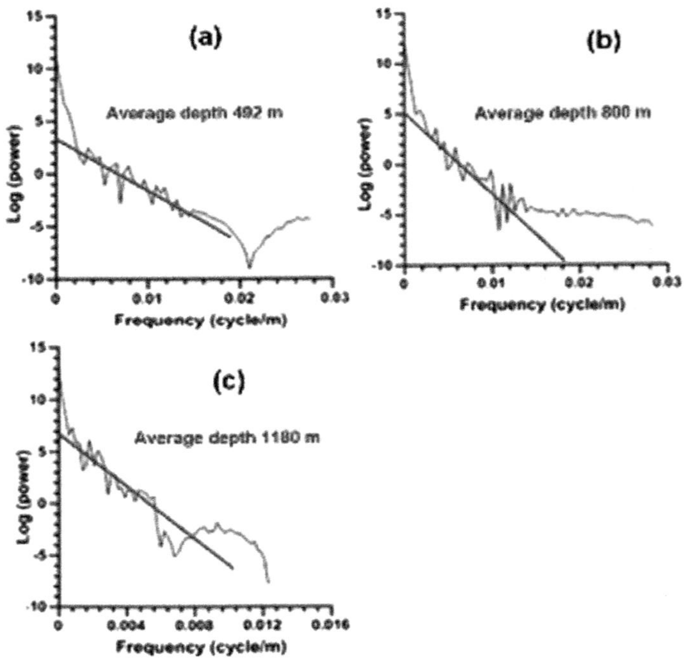

Fig. 8 Depth estimation using power spectrum curves along the magnetic profiles carried out at **a** N–S profile at NE Jabal Kamel, **b** E–W Jabal Kamel profile, **c** N–S Eight Bells profile

single ensemble associated to a local source (Fig. 8a, b). This pronounced manifestation is attributed to the coarser sampling interval of acquiring magnetic data along with the two former investigated sites and consequently more depth penetration was achieved. Moreover, the deeper inspection of the radially averaged power spectrum curves shows a lower frequency of less than 0.00208 cycle/m (regional component corresponding to about 2000 m depth) reflecting the deeper bodies. Meanwhile, the slope of the higher frequency portion from 0.00208 to 0.03167 cycle/m (red line) was used to estimate the average depth to the basement rocks and/or igneous intrusions. The minimum estimated depth to the basement (90 m) was recognized at Nosb Al Amir site in the West whereas the maximum (1180 m) to East of Eight Bells site (Fig. 8c). Consequently the thickness of sedimentary cover increases towards the east. Depths obtained from the power spectrum technique are well correlated with depths interpreted from the geoelectric cross-sections.

In addition to spectral analysis, 2D Euler Deconvolution (ED) was applied on selected profiles to calculate the depth and position of sources of an assumed structural index. 2D Euler solutions revealed subsurface faulting zones and the presence of sub-basins within the survey area. Traditionally, ED has been applied to the total magnetic field intensity, but recently it has been also applied to the vertical gradient of the field with much success [32]. Euler solutions were obtained for structural index values of 1, 1.25, 1.5, 1.75 and 2 since it best represents dikes, faults and contacts. Also, it is worth mentioning that the structural index is increased by (n) when using a derivative of order (n) to reduce the sensitivity of the solutions to the structural index [32]. Depth limits, obtained through a geoelectric method and downward continuation of the magnetic data and spectral analysis technique have been used here to constrain the Euler results obtained.

For DC resistivity data, quantitative interpretation was performed to obtain the true resistivities and thicknesses of subsurface layers using IPI2win software [33]. This program is based mainly on the linear filtering algorithm, which provides a fast and accurate direct problem solution for a wide range of models, covering all the reasonable geologic situations. In which, the resistivity data for a profile are treated as a unit representing the subsurface geological structures rather than a set of independent objects. During the inversion procedures, the principle of equivalence was considered, where the same resistivity curve may produce many interpreted curves [9]. The resulting 1D resistivity models were compared with the original input geological information from the prior geological and hydrogeological data. Figure 7 shows an example of the geological interpretation of sounding points 1 and 3 at NE Jabal Kamel site.

Results and Discussion

The obtained results from magnetic survey are integrated with resistivity distribution along the measured profiles in order to visualize the subsurface structures including normal faults (E–W trend) and igneous intrusions which are divided the area south to Gilf Kebir into windows or sub-basins in the basement surface and act as preferential flow and transport paths of groundwater from the southern parts of NSAS. In this situation, the connected sand windows and fluvial features can be considered as recharge sources in the area east to Gilf Kebir-Uweinat [16, 21].

Eight Bells Site

Eight Bells are conical high hills and located directly southern of the Gilf Kebir plateau along NW–SE line. From a geological point of view, the Eight Bells have a connection with the wide-spread intrusive magmatism in the region, which has produced numerous craters and dykes [34]. The southeast area of Eight Bells is covered by sands and rock fragments with different subsurface structural features, faults, fractures, and dikes as indicated from the topography and field observations as well as the Landsat images. Each of these structures has distinctive effects on the total magnetic field. Therefore, the area is considered as targets of a magnetic survey. Accordingly, the variations of the total magnetic field were measured along N–S profile with 35 km length and 850 m station interval towards Uweinat uplift in the south (base map in Fig. 9). The magnetic measurements were followed by conducting eight Schlumberger soundings for assessing the suitability of the region for groundwater potentiality and delineate the role of the subsurface structures in groundwater movements as discussed before from radar-sat and remote sensing data [18, 22].

Figure 9a illustrates the total magnetic intensity field along N–S profile at Eight Bells site. The profile was subjected to horizontal and vertical gradient combined with analytic signal technique to locate the tops of isolated geological contacts or faults from the reduced-to-pole magnetic field. Inspection of horizontal and vertical gradient profiles reveals that the area is affected by dikes/magnetic contacts such as faults. Four major vertical magnetic contacts/faults (E–W) are recognized at a distance of 1000, 1200, 2000 and 2500 m from the start point of the profile in the south. The depth to the basement surface varies between 30 m towards the south and more than 800 m as maximum depth in the middle part. On the other hand, the analytical signal profile (red dashed line, Fig. 9a) shows a bell-shaped symmetric amplitude with two closed spikes located directly above igneous intrusion at 500 m from the start point in the south. Additionally, the four recognized faults from horizontal and vertical gradient profiles have appeared here as one spike anomalies confirming

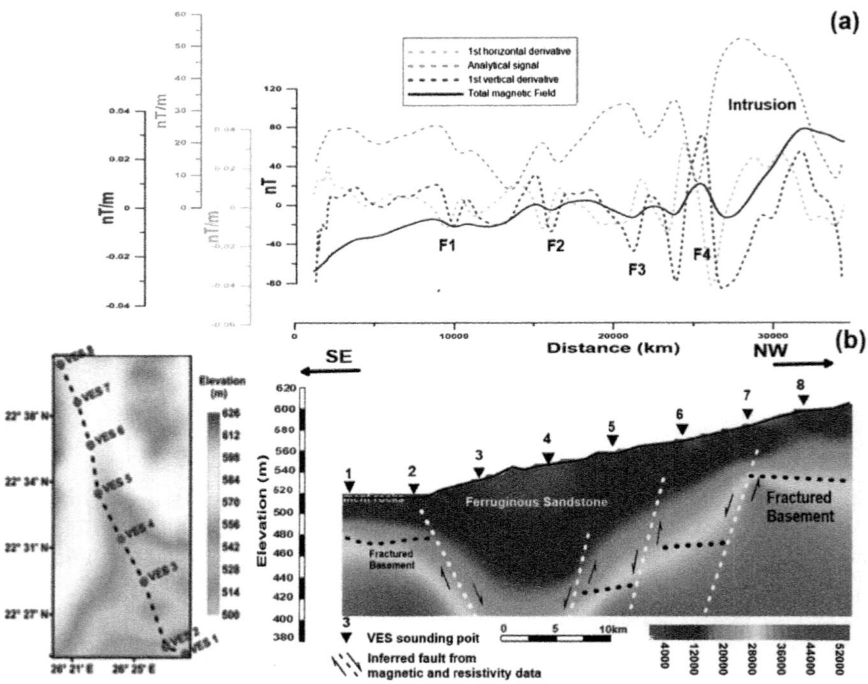

Fig. 9 Integration of magnetic and DC resistivity results along N–S profile at Eight Bells site, for location see, Fig. 1. **a** Vertical Derivative (VD), Horizontal Derivative (HD) and Analytical Signal (AS) along the magnetic profile. **b** Resistivity distributions along the measured profile

the E–W fault structures which are common in the area as mentioned earlier using satellite radar data [18] (see Fig. 3).

The combined use of the three methods is generally considered as the best approach for accurate detection of the magnetic sources. For example, the horizontal gradient (HG) anomaly that overlies analytic signal (AS) at places of F3 and F4 (Fig. 9a) indicates the edges of horizontal sheets or vertically dipping contacts due to the faults [30]. While, the HG anomaly is offset from the (AS) overlie the southern igneous intrusion (at 500 m distance) reflects dipping contacts, with the true location close to the analytic signal and the dip in the direction of the horizontal gradient solution.

To visualize the sedimentary cover over the basement bedrock and the impact of the detected vertical magnetic contacts in aquifer configuration in the area, the interpreted resistivities from the surface resistivity soundings were used to create a 2D integrative model to delineate the suitability of the region for water accumulation as seen in Fig. 9b. The input data of this model was created by sampling of the true resistivities and elevations of the geoelectrical layers. In the case of considering the basement surface in the area as markers, the accuracy of the model can be assessed by comparing the predicted structural features with the surface elevations and magnetic data [35].

Inspection of the resistivity patterns with depth along the Eight Bells profile (Fig. 9b) indicates four resistivity characters are observed which are considered as relatively high near the surface in middle part, low (dry ferruginous sandstone), moderate (fractured basement), and very high resistivity (basement complex) zones, respectively. Therefore, the very high resistivity values (more than 5000 Ω m) over the profile is considered the region as dry ferruginous sandstone accumulation covering the basement rocks with 30 m thickness at both ends (VES 1 and VES 8) and enormously increases to reach 160 m at VES4 in the middle lowland. As appeared in the resistivity section (Fig. 9b), the region along the profile is considered as a typical sub-basin, however the presence of the shallow basement uplifts in the south at soundings 1 and 2 acts as a barrier in the N–S direction of groundwater flow and the area appears to be closed dry basin and is not connected with southern NSAS. The comparison between DC resistivity and magnetic results shows a significant agreement in the detection of E–W faults (sub-parallel to Uweinat-Aswan uplift) that disrupt the continuity of groundwater aquifer in this site.

NE Jabal Kamel Site

Magnetic and DC resistivity measurements were collected along two crossing profiles in northeast of Jabal Kamel (for location see, Fig. 1). This site lies east to the earlier site of Eight Bells. Figures 10 and 11 illustrate the combination of magnetic and DC resistivity along E–W and N–S profiles in the area NE Jabal Kamel, respectively. In E–W profile (Fig. 10a), the analytical signal profile reveals intrusion bodies with shallow depth in at 3000 m from the start point to the west. The good matching between the

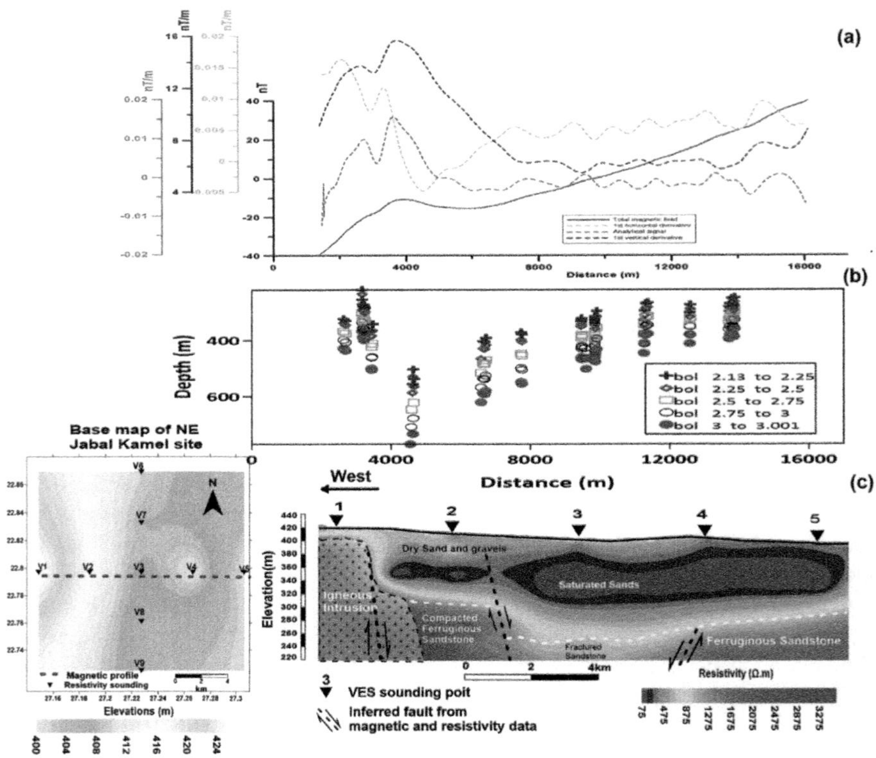

Fig. 10 Integration of magnetic and DC resistivity results along E–W profile at EN Jabal Kamel site, for location see, Fig. 1. **a** Vertical Derivative (VD), Horizontal Derivative (HD) and Analytical Signal (AS) along the magnetic profile. **b** Euler depth solutions along magnetic anomalies. **c** Resistivity distributions along the measured profile

horizontal gradient and analytical signal anomaly at that place indicates the sharp vertical contacts of shallow intrusion with sedimentary cover along the profile. All three methods (AS, HG and VG) show great similarities of low amplitude anomalies towards the east side of the area to indicate deep basement refiles. Accordingly, the depth to basement surface varies between 350 and 500 m where the area is affected by deep faults. The most relevant observation is that the area east to the detected basement intrusion is clearly connected with the south and southeast NSAS aquifer to increase the suitability of the sedimentary cover for groundwater accumulations as indicated from N–S profile (Fig. 11). Also, the negative polarity of the first vertical derivative anomalies (Fig. 11a, blue dashed line) could be attributed to the presence of deeper sedimentary sub-basin.

From the calculated Euler solutions of Fig. 10b, the clusters are observed at 4500, 6500, 9000, 11,000 and 1500 m along the profile reflected major discontinuities. The deepest part of the profile was 500 m at the distance of 5 km from the start

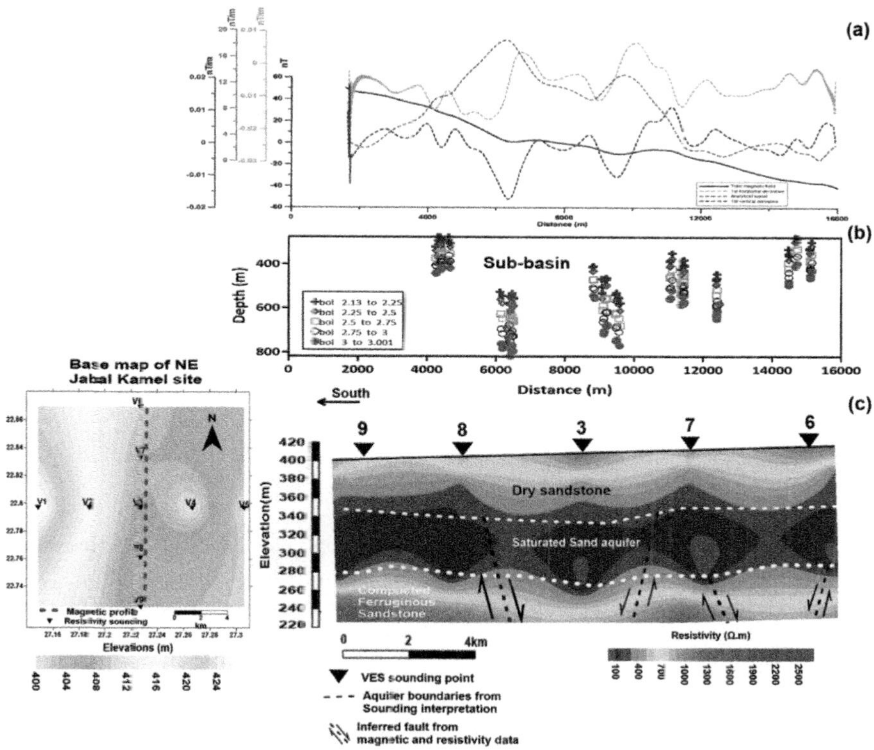

Fig. 11 Integration of magnetic and DC resistivity results along N–S profile at EN Jabal Kamel site, for location see, Fig. 1. **a** Vertical Derivative (VD), Horizontal Derivative (HD) and Analytical Signal (AS) along the magnetic profile. **b** Euler depth solutions along magnetic anomalies. **c** Resistivity distributions along the measured profile

point and extends to 10 km which is considered as promising sub-basin for water accumulations. While the calculated solutions along the N–S profile of Fig. 11b, indicate an abrupt change in the basement surface at a distance between 5 and 8 km with great sediments (550 m thick). Also, the horizontal and vertical gradients highly fluctuate over a distance of 6 km to 14 km along the profile which represents rocks of high and low magnetic susceptibility respectively relative to the host rocks due to the E–W faults.

The integrative resistivity models along both profiles are compatible with magnetic results where the western border of the region is suffered by exceedingly increase in resistivity to reach 8000 Ω m at 15 m depth (VES1 in Fig. 10c) to indicate shallow basement intrusion to confirm the magnetic interpretation. Based on the resistivity characters in the east and south of both profiles, the region is totally different from the Eight Bells site. Hence, the subsurface layer distributions here have significant low resistivity values compared with exciting high resistivities to the west at the Eight Bells site. The lowering of resistivity values in this site is attributed to the presence of saturated sands in the low basement reliefs with resistivity varies between 70–200 Ω m and thickness range 60–100 m. The comparison of geophysical results in this promising site with the regional geological and hydrogeological results from RS and Radarsat-1 data indicate that the area northeast and east of Jabal Kamel and Uweinat is considered as favorable sites for exploration drilling. As seen in radar data [18] (see Fig. 5), this region is covered by sand sheets and directly connected with southern NSAS through NE–SW and nearly N–S buried channels in the sedimentary cover over the basement rocks [18, 21–23]. Therefore, as seen in the resistivity sections (Figs. 10c and 11c), considering the subsurface structures and basement reliefs is very important in order to make a better understanding of aquifer distribution and the groundwater flow in the area.

North of Jabal Kamel Site

The region north of Jabal Kamel was selected for geophysical survey to delineate the continuity of the promising aquifer in the former site in the west direction (for location see, Fig. 1). The geophysical exploration program was started by measuring the total magnetic field along 8 km profile with 300 m station interval. Figure 12 illustrates the results of the HG, VG and AS anomalies after treatment of total magnetic intensity along the measured profile. The preliminary qualitative interpretation of the HG and VG profiles reveals three magnetic sources; the middle low amplitude anomaly is related to deep-seated intrusion bodies, while the two high frequency anomalies at both ends are associated with shallow magnetic contacts related to intrusion structures along faults. Figure 12a, b compares the results of ED and different derivatives of the magnetic field. The Euler solutions are well clustered much more accurately over the vertical contacts and well correlated to the fluctuations of the derivatives. The major faults located at horizontal distances of 1700, 2900, 4400 and 6600 m from the start point. The basement surface depth along the profile ranges between 150 and to a maximum of 250 m.

Based on the magnetic results, the region is considered as closed sub-basin and structurally complicated due to the presence of isolated intrusions controlled by NE–SW and N–S fault trends. Additionally, the area is closed totally from the south by the sub-parallel Uweinat basement complex. These results are conformity with those of previous regional-scale investigations using SRTM radar and GRACE data [22, 23]. Therefore, it is expected that the sedimentary cover in these closed sub-basins is dry and there is no aquifer in that site due to the presence of shallow basaltic intrusions and Uweinat uplift in the front of groundwater flow from the south.

According to the predicated results from the magnetic data, two resistivity soundings were carried out in the eastern side of the magnetic profile to confirm the results. In contrast with the NE Jabal Kamel site, the interpretation of both soundings shows high resistivity values of more than 10,000 Ω m for the sedimentary cover indicating dry and compacted ferruginous sandstones over basement bedrocks (Fig. 12c).

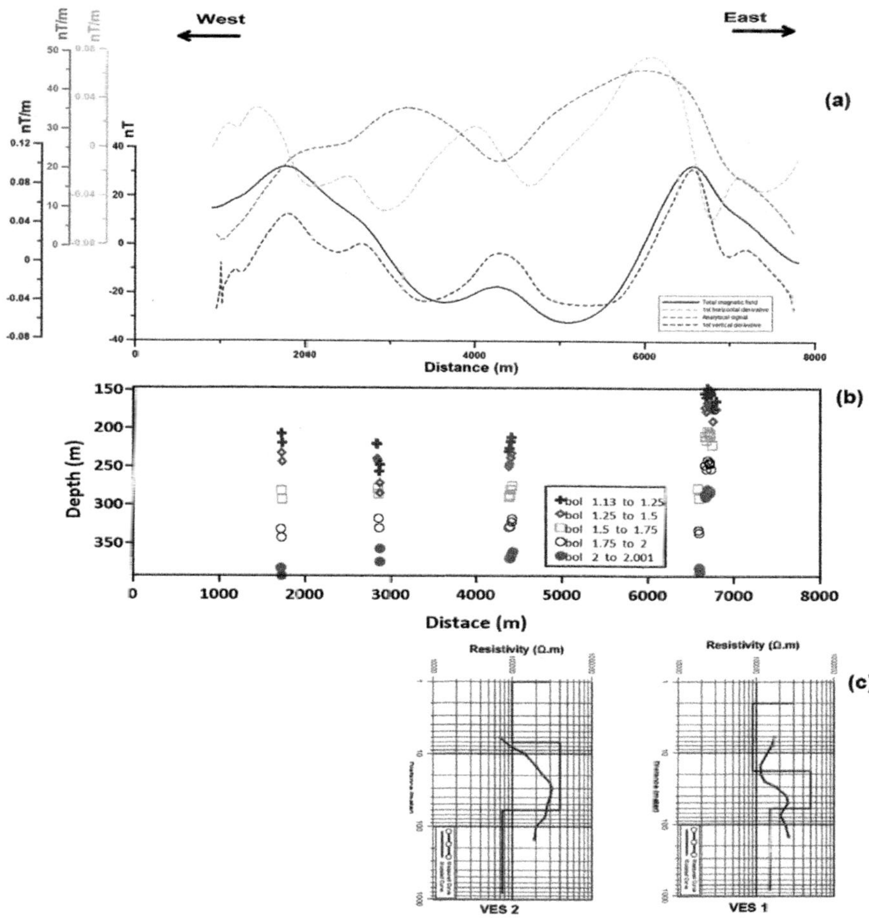

Fig. 12 Integration of magnetic and DC resistivity soundings at north of Jabal Kamel site, for location see, Fig. 1. **a** Vertical Derivative (VD), Horizontal Derivative (HD) and Analytical Signal (AS) along the magnetic profile. **b** Euler depth solutions along magnetic anomalies. **c** Resistivity soundings in the eastern side of the measured magnetic profile (no horizontal scale for the sounding points)

Nosb El-Amir Site

Nosb El-Amir site lies to the west Jabal Kamel and the Eight Bells sites. The site was selected based on the regional scale investigation where the area is characterized by buried channels controlled by E–W faults [22]. To delineate the impact of E–W faults, a magnetic profile with 20 km length and 1300 m station interval was measured in N–S direction. Additionally, five DC resistivity sounding points were carried out to detect the subsurface layer distribution in the sedimentary cover. The visual inspection of the magnetic anomalies of horizontal and vertical gradients (Fig. 13) indicates the presence of several E–W faults which are recognized in the Eight Bells and Jabal Kamel sites. The shape of AS anomalies shows the presence of basaltic intrusions accompanied by the E–W faults. Horizontal gradient contacts that are offset from analytic signal contacts indicate dipping basement contacts along the profile, with the true location close to the analytic signal [30] and the dip in the north–south direction due to E–W normal faults (Fig. 13).

The depths to the basement along the profile vary from 60 to 85 m. The estimated depth is conformity with the field observations where the surface intrusions have appeared in the area as conical hills. On the other hand, the resistivity values of

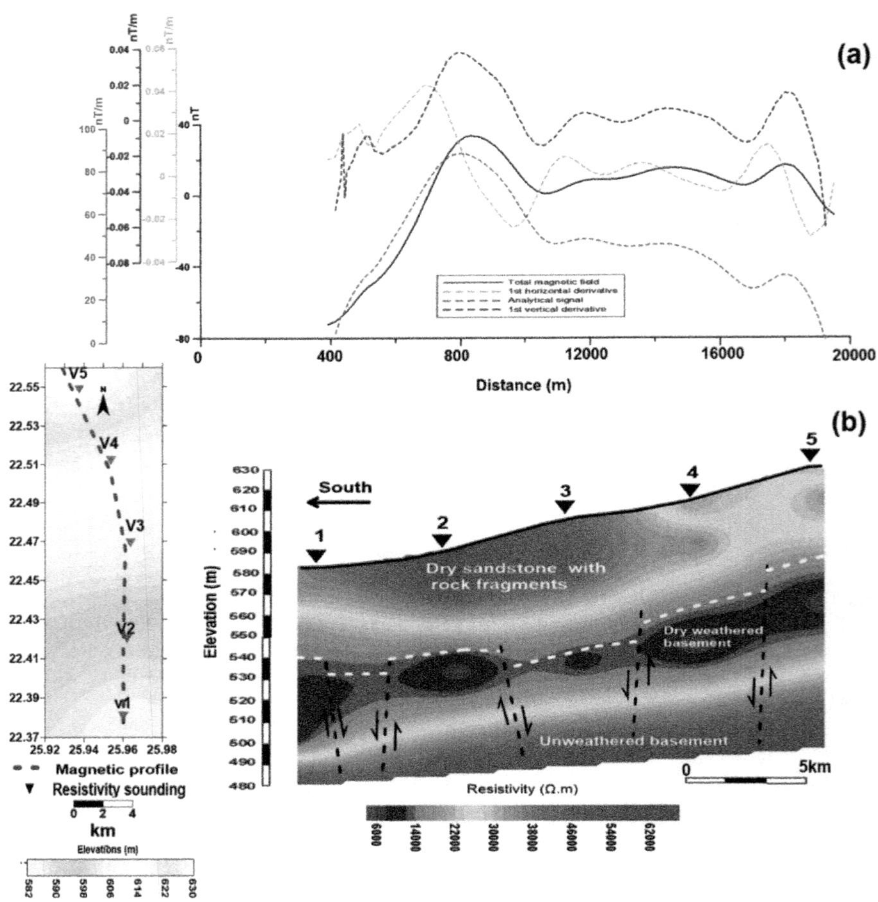

Fig. 13 Integration of magnetic and DC resistivity results along N–S profile at Nosb El-Amir site, for location see, Fig. 1. **a** Vertical Derivative (VD), Horizontal Derivative (HD) and Analytical Signal (AS) along the magnetic profile. **b** Resistivity distributions along the measured profile

subsurface layers under all VES show a continuous increase with depth until to basement rocks. The high resistivities of the sedimentary cover and bottom layers along the profile are due to the presence of dry sandstone layers and basement rock, respectively. Based on the resistivity characters as well as shallow basement, there is no evidence for groundwater accumulations in the area.

Integrative Approach

The collected and acquired data were employed to establish how the subsurface structures and basement blocks are influenced on groundwater movements and to fully understand the linkage between the common faults and groundwater potentialities in the area which is structurally complicated. Hence, the target of integrating multi-scale responses is to improve the efficiency of the applied sequential methodology and establishing a conceptual groundwater flow model considering the site characterization and the common structures. To achieve this concept, an integrated approach was implemented using the foregoing regional scale radarsat-1 data [22] geological and hydrogeological information [20], with land-based magnetic and DC resistivity data for delineating promising areas for further groundwater exploration and drilling. Accordingly, the groundwater occurrence was assessed based on the detected considerable buried channels from radar, great thickness of the sedimentary cover, valuable electrical resistivity and interconnection of the expected aquifer with NSAS from the south where the aquifer recharge occurs in the southern sections in Abdelmohsen et al. [24].

As illustrated in Fig. 14, the eastern side of the area (dashed line) is considered as a high potential region for groundwater accumulations based on the significant resistivity values (70–200 Ω m) and considerable aquifer thickness (about 100 m) at NE Jabal Kamel site. The depth to the basement rocks, and corresponding thickness of the Nubian aquifer, increases northeastward. This region is characterized by interconnection with the NSAS from the south through N–S buried channel as indicated by SRTM (Fig. 14). This promising area is considered as the extension of east Uweinat aquifer where the western part of the East Uweinat area may have a considerable reserve of groundwater with a good quality [17]. By referring to Fig. 14, the rest area to the west is considered as poor groundwater potential where there is no aquifer in the region as interpreted from geophysical measurements at Eight Bells, Nosb El-Amir and west of Jabal Kamel sites. Where these sites are located in shallow basement zones which are suffered by E–W faults and basaltic intrusions. The complex structures have divided the region into closed sub-basins and there is no evidence for groundwater flow due to the presence of Uweinat uplift in the southern border of the area. In addition, the radar map (Fig. 14) shows isolated buried channels in E–W and NE-SW directions to confirm the magnetic and DC resistivity results.

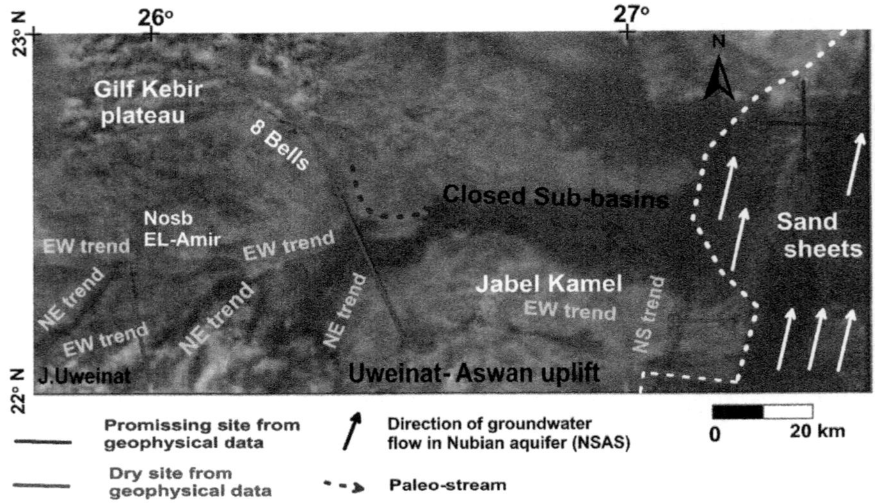

Fig. 14 Integration of multi-scale results of Radarsat, DEM and geophysical data. The radar data from Ghoneim et al. [22] projected onto DEM data showing the paleo-rivers. The dot white line is the promising site for groundwater while, the red solid line is the dry site as indicated from high resistivity values (for more details, see Figs. 9, 12 and 13)

Summary and Conclusions

In the area south of Gilf Kebir, the southwestern corner of Egyptian Western Desert, land magnetic and resistivity surveys as local scale were carried out to recognize and locate the subsurface structures that control the groundwater flow and the more promising sites for the groundwater accumulation in this arid region. Land-based magnetic survey was used to determine the depth to the basement rocks and the subsurface discontinuities such as faults, igneous intrusions and dikes that are configured the sands aquifer in the area. Accordingly, the trend analyses show that the dominant tectonic trends are E–W and NE–SW in the western side from the area and NNE–SSW and N–S trends in the east. These structures play an important role in controlling the aquifer depths and recharge system where the E–W igneous intrusions and shallow basement blocks in the west act as barriers in the direction of water flow from the south or deep aquifer layers. The depths of the basement surface were found to be widely variable where the maximum depth of 500 m was estimated in the eastern border at NE Jabal Kamel while the minimum was 90 at Nosb El-Amir in the western border.

The inferred results from the land magnetic and resistivity surveys were compared and validated with foregoing regional geological, hydrogeology obtained from remote sensing and radarsat data. The integration of magnetic and DC resistivity methods in this study approves their capabilities as useful tools for groundwater exploration in arid regions. The applied integrative procedure as time and cost-effective tool demonstrates the importance of multi-scale investigations including proposed land geophysical surveys in combinations with regional-scale data, for delineations and identification of high groundwater potentials in the extremely arid regions as in the case of Egyptian Western Desert.

Recommendations

The obtained results in the present works can help further groundwater exploration program in the area. Thus, it is recommended to encourage more detailed investigations in the eastern region around NE and East of Jabal Kamel using small scale geophysical surveys which could easily lead to determine the aquifer characterization and to study the linkage between the detected aquifer zone in the area and the aquifer in the east of Uweinat area.

Acknowledgements The authors would like to highly appreciate to Research Institute for Groundwater (RIGW) and the National Water Research Center (NWRC) for their supporting to achieve this work.

References

1. Reynolds JF, Stafford Smith DM, Lambin EF, Turner BL, Mortimore M, Batterbury SPJ, Downing TE, Dowlatabadi H, Fernandez RJ, Herrick JE (2007) Global desertification: building a science for dryland development. Science 2007(316):847–851. https://doi.org/10.1126/science.1131634
2. Botai CM, Botai JO, Muchuru S, Ngwana I (2015) Hydrometeorological research in South Africa: a review. Water 7:1580–1594. https://doi.org/10.3390/w7041580
3. Masoud AA, El-Horiny MM, Atwia MG, Gemail K, Koike K (2018) Assessment of groundwater and soil quality degradation using multivariate and geostatistical analyses, Dakhla Oasis, Egypt. J Afr Earth Sci 142:64–81
4. Huang F, Zhang YD, Zhang DR, Chen X (2019) Environmental groundwater depth for groundwater-dependent terrestrial ecosystems in arid/semiarid regions: a review. Int J Environ Res Public Health 16:763. https://doi.org/10.3390/ijerph16050763
5. Darius F (2013) Wadi Sura in its environmental setting. In: Kuper R (ed) Wadi Sura—the Cave of Beasts a rock art site in the Gilf Kebir (SW-Egypt). HEINRICH-BARTH-INSTITUT e. V., Köln, 542 pp. ISBN 978-3-927688-40-7
6. Masoud AA, El-Horiny MM, Atwia MG, Koike K, Gemail K (2019) Mapping soil salinity using spectral mixture analysis of landsat 8 OLI images to identify factors influencing salinization in an arid region. Int J Appl Earth Obs Geoinf 83:101944. https://doi.org/10.1016/j.jag.2019.101944
7. Ghoneim E, El-Baz F (2020) Satellite image data integration for groundwater exploration in Egypt, Chapter 8. In Elbeih et al (eds) Environmental remote sensing in Egypt. Springer, ISSN 2364-9127 (electronic) https://doi.org/10.1007/978-3-030-39593-3
8. Abu Salem H, Gemail Kh, Nosair A (2021) A multidisciplinary approach for delineating wastewater flow paths in shallow groundwater aquifers: a case study in the southeastern part

of the Nile Delta, Egypt. J Contam Hydrol 103701. https://doi.org/10.1016/j.jconhyd.2020.103701
9. Gemail Kh (2015) Application of 2D resistivity profiling for mapping and interpretation of geology in a till aquitard near Luck Lake Southern Saskatchewan, Canada. Environ Earth Sci 73:923–935
10. Abdel Rahman SI, Gad A, Younes HA (1990) Monitoring of the drought on Lake Nasser region using remote sensing technology. In: International conference for disaster management, Cairo, Sep. 24–27, Academy of Scientific Research & Technology, pp 161–171
11. ACC, "The American Chamber of Commerce in Egypt" (1998) The South valley development project. Business studies and analysis center, five year plan (1997–1998 to 2001–2002), Ministry of Planning
12. Issawai B, Sallam E (2018) Stratigraphy and facies development of the pre-Cenozoic sediments in southern Egypt: a geodynamic approach. Arab J Geosci 11:271. https://doi.org/10.1007/s12517-018-3626-z
13. Issawi B (2002) Egypt during the Phanerozoic. In: 6th international conference on the geology of the Arab World, Cairo University, pp 401–450
14. USGS (2002) Map showing geology, oil and gas fields and geological provinces of Africa. U.S. Geological Survey: Open File Report 97-470A, version 2.0. http://pubs.usgs.gov/of/1997/ofr-97-470/OF97-470A/. Accessed July 2013
15. Mansour MM, Baniamin NY, Mostafa MBEO, Mousa (2007) Main tectonic features of the southeastern part of the Western Desert, Egypt. Ann Geol Surv Egypt, Cairo, Egypt, Abstr 29:584–585. SE-W Desert; Lake Nasser; Gilf Kebir Plateau; Gabal Uweinat
16. El Alfy M (2014) Numerical groundwater modelling as an effective tool for management of water resources in arid areas. Hydrololo Sci J 59(6):1259–1274. https://doi.org/10.1080/02626667.2013.836278
17. GARPAD (General Authority for Rehabilitation Projects and Agricultural Development) (1994) Internal report for projects resource evaluation of (3000 ft). Ministry of Agriculture and Land Reclamation, Cairo
18. Laake A (2011) Integration of satellite imagery, geology and geophysical data. Earth Environ Sci. Imran Ahmad Dar and Mithas Ahmad Dar, IntechOpen. https://doi.org/10.5772/27613. https://www.intechopen.com/books/earth-and-environmental-sciences/integration-of-satellite-imagery-geology-and-geophysical-data
19. El Nahry H, Elewa H, Qaddah A (2010) Soil and groundwater capability of East Oweinat area, Western Desert, Egypt using GIS spatial modeling techniques. Nat Sci 8(8):1–17
20. Foster S, Loucks D (2006) Non-renewable groundwater resources. UNESCO, IHP-VI, Series on Groundwater, no 10, Paris, 81 pp
21. Robinson C, Werwer A, El-Baz F, El-Shazly M, Fritch T, Kusky T (2007) The Nubian aquifer in the southwestern Egypt. Hydrogeol J 15:33–45. https://doi.org/10.1007/s10040-006-0091-7
22. Ghoneim E, Robinson C, El-Baz F (2007) Radar topography data reveal drainage relics in the eastern Sahara. Int J Remote Sens 28(8):1759–1772. https://doi.org/10.1080/01431160600639727
23. Mohamed A, Sultan M, Ahmed M, Yan E, Ahmed E (2017) Aquifer recharge, depletion, and connectivity: inferences from GRACE, land surface models, and geochemical and geophysical data. Geol Soc Am Bull 129(5/6):534–546. https://doi.org/10.1130/B31460.1. R (ed) The geology of Egypt. Balkema, Rotterdam, pp 259–292
24. Abdelmohsen K, Sultan M, Ahmed M, Save H, Elkaliouby Emil M, Yan E, Abotalib AZ, Krishnamurthy RV, Abdelmalik K (2019) Response of deep aquifers to climate variability. Sci Total Environ 677:530–544. https://doi.org/10.1016/j.scitotenv.2019.04.316
25. RIGW (2009) Research Institute for Groundwater "Geophysical Studies in Gilf Kebir area", Internal Technical Report Submitted to National Water Research Center (NWRC), Cairo, Egypt
26. Sultan SA (1998) Shallow geophysical investigation for detecting the characteristics of near surface section of Inshas Area. PhD thesis, Ain Shams University, Cairo, Egypt, 186 pp
27. Gemail Kh, Abd-El Rahman NM, Ghiath BM, Aziz RN (2016) Integration of ASTER and airborne geophysical data for mineral exploration and environmental mapping: a case study,

Gabal Dara, North Eastern Desert, Egypt. Environ Earth Sci 75:592. https://doi.org/10.1007/s12665-016-5368-0
28. Gemail KhS, Attwa M, Eleraki M, Zamzam S (2017) Imaging of wastewater percolation in heterogeneous soil using electrical resistivity tomography (ERT): a case study at east of Tenth of Ramadan City, Egypt. Environ Earth Sci 76:666. https://doi.org/10.1007/s12665-017-7013-y
29. Milsom J (2003) Field geophysics, 3rd ed. Wiley Ltd., the Atrium, Southern Gate, Chichester, West Sussex PO19 8SQ, University College London, England, 232 pp
30. Mendonca CA, Silva JBC (1993) A stable truncated series approximation of the reduction-to-the-pole operator. Geophysics 58:1084–1090
31. Phillips JD (2000) Locating magnetic contacts: a comparison of the horizontal gradiant, analytical signal, and local wavenumber methods: Society of Exploration Geophysicists, Abstracts with programs, Calgary, 2000. https://doi.org/10.1190/1.1816078
32. Cooper GJ (2004) (2004) Euler deconvolution applied to potential field gradients. Explor Geophys 35:165–170
33. Bobachev A, Modin I, Shevinin V (2008) IPI2Win V2.0: user's guide. Moscow State University, Moscow. http://geophys.geol.msu.ru/ipi2win.htm
34. Brűgge N (2017) Remarks to the geological origin of the hills "Eight Bells" at the southern Gilf Kebir (Egypt). http://www.b14643.de/Sahara/Eight_Bells/index.htm. Accessed 10 May 2017
35. Jessell MW, Valenta RK (1996) Structural geophysics: integrated structural and geophysical modelling. In: De Paor DG (ed) Structural geology and personal computers. Pergamon, pp 303–324, 10014, 10016, 10018, 10046

Ground Water Potentiality in Siwa and Baris Oases, Western Desert, Egypt

Hosny Mahmoud Mahmoud Ezz El-Deen

Abstract The area selected for the present investigations are Siwa and Baris Oases. Siwa Oasis lying below the zero contour level elevation. It is usually considered as Siwa Depression. It lies about 65 km east of the Libyan borders. It attains about 75 km length and variable width ranging from 5 to 25 km. Baris Oasis comprises the southern part of El-Kharga Oasis. It covers about 900 km^2. Generally, the climate of the Western Desert especially in Siwa and Baris Oases is torrid and arid. Geologically, Siwa and Baris Oases are part of the Western Desert. They are occupied by deposits belonging to Pre-Cambrian (Basement complex), Upper Cretaceous (Nubian Sandstone), Middle Miocene, and Quaternary rocks (Lake and Aeolian deposits). A grid pattern of 85-deep Vertical Electrical Sounding Stations (VES) are carried out in Baris Oasis to evaluate the geologic succession, lateral and vertical variations of surface and subsurface beds, delineation of the cap rock, and evaluation of the water-bearing formations and geologic structures which affects the groundwater potentialities. The geoelectrical succession in Baris Oasis consists of 7-geoelectrical zones {surface zone, cap rock zone, zone "A", zone "B", zone "C", zone "D", and basement rock zone}. The geoelectrical zones ("A, B, C, and D") are Nubian Sandstone water-bearing zones. Hydrogeologically, the groundwater system that underlies Siwa Oasis consists of two productive aquifers. These are the Lower Cretaceous Nubian Sandstone and the Middle Miocene fractured limestone. Besides the Quaternary sand and clay uppermost layer is water-bearing due to water logging. In Baris area the Nubian Sandstone aquifer represents the main aquifer. It is classified into four productive zones (A, B, C, and D) andcomposed of sand and sandstone separated by water confining interbeds of clay and shale. In Siwa, the total salinity groundwater of the Nubian Sandstone aquifer, in terms of TDS, varies from 200 ppm to more than 1,500 ppm, while in Baris Oasis, the total salinity groundwater of the Nubian Sandstone aquifer is fresh and ranging from 353 to 694 mg/l and the chemical eater type is Sodium chloride.

H. M. M. E. El-Deen (✉)
Department of Geophysical Exploration, Desert Research Center, Mathaf El Matriya St, El Matariya 11753, Cairo, Egypt

A. Negm and A. Elkhouly (eds.), *Groundwater in Egypt's Deserts*, Springer Water,
https://doi.org/10.1007/978-3-030-77622-0_6

Keywords Siwa oasis · Baris oasis · Groundwater · Aquifer · Salinity · Nubian sandstone · Quaternary · Miocene · Cretaceous · Vertical electrical sounding stations · Water logging

Introduction

In the last few decades, great attention has been paid for the establishment of new settlements in the Egyptian desert areas that have good groundwater potentialities to preserve of our land in the Nile Valley and Nile Delta. For these purposes, favorable effort has been given in several integrated fields of study. As Egypt is essentially a desert land (about 96%), great consideration is given for its reclamation and use. In the last two decades, studies have been conducted in the fields of geology, hydrogeology, and geophysics for selected Egyptian desert areas by several governmental organizations and private sectors. These studies had been done to secure proper evaluation of new settlements. Examples of these areas are the Oases which lie in the Western Desert as Siwa and Baris Oases. The present work deals with the employment of geological and geoelectrical and hydrogeological techniques for evaluation the potentiality of groundwater in Siwa and Baris Oases.

Groundwater Potentiality in Siwa Oasis

Location of Siwa Oasis

The areas (Siwa Oasis) selected for the present investigations comprises part of the desert area, Siwa Oasis lies between longitudes 25° 16′ and 26° 12′ E and latitudes 29° 06′ and 29° 24′ N. The area lies below the mean seawater level and is usually referred as Siwa Depression. It lies at 330 km southwest of Sallum and 65 km east of the Libyan borders (Fig. 1). It attains about 75 km length and a variable width ranging from 5 to 25 km.

Climate and Geomorphology of Siwa Oasis

Siwa Oasis is characterized by desert climate. The average temperature ranges from 5.8 °C in January to 37.8 °C in July. The precipitation is scarce and does not exceed 10.44 mm/yr.

Topographically, in Siwa Oasis, the ground surface elevation ranges from 10 to 20 m below mean sea level. The total surface area inside the zero contour line reaches 1088 km^2 [1].

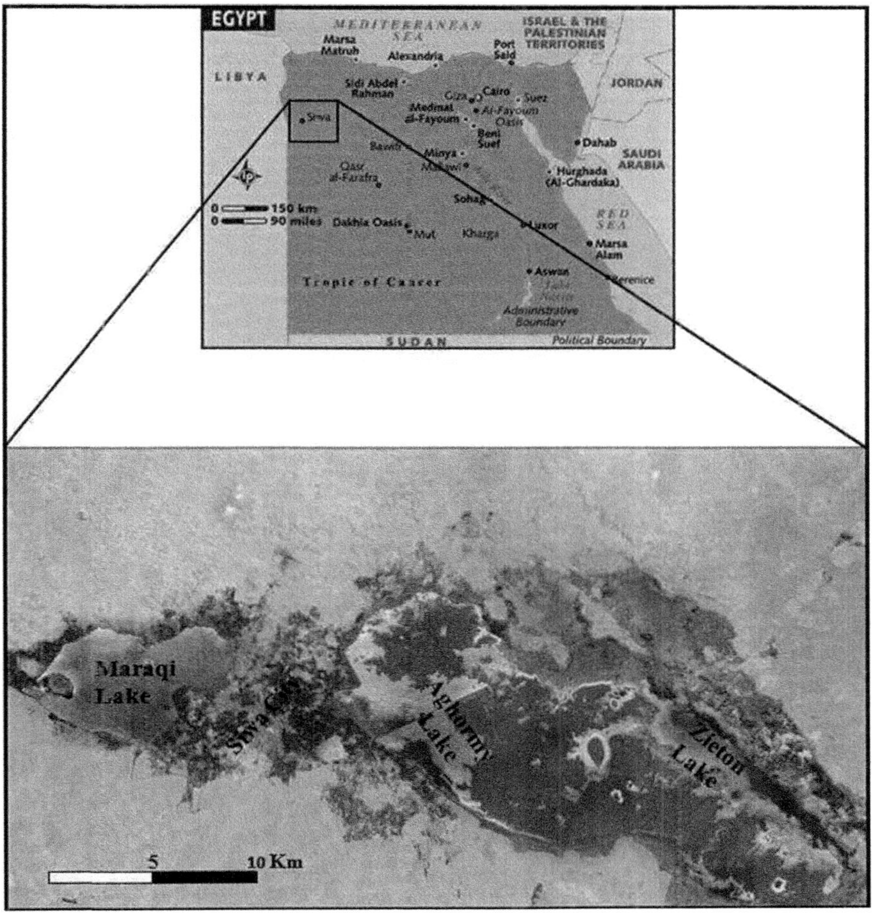

Fig. 1 Map of Egypt showing the location map of Siwa Oasis

Geomorphologically, Siwa oasis occupies a large depression opened from the border of the Tamir and Libyan plateaus to the large desert area of the Great Sand Sea. Proceeding from the north, the plateau is fragmented into numerous small isolated relieves (outlier hills), representative of erosion remnants of the depression escarpment [2, 3]. These relieve with more or less steep concave slopes reach the bottom of the basin occupied by large water stretches and by emerged zones, gradually moving towards south to the vast extensions of the dunes of the desert. Within the oasis some geomorphological features may be identified and grouped as follows (Fig. 2) [1].

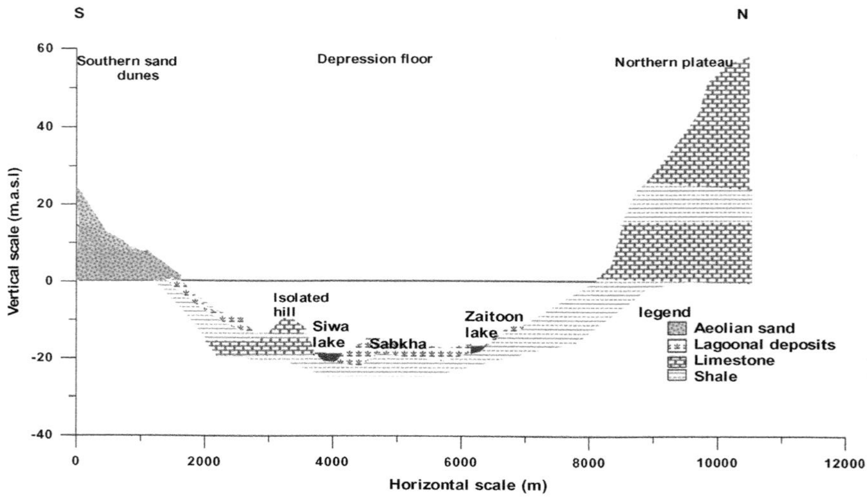

Fig. 2 Geomorphologic cross-section in N-S direction in Siwa Oasis [1] showing types of rocks and deposits and its elevation with sea level

Geology of Siwa Oasis

Geologically, sedimentary rocks belong to Middle Miocene and Quaternary occupy the surface in Siwa Oasis [4–7]. The Middle Miocene section is composed of chalky limestone, marl, shale and dolomite having a thickness of about 94 m. The Quaternary section is differentiated into the aeolian sand and alluvial and lagoonal deposits of variable thickness. In the subsurface, the stratigraphic section attains a thickness of about 3527 m and rests directly on the basement rocks (cite a reference). This section belongs, from top to base, to Miocene (250 m), Eocene (350–400 m) Cretaceous (600 m), Carboniferous (765 m), Devonian (1038 m), Silurian (470 m) and Carbo-Ordovician (320 m) [1].

Hydrogeology of Siwa Oasis

Hydrogeologically, the groundwater system that underlies Siwa Oasis consists of two productive aquifers; the Lower Cretaceous Nubian Sandstone and the Middle Miocene fractured limestone beside the Quaternary sand and clay water-bearing beds. The former represents the only sustainable fresh groundwater source in the whole Western Desert. Its thickness generally decreases from south to north till it reaches about 200 m at Siwa Oasis [8]. The groundwater flow is from SW to NE direction. The hydraulic pressure differs from 5.5 to 11 atm [6, 9] and increases towards south and east. The permeability coefficient of the saturated zone ranges between 0.93 and

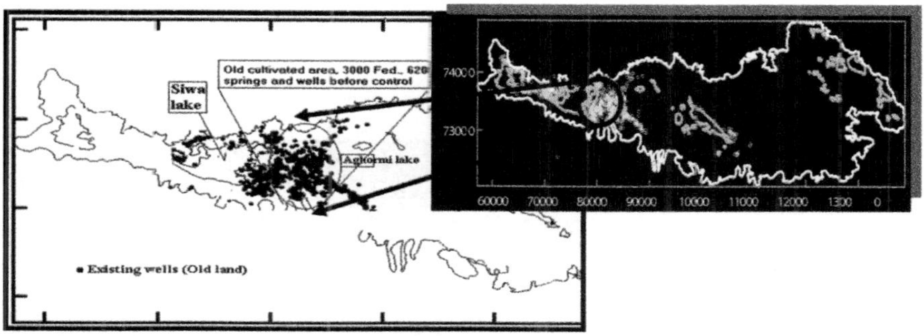

Fig. 3 Map showing the location map of the different flowing wells and springs in Siwa Oasis [1]

24.6 m/day. The extracted water from this aquifer for irrigation purposes reaches 15 $\times$ 10^6 m^3/year (Fig. 3) [1, 10].

In Siwa Oasis two aquifers exist. The lower one is the Nubian Sandstone aquifer which consists of sandstone and shale intercalations of Paleozoic and Mesozoic age and the upper one is the fractured carbonate aquifer which composed of limestone and dolomite beds of Miocene age. The thickness of the Nubian Sandstone is about 2600 m whereas the thickness of the fractured carbonate aquifer is 400–600 m. The two aquifers are separated by an aquitard composed of low permeability shale and clay layer. This aquitard has a thickness varies from 60 m in the west to 250 m in the east (Fig. 4). The thickness of the upper zone of the Nubian Sandstone aquifer is about 500 m. It contains fresh water with total salinity of about 500 mg/l. The lower part of the aquifer contains groundwater with relatively higher salinity up to 1500 mg/l. The net sand percent of this zone decreases in the northwestern direction. The potentiometric level of the Nubian Sandstone ranges from 80 masl in west Siwa to 120 masl in the east. The fractured carbonate aquifer is recharged by the upward leakage of groundwater from the Nubian Sandstone. The change of facies from silicate to carbonate has impact on the groundwater chemistry. This impact is represented by the increase of the groundwater salinity of the carbonate aquifer. Hence the total salinity of groundwater in this aquifer varies from 1500 to 8000 mg/l [11].

The Factors Constraining Sustainable Development of Siwa Oasis

Groundwater Quality

The estimated groundwater extraction from the aquifers in Siwa oasis is about 500,000 m^3/d [12]. This amount is obtained from wells taping either the Nubian Sandstone aquifer only or both aquifers in the area [11, 13].

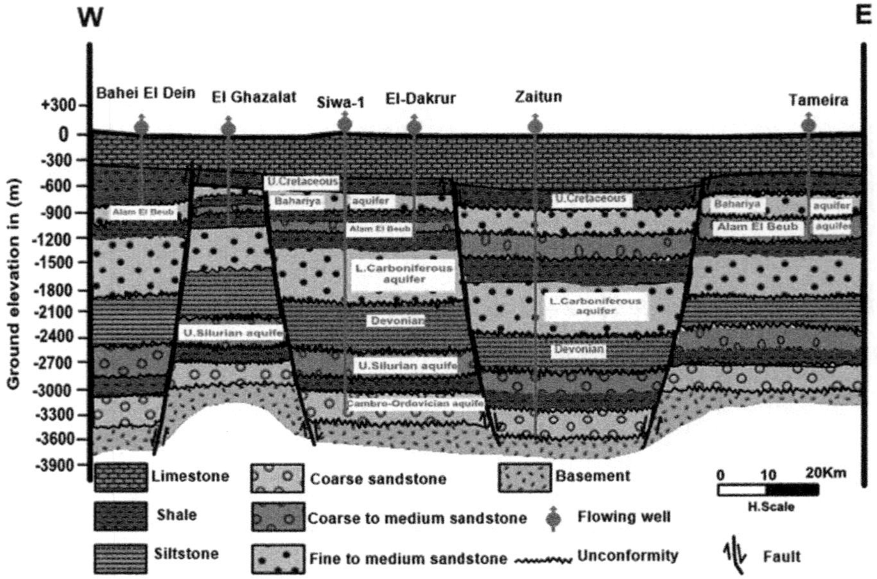

Fig. 4 Hydrogeological cross section of Siwa area [11] showing the types of aquifers in Nubian sandstone and carbonate rocks, its elevation with respect to mean sea level, different types of faults affecting different aquifers and total depth of each well (meter)

Due to the high groundwater flow rate, high temerature, and iron concentration, the pipes of a great number of well were corroded. Therefore, 400 wells were rehabilitation due as a result of the rehabilitation,which took place during the period 1996–2001, the total groundwater discharge raised up to 130 million m^3/year, However, due to the continuous random drilling of uncontrolled wells, 51 million m^3/year are lost. The lost water has created lakes and swamps which represent an environmental issue. The salinity of the deepest part of the Nubian Sandstone reaches more than 55,000 ppm [13, 14]. The thickness of the Middle Miocene carbonate aquifer ranges from 400 to 700 m. Two salinity zones were distinguished in this aquifer. The maximum depth of the upper zone is about 200 m. The groundwater salinity of this zone ranges from 1500 to 7000 ppm. The lower zone is separated from the upper one by a thin clay layer. Its groundwater salinity is high, up to 12,000 mg/l [11, 15].

Sustainable Groundwater Flow

It could summarize the present discharge of groundwater, cultivated area,water requirements and percent losses to the drainage lakes. It can be concluded from the table that, irrespective of the rehabilitation program and replacement of wells, still the percent of losses to the lakes is high. Due to the habits of the oasis people, a quick change can be harmful and thus resisted by the majority. However, introducing the sustainability factor and the effect of the uncontrolled flow of wells can be understood, especially that the people have monitored the continuous decrease in flow and increased salinity [11, 15].

Water Resources Management

The most striking feature of Siwa is certainly its springs. At one time it was said to possess 1,000, but there are now less than 200 and of these only about 80 are of use for drinking and irrigation purposes. The water in them is particularly clear and sparkling; continuous streams of bubbles are always ascending to the surface, in some cases with such rapidity and violence as to give the impression that the water is boiling. Some of the springs are hot, which are in constant use by the people in bathing and washing clothes such as Ain el-Hammam and Ain Tamousa [11].

The average ground water level rose from 150 cm below ground surface in 1962 to 75 cm below ground surface in 1987. This affects the agriculture by damaging the roots. The nature of irrigation water plays an important role in determining soil characteristics and productivity of the land. The only source of irrigation water in Siwa oasis is the springs which are spread all over the oasis. The water flow rate is about 200,000 m^3/day. Therefore, it was important to study the quality of this water. Metals including Fe, Mn. Cu, Cd, Cr, Pb, Co, and Zn were determined in water, soil and plant samples. It was concluded from the results obtained and from the given classifications that these water sources are relatively poor quality, since they contain large amounts of soluble salts. Such water is suitable for the irrigation of tolerant and semi-tolerant plants. Tolerant crops, date palm or olive trees, are suitable for growing in the Siwa oasis, because they have great salinity resistance. 5 Settling basins Irrigation channels Bier Wahed—Hot water spring Cold Water Spring Field Visits of Water and irrigation features in Siwa, Nov 2012. Source: the common irrigation method in Siwa oasis is a traditional surface irrigation system and in the east of the oasis is crystal clear that there are some new farmers used drip irrigation systems [11].

The current annual groundwater abstraction in the Western Desert is about 0.7 bcm, most of which is being utilized in irrigated agriculture and for domestic purposes [11].

Controlling the flow from the wells tapping the Nubian Sandstone aquifer is difficult because of their high hydrostatic pressure (Fig. 5). The sudden shut of the wells may cause back pressure into the aquifer. The frequent rapid opening and shutting of wells may result in the collapse of aquifer beds and well sides leading to the abandonment of the wells. The continuous flow in the shorter term produces water in excess of demand. However, these wells can be controlled on a longer term cycle (weekly or seasonally, for example) [16]. During the irrigation period, unused flow during the night occurs (the non- irrigation period) and, consequently, water logging, drainage problems, and soil salinization developed. These environmental effects may

Fig. 5 Features associated with the groundwater flow from artesian wells and springs [16] showing the different resources of water

seriously reduce agricultural productivity and mining sectors. Deep wells in some reclaimed areas (200 m in Siwa and 800–1,000 m in El-Farafra and El-Dakhla) are free flowing at high rates (5,000–30,000 m^3 per day) and high pressures (5–8 atm. at the wellheads).

Salt Lakes in Siwa Oasis

Siwa Oasis is the most westerly and remote of Egypt's major inhabited oases. It is a microcosm of recent economic, social, and environmental changes. New and improved irrigation facilities have conquered vast swathes of desert, but profligate groundwater use has resulted in the expansion of naturally occurring salt lakes (Fig. 6) as well as the loss of arable land through water-logging and land salinization [17, 18]. The lakes are mostly fed by springs. The continuous evaporation of the water of the lakes increases its salinity to levels not suitable for irrigation of many crops. Therefore, the Siwans focus on cultivating crops and fruits that can tolerate salty water such as plam trees and olives. The salty soil of the Oasis is called karshif in Arabic, and ererig by the Siwans. It is used by the locals to build their traditional mudbrick houses, which creates another problem. While the salt helps strengthen the walls of the houses, it also is unhealthy, and melts in the rain, making the houses potential death traps. Although Siwa does not receive much rain, it doesn't take long time to destroy these homes. In 1928, a major storm resulted in the local inhabitants abandoning their ancient town. These days, there are more and more prefab houses in the area. There are many hills in the oasis. The most important four are Gebel al-Mawta, Gebel al-Dakrur, Gebel Hamra, and Gebel Baylin. These carbonate hills are riddled with karst and caves, which were sometimes used as tombs, and at times inhabited both in antiquity and during more modern times [19].

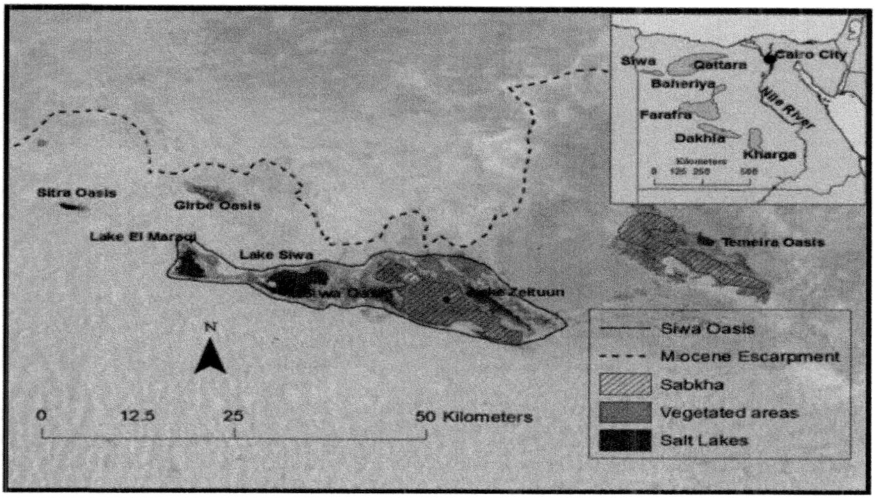

Fig. 6 Salt lakes in Siwa oasis [11] showing the result of excessive use of groundwater as well as the loss of arable land through water-logging and land salinization and the lakes are mostly fed by springs. Also, the continuous evaporation of the water of the lakes increases its salinity

Salinization in Siwa Oasis

In 2002, the environmental changes in the total surface area of saline lakes and associated forms of it such as marshes, Salinas and Sabkha (salt flats) were assessed during the period 1933–2002. The total surface area of saline lakes in 1933 is 62.97 km^2 whereas it decreased to 52.31 km^2 in 2002. The total surface area of marshes, salinas and sabkhas around the saline lakes increased from 30.78 km^2 in 1933 up to 51.33 km^2 in 2002. The total surface area of the salt flats changed by 20.55 km^2 representing 5.3% of the total surface area of the Oasis. The results of detecting changes in the Siwa Oasis explain the rapid increase of the salt flats area, high activity of salt weathering processes and increasing its hazards, where it threatens cultivated lands, roads, archaeological sites and urban areas.

There are two main types of salinity in Siwa oasis, primary (naturally) and secondary salinity resulting from human activities. Salt lakes, salt pans, salt marshes and salt flats are examples of naturally occurring saline areas [20] (Fig. 7). Examples of secondary salinity include the soil salinization due to the flood irrigation with brackish water (Fig. 8) and/or the water logging due to excess irrigation water that cannot infiltrate deep into the ground due to the occurrence of shallow clay layers.

Fig. 7 A naturally occurring saline lake and Black and gray crust near of Zeton Lake showing the salinity resulting from human activities, salt lakes, salt pans, salt marshes and salt flats as an examples of naturally occurring saline areas

Fig. 8 Soil salinization as secondary salinity of soil in Siwa Oasis due to the flood irrigation with brackish water and/or the water logging due to excess irrigation water that cannot infiltrate deep into the ground due to the presence of shallow clay layers

The Springs, Wells and Water Resources

More than one thousand springs were active in Siwa Oasis. Nowadays, 200 springs are active. Only 80 springs discharge groundwater suitable for drinking and irrigation. Most of these springs were surrounded by circular basins [21]. Many of them have some fame and glory derived from mythical association due to symbolic and the phenomenological significance. Three types of springs can be distinguished in Siwa Oasis. These are (1) common springs for the public use of the community, (2) springs shared among groups of owners, and (3) private springs. The first type, was includes the springs which are close to the Shali hill and have a good quality

Fig. 9 Types of springs [22, 23] showin the different types of springes and its uses in several fields

of drinking water as well as the springs which carry significant mythical meaning in the community,such as *Tamusi and Tiiihram* [22, 23].

The second type follows a very strict irrigation scheduel which divides the day into 16 parts and according to these parts the owners may take the amount of the water they need for irrigation. The third type comprises the springs which are owned by the tribes before the establishment of Shali in 1203. These are fairly far from Shali hill such as the spring of Quryshat [22, 23] (Fig. 9).

Ground Water Potentiality in Baris Oasis

Location of Baris Oasis

The area selected for the present investigations comprises part of the desert area to the south of El-Kharga area (Baris Oasis) and covers about 900 km^2 (Fig. 10). It is bounded southward by group of Abu-Bayan ridges (Lat. 24° 10′ to 24° 15′ N), northward by Garmashine area (Lat. 25° 00′ N) and is limited by steep escarpment in the east (30° 40′ to 30° 45′ E) whereas its western part is bounded by belt of sand dunes (Long.30° 30′ E).

As to its accessibility, the area is reached by air plane to Kharga Oasis, also there is a new asphaltic road (225 km) between Assiut and El-Kharga. El-Kharga area is connected with Baris Oasis by an asphaltic road about 90 km long. Also, Darb El-Arabian is the old Camel road from Darfour to Assiut passing through the Oases. Part of this road from south Baris to El-Kharga is now asphalt.

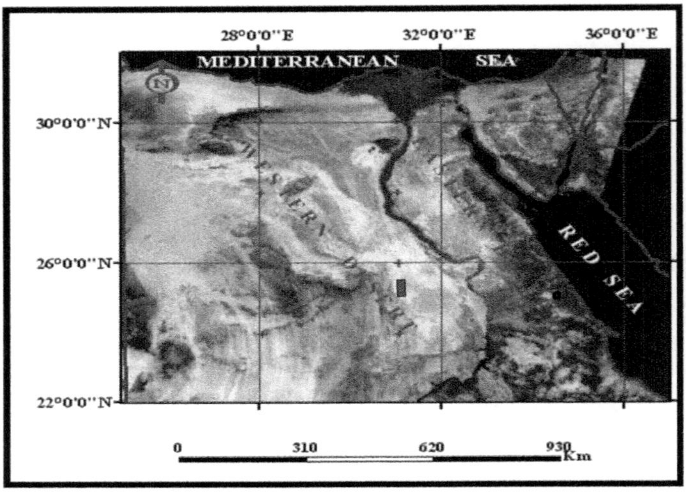

Fig. 10 Map of Egypt showing the location map of Baris Oasis

Topography of Baris Oasis

Topographically, the study area is generally as low relief. Maximum elevation is in the southern part of the area in Abu-Bayan area where it reaches 290 m above sea level at Abu-Bayan El-Kebli and decreases northward to reach <50 m at Garmashine village and westward at Baris Oasis (50–100 m).

The area can be divided into three parts. These are:

- High land (Group of Abu Bayan ridges).
- Moderate land (belt of sand dunes).
- Low land (cultivated area or depressions).

This low land comprises El-Kharga depression as morphotectonic depression El-Kharga depression. Its long axis oriented N-S, and extending southwards from the escarpment of the Abu-Tartur plateau to granite hills of Abu-Bayan. Generally the ground of this depression rises gradually to the west and the south. This depression slopes gently (5 m/km) from south to north. Baris area represents the southern part of the depression and has an area about 900 km^2 from Abu-Bayan area to Garmashine village. The elevation of this depression is about +160 m in the southern part and it is about +30 m in the northern part. The lower portion of the depression is the most suitable area for infiltration from the surface runoff water and it can store groundwater. This depression is filled with detrital materials and its floor is covered by alluvial deposits. The surface deposits are dominated by calcareous and ferruginous loamy sand and gravels in the south and by clay and shale with sand in the north. Underlying these deposits is the Nubian Sandstone.

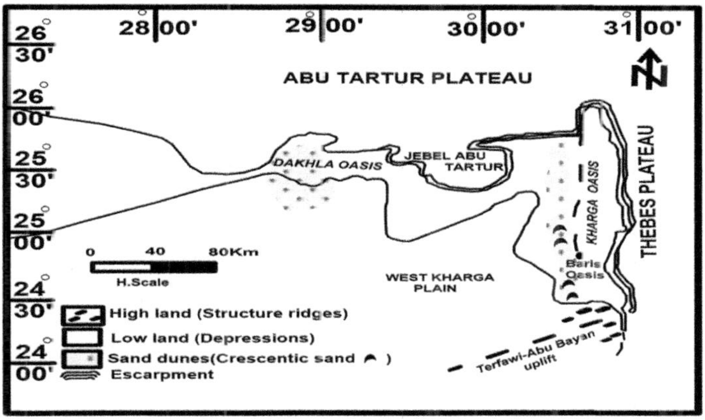

Fig. 11 Geomorphological map of the Baris area [5, 24] showing the types of land, ridges, depressions and escarpments

The surface of this depression is covered partially with sand accumulations in the form of crescentic sand dunes and sand sheets. The most conspicuous dune in the depression is that deposited in the western part of the area. It is widely distributed over a large portion (Fig. 11). The maximum elevation of these dunes is 50 m in the north and 20 m in the south. These dunes occupy the low area and are composed of loose quartz grains which are of the moving types. The sand movement is towards the SSW in the direction of prevailing winds and it is represented the young aeolian deposits. This depression with its fillings is nearly flat plain.

Another feature, which characterizes this depression, is the lake deposits (playas). These deposits are occupying the topographically low areas at the foot of the scarp. Baris playa covers an area of about 125 km^2 extending from Gada village in the north to El-Max area in the south (El-Max area 15 km south Baris Oasis). The playas consist of sand and silt in Baris area and are highly affected by wind erosion which abroad the upper parts leaving isolated hummuks representing relies on the old exhumed surface [25]. The climate of the Western Desert especially in the New Valley (El-Kharga area, Baris Oasis) is torrid and arid. The maximum recorded monthly temperature is about 41.7 °C in June, while the minimum is about 4.4 °C in January. Evaporation has its maximum value during June (22.9 mm/day) and the minimum value in January (4.35 mm/day). Rainfall is rarely and rareness in some months from November to May. Generally, the rainfall decreases southward in the area.

Age		Rock units	Lithic description
Quaternary	*Holocene*	Aealian deposits	Sand dunes and sand sheets
	Pleistocene	Lake deposits	sand and silt affacted by wind erosion
Tertiary	*Eocene*	Thebes Formation	Limestone intercalation with few layers of shale and marls
	-	Esna Formation	Shale layers convered by sand and rubble
	Paleocene	Tarawan Formation	Chalk to chalky limestone, limestone or siliceous limestone
Upper Cretaceous	*Meastrichtin*	Phosphate Formation	Phosphate shale and fossiliferous mudstone beds
	-	Nubia Formation	Variegated shale: grey, greenish grey shale together with bands of brown and purple clays
	Cenomanian		Nubian Sandstone: it consists of alternating bed sandstone, shale and clay
Pre-Cambarian	*Basement complex*	Basement rock	Pink granits, granodiorites, Pegmatite and Xenoliths rocks

Fig. 12 Idealized compiled stratigraphic columnar section, Baris Oasis, El-Khraga area, New Valley, Egypt [5, 24, 29, 30] showing the rock units and its lithological description

Geology of Baris Oasis

The southern part of the Western Desert is occupied by deposits belonging to Pre-Cambrian (Basement complex), Upper Cretaceous (Nubian Sandstone and Quaternary rocks (Lake and Aealian deposits).

The stratigraphy of the Kharga area (Baris Oasis) has been discussed by many authors since the beginning of the 19th Century, such as [13, 24–36]

The idealized stratigraphic section of the study area is shown in (Fig. 12).

Geologic Structures of Baris Oasis

Generally, the area was affected by a fod and normal faults. El-Khraga area is a plunging anticline fold with its axis extends in NE-SW direction and is dissected by hinged faults forming a graben in N-S direction. Baris Oasis is a part of El-Khraga area.

It is noticed that the N-S and E-W faults are most common in the area, while the NE-SW fault set is numerous but of small extent. There are also minor faults which are recorded in the area following the same trends of major faults. The surface faults and their concealed extensions below the surface sediments in the area have important effects on the interpretation of the geoelectrical data and hydrogeological setting as well as the evaluation of the water aquifer.

Hydrogeology of Baris Oasis

The Nubian Sandstone aquifer represents the main aquifer in Baris Oasis, it is composed of sand and sandstone separated by water confining interbeds of clay and shale. According to [36, 37] who classified the Nubian Sandstone aquifer into four productive zones (A, B, C, and D). the groundwater exists under confined condition. The depth to water varies from 35 m below ground surface at the south to 16 m at the north, so the piezometric water level varies from +72 to +26 m from south to north. This indicates that the general trend of groundwater movement is from south to north. The total salinity groundwater of the Nubian Sandstone aquifer is fresh and ranging from 353 to 694 mg/l and the chemical eater type is Sodium chloride.

Hydro-geophysical Aspect of Baris Area

A grid pattern of 85-deep Vertical Electrical Sounding stations (VES) are carried out in the area (Fig. 11) between Abu-Bayan El-wastani and Garmashine. The distribution of the Vertical Electrical Sounding stations are carried out through 3-geoelectrical profiles extends from south at Abu-Bayan El-wastani area to the north at Garmashine area for a distance about 88 km. Some of these vertical electrical soundings are conducted close to water wells (Fig. 13) [38, 39].

Schlumberger configuration Fig. 14 is used in the vertical electrical sounding (VES) measurements. The field potentiometer type (ERS) is employed for measuring the current intensity between current electrodes A & B in milliamperes and potential difference created between the potential electrodes M&N in mill volts with high resolution. The distance between the current electrodes began from 2 m until 6000 m. while the distance between the potential electrodes began from 0.5 m until 500 m. These deep vertical electrical soundings with distance 6000 m between the current electrodes were used for the first time in Egypt. The spreading is sufficient to reach to the base of the aquifer especially in the southern part of the study area.

The data of the Vertical Electrical Sounding (VES) curves was interpreted qualitatively and quantitatively by construction of a model for every curve using computer programs [40–43] for estimating the resistivities and thicknesses for every geoelectrical zones (A, B, C, and D). These zones are attributed to the four productive zones (A, B, C, and D) of the Nubian Sandstone aquifer.

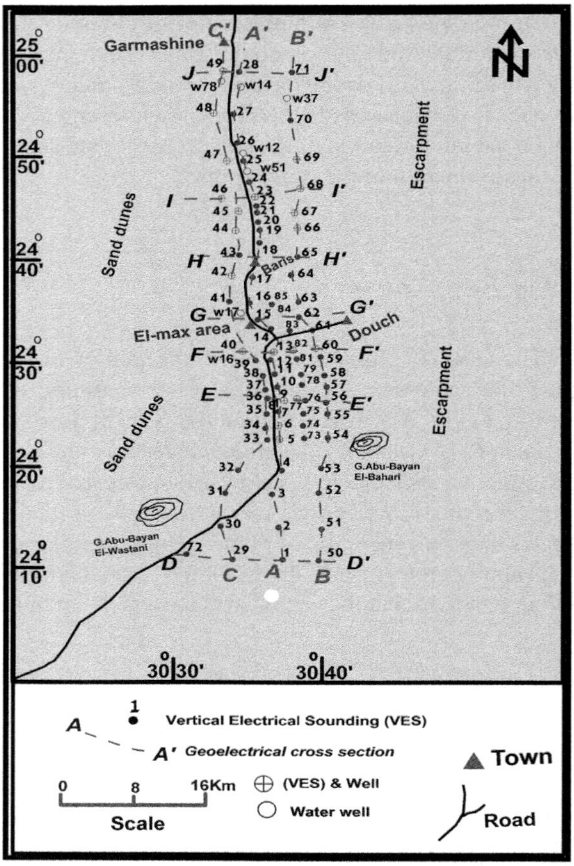

Fig. 13 A map showing the locations of the grid pattern of VES stations, geoelectrical profiles and water wells in Baris area

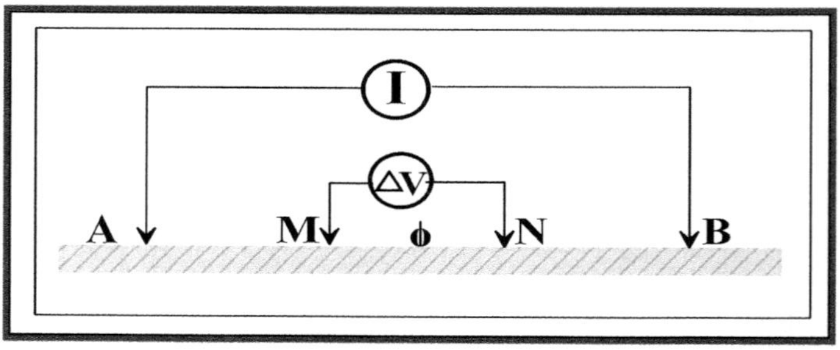

Fig. 14 Schlumberger configuration (4-electrodes array, two current electrodes A & B & two potential electrodes M&N) is used for the vertical electrical sounding (VES) measurements and for measuring the apparent resistance of rocks with high accuracy at different electrode spacing

Geoelectrical Successions

The comparison between the interpreted data for the vertical electrical sounding and the lithologic logs of some wells indicate that the geoelectrical succession of the Baris area consists of 7-geoelectrical zones {surface zone, cap rock zone, zone "A", zone "B", zone "C", zone "D" and basement rock zone}.

Geoelectrical Cross Sections

Geoelectrical cross sections illustrate the geoelectrical sequence, lateral and vertical variations for every zone and the subsurface structures in its directions. Ten geoelectrical cross sections have been constructed from the interpreted data of the vertical electrical sounding curves and lithologic data of wells, three of them in S–N direction and the other in W-E direction (Fig. 13) [38, 39].

Such geoelectrical cross sections would complete the hydro-geophysical picture and determine the distribution of the different water-bearing zones "A, B, C, D". The sequence of the geoelectrical zones and some geophysical features differ from south to north and from west to east.

These cross sections (A-A′), (B-B′), and (C-C′) extend in Baris area from Abu-Bayan El-Wastani area in the south to Garmashine area in the north for a distance of about 88 km (Fig. 11). The geoelectrical cross section (A-A′) runs parallel to the asphaltic road of Baris-El-Kharga, the (B-B′) extends parallel to cross section (A-A′) from the western side of the area, while the (C-C′) strikes in the same direction of the cross section (A-A′) from the eastern side of the area (Fig. 13), while the geoelectrical cross sections (D-D′), (E-E′), (F-F′), (G-G′), (H-H′), (I-I′), and (J-J′) are extended from sand dunes in the west to escarpment in the east for a distance ranging from 7 to 16 km (Fig. 13).

Two geoelectrical crosses as an example of ten geoelectrical cross sections in the Baris area. The first geoelectrical cross section A-A′ and the second geoelectrical cross section J-J′ (Figs. 15 and 16) respectively [38, 39].

The main observations and conclusions from these cross sections in Baris Oasis are

1. Generally, the geoelectrical cross sections consist of complete geoelectrical successions of 7-geoelectrical zones {surface zone, cap rock zone, zone "A", zone "B", zone "C", zone "D" and basement rock zone} intercalated with thin beds of clay, shale and silt (Figs. 15 and 16).
2. It is clear that the whole succession of 6-geoelectrical zones belonging to Upper Cretaceous are resting unconformably over the basement rock.
3. This succession of 7-geoelectrical zones is represented in the northern part of the area, while it is incomplete in the southern part of the area, because the sequence of the surface zone, cap rock zone, zone "A" and subzone "B1" are resting directly over the basement rocks. This is due to the deposition of the post-

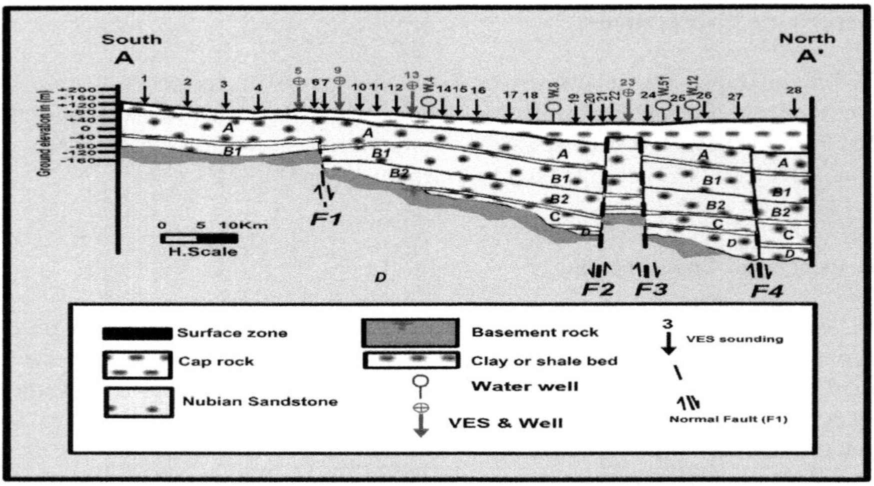

Fig. 15 Geoelectrical cross-section along the profile (A-A′) in Baris area showing the types of zones or aquifers in Nubian Sandstone, thickness and extensions of these zones, its elevation with respect to mean sea level, depth to water bearing zones or formations and impact of different types of faults on the aquifers

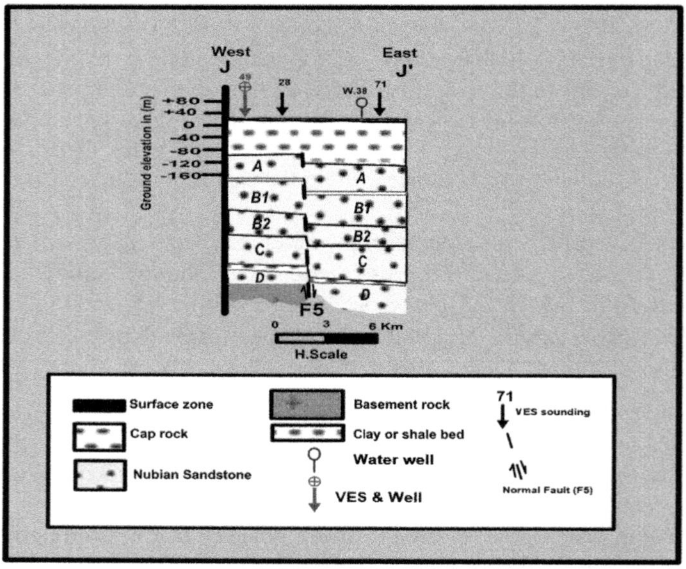

Fig. 16 Geoelectricalcross section along the profile (J-J′) in Baris area showing the types of zones or aquifers in Nubian Sandstone, thickness and extensions of these zones, effect of fault on aquifers, its elevation with respect to mean sea level and depth to water bearing zones or formations

basement rocks on an undulating surface where E-W faults system (F1, F2, F3, and F4) exist as indicated in (Figs. 15 and 16).

4 The thicknesses of the geoelectrical zones especially the Nubian Sandstone aquifer increase towards north and east as a result of E-W faults which have their down thrown sides to the north and east respectively.

5 Hydrogeologically, the geoelectrical zones ("A, B, C, and D") could be water bearing zones of Nubian Sandstone.

6 The depth to basement rock increases from 200 m at the southern part and to more than 600 m at the northern part of the study area (Fig. 15).

7 The Nubian Sandstone aquifer represents the main aquifer in Baris area, it is composed of sand and sandstone separated by water confining interbeds of clay and shale.

Discussion

Siwa Oasis lies between longitudes 25° 16′ and 26° 12′ E and latitudes 29° 06′ and 29° 24′ N. The area is lying below the zero contour level elevation is usually considered as Siwa depression, while Baris Oasis selected for the present investigations comprises part of the desert area to the south of El-Kharga area (Baris area) and covers about 900 km^2. It is bounded southward by a group of Abu-Bayan ridges, northward by Garmashine area and is limited by steep escarpment in the east whereas its western part is bounded by a belt of sand dunes.

Geologically, Siwa and Baris Oases are part of the Western Desert. It is occupied by deposits belonging to Pre-Cambrian (Basement complex), Middle Miocene, Upper Cretaceous (Nubian Sandstone), and Quaternary rocks (Lake and Aeolian deposits).

Hydrogeologically, the groundwater system that underlies Siwa Oasis consists of two productive aquifers. These are the Lower Cretaceous Nubian sandstone and the Middle Miocene fractured carbonates. The Nubian Sandstone aquifer is the main aquifer in Baris area where it consists of four productive zones. It is composed of sand and sandstone separated by water confining interbeds of clay and shale. In Siwa, the total groundwater salinity of the Nubian Sandstone aquifer varies from 200 to more than 1,500 mg/l. In Baris Oasis, the total salinity groundwater of the Nubian Sandstone aquifer is fresh and ranging from 353 to 694 mg/l and the chemical water type is Sodium-chloride.

Conclusions

It can be concluded from this work that the groundwater potentiality in Siwa and Baris Oases are excellent. However, the groundwater in the Nubian Sandstone in these oases was recharged during the Pleistocene pluvial periods tens to hundreds of thousands of years ago. The recharge of the aquifer in modern times is negligible due to the sever aridity prevailing the region. At Siwa Oasis, the fractured carbonate aquifer is recharged mainly from the underlying artesian Nubian Sandstone aquifer by upward leakage that finds its ways through the vertical and sub vertical open fractures. In Siwa, the total groundwater salinity of the Nubian Sandstone aquifer varies from 200 to more than 1,500 mg/l, while in Baris Oasis, the total salinity groundwater of the Nubian Sandstone aquifer is fresh and ranging from 353 to 694 mg/l and the chemical water type is Sodium-chloride.

Recommendations

It is highly recommended to set up a groundwater allocation plan that include both legal and pricing rules to organize the pumping and usage of the Nubian aquifer groundwater. Modern drip and spray irrigation techniques must be used in these areas. No flood irrigation should be allowed. Scientifically, there are some outstanding issues raising debate regarding the apparent reversed pattern of groundwater salinity both laterally and vertically. Future studies should focus on the pale recharge from both local and distant areas. Also, the following points are recommended and very important:-

1 Convert the available traditional surface irrigation to a developed irrigation system like (gated irrigation pipes, siphon tubes, alternative furrow irrigation and surge flow furrow irrigation systems).
2 The recommendation to reuse the drainage water in the new reclamation projects will also mitigate the groundwater level increase.
3 Stop drilling of random wells and search and study the best solutions to the problem.
4 Managing salinity involves striking a balance between the volumes of water entering (recharge) and leaving (discharge) the groundwater system.
5 Maintenance of water and soil resources.
6 In Siwa and Baris oases, Recommendations concerning mitigation of water logging problem through improvement of irrigation and drainage.
7 The increase of lifting stations' efficiency is highly recommended since it will decrease the groundwater demand.

References

1. Gad MIM, Abdel-Baki AA (2002) Estimation of salt balance in the soil-water of the old cultivated land in Siwa Oasis, western Desert, Egypt. In: Proceedings of the international symposium on optimum resources utilization in salt-affected ecosystems in arid and semi-arid regions, 8–11 April 2002, Cairo, Egypt, pp 16–23

2. Embabi NS (1977) Morphological map of Siwa depression. In: Sym. record of Siwa Oasis, record 4-d. Ain Shams University, pp 175–191
3. Mohamed HMH (1994) The morphological map of Siwa depression. M. Sc. thesis, Fac. Liter., Monofia University 220
4. Shata AA (1953) New light on the structural development of the Western Desert of Egypt. Bull Inst Desert Egypt
5. Said R (1962) The geology of Egypt. Elsvier Publishing Co., Amsterdam
6. Zakaria MH (1972) Geological and hydrogeological studies of Siwa Oasis. M. Sc. thesis, Fac. Sci., Alex University
7. Sallam ES, Ruban DA (2017) Palaeogeographical type of the geological heritage of Egypt: a new evidence. J Afr Earth Sci 129:739–750
8. REGWA (1997) Technical report for Siwa deep wells. A report submitted to Research Institute for groundwater. RIGW, p 115
9. Research Institute for groundwater RIGW (1996) Project on development of Siwa Oasis, well inventory, p 79
10. Gad MIM (1999) Water budget analysis for the topmost aquifer for some cultivated areas in the central part of Siwa Oasis, Western Desert, Egypt. PhD. thesis, Faculty of Science, Al-Azhar University, p 228
11. El Hossary MFM (2013) Investigating the development challenges to Siwa Oasis, Northwestern Desert, Egypt. N Y Sci J 6(4):55–61. ISSN: 1554-0200
12. El Hossary MFM (1999) Investigating the development challenges to Siwa Oasis, Northwestern Desert, Egypt. N Y Sci J
13. Powell O, Fensham R (2016) The history and fate of the Nubian sandstone aquifer springs in the oasis depressions of the Western Desert Egypt. Hydrogeol J 24:395–406
14. The General Petroleum Company GPC (1991) Salinity study on Pre-Upper-Cenomanian (PUC) sediments in Siwa. Paper presented during the workshop on Siwa, Alexandria, November 1991
15. Aly AA, Kishk FM, Gaber HM, Al-Omran AM (2016) Long-term detection and hydrochemistry of groundwater resources in Egypt: case study of Siwa Oasis. J Saudi Soc Agric Sci 15:67–74
16. Badawy HA (2005) Exchange of irrigation water between farmers in Egypt. In: Ninth international water technology conference, IWTC9 2005. Sharm El-Sheikh, Egypt, pp 471–484
17. Misak RF, Baki AA, El-Hakim MS (1997) On the causes and control of the waterlogging phenomenon, Siwa Oasis, Northern Western Desert, Egypt. J Arid Environ 37(1):23–32
18. Masoud AA, Koike K (2006) Arid land salinization detected by remotely-sensed landcover changes: a case study in the Siwa region NW Egypt. J Arid Environ 66 151–167
19. http://www.touregypt.net/featurestories/siwa1.htm
20. https://www.qld.gov.au/environment/land/management/soil/salinity/types
21. https://www.coursehero.com/file/pf1rqb/The-following-sections-analyze-the-Siwan-built-environment-to-explore-possible/
22. http://www.ijens.org/2F106604-3131/2520IJCEE-IJENS.pdf
23. El-Shafie M (2010) Human-environment interactions: phenomenal relationships. Int J Civil Environ Eng IJCEE-IJENS 10(04)
24. El Shazly MM (1965) Geological studies of kharga and Dakhla area. Mataria, Cairo
25. El-Hinnawi M (1976) Geology of Abu Bayan—Balag stretch, western desert. M.Sc. thesis, Cairo University, Egypt
26. Abdulaziz AM, Faid AM (2015) Evaluation of the groundwater resources potential of Siwa Oasis using three-dimensional multilayer groundwater flow model, Mersa Matruh Governorate Egypt. Arab J Geosci 8(2):659–675
27. Said R (ed) (1961) The geology of Egypt. Elsevier, The Netherlands, p 377
28. Shata A (1951) Remarks on the geological structure of the groundwater reservoir at El-Kharga and Dakhla Oases (Western Desert of Egypt). Bull Soc Geograph Egypte 34:165
29. Moghazy NH, Kaluarachchi JJ (2020) Assessment of groundwater resources in Siwa Oasis, Western Desert, Egypt. Alex Eng J 149–163
30. Hermina MH (1967) Geology of the North-Western approaches of kharga Oasis. Geol. Surv. and Min Research Dept., Paper no. 44 pp 87

31. Youssef MI (1968) Structural pattern of Egypt and its interpretation. Bull Am Assoc Petrol Geol 52:601–614
32. Ezzat MA (1974) Exploration of groundwater in El Wadi El Gedid project area (NewValley), Part I, hydrogeological conditions, kharga and Dakhla area. Ministry of Agriculture and Land Reclamation, Cairo
33. Ezzat MA (1976) Groundwater series in the Arab Republic of Egypt. Exploitation of groundwater in ElWadi El-Gedid project area (NewValley), Part IV, Exploitation of groundwater. Kharga Oasis Cairo–Egypt, p 203
34. Issawi B (1971) Geology of Darb El Arbain, WesternDesert. Ann Geol Surv Cairo Egypt 1:53–92
35. Abu Risha UA (2013) Geological controls upon groundwater occurrence in Kharga and Dakhla Oases, western desert. Egypt J Sediment Soc Egypt 21:155–170
36. Aggour TA, El Abd E (2009) New lights on the hydrogeology of Darb El Arbaein Area Western Desert, Egypt. Sedimentol Egypt 17:167–184
37. Ezzat MA, Abul Atta AA (1974) Exploration of groundwater in El Wadi El Gedid project area (NewValley), Project area. UAR Series government press, Cairo
38. Ezz El Deen HM (1996) Uses of geophysical, hydrogeological methods and application of geographic information systems (GIS) for the development of Paris area, New Vally, Egypt. PhD thesis, Geophysics Dept., Fac. of Science, Ain Shams University, p 212
39. Mostafa Said MB (2016) Groundwater exploration by using geophysical techniques in the area between Gable Abu Byan El Qibli and El Tamaneen Village, Darb el Arbian, Western Desert, Egypt. Egypt J Appl Geophys 15(2):1–23
40. Zohdy AAR, Bisdorf RJ (1989) A new method for the automatic interpretation of Schlumberger and Wenner sounding curves. Geophysics 54(2):245–253
41. Van der Velpen BPA (1988) Computer program "Resist" ,Ver.1. M.Sc.thesis, ITC, Delft, Netherlands
42. Interprx Limited (1996) "RESIX- PLUS". Resistivity data interpretation software, v. 2.39. Golden, Colorado, USA
43. IPI2Win Software (2003) Resistivity sounding interpretation Version 3.0.1.a. Moscow State University
44. Shata A (1961) Remarks on the regional geological structural of the groundwater reservoir of El-kharga and Dakhla Oases. Bull Soc De Geogr D Egypt 34:177–186

Groundwater and Characteristics of the Tertiary-Quaternary Aquifer System West of Mallawi, Upper Egypt

Sawsan Moselhy Mohamed Ibrahem

Abstract The desert areas in Egypt are suffering from the water shortage, so the groundwater exploration is the main solution for water demands in such areas. The area west of Mallawi, Upper Egypt is considered one of the promising areas for sustainable development. Mallawi area is located to the west of the Nile valley and is bounded by latitudes 27° 23′ 00″ and 28° 00′ 00″ N, and longitudes 30° 25′ 47″ and 31° 00′ 00″ E. The area is arid to semi arid, hot climate, dry, rainless in summer, and mild with rare precipitation in winter. In such area, the groundwater is the only source for sustainable development. So, the evaluation of the groundwater is necessary, where it is characterized by good water potentialities through the existing of three aquifers. These aquifers are; Quaternary sand and gravels, Oligocene-Pleistocene gravels and Middle Eocene fractured limestone. The Quaternary aquifer in the flood plain is considered as highly productive where the aquifer has an average transmissivity value of the order of 8906 m^2/day and average specific capacity of 25 m^3/h/m. The water salinity of this aquifer ranges between 203 and 549 mg/l and increases westward. The second aquifer; Oligocene-Pleistocene aquifer is moderately productive aquifer having an average transmissivity value of the order of 3291 m^2/day and hydraulic conductivity ranges between 13.22 and 59.9 m/day and average specific capacity of 9.6 m^3/h/m. Its saturated thickness ranges between 91.5 and 113.5 m and its salinity ranges between 719 and 801 mg/l. The third aquifer; the Eocene fractured limestone aquifer is considered as highly productive aquifer where it attains an average transmissivity value of the order of 6091 m^2/day and average specific capacity value of 208 m^3/h/m. In this aquifer, the water salinity ranges between 462 and 845 mg/l. The geologic structures play an important role in the direct hydraulic connection between the three aquifers, where; the Eocene fractured limestone aquifer is recharged directly from the Quaternary aquifer and then acts as a rechargeable source for the Oligocene-Pleistocene aquifer. The general flow direction of the groundwater is from east to west with some local change in the Eocene fractured limestone aquifer where there is a flow direction from southwest to the northeast direction which may be attributed to the over-exploitation in this locality or the effect of the fractures orientation in

S. M. M. Ibrahem (✉)
Hydrology Department, Desert Research Center (DRC), 1 Mathaf El-Mataryia Street, El-Mataryia, Cairo, Egypt

A. Negm and A. Elkhouly (eds.), *Groundwater in Egypt's Deserts*, Springer Water,
https://doi.org/10.1007/978-3-030-77622-0_7

this aquifer system. Groundwater salinity in the three dominant aquifers generally increases westwards; i.e. in the same water flow direction.

Keywords Groundwater · Deserts · Egypt · Hydrology · Hydraulic parameters · Granular rocks · Fractured rocks

Introduction

Water is the important component of the development of any area. The human settlement depends on a large extent on the availability of water resources. In the recent years, the consumption of water is greatly increased due to the increase in human population in the study area.

In the last decades, the development of the Upper Egypt governorates desert areas attracted the attention of the decision makers and the investors which is achieved by reclamation of more desert lands and building up new communities. This natural expansion for agricultural, industrial and civil activities in the Western Desert of Egypt necessitates more exploration activities for groundwater resources.

The pilot area is located to the west of the Nile valley occupying the floodplain and the desert fringes between latitudes 27° 23′ 00″ and 28° 00′ 00″ N, and longitudes 30° 25′ 47″ and 31° 00′ 00″ E (Fig. 1). The area is arid to semi arid, hot climate,

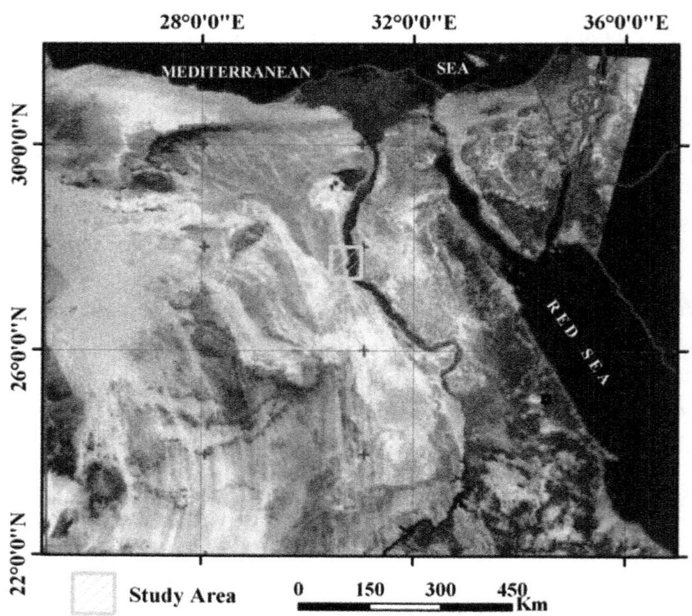

Fig. 1 Satellite image of Egypt showing the location of the study area

dry, rainless in summer, and mild with rare precipitation in winter. The rainfall average value for the last 15 years ranged from 23.05 to 33.15 mm/year, while the evapotranspiration at El Minia is 4897.91 mm/year [1, 2]. The average temperatures during January are 4.5–20.5 and 20.5–37.7 °C during August.

In this area, the water resources are represented by the Nile River as well as the groundwater of the existed three aquifers.

The main objectives of the present study are to evaluate the characteristics of the existed aquifers and assessment of the groundwater potentialities and quality at the west of Mallawi area aiming to realize the sustainable development of such area.

Geomorphological and Geological Setting

The geomorphological and geological setting of the concerned area plays an important role in the groundwater occurrences through the geomorphological features and the geological structures.

Geomorphological Setting

The study area includes three geomorphological units [3]. The first unit is young alluvial plain of the Nile which occupies the area adjacent to the Nile bank from west to the eastern scarp of the limestone plateau The second unit is Nile terraces which lay adjacent to the cliff of the limestone plateau and to the west of the young alluvial plain forming a gently undulated dissected surface made of several broad bench differentiated into old and young terraces. The third one is western limestone plateau which bounds the Nile valley from the east and the west.

Generally, the ground elevation varies widely from less than 38 m in the flood plain area to more than 160 m at the limestone plateau at west as shown in the Digital Elevation Model (Fig. 2).

Geological Setting

The geological map (Fig. 3) [4], shows that the surface of the study area is covered by the Middle Eocene limestone, Oligocene-Pleistocene gravel and sand and Quaternary deposits. The recorded rock units from base to top are [3, 5]:

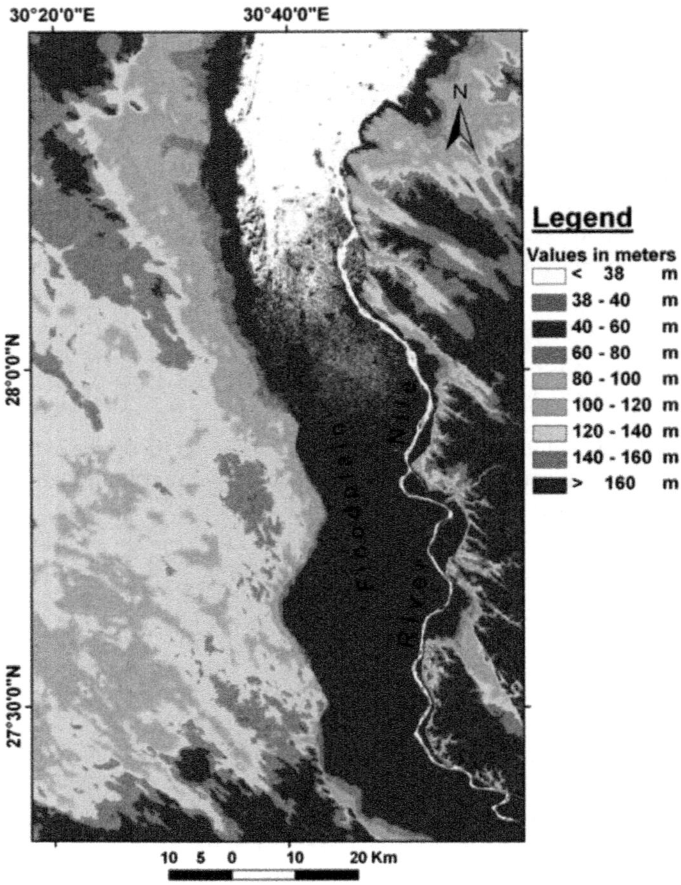

Fig. 2 Digital Elevation Model (DEM) of the study area showing the ranges of ground elevations along the study area

Middle Eocene Limestone (Samalut Formation)

This Formation is the oldest exposed rock unit in the vicinities of the study area overlying El-Minia Formation of Lower Eocene. It consists mainly of massive limestone assigned to the Early-Middle Lutetian times. This Formation is exposed at the northern, southern and eastern portions, while in the western portion; it is covered by a great thickness of Oligocene-Pleistocene gravels and sands. Samalut Formation represents one of the main water bearing formations in the study area through its highly fractured systems.

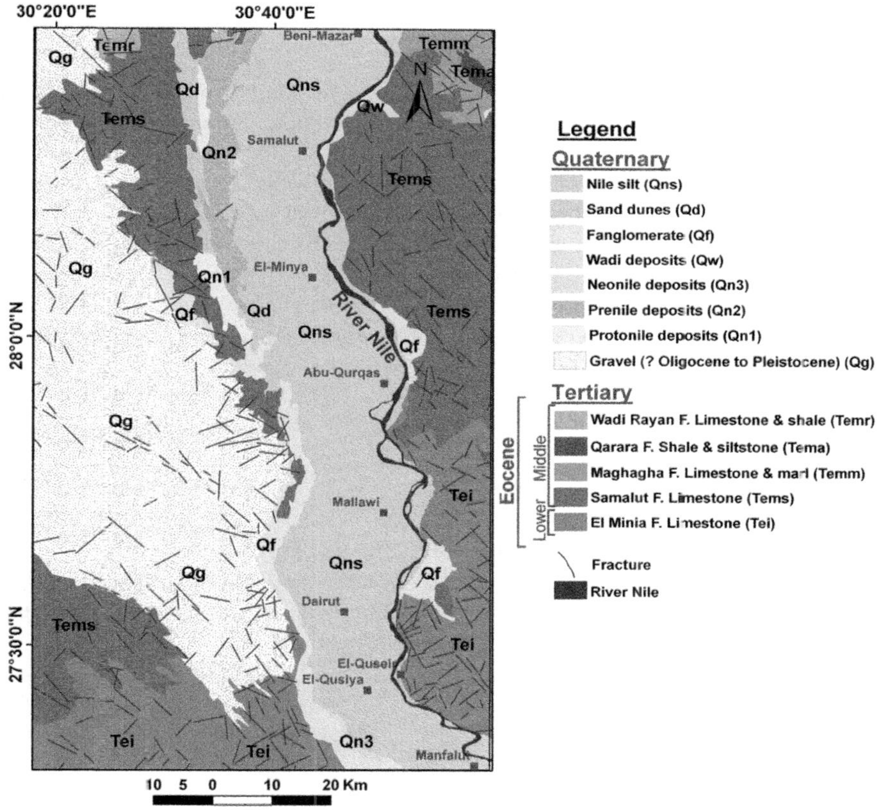

Fig. 3 Geologic map showing the different exposed geologic units in the study area and its vicinities (After CONOCO, 1987)

Oligocene-Pleistocene Gravel and Sand

It covers a wide area and composed mainly of gravel, sand, clay and limestone fragments, varying in thickness from few meters in the northern, southern and eastern portions to more than 200 m in the western portion.

Quaternary Deposits

It has a wide distribution in the study area including the Nile silt, sand dunes, and Fanglomerates.

Geologic Structure

Structures as hydrodynamic contacts impact on the groundwater flow pattern of an aquifer, as well as, the major structural features impacting on groundwater are fractures and folds.

From the structural point of view, the study area affected by a series of fractures and faults which are formed by brittle fracturing of rocks. Fractures are not homogeneously distributed in the rock mass, and because the permeability of the fracture system is mainly depended on the width and degree of fracture connectivity, it is very difficult to predict the yield of a well or borehole in crystalline aquifer [6].

The Nile Valley is essentially bounded by wrench faults that are more or less parallel either to the Gulf of Suez (NW–SE) or Gulf of Aquba (NE-SW) [3, 7].

Aquifer Systems

The investigation of 31 wells (Fig. 4 and Table 1) tapping the three existed aquifers was carried out both in the field and in the lab to assess the groundwater potentialities. Water depths, long term and phase discharge pumping studies, and water sampling for chemical analysis are examples of such investigations.

Quaternary Aquifer (The Nile Alluvium Aquifer)

This aquifer occupies the central strip of the Nile valley system forming the old cultivated lands on both sides of the Nile (floodplain). It is composed of Pleistocene clay, graded sand and gravel and capped by a Holocene silty clay layer which acts as an aquitard. The thickness of this aquifer ranges from 25 m at the desert fringes to 300 m at the central Nile Valley [8] and is recharged mainly from Nile water, irrigation system, drains, and agricultural wastewater [9].

Oligocene-Pleistocene Aquifer

This aquifer is composed of clay, sand and gravel. It occupies the outer fringes of the Nile aquifer system adjacent to the floodplain.

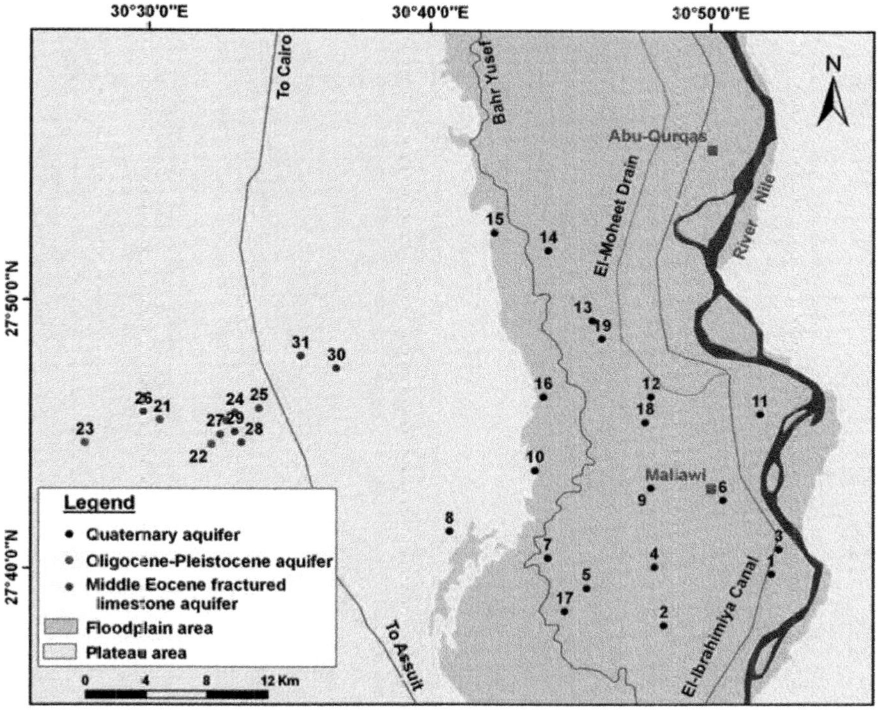

Fig. 4 Well location map showing the distribution of the different existing wells of the three aquifers (Quaternary, Oligocene-Pleistocene and Middle Eocene Limestone Aquifer)

Middle Eocene Limestone Aquifer

This aquifer is not outcropping in the study area but it exists in the subsurface as a fractured system and constitutes the main aquifer through its highly fractured systems which increase its permeability and its productivity. The Middle Eocene limestone aquifer was subjected to different studies including its potentials and its geochemical characteristics. Abdel Moneim et al. [10] studied the groundwater management at west El-Minya desert area, [11] studied the water resources management of El-Minya Governorate, [12] studied the assessment of the hydrogeochemical processes affecting groundwater quality in the Eocene limestone aquifer at the desert fringes of El-Minya governorate, [13] discussed hydrological evaluation of the tertiary-quaternary aquifer system west Mallawi, [14] studied the hydrogeology of the shallow aquifer in the area west Samalot, [15] utilized the geological data and remote sensing applications for investigation of groundwater occurrences in West El Minia governorate and [16] evaluated the groundwater potentials of Eocene limestone aquifer in West-West El-Minya area.

Table 1 Depth to water and water salinity of the existed wells of the three aquifers (Quaternary, Oligocene-Pleistocene and Middle Eocene)

Well No	Aquifer	Depth to water m	Water salinity mg/l
1	Quaternary aquifer		323.3
2			369.4
3			314.8
4			445.2
5			510.2
6			203.5
7			351.2
8			549.4
9			367.2
10			
11			325.4
12			380.0
13			363.1
14			429.0
15			315.5
16		3.8	
17		3.8	
18		2.67	
19		2.2	
20	Oligocene-Pleistocene aquifer	85.73	785.0
21		92.6	794.4
22		88.49	801.0
23		87	719.0
24	Middle Eocene aquifer		715.0
25		86.2	626.5
26		86.5	844.9
27		83.4	
28		78.1	487.9
29		80.63	490.1
30		113.6	461.8
31			475.8

Hydrogeological Cross Sections

The lateral and vertical distributions of the aquifers mentioned above, as well as their interrelationships, are depicted by constructing two hydrogeological parts A-A′ (Fig. 5) and B-B′ (Fig. 6) that cross the study region in North–South and West–East directions, respectively. The following is a discussion of these hydrogeological cross sections based on these cross sections:

Cross-Section A-A′

With a length of around 20 km, this cross-section runs nearly parallel to the Nile River and runs nearly parallel to the flood plain (Fig. 5). The following conclusions can be drawn from this section:

1. The Quaternary aquifer, which consists of sand, gravel, and clay and has a penetrating saturated thickness of about 97 m, is the water-bearing formation along this line. A capping layer of semi-pervious silty clay with a thickness of 10–15 m sits on top of the aquifer. As a result, the aquifer around the Nile River is in a semi-confined state. This makes the aquifer to exist under a semi-confined condition around the River Nile.

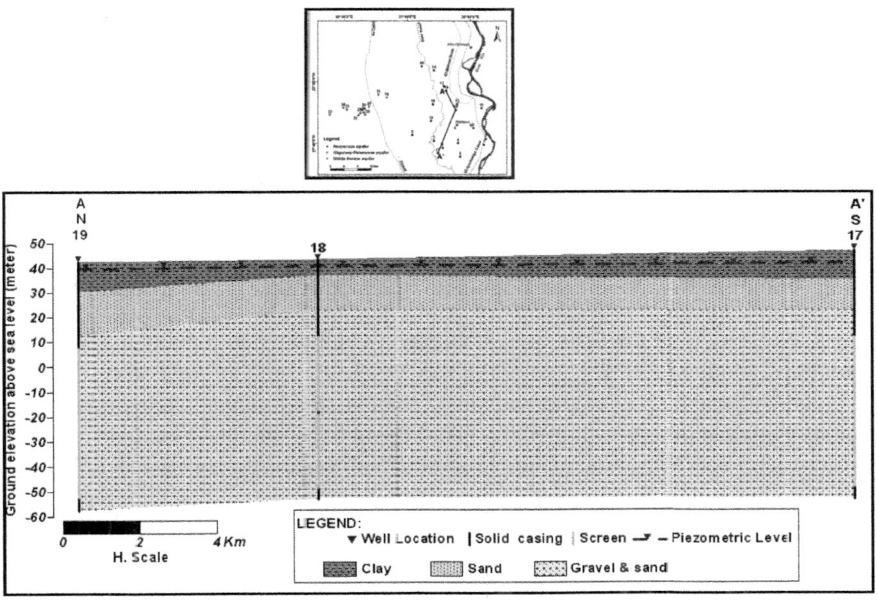

Fig. 5 Hydrogeological cross section A-A′ showing the vertical and horizontal distribution of the lithological succession, the design of the wells and the piezometric level of the surface of the groundwater

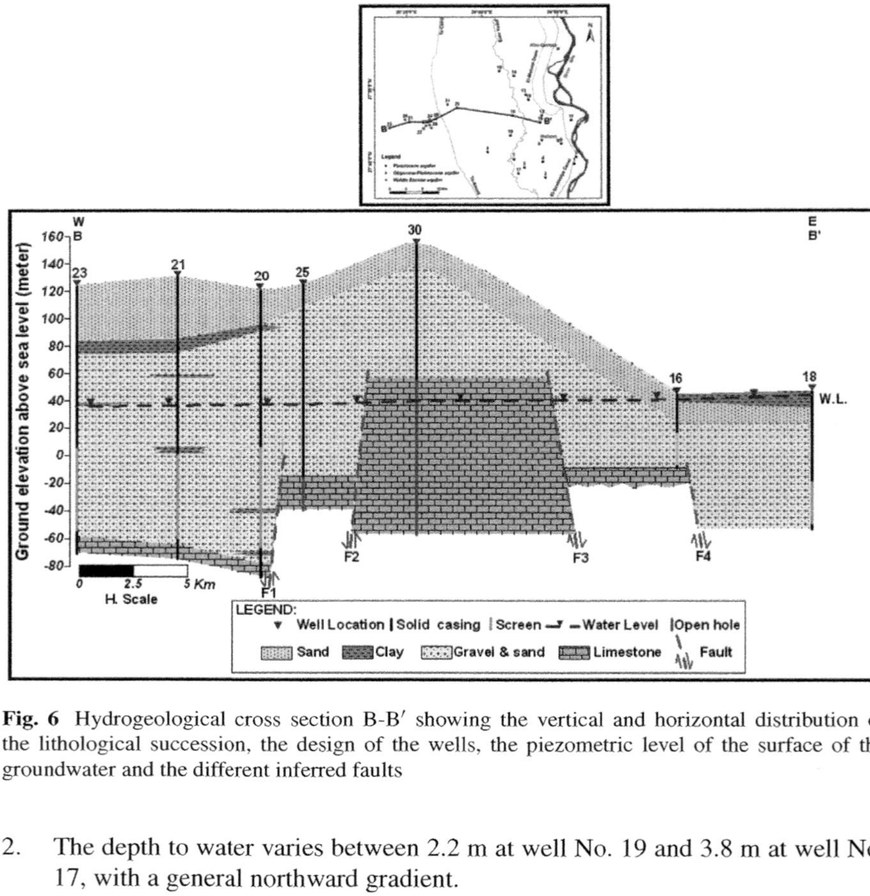

Fig. 6 Hydrogeological cross section B-B′ showing the vertical and horizontal distribution of the lithological succession, the design of the wells, the piezometric level of the surface of the groundwater and the different inferred faults

2. The depth to water varies between 2.2 m at well No. 19 and 3.8 m at well No. 17, with a general northward gradient.

Cross-Section B-B′

This section has W–E direction which is nearly perpendicular to the River Nile with a length of about 33 km. It passes from the west to the east through wells Nos. 23, 21, 20 (fully penetrating the Oligocene-Pleistocene aquifer), 25, 30 (penetrating the fractured Eocene limestone aquifer), 16 and 18 (penetrating the Quaternary aquifer) respectively. From this section some important features can be concluded as follows:

1. Faults are common along the whole section causing a pronounced fault rising for the fractured limestone aquifer in the middle part of the section (well 30). This rising brought the fractured aquifer in juxtaposition with the Quaternary aquifer from the eastern side and the Oligocene-Pleistocene aquifer from the other side. This situation favours the hydraulic interconnection between these aquifers forming one hydraulic continuous system. On the other hand, the comparison between the water levels in wells 16 and 18 of the Quaternary aquifer with those

of the Oligocene-Pleistocene aquifer shows slight difference of about 8.33 m higher at the east than at west indicating slow motion of groundwater from east to west.

2. The depth to the Eocene fractured limestone ranges between 98 m at well No. 30 and 199 m at well No. 20. Through the present study, the base of this aquifer was not reached; accordingly, its thickness is not recorded.
3. In the western part of the cross-section, the Oligocene-Pleistocene aquifer is fully penetrated through the wells Nos. 23, 21 and 20, which is mainly composed of sand and gravel with few clay intercalations having a saturated thickness ranges between 95 m at well No. 23 and 113.5 m at well No. 20.
4. In the study area, the groundwater in the Oligocene-Pleistocene and the fractured Eocene limestone aquifers are generally under phreatic conditions.

Groundwater Movement

The groundwater movement in the study area is discussed through the construction of water level contour maps for the studied aquifers (Figs. 7 and 8). The water level ranges between 44.33 m above mean sea level at the eastern part of the study area (Quaternary) and 36 m above mean sea level at the western part (Oligo-Pleistocene) with general trend of movement from east to west.

The intervals and the spacing between the contour values show different gradient values ranging from 0.3 m/km at the east where the groundwater flow is through the granular aquifers to 0.47 m/km in the fractured limestone aquifer. The groundwater flow through that aquifer is faster than the flow in granular media due to the width of existing fractures (Non Darcyan flow). This could be attributed to variation in fractures orientation where they can form a well-connected network for fluid flow [17].

Aquifer Parameters

The determination of the hydraulic parameters of the aquifers is essentially in evaluation of groundwater resource in the study area. Better and more reliable data are obtained if pumping continues until steady or pseudo-steady flow has been attained.

Long Duration Pumping and Recovery Tests

To evaluate the groundwater of the different three aquifers, 15 constant discharge pumping and recovery tests were carried out including; 3 in the Quaternary, 4 in

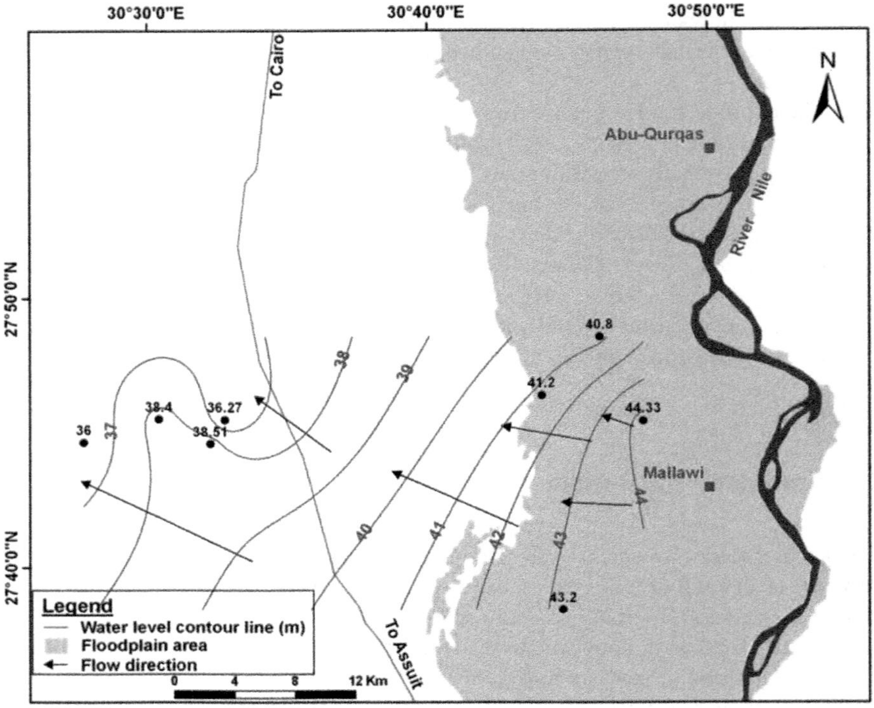

Fig. 7 Hydrogeological cross section B-B′ showing the vertical and horizontal distribution of the lithological succession, the design of the wells, the piezometric level of the surface of the groundwater and the different inferred faultsWater-level contour map of the Quaternary and Oligocene-Pleistocene aquifer showing the water level with respect to mean sea level

the Oligocene-Pleistocene and 8 in the Eocene fractured limestone aquifers. The analyzing of the pumping tests for both the Quaternary and Oligocene-Pleistocene aquifers (porous aquifers) was carried out according to [18]. On the other hand, the transmissivity of the Eocene fractured limestone can be approximated estimated from the specific capacity according to the following equation [19]:

$$T = C\,(Q/s)\quad [19]$$

where,

T is the transmissivity m^2/day.

Q/s is the specific capacity m^3/day/m.

C is a constant value and is found to vary from 0.9 to 1.5 with an average of 1.2 [20] or 1.22 [21].

The results of these tests (Table 2) reveal the followings:

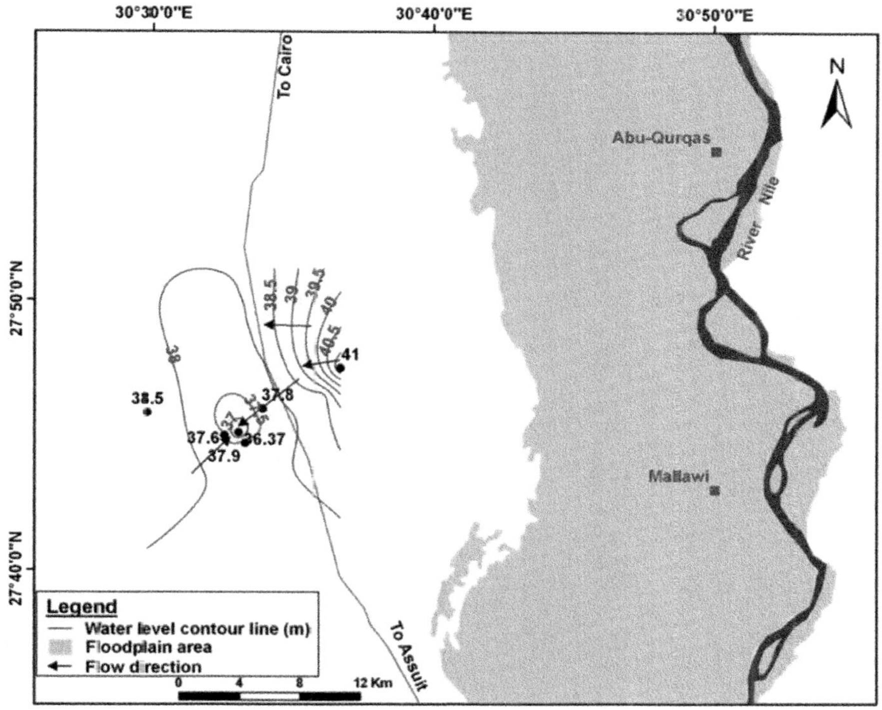

Fig. 8 Water-level contour map of the Eocene fractured limestone aquifer showing the water level with respect to mean sea level

- The Quaternary aquifer shows high values of transmissivity range between 6592 and 12,700 m^2/day with an average value of 8906 m^2/day, while the Oligocene-Pleistocene aquifer has transmissivity values ranging between 1256 and 6800 m^2/day with an average value of 3291 m^2/day. The approximated transmissivity values for the Eocene fractured aquifer range between 1083 m^2/day and 14,054 m^2/day with an average value of 6091 m^2/day.
- In the Quaternary aquifer, the specific capacity values (Q/s) range between 11.8 and 45 m^3/h/m and according to well productivity classification [22], the Quaternary aquifer can be classified into highly productive (Q/s more than 18 m^3/h/m) and moderate productive (Q/s from 1.8–18 m^3/h/m) while all the Oligocene-Pleistocene aquifer have moderate productivity. In the fractured limestone aquifer, the specific capacity (Q/s) ranges between 37 m^3/h/m (Well No. 31) and 480 m^3/h/m (Well No. 24), reflecting the high productivity of the aquifer [22].
- Drawdowns in the wells tapping the granular aquifers (Quaternary and Oligocene-Pleistocene) are always greater than drawdowns in the wells tapping the fractured rock aquifer (Eocene fractured limestone) where the former type achieving final drawdowns ranges between 14.34 m in well No. 22 and 17.28 m in well No. 21 with

Table 2 Analysis of long duration pumping test data for selected wells tapping different aquifers

Final No	Aquifer	Aquifer type	Q (m^3/h)	Drawdown S (m)	Specific capacity ($m^3/h/m$)	b (m)	K (m/day)	T (m^2/day)	Well productivity classification [15]
17	Quaternary	Semi-conf	195	16.57	11.8	Partial penetration	Undefined	6592	Moderately productive
18	Quaternary	Semi-conf	338	18.27	18.5	Partial penetration	Undefined	7427	Highly productive
19	Quaternary	Semi-conf	300	6.52	46.0	Partial penetration	Undefined	12,700	Highly productive
20	Oligo-Pleist	Unconfined	93	5.86	15.9	113.5	59.9	6800	Moderately productive
21	Oligo-Pleist	Unconfined	122	17.28	7.0	102.4	30.8	3154	Moderately productive
22	Oligo-Pleist	Unconfined	120	14.34	8.4	91.5	21.3	1953	Moderately productive
23	Oligo-Pleist	Unconfined	120	17.23	6.9	95	13.22	1256	Moderately productive
24	Eocene	Unconfined	120	0.25	480.0	Partial penetration	Undefined	14,054	Highly productive
25	Eocene	Unconfined	125	0.63	198.0	Partial penetration	Undefined	5797	Highly productive
26	Eocene	Unconfined	125	0.5	250.0	Partial penetration	Undefined	7320	Highly productive
27	Eocene	Unconfined	125	0.52	240.4	Partial penetration	Undefined	7039	Highly productive
28	Eocene	Unconfined	125	1.5	83.3	Partial penetration	Undefined	2439	Highly productive
29	Eocene	Unconfined	122	0.9	135.6	Partial penetration	Undefined	3970	Highly productive
30	Eocene	Unconfined	120	0.5	240.0	Partial penetration	Undefined	7027	Highly productive
31	Eocene	Unconfined	110	2.97	37.0	Partial penetration	Undefined	1083	Highly productive

pumping rate approximately of 120 m^3/h while in the latter, representing fractured type, the drawdowns values range between 0.25 m in well No. 24 and 2.97 m in well No. 31 with approximately the same rate of pumping (120 m^3/h). The low values of drawdowns in the Eocene fractured aquifer is attributed to the large volumes of water that move very rapidly through the karst features that develop in rocks where groundwater has widened fractures and porous zones into solution cavities by dissolving soluble minerals. References [23–26] concluded that, secondary porosity features such as conduits and caves are formed by dissolution and act as highly permeable pathways for water to flow through karst aquifers. The presence of these solution cavities are observed during the well drilling process in the Eocene fractured limestone aquifer where complete loss is occurred at different depths in the wells. Generally, this widening is limited in the carbonate rocks such as limestone and dolomite.

- For the wells tapping the granular aquifers, unsteady-state flow occurs from the moment pumping starts until steady-state flow is reached where the flow is considered to be unsteady as long as the changes in water level in the well is measurable. In these aquifers the time taken to reach the steady-state varies in the tested wells and ranges between 1000 min in well No. 18 and 1260 min in well No. 17 from starting well pumpage in the Quaternary aquifer while in the Oligocene-Pleistocene aquifer, this time ranges between 160 min in well No. 22 and 720 min in well No. 21.
- In the Eocene fractured limestone wells, flow attains its steady-state after few minutes (2–6 min) from starting well pumpage and the aquifer shows a pseudo-steady-state where the changes in the water level in the wells have become so small with time which can be neglected and this indicates that the recharge boundary is able to provide enough water to compensate for discharge.
- The difference between the complete recovery times for the wells tapping the granular aquifers (Quaternary and Oligocene-Pleistocene) and that for the fractured aquifer (Eocene fractured limestone) is attributed to the nature of the karst aquifers which have multiple types of porosity and two flow types of water: laminar and turbulent flow. The laminar flow occurs within the matrix domain of the karst aquifer with slow velocities and the turbulent flow occurs in the large diameter void spaces that are well connected forming conduit domain with high permeability and fast water movement or may be attributed to a direct connection between the Nile and the karst Eocene aquifer through step faulting.

Well Performance Test (Step Drawdown Test)

A number of seven step-tests and one step-test were carried out for the porous sands & gravels and fractured limestone aquifer, respectively. Three rates of well discharge were applied during the conducting of these tests. The parameters of aquifer loss, well loss, and well efficiency (Table 3) were calculated through the application of the **GWW Software** (Ver. 1.10).

Table 3 Calculated parameters of analyzing step-tests data for the porous and fractured limestone aquifers in the study area

Well No	Aquifer type	Maximum rate of pumping (Q_{max}) m^3/h	Total drawdown (S_t) corresponding to Q_{max} M	Aquifer loss S_a %	Well loss S_b %	$A * 10^{-2}$ day/m^2	$B * 10^{-6}$ day^2/m^5
17	Porous	190	15.56	63.8	36.2	0.2176	0.2786
18		300	16.45	75.9	24.1	0.1733	0.0781
19		300	6.49	70.6	29.4	0.0636	0.0366
20		90	5.74	67.9	32.1	0.1805	0.3960
21		121	16.77	85.7	14.3	0.4950	0.2825
22		120	14.11	72.4	27.6	0.3545	0.4711
23		120	17.26	82	18	0.4918	0.3753
31	Fractured limestone	110	2.86	8.4	91.6	0.0093	0.3782

The total drawdown (s_t) in the 8 pumped wells consist of the drawdown (s_a) in the aquifer and the drawdown (s_b) that occurs through the moving of water from the aquifer into the well and up the well bore to the pump intake. The total drawdown (s_t) can be defined in the following equation:

$$s_t = s_a + s_b$$

where s_a is the drawdown in the aquifer at the effective radius of the pumping well, s_b is well loss.

From Table 3, it can be concluded that, the aquifer loss percentage (S_a%) in the total drawdown (S_t) ranges between 63.8% in well No. 17 and 85.7% in well No. 21 which are representing the granular aquifers (Quaternary and Oligocene-Pleistocene). On the other hand, it is 8.4% in the well No. 31 which represents the Eocene fractured limestone aquifer and its well loss percentage (S_b%) is 91.6%. The low percentage of the aquifer loss in the well tapping the fractured aquifer could be attributed to the dominance of large diameter void spaces reflecting high permeability and subsequent fast water flow through aquifer.

Hydrochemical Characteristics of Different Aquifers

To study the variation in groundwater quality of the different dominant aquifers and its relationship with surface water, 25 samples were collected (2010) to represent the different aquifers besides 4 surface water samples from Nile River, canals and drains.

Groundwater Salinity

The water salinity of the three aquifers (three parts of one hydrogeologic system) is less than 1000 ppm (the maximum recorded salinity is 845 ppm of well 26 (Middle Eocene aquifer)). This means that groundwater of the study area (in the three aquifers) is fresh to fairly fresh water (Figs. 9 and 10).

Variation in salinity may be attributed to several factors, among them; local change of lithology, occurrence of clay adjacent to the well site, depth of the well (where salinity increases by depth) and arid conditions.

Sodium/Chloride ratio (rNa^+/rCl^-) indicates that, for surface water samples, the ratios are more than unity and range between 1.37 and 2.31.

The comparison of the rNa^+/rCl^- ratio of groundwater of all the dominant aquifers in the study area with those of River Nile and sea waters (>2 and <1, respectively) shows that, the sources of recharge for the Quaternary aquifer, Oligocene-Pleistocene and Eocene limestone aquifers are directly from the Nile water and the seepage from irrigation canals and drains. Also, the percolation of the excess irrigation water may contribute significantly in recharging the aquifers.

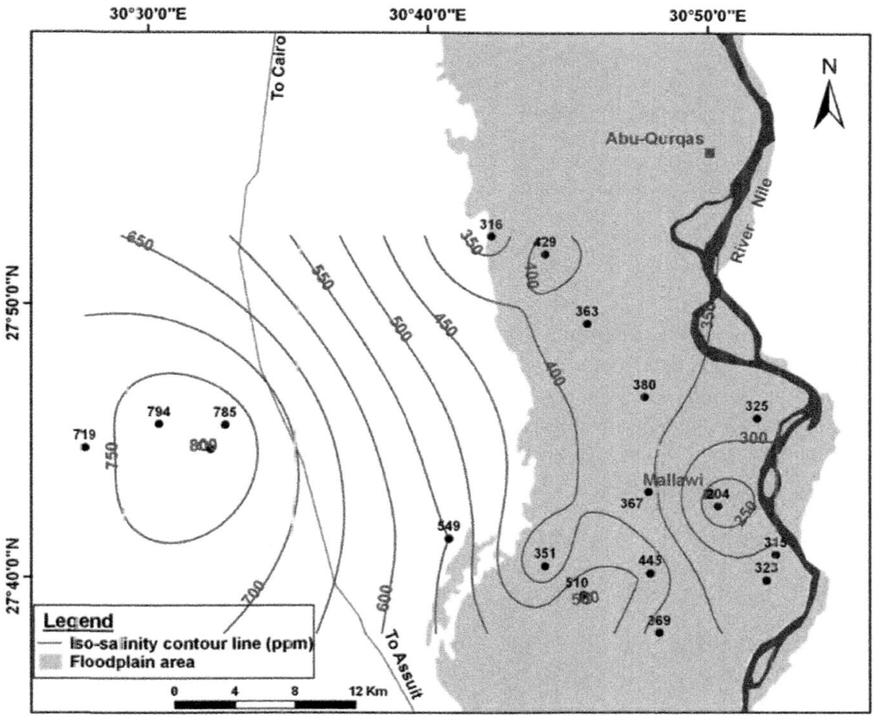

Fig. 9 Iso-salinity contour map for the Nile aquifer system (Quaternary and Oligocene-Pleistocene aquifers) showing the water salinity in ppm at each well and its areal distribution

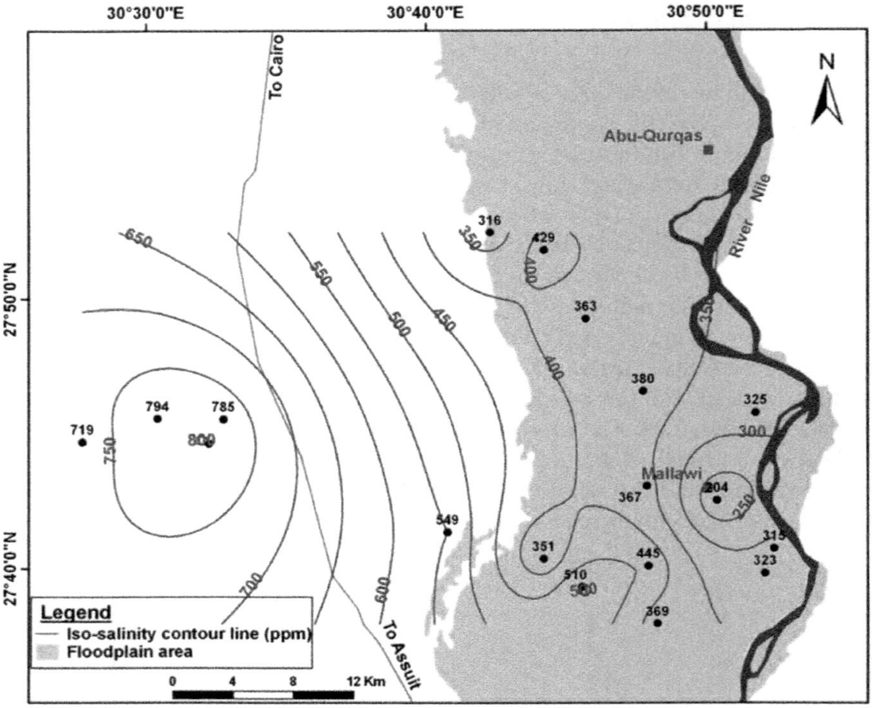

Fig. 10 Iso-salinity contour map for the Eocene fractured limestone aquifer system showing the water salinity in ppm at each well and its areal distribution

The decrease in the rNa^+/rCl^- ratio of the Oligocene-Pleistocene (0.88–1.04) and Eocene limestone (1–1.43) aquifers than the Quaternary aquifer (1.17–2.28) may be attributed to possible contamination, change of facies, depth of wells and hardly to consider it to marine effects.

The Quaternary aquifer shows one assemblage of hypothetical salts for all samples: NaCl, Na_2SO_4, $NaHCO_3$, $Mg(HCO_3)_2$ and $Ca(HCO_3)_2$.

For the Oligocene-Pleistocene aquifer, two assemblages of hypothetical salts combinations are recognized:

- NaCl, Na_2SO_4, $MgSO_4$, $Mg(HCO_3)_2$ and $Ca(HCO_3)_2$ in wells Nos. 20, 21 and 22.
- NaCl, $MgCl_2$, $MgSO_4$, $Mg(HCO_3)_2$ and $Ca(HCO_3)_2$ in well No. 23.

The Eocene fractured limestone aquifer shows two assemblages of hypothetical salts combination as follows:

- NaCl, Na_2SO_4, $NaHCO_3$, $Mg(HCO_3)_2$ and $Ca(HCO_3)_2$ in wells Nos. 30 and 31.
- NaCl, Na_2SO_4, $MgSO_4$, $Mg(HCO_3)_2$ and $Ca(HCO_3)_2$ in wells Nos. 24, 25, 26, 28 and 29.

From the above mentioned assemblages of the hypothetical salts combinations for the different aquifers, it can be concluded that:

1. The Quaternary aquifer has chemical composition similar to the chemical composition of the Nile River water where the hypothetical salts of the Nile water is NaCl, Na_2SO_4, $NaHCO_3$, $Mg(HCO_3)_2$, and $Ca(HCO_3)_2$ which indicate that the Nile water is the main recharging source for this aquifer.
2. It is obvious that the two assemblages of the Eocene fractured limestone aquifer are represented in the Quaternary and Oligocene-Pleistocene aquifers which mean that the Eocene water is a mixed water of these two aquifers i.e. the Eocene fractured limestone aquifer is recharged from the Quaternary aquifer through their direct connection and then recharges the Oligocene-Pleistocene aquifer as illustrated by the hydrogeological section B-B′ (Fig. 6).
3. From the studied 25 groundwater samples of the three aquifers, only sample of well 23 (Oligocene-Pleistocene aquifer) which reflects a marine conditions (due to the occurrence of $MgCl_2$). This may be attributed to a limited local condition around this well.

Groundwater Types

The Piper diagram [27] has been used to illustrate the possible relationship among the investigated aquifers and the surface water. The plotted data on the diamond field shape (Fig. 11) could be differentiated into two groups:

1. The first group occupies the lower part of the diamond shape, where water is characterized by sodium- bicarbonate and calcium/magnesium- bicarbonate water types. It includes the surface water of the Nile River together with almost all samples of the Quaternary aquifer and some samples of the Eocene fractured limestone aquifer (25, 28, 29, 30 and 31).
2. The second group which represents all the Oligocene-Pleistocene aquifer water samples and some samples of the Eocene aquifer (24 and 26) occupies the upper part of the diamond shape. Its chemical properties are dominantly characterized by sodium-chloride water type.

Conclusions

This chapter is mainly focused on the following main approaches related to the potentials and relationships of the dominant aquifers in the study area. From the abovementioned discussion, the followings are the main conclusions:

1. The dominated aquifers in the study area fall into two broad categories: The unconsolidated aquifers (granular) represented by Quaternary and Oligocene-Pleistocene. The consolidated fractured aquifer (fractured rock) represented by Eocene fractured limestone which considered as karst aquifer.

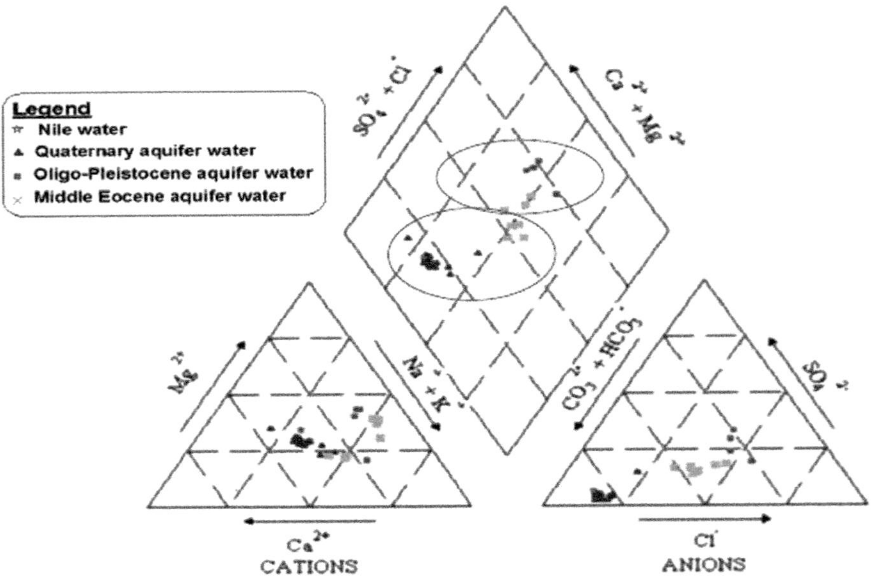

Fig.11 Piper's diagram showing the plotted data on the diamond field shape, where the water of the first group occupies the lower part of the diamond shape characterized by sodium-bicarbonate and calcium/magnesium-bicarbonate water types representing the surface water of the Nile River together with almost all samples of the Quaternary aquifer and some samples of the Eocene fractured limestone aquifer and the water of the second group occupies the upper part of the diamond shape characterized by sodium-chloride water type representing the all samples of the Oligocene-Pleistocene aquifer and some samples of the Eocene fractured limestone aquifer

2. The Quaternary aquifer in the flood plain is considered as highly productive where the aquifer has transmissivity values ranging between 6592 and 12,700 m^2/day and specific capacity ranging between 11.8 and 46.0 m^3/h/m. The water salinity of this aquifer ranges between 203.5 and 549.4 ppm increasing from east to west. The main source of recharge of this aquifer is the Nile water as indicated by the ion relationships.
3. The Oligocene-Pleistocene aquifer is classified as moderately productive having transmissivity values ranging between 1256 and 6800 m^2/day and hydraulic conductivity ranges between 13.22 and 59.9 m/day with saturated thickness ranges between 91.5 and 113.5 m. This aquifer has specific capacity values between 6.9 and 15.9 m^3/h/m. The water salinity of this aquifer ranges between 719 and 801 ppm. rNa^+/rCl^- ratio, hypothetical salts combinations, Piper (1944) and Schoeller (1955) diagrams show that the groundwater of this aquifer is in the progress stage of evolution than the other two aquifers (Quaternary and Eocene fractured) where it has sodium- chloride chemical type.
4. For the Eocene fractured limestone aquifer, the aquifer is considered as highly productive aquifer having transmissivity values range between 1083 m^2/day and 14,054 m^2/day and attaining specific capacity ranges between 37.0 and 480

$m^3/h/m$. Its water salinity ranges between 462 and 845 ppm. The groundwater of this aquifer is, chemically, in the middle stage of evolution as its groundwater is a mixture from the other two aquifers (Quaternary and the Oligocene-Pleistocene) due to the direct hydraulic connection between the Quaternary aquifer and the Eocene fractured limestone aquifer through faulting. The conclusion of this situation is that, the Eocene fractured limestone aquifer is recharged directly from the Quaternary aquifer and then acts as a rechargeable source for the Oligocene-Pleistocene aquifer.

5. The general flow direction of the granular aquifers (Quaternary and Oligocene-Pleistocene) is from east to the west as the Eocene fractured aquifer with some local change where there is a flow direction from southwest to the north east direction which may be attributed to the over-exploitation in this locality or the effect of the fractures orientation in this aquifer system.
6. The groundwater salinity in the three dominant aquifers increases generally in the west direction since the groundwater movement is towards west.
7. A direct connection between the Nile and the karst Eocene aquifer through step faulting is quite believable as indicated from the pumping test data carried out in the present study. Such important conclusion needs to be verified and emphasized by geological and geophysical surveys to establish a new approach about the hydro-structural aspects affect groundwater occurrence and potentials along the Nile valley.

Recommendations

From the above mentioned discussions, the followings can be recommended:

1. Generally, the whole area should be subjected to a monitoring network of wells representing the three aquifers. The monitoring system includes both water levels and salinity on periodical basis.
2. It is recommended to increase the drilling depth in the Eocene aquifer in order to maximize, as much as possible, the groundwater potentials of this karstic phenomenon.
3. Since the study area is a new development area, it is strongly recommended that a mathematical model be built in order to forecast future conditions based on anticipated heavy pumping programs by investors. The management should take into consideration the safe distance between drilled wells as well as the pumping rates.
4. The safe distance between drilled wells, as well as the pumping speeds, should be considered by the management.
5. Since the fractured limestone (Eocene) has good groundwater quality and availability, water extraction should be based on this aquifer, which would directly improve the reclamation of new areas.
6. It is recommended that a modern irrigation system be used.

7. To make the necessary management on groundwater extraction in the study area, cooperation between the Ministry of Irrigation & Water Resources and the Ministry of Agriculture and Land Reclamation is critical.

References

1. Korany E (1980) Peak runoff calculations and preventing the risk of occasional flooding in Sannur drainage basin, Estern Desert, Beni Suef Governorate, Egypt. 5th Intern Congr Statist Comput Sci, Cairo 505–534
2. Korany E, Sakr S, Darwish M, Morsy S (2008) Hydrogeologic modeling for the assessment of continuous rise of groundwater levels in the quaternary aquifer, Nile Valley, Egypt: case study. In: International conference of Geological Arab World (GAW8), Cairo University, pp. 703–711
3. Shabana AR (2010) Hydrogiological studies on the area West Deir Mouas-Mallawi. El Minia governorate-Egypt. Egypt J Geol 54:61–78
4. CONOCO Coral Egypt (1987) Geologic map of Egypt (Scale 1: 500000). General Petroleum Company, Cairo
5. El Boukhary MA, Abdel-Malik W (1983) Revision of the stratigraphy of the Eocene deposits in Egypt, N Jb Geol Palaont Mh. Stuttgart 6:321–337
6. David Banks and Nick Robins (2002) An introduction to Groundwater in Crystalline Bedrock, Published by Norges geologiske undersøkelse (Geological Survey of Norway) N-7491 Trondheim. Norway
7. Youssef MI (1968) Structural pattern of Egypt and its interpretation. Bull Am Petrol Geol 52(4):601–614
8. Sadek M (2001) Istopic criteria for upward leakage in the alluvial aquifer in north El Minia district, Egypt. Ann Geol Surv Egypt XXIV:585–596
9. E Korany 1984 Statstical approach in the assessment of the geohydrologic profiles. In: 9th international congress. Statistics, computer. sciences. Social and demerges research. Ain Shams University Press, Cairo, pp. 161–176
10. Abdel Moneim AA, Fernández-Álvarez JP, Abu El-Ella EM, Masoud AM (2016) Groundwater management at west El-Minya desert area, Egypt using numerical molding. J Geosci Environ Prot 4:66–76
11. El-Deeb H, El Rawy M, Habib E (2015) Water resources management: case study of El-Minya Governorate. Egypt Int J Sci Eng Res 6(6):48–55
12. Ibrahim RGM, Lyons WB (2017) Assessment of the hydrogeochemical processes affecting groundwater quality in the Eocene limestone aquifer at the desert fringes of El-Minya governorate. Egypt Aquat Geochem 23:33–52
13. Ibrahim SMM (2013) Hydrological evaluation of the tertiary-quaternary aquifer system west Mallawi, Upper Egypt: case study. Egypt J Geol 57:1–30
14. Salem AAA (2015) Hydrogeological studies on the shallow aquifer in the area west Samalot, El-Minya governorate. Egypt Egypt J Pure Appl Sci 53(4):49–60
15. Yousif M, Sabet H, Ghoubachi S, Aziz A (2018) Utilizing the geological data and remote sensing applications for investigation of groundwater occurrences. West El Minia, West Desert Egypt, NRIAG J Astron Geophys 7(2):318–333
16. Ibrahem SMM, Elalfy M, Hagras MA (2020) Groundwater potentials of Eocene limestone aquifer in West-West El-Minya area. Egypt Egypt J Desert Res 70(1):59–82
17. Martel SJ (1999) Analysis of fracture orientation data from boreholes. Environ Eng Geosci 2:213–233
18. Jacob CE (1950) Flow of groundwater in engineering hydraulics. John Willey and Sons, New York

19. Verbovsek T (2008) Estimation of transmissivity and hydraulic conductivity from specific capacity and specific capacity index in Dolomite aquifers. J Hydrol Eng 13(9)
20. Jalludin M, Razack M (2004) Assessment of hydraulic properties of sedimentary and volcanic aquifer systems under arid conditions in the Republic of Djibouti (Horn of Africa). Hydrogeol J 12(2):159–170
21. Misstear BDR (2001) The value of simple equilibrium approximations for analyzing pumping test data. Hydrogeol J 9(2):125–126
22. Sen Z (1995) Applied hydrogeology for scientists and engineers. C.R.C. Press, Inco, Florida, U.S.A, Lewis Publishers, p 310p
23. White WB (1977) Role of solution kinetics in the development of karst aquifers. In: Tolson JS, Doyle FL (eds) Karst hydrogeology, congress of the international association of hydrogeologists, 12th, Memoirs, 503–517
24. White WB (1988) Geomorphology and hydrology of karst terrains. Oxford University Press, New York
25. Dreybrodt W (1990) The role of dissolution kinetics in the development of karst aquifers in limestone—a model simulation of karst evolution. J Geol 98(5):639–655
26. Palmer A (1991) Origin and morphology of limestone caves. Geol Soc Am Bull 103(1):1–21
27. Piper AM (1944) A graphic representation in the geochemical interpretation of groundwater analyses. Am Geophys Union Trans 25:914–923

Groundwater Characterization and Quality Assessment in Nubian Sandstone Aquifer, Kharga Oasis, Egypt

Mustafa El-Rawy, Fathy Abdalla, and Abdelazim M. Negm

Abstract The present study integrates groundwater hydrochemistry to assess groundwater quality under expanding agricultural activities in Kharga Oasis, Egypt using field data, GIS technique and the multivariate statistical analysis (Factor analysis). A total of 144 groundwater samples were collected and analyzed for major ions and trace elements and the spatial distribution of physicochemical parameters was visualized with GIS. The result has been used to evaluate the groundwater quality in the area for drinking and irrigation purposes comparing with the World Health Organization WHO and Egyptian water standards (EHCW). Results revealed that 93.8% of the groundwater samples are freshwater with TDS < 1000 mg/l, whereas the rest 6.2% are slightly saline (TDS 1000–3000 mg/l). 35.4% of the groundwater samples are unsuitable for domestic purposes due to the high level of hardness. The calculated water quality index WQI indicates that 88.2% of the samples are excellent for drinking use. Salinity hazard, sodium percent (Na %), sodium adsorption ratio (SAR), residual sodium carbonate (RSC), magnesium hazard (MH), Kelly's ratio (KR), and soluble sodium percentage (SSP) were calculated to evaluate groundwater quality for irrigation purposes. According to the SSP values, 46% of the samples are good quality water for irrigation purpose, 71% of the samples are good to permissible for irrigation based on Na%, 79.2% of the samples are good for irrigation based on PI values and 91.1% of the samples are suitable for irrigation based on MH values.

M. El-Rawy (✉)
Faculty of Engineering, Civil Engineering Department, Minia University, Minia 61111, Egypt
e-mail: mustafa.elrawy@su.edu.sa; mustafa.elrawy@mu.edu.eg

Civil Engineering Department, College of Engineering, Shaqra University, Ar Riyadh 11911, Dawadmi, Saudi Arabia

F. Abdalla
Deanship of Scientific Research, King Saud University, Riyadh, Saudi Arabia
e-mail: fathy.abdallah@sci.svu.edu.eg

Faculty of Science, Geology Department, South Valley University, Qena, Egypt

A. M. Negm
Faculty of Engineering, Water and Water Structures, Engineering Department, Zagazig University, Zagazig 44519, Egypt
e-mail: amnegm@zu.edu.eg

A. Negm and A. Elkhouly (eds.), *Groundwater in Egypt's Deserts*, Springer Water,
https://doi.org/10.1007/978-3-030-77622-0_8

Based on the above results, most of the collected samples in the Kharga Oasis are unsuitable for drinking and agriculture under ordinary conditions without treatment while it is suitable for many industrial purposes. The factor analysis produced three factors which described 83.13% of the total variance. The results show that the concentrations of pH and HCO_3^- do not contribute to other chemical parameters. This study demonstrated that the hydrochemical analysis and GIS with statistical factor loading could deliver an influential tool to distinguish factors controlling the groundwater quality in Kharga Oasis area.

Keywords Groundwater quality assessment · Hydrochemistry · Nubian sandstone aquifer · Kharga Oasis · GIS · Factor analysis

Introduction

Worldwide, groundwater becomes very important where many people rely upon groundwater for domestic, industrial, and agricultural purposes, especially in desert areas, where there are no adequate surface water supplies. Groundwater provides about 50% of the potable water supplies, 30% of the industrial water, and 20% of the irrigation water worldwide. Some desert countries like Saudi Arabia, groundwater contributes to nearly 79% of the total water supply [1]. Groundwater contamination decreases the amount of available groundwater, chemical and biological contamination of water was behind 80% of the diseases and deaths in the developing countries [2]. According to the World Health Organization (WHO) more than 1×10^9 people lack access to safe drinking water and 4000 children die every day from waterborne disease [3]. Moreover, plants growth and metabolism and soil fertility may adversely affect by using irrigation water with a high level of dissolved ions [4]. In Egypt and in many areas in the Middle East, groundwater is facing increasing environmental problems, especially superficial aquifers due to anthropogenic activities [5]. Many factors may cause groundwater quality deterioration such as rock–water interaction, urbanization, population growth, water-borne pathogens, agriculture runoff, and the anticipated climate change [6]. Furthermore, groundwater quality is adversely affected by geogenic sources or percolating water interaction with aquifer media. Accordingly, it is important to know the characteristics of the dissolved constituents in groundwater [7]. Raju et al. [8] mentioned that overexploitation and excessive pumping have detrimentally affected groundwater availability in the form of quality and quantity. Groundwater suitability for domestic and irrigation purposes was determined based on several chemical and bacteriological parameters, where parameters exceeded the maximum permissible levels of WHO have many adverse health effects [9].

This chapter evaluates the groundwater quality in Kharga Oasis area using hydrochemistry, GIS, and Factor analysis methods to determine the factors controlling groundwater quality in the study area. The obtained data were interpreted according to the WHO standard and the Egyptian water standards (EHCW). Geographic information system (GIS) is an important tool for plotting. In this study, GIS may help as a database system to generate distributed maps of ions concentration and assessment of groundwater quality assessment [10–16].

Statistical factor analysis has been widely used to study groundwater geochemistry [15–24]. Factor analysis is a multivariate statistical method, used to classify the main factors influencing the groundwater quality and water types in the study area. The factor analysis differentiates between dependent and independent variables. Factor analysis, generating the overall relationship between measured factors by rearranging them in a method that better explains the basic structure [17].

This chapter aims to assess the hydrogeochemical characteristics by identifying major variables affecting the quality of groundwater in the Nubian Sandstone Aquifer (NSSA) in Kharga Oasis and to evaluate the suitability of groundwater for domestic and agricultural purposes. These can be achieved by preparing thematic maps through a GIS for the most important parameters indicating groundwater quality.

The Study Area

The study area, Kharga Oasis, is a depression and situated in southern part of the Western Desert of Egypt, at approximately 200 km west of the Nile (Fig. 1). It covers about 7500 km^2. Its lies between 22°30′14″ and 26°00′00″ N, and between 30°27′00″ and 30°47′00″ E. Kharga Oasis is part of the NSSA. The eastern and western boundaries of Kharga Oasis are bounded by Eocene limestone plateau. The temperature ranges between 22.1 and 45 °C with an average 31.2 °C. August is the warmest month of the year and January is the coldest month. The annual rainfall is 0 mm, and the relative humidity ranges between 27 and 52% with an average of 39%. The average potential evaporation rate is about 18.4 mm/day. Nubian Sandstone Aquifer (NSSA) in Kharga Oasis is formed by predominantly continental sandstone of Mesozoic and Palaeozoic age (pre-Senonian strata). Its main structural components are the Kufra Basin in Libya and the Dakhla Basin in south-western Egypt, with an aquifer thickness system up to 4000 m [25]. CEDARE [26] reported that the total groundwater storage amounts in the NSSA at 1960 is 150 000 km^3, which represents a huge groundwater resource. The NSSA is located in the Eastern Sahara, in south-eastern Libya, north-eastern Chad, northern Sudan and Egypt with an area which covers about 2×10^6 km^2 [27]. The NSSA is therefore a generally closed system, except in the south-east where there is a groundwater discharge from the Blue Nile/Main Nile Rift System [28]. The NISSAN is non-renewable aquifer and the total water storage in the Nubian sandstone aquifer in Egypt was estimated to be around 40,000 BCM [27, 29]. The transmissivity values of the NSSA at Kharga Oasis range from 100–1475 m^2/d, with a geometric average of 400 m^2/d [29].

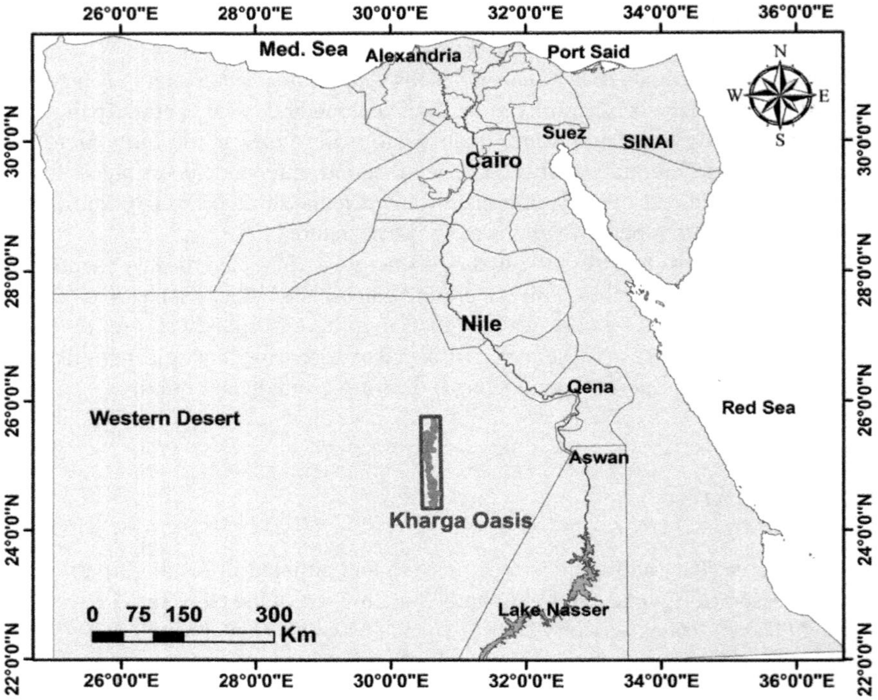

Fig. 1 Location of Kharga Oasis area in Egypt

Material and Methods

Sampling Protocol

The data of groundwater samples have been obtained from General Directorate of Groundwater in Al-Kharga (Ministry of Water Resources and Irrigation (MWRI)). The Samples have been collected from 144 representative wells in Kharga Oasis, in 2016 (Fig. 2) were collected in pre-cleaned polyethylene bottles (1L) and transported to the laboratory in an ice-box for analysis based on the standard American Public Health Association methods [30]. The field surveys provided wells location, water depth, ground elevation, and physicochemical properties of water samples with field observations. Total dissolved solid (TDS), total hardness (TH), electrical conductivity (EC), and (pH) of the collected samples were measured in situ using Eutech digital portable meters.

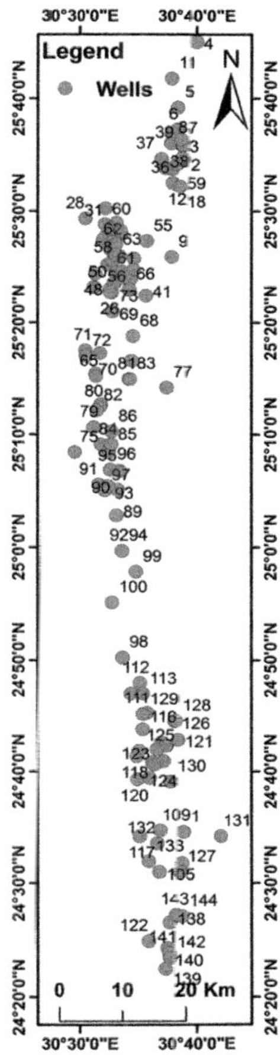

Fig. 2 Location of the wells in Kharga Oasis area

Laboratory Analysis

The collected samples have been analyzed in the laboratory of MWRI in Al-Kharga area to determine major ions including Ca^{2+}, Mg^{2+}, Na^{+}, K^{+}, HCO_3^-, Cl^-, and SO_4^{2-} and trace metals. Calcium (Ca^{2+}), magnesium (Mg^{2+}), carbonate (CO_3^{2-}), bicarbonate (HCO_3^-), and chloride (Cl^-) were examined by volumetric analysis methods, sodium (Na^+) and potassium (K^+) were measured using the flame photometer, and sulfate (SO_4^{2-}) was determined by spectrophotometric technique. Statistical analysis to get the minimum, the maximum, and the average for each parameter was applied and the results were compared with World Health Organization [31] and Egyptian water standards [32]. The spatial distribution maps of groundwater quality parameters have been prepared using ArcGIS. Table 1 shows the standards of water for drinking, according to WHO and EHCW.

Table 1 Statistical analysis of different water quality parameters of the collected groundwater samples of the study area

Parameter	Min	Max	Mean	St. dev	WHO desirable limit (mg/l)	WHO allowable limit (mg/l)	EHCW Egypt limit (mg/l)
pH	3.5	8.6	6.96	0.6	6.5–8.5	8.5	7–8.5
Electrical conductivity (EC) (μS/cm)	276.5	2256.3	794.8	344.2	1500	1500	–
Total dissolved solids (TDS) (mg/l)	177	1444	422.91	215.8	500	1000	500
Calcium (Ca^{2+}) (mg/l)	11	115.1	30.36	18.94	75	75	75
Magnesium (Mg^{2+}) (mg/l)	5.3	59.2	17.53	9.15	30	30	50
Sodium (Na^{+}) (mg/l)	13.6	308.8	58.01	48.8	200	200	200
Potassium (K^{+}) (mg/l)	10.5	65.9	25.72	7.65	10	10	–
Bicarbonate (HCO_3^{-}) (mg/l)	0	300	115.8	59.3	100	100	–
Chloride (Cl^{-}) (mg/l)	34.9	454.8	105.9	71.64	200	200	200
Sulfate ($SO4^{2-}$) (mg/l)	9.5	338.7	77.01	70.45	200	200	400
Iron (Fe^{-}) (mg/l)	0	35.5	2.5	4.34	0.3	1	0.3
Manganese (Mn) (mg/l)	0	3.0	0.2	0.4	0.4	0.4	0.1
Total hardness (TH) (mg/l)	54.9	439.9	147.7	75.95	300	600	500
Sodium adsorption ratio (SAR) (meq/l)	3	36	12	7	–	–	–
Sodium percent (Na %)	38.33	83.60	63.33	8.6	–	–	–

(continued)

Table 1 (continued)

Parameter	Min	Max	Mean	St. dev	WHO desirable limit (mg/l)	WHO allowable limit (mg/l)	EHCW Egypt limit (mg/l)
Permeability index (PI) (meq/l)	27.4	92.6	64.01	13.6	–	–	–
Kelley ratio (KR) (%)	0.3	4.4	1.3	0.8	–	–	–
Magnesium hazard (MH) (%)	11.5	69.5	37.9	9.6	–	–	–
Soluble sodium percentage (SSP) (%)	20.1	81.6	52.1	12.7	–	–	–

Spatial Distribution Analysis

Data integration and sampling locations were digitizing using the spatial analyst tools of ArcGIS 10.3 (Arc Map) [33] to prepare the spatial distribution maps of groundwater quality parameters with the inverse distance weighted (IDW) technique.

Results and Discussions

Major Hydrochemical Parameters

The groundwater samples were analyzed for pH, EC, total hardness (TH), TDS, Cl^-, SO_4^{2-}, NO_3^-, and Fe as represented in Table 1. The results of the hydrochemical analyses showed that groundwater is slightly alkaline reflecting largely various in chemical composition. The plenty of the major ions is as follows: $Na^+ > Ca^{2+} > Ma^{2+} > K^+$ and $Cl^- > SO_4^{2-} > HCO_3^-$ with three major hydrochemical facies (Ca–Cl, Na–Cl, and mixed Ca–Mg–Cl).

Hydrogen Ion (pH)

pH is the expression of acidity, the pH values of the collected samples varied from 3.5 to 8.6 with an average of 7, which is indicative of a little acidic to a little alkaline condition (Table 1).

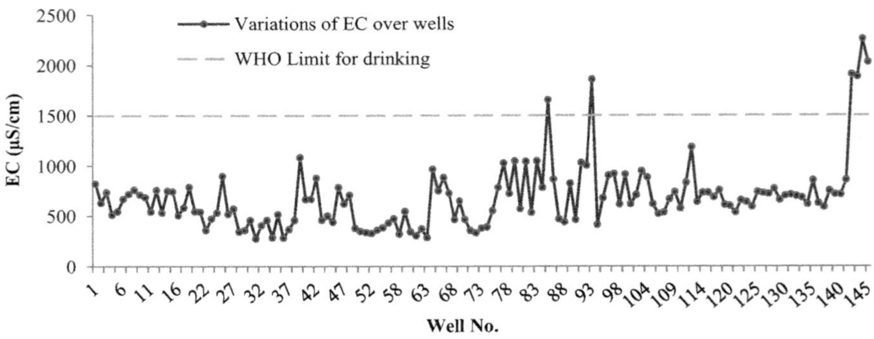

Fig. 3 The variations of EC over wells

Electrical Conductivity (EC) and Total Dissolved Solids (TDS)

The EC values vary from 276.5 to 2256.3 mg/l with an average of 794.8 mg/l (Table 1). The variations of EC over wells reflect that groundwater in the area is largely various in chemical composition (Fig. 3). The large variations in EC may be due to geochemical processes such as ion exchange, evaporation, and rock–water interaction [34]

TDS in water can result from natural sources, urban runoff, livestock waste, industrial wastewater, sewage, and nature of the soil. The spatial distribution of the TDS in the Kharga Oasis area is shown in Fig. 4. In the study area, TDS values range from 177 to 1444 mg/l, with an average of 422.91 ppm (Table 1 and Fig. 4). TDS was under the maximum permissible limit of drinking (500 mg/l) in 74.8% of the groundwater wells stations as per WHO standards (Table 1). According to Hem [35], 93.8% of the groundwater samples are freshwater with TDS less than 1000 mg/l (Table 2) whereas 6.2% are slightly saline (TDS 1000–3000 mg/l).

Total Hardness (TH)

The TH of water depends on the amount of dissolved calcium and magnesium [31]. Hard water causes scales in the boilers, pipes, and other domestic appliances, while soft water is more corrosive and contains more metal contaminants from the water pipes [36]. The TH values in the study area vary from 54.9 to 439.5 mg/l with an average value of 147.7 mg/l (Table 1), which is within the safe limit of 500 mg/l as recommended by [31].

The TH classification (Table 3) [37] indicates that most of the groundwater samples are from the hard to moderate hard class. 35.4% of the groundwater samples are hard to very hard, while 57.6% of the groundwater samples are moderate hard. Only about 6.9% of moderate groundwater samples are soft. The soft water unit having exchange resins may remove the hardness of water.

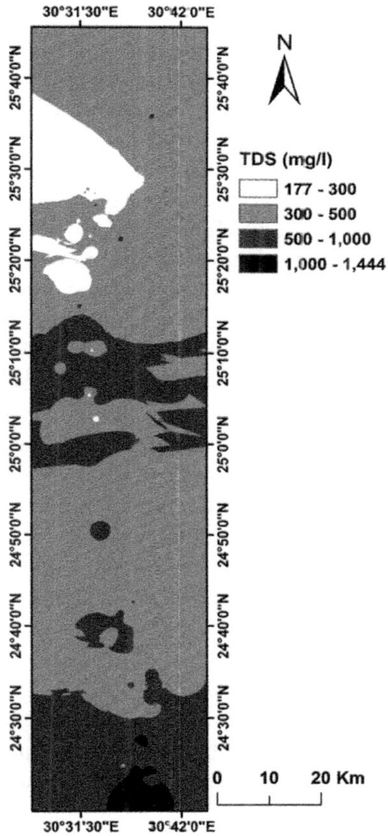

Fig. 4 Total dissolved solids distribution map for the study area

Table 2 Classification of water types according to Hem [35]

Water type	TDS (mg/l)	Number of samples	Percentage of samples (%)
Freshwater	<1,000	135	93.8
Slightly saline	1,000–3,000	9	6.2
Moderately saline	3,000–10,000	–	–
Very saline	10,000–35,000	–	–
Brine	>35,000	–	–

Table 3 Groundwater quality classification based on TH [31]

Total hardness (mg/l)	Classification	Number of samples	Percentage of samples (%)
<75	Soft	10	6.9
75–150	Moderate hard	83	57.6
150–300	Hard	43	29.9
>300	Very hard	8	5.5

Chloride and Sulfate (Cl^- and SO_4^{2-})

Chloride (Cl^-) is the mixture of metal gas chlorine and is a slight component of the Earth's crust but a significant dissolved component of most natural waters. Water contains several sources of chloride, including agricultural runoff, rocks, industrial wastewater, street salting, and wastewater from wastewater treatment plants.

The Cl^- values in the study area ranges from 34.9 to 454.8 mg/l with an average of 105.9 mg/l (Table 1).

Sulfates exist in various minerals and are utilized commercially within the chemical industry. Sulfates released into the water in industrial wastes and through atmospheric deposition. Anyhow, the highest levels of the sulfates ordinarily are in groundwater and are from natural sources [31].

Sulfate particle focuses are most likely gotten from the enduring of sulfate and gypsum-bearing sedimentary rocks [38]. The sulfate concentration is safe for human if the value will not exceed the maximum allowable limit of 400 mg/l. In the study area, the sulfate values are between 9.5 and 338.7 mg/l with an average of 77.01 mg/l, which is within the maximum allowable limit (400 mg/l) in all sampling locations as per WHO specification (Table 1).

These variations may show the breaking of natural substance from topsoil/water, leachable sulfate present in compost, and other human impacts [39].

Trace Elements

Trace metals happen in dissolved form in water, as precipitates, or sorbed onto aquifer materials and they may present a serious threat to water quality due to their toxicity. Trace metals above the detection boundary in the study area include only Fe and Mn, where the others were below the detection boundaries. Iron is a necessary ingredient in the marine ecosystem, making it one of the most abundant ingredients in water and sediments. The association of manganese and iron is well known and belongs to the iron-cobalt group metals. The iron concentration i the groundwater of the study area range from 0.0 to 35.5 with an average of 2.5 mg/l, while manganese concentration ranges from 0.0 to 3 with an average of 0.2 mg/l.

The distribution of the major cations (Ca^{2+}, Mg^{2+} , Na^+, K^+) and major anions (HCO_3^-, Cl^-, SO_4^{2-}) is shown in Figs. 5 and 6, respectively.

Groundwater Quality Evaluation

The quality of groundwater provides a clear description of water usability for various uses. In this research, groundwater quality was assessed for drinking, irrigation and industrial purposes, according to WHO [31] and EHCW [32].

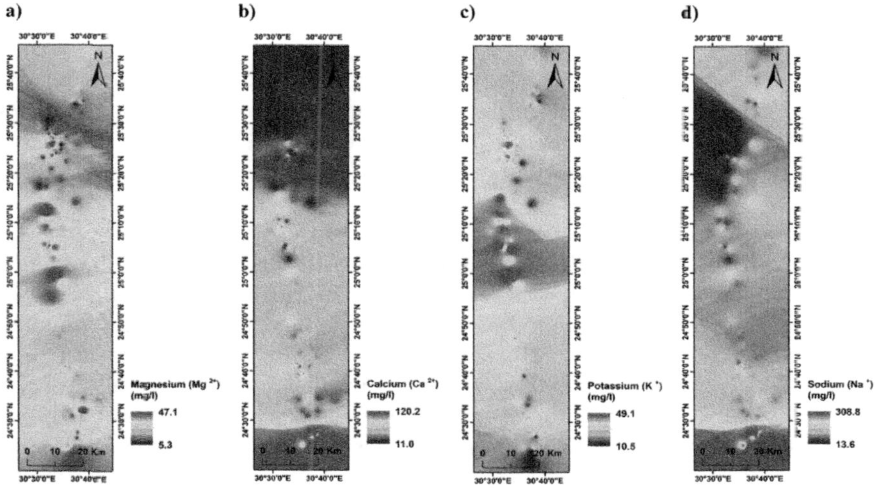

Fig. 5 The distribution of the major cations Mg^{2+}, Ca^{2+}, K^+ and Na^+ in the Kharga Oasis

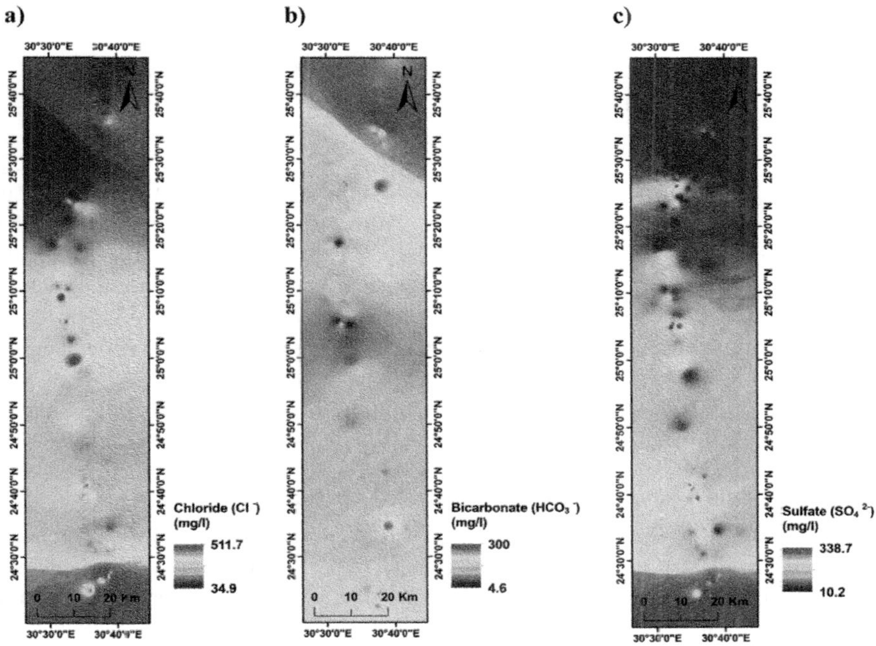

Fig. 6 The distribution of the major anions Cl^-, HCO_3^- and SO_4^{2-} in Kharga Oasis

Domestic and Drinking Uses

The range of hydrochemical parameters of groundwater wells in the study area and their comparison with WHO [31] are represented in Table 1, to assess the groundwater suitability for drinking and domestic uses. The groundwater suitability for drinking and domestic uses was assessed based on different parameters such as pH, EC, TDS, TH, CL, SO_4, NO_3, and Fe.

Chloride concentration ranged from 34.9 to 454.8 mg/l, with an average of 105.9 mg/l (Table 1). The WHO [31] standard of Cl^- acceptable limit in drinking water is 150 mg/l. The chloride concentration in the study area is more than the allowable limit of 150 mg/l in 13 wells. Cl^- is the most dominant anion in most samples, followed by SO_4^- and HCO_3^-. High levels of chloride may result from leaching of soluble salts by irrigation water (e.g., NaCl) precipitated in soil due to high evaporation rates [34]. Sulfate concentrations in the study area vary greatly between 9.5 and 338.7 mg/l with an average of 77.01 mg/l which is less than the maximum allowable limit in all sampling locations with respect to the WHO standards (Table 1). The Fe content in the groundwater samples is between 0.0 and 35.5, with an average of 2.5 (Table 1). Fe content of 81.1% of the wells exceeds the standard value of 0.5 mg/l for drinking, and 18.1% of samples exceed the limitation of 3 mg/l for drinking. The manganese content in groundwater samples varies from 0.0 to 3, with an average of 0.2 mg/l (Table 1). Mn contents exceeded the limitation of drinking water in 31.7% of the groundwater samples.

Evaluating groundwater quality index (WQI)

The water quality index (WQI) was calculated according to [16, 40–42]. The WQI is described as a score that reflects the effect of various water quality parameters on the composite [43]. The Water Quality Index (WQI) established by Saeedi et al. [44], is a mathematical tool used to convert big amounts of water quality information into a single number representing the water quality level. Also, it is defined by Yidana et al. [45] as a parameter assigning a weight to the sampling points based on the water's physicochemical parameters and/or biological constituent's levels.

In this study, WQI calculated in three steps. In the first step, the following 11 physicochemical parameters (n = 10) namely, pH, TDS, F^-, Cl^-, SO_4^-, HCO_3^-, Ca, Mg, Na, and K have been given a weight (w_i) according to its importance in the overall groundwater quality for drinking purposes. The maximum weight of 5 has been given to TDS, Na, F^-, Cl^-, and SO_4^-. The maximum weight of 1 has been given to HCO_3^-, as it plays an insignificant role in the water quality assessment and other parameters, such as pH, Ca, Mg, and K were given weights between 1 and 5 according to their importance in water quality calculations [46, 47]. The second step is the calculation of the relative weight (W_i) using the following equation:

$$W_i = w_i / \sum_{i=1}^{n} w_i \quad (1)$$

where W_i is the relative weight, w_i is the weight of each parameter, and n is the number of parameters. The third step is a quality classification scale (q_i) for each parameter. It can be calculated by the following equation [46]

$$q_i = \left(\frac{Ci}{Si}\right) * 100 \tag{2}$$

where q_i is the quality classification, Ci is the concentration of each chemical parameter in each water sample in mg/l. For computing the WQI, the Si is first calculated for each chemical parameter, which is then used to calculate the WQI as per the following equation [47]:

$$SI_i = W_i * q_i \tag{3}$$

$$WQI = \sum SI_i \tag{4}$$

where SI_i is the sub-index of ith parameter, q_i is the rating based on the concentration of the ith parameter, and n is the number of parameters. If WQI equals 100, the water is saturated regarding the measured parameters, with respect to their standards. If WQI < 100, the water is suitable for drinking and domestic purposes. If WQI > 100 means that, the water is polluted and it is not suitable for drinking and domestic purposes.

The WQI in the study area varies from 16.97 to 113.71, with an average of 37.28. For all the collected groundwater samples, the WQI percentages classifications are excellent (88.2%), good (10.4%), and poor (1.4%) for domestic uses. It is noted that most of the hydrochemical groundwater parameters in the study area are within the allowable limits of the WHO, while a number of samples have values more than the maximum allowable limits.

Irrigation Purposes

The general criteria and the most important hydrochemical parameters of groundwater used to determine its suitability for irrigation are salinity (EC), Cl, percent sodium (Na%), sodium adsorption ratio (SAR), residual sodium carbonate (RSC), permeability index (PI), chloroalkaline indices (CAI), and magnesium ratio.

The groundwater suitability for irrigation depends on the effects of the mineral ingredients in the water the plants and soil [48]. The collected groundwater samples in the study area are classified based on the chlorinity index as shown in Table 4 [16, 49, 50]. From Table 4, it observed that 95.1% are suitable for all crops and 3.1% are suitable for low to high tolerant crops.

The EC values indicate that 99% of the groundwater wells in the study area are suitable for irrigation with 72% being good and 27% permissible (Table 5). Only 1%

Table 4 Classification of irrigation water quality based on electrical conductivity (EC) [49]

Class	Chloride (mg/l)	suitability for irrigation	% of location
I	<350	Suitable for all crops	95.1
II	350–700	Suitable for high-, medium-, and low-salt-tolerant crops	3.1
III	700–900	Suitable for high- and medium-salt-tolerant crops	0
IV	900–1300	Suitable for high-salt-tolerant crops	0
V	>1300	Not suitable for any crops	0

Table 5 Quality of irrigation water based on electrical conductivity [51]

EC	Classification	Number of location	% of location
<250	Excellent	0	0
250–750	Good	104	72
750–2250	Permissible	39	27
2250–5000	Doubtful	1	1
>5000	Unsuitable	0	0

is doubtful.

Sodium hazard (SR):

Sodium concentration is an essential parameter in water classification for irrigation purposes because sodium concentration can decrease the soil permeability [51, 52]. SAR is an important parameter for calculating the groundwater suitability for irrigation uses. AR is a measure of alkali/sodium hazard to plants. It can be calculated by Karanth [53] as:

$$SAR = \frac{Na}{\sqrt{(Ca + Mg)/1}} \tag{5}$$

where the concentrations are reported in milliequivalents per liter (meq/L). The SAR values range from 3 to 35.9, with an average value of 12 (Table 1). The classification of groundwater quality in the study area based on SAR is presented in Table 6. About 45% of the groundwater in the study area falls in the low-sodium class (S1) and 40%

Table 6 Quality of irrigation water based on SAR values in Kharga Oasis [51]

SAR	Alkalinity hazard	Classification	Number of locations	Percentage of locations (%)
<10	S1	Excellent	65	45
10–18	S2	Good	57	40
18–26	S3	Doubtful	14	10
>26	S4	Unsuitable	8	6

in good class (S2). There are 8 samples unsuitable for irrigation purposes, and 14 samples are doubtful.

The sodium percentage (Na %): Sodium percentage is defined by Raghunath [54]. It is calculated using the formula given in the following:

$$Na\,\% = \frac{(Na + K) * 100}{(Ca + Mg + Na + K)} \tag{6}$$

where all the concentrations are reported in meq/l. The Na% values vary from 38.33 to 83.6, with an average of 63.33 (Table 1). The Na% indicates that about 71% of the collected samples are good to permissible for irrigation, while the 29% is doubtful to unsuitable water (Table 7).

Permeability index (PI): The PI values indicate groundwater suitability for irrigation is the PI can be calculated by the following formula:

$$PI = \frac{Na + \sqrt{HCO_3}}{Ca + Mg + Na} * 100 \tag{7}$$

where all the ions are reported in meq/l. According to PI values, water can be classified into three classes: class I, class II, and class III [55]. The PI values vary from 27.4 to 92.6 with an average of 64.01. Most of the groundwater wells (79.2%) are considered as good suitability for irrigation purposes (Table 8).

Kelly's Ratio (KR): KR is using for evaluating the effect of sodium on water quality for irrigation water. Sodium measured against calcium and magnesium was measured by Kelly [56] for calculating KR. KR is calculated by the following equation:

Table 7 Quality of irrigation water based on Na% [54]

Na%	Classification	Number of locations	% of locations
<10	Excellent	0	0
Oct-40	Good	56	39
40–60	Permissible	46	32
60–80	Doubtful	20	14
>80	Unsuitable	22	15

Table 8 Classification of irrigation water quality based on PI [55]

PI	Standard of irrigation	Class	Number of locations	Percentage of location
<25	Unsuitable	III	0	0
25–75	Good	II	114	79.2
>75	Permeability	I	30	20.8

Table 9 Quality of irrigation water based on MH % [57]

Mg hazard	Quality	Number of location	Percentage of location
<50	Very good for irrigation	131	91.0
>50	Poor for irrigation	13	9.0

$$KR = \frac{Na^{2+}}{(Ca^{2+} + Mg^{2+})} \tag{8}$$

where all the ions are reported in meq/l. If KR more than one is generally categorized as unfit for irrigation. KR values in the study area range from 0.3 to 4.4 with an average of 1.3 (Table 1). About 23% of the groundwater wells are unsuitable for irrigation (KR > 1).

Magnesium hazard (MH): The MH of irrigation water is considered by Szabolcs and Darab [57] with the following equation:

$$MH = \frac{Mg^{2+}}{(Ca^{2+} + Mg^{2+})} \times 100 \tag{9}$$

where all the concentrations are reported in meq/l. The MH values more than 50 are considered as unsuitable for irrigation use. In the study area, it is observed that the MH values are between 11.5 and 69.5, with an average of 9.6 (Table 1). The analyzed water samples indicate that about 91.1% have MH < 50 and hence are suitable for irrigation, and 9% of the samples are poor for irrigation (Table 9).

The Soluble Sodium Percent (SSP): It has been determined by the following equation [51]

$$SSP = \frac{Na^{+} + K^{+}}{(Ca^{2+} + Mg^{2+} + Na^{+} + K^{+})} \times 100 \tag{9}$$

where all the concentrations are reported in meq/l. **SSP** values less than 50% or equal to 50% indicate good quality water and if it is more than 50% shows the samples unfit for irrigation. The values of SSP ranges from 20.1 to 80.6% with an average value of 51.8% (Table 1). About 46% of SSP values of the samples are good quality water for irrigation and 54% indicate the unsuitable water quality for irrigation.

Industrial Purposes

Kharga Oasis is observed as one of Egypt's most promising industrial extension regions. It is therefore very essential to evaluate groundwater quality for industrial reasons. Water with elevated chloride, sulphate and sodium concentrations or other ions will improve water conductivity and may encourage corrosion [58]. Groundwater in the study area is appropriate for many industrial reasons, according to the National Academy of Science [59].

Statistics Analysis

Correlation Coefficient Analysis

The coefficient of correlation is a frequently used metric for establishing the connection between two factors. The correlation coefficients for EC, TDS, TH, and major ions were estimated and presented in Table 10. It is observed that EC and TDS illustrate high positive correlation with TH, Ca^+, Mg^+, Na^+, $SO4^{2-}$and Cl^-. Particularly, Na^+ and Cl^- represent a strong correlation with TDS and among themselves. A high positive correlation between TH and Ca^{2+} (r = 0.934), Mg^{2+} (r = 0.83), and SO_4^{2-} (r = 0.891) shows that the hardness in the groundwater is regarding to these parameters. pH shows a small negative correlation with other parameters. K^+–Na^+, SO_4^{2-}–a^{2+}, $SO4^{2-}$–Mg^{2+}, Cl^-–K^+, HCO_3^-–Na^+, and HCO^{3-}–SO_4^{2-} are also the more important correlation pairs. The close positive relationship between Ca–Mg–SO_4 can indicate the source of Ca, Mg, and SO_4 and maybe regard to gypsum, and the relationships between Na–K–Cl are due to the dissolution of chloride metals, while the relationships between Na–K–HCO_3 are due to for the weathering of silicate metals.

Factor Analysis

Liu et al. [17] classified the factor loading to three clusters: strong positive loadings when factor loading is more than 0.75, moderate positive loadings when factor loading between 0.75–0.5 and weak positive loadings when factor loading between 0.5–0.3. Table 11 shows the factor loading analysis, eigenvalue, specific variance, cumulative variance for water quality parameters in the study area. The factor analysis results of the groundwater quality indicate three factors, which described 83.13% of the total variance. Factor 1 describes 56.34% of the total variance, has a strong positive loading on Na^+, Ca^{2-}, Mg^{2+}, Cl^-, SO_4^{2-}, EC, TDS, and TH, with loading values of 0.79, 0.93, 0.78, 0.92, 0.89, 0.96, 0.96, and 0.96, respectively. It has a negative loading on pH and HCO_3^-. This means that the values of pH and HCO_3^- do not contribute to other chemical parameters. Factor 2 describes 18.16% of the total variance, has a strong positive loading on HCO_3^- with loading value of 0.78 and a moderate positive loading on pH with a loading value of 0.52. This factor has a negative loading of Fe and Mn. Factor analysis 3 describes 8.63% of the total variance, has only strong positive loading on Mn.

Table 10 Correlations coefficient of the parameters for the samples of the study area

Variables	pH	EC	TDS	TH	Ca^{2+}	Mg^{2+}	Na^+	K^+	SO_4^{2-}	Cl^-	HCO_3^-
pH	**1**	−0.114	−0.105	−0.130	−0.104	−0.364	−0.004	−0.443	−0.155	−0.178	0.143
EC		**1**	0.994	0.898	0.916	0.614	0.919	0.049	0.811	0.945	0.081
TDS			**1**	0.901	0.914	0.637	0.916	0.037	0.811	0.940	0.090
TH				**1**	0.934	0.830	0.715	0.310	0.891	0.864	−0.197
Ca^{2+}					**1**	0.576	0.786	0.111	0.873	0.881	−0.131
Mg^{2+}						**1**	0.410	0.510	0.676	0.601	−0.145
Na^+							**1**	−0.104	0.681	0.860	0.311
K^+								**1**	0.180	0.163	−0.508
SO_4^{2-}									**1**	0.718	−0.116
Cl^-										**1**	−0.090
HCO_3^-											**1**

Table 11 Factor loading analysis, and eigenvalue, specific variance, the cumulative variance for water quality parameters in the study area. Bold values denote high correlation between variables

Variable	Factor 1	Factor 2	Factor 3
Na^+	**0.79**	0.55	–0.10
Ca^{2+}	**0.93**	0.06	0.10
Mg^{2+}	**0.78**	−0.37	0.13
Cl^-	**0.92**	0.12	−0.20
HCO_3^-	−0.08	**0.78**	0.05
SO_4^{2-}	**0.89**	−0.12	0.14
EC	**0.96**	0.26	−0.01
TDS	**0.96**	0.27	0.00
pH	−0.30	**0.52**	0.51
TH	**0.96**	−0.15	0.13
Fe	0.42	–0.65	−0.34
Mn	0.16	–0.50	**0.74**
Eigenvalue	6.76	2.18	1.04
Total variance (%)	56.34	18.16	8.63
Cumulative variance (%)	56.34	74.50	83.13

Conclusions

The main objective of this study was to assess the groundwater quality in Kharga Oasis using traditional hydrochemical analysis, GIS technique. The water samples were analyzed for various physiochemical attributes. Groundwater is somewhat alkaline and largely various in chemical composition; e.g., electrical conductivity (EC) ranges from 276.5 to 2256.3 ppm. The analysed groundwater samples have TDS values under the allowable limit. The plenty of the major ions is as follows: $Na^+ > Ca^{2+} > Ma^{2+} > K^+$ and $Cl^- > SO_4^{2-} > HCO_3^-$. The study area includes three major hydro chemical facies (Ca– Cl, Na–Cl, and mixed Ca–Mg–Cl). The results of WQI indicate that 88.2% are Excellent for drinking. Based on TDS, TH, Na%, PI%, MH%, SAR, it is clear that the majority of the collected groundwater samples are suitable for irrigation.

It is also observed that the Iron element value exceeds the standard value 0.5 ppm permitting for Drinking (88.2% of samples), and according to water quality index, there is 4 wells are categorized as poor wells for drinking and 88.2%, 10.4% are Excellent and Good respectively.

The hydrogeochemical analysis reveals that the groundwater of the study area is valid for Drinking and irrigation except for some wells which are considered Poor for Water (4 wells) according to the water quality index and PSS. Almost all the groundwater wells of the study exceeded the permissible limits of Fe (81.8%) and K (98.5%).

The factor analysis results of the groundwater quality indicate three factors, which described 83.13% of the total variance. Factor 1 describes 56.34% of the total variance, has a strong positive loading on Na^+, Ca^{2+}, Mg^{2+}, Cl^-, $SO_4{}^{2-}$, EC, TDS, and TH and a negative loading on pH and $HCO_3{}^-$. Factor 2 describes 18.16% of the total variance, has a strong positive loading on $HCO_3{}^-$ and a moderate positive loading on pH. Factor analysis 3 describes 8.63% of the total variance, has only strong positive loading on Mn. The factor analysis results show that the concentrations of pH and $HCO_3{}^-$ do not contribute to other chemical parameters.

Recommendations

Some suggestions and recommendations should be taken into consideration to maintain the groundwater quality in the study area:

- Controlling the groundwater overpumping to avoid salinity hazards
- Continuous monitoring of groundwater quality to avoid degradation
- To overcome the problem of iron and manganese in groundwater, aeration method (or using compact unites in case of low concentration for domestic purposes).

Acknowledgements The authors are thankful to the General Directorate of Groundwater in Al-Kharga and to the Ministry of Water Resources and Irrigation, Egypt, for providing access to the data.

Also, Abdelazim Negm want to acknowledge the support of the Science, Technology and Innovation, Funding Authority (STIFA) of Egypt in the framework of the grant no. 30771 for the project titled "A Novel Standalone Solar-Driven Agriculture Greenhouse—Desalination System: That Grows Its Energy and Irrigation Water" via the Newton-Mosharafa funding scheme, Call 4.

References

1. Abdalla F, Al-Turki A, Amri AA (2015) Evaluation of groundwater resources in the Southern Tihama Plain, Saudi Arabia. Arab J Geosci 8:3299–3310. https://doi.org/10.1007/s12517-014-1401-3
2. UNESCO (2007) Water portal newsletter no. 161: water-related diseases. http://www.unesco.org/water/news/newsletter/161.shtml
3. Bacha AA, Durrani MI, Paracha PI (2010) Physical and bacteriological characteristics of drinking water of Peshawar. Pak J Nutr 9:1028–1033
4. Yesilnacar MI, Gulluoglu MS (2007) The effects of the largest irrigation of gap project on groundwater quality, Sanliurfa–Harran plain, Turkey. Fresenius Environ Bull 16(2):206–211
5. Yesilnacar MI, Gulluoglu MS (2008) Hydrochemical characteristics and the effects of irrigation on groundwater quality in Harran Plain, GAP Project, Turkey. Environ Geol 54:183–196
6. Nalbantis I, Tsakiris G (2009) Assessment of hydrological drought revisited. Water Resour Manag 23:881–897

7. Nagaraju A, Kumar SK, Thejaswi A, Sharifi Z (2014) Statistical analysis of the hydrogeochemical evolution of groundwater in the Rangampeta area, Chittoor district, Andhra Pradesh, South India. Am J Water Res 2:63–70
8. Raju NJ, Ram P, Dey SJ (2009) Groundwater quality in the lower Varuna river basin, Varanasi district. Uttar Pradesh. Geol Soc India 73:178. https://doi.org/10.1007/s12594-009-0074-0
9. Vijay R, Samal DR, Mohapatra PK (2011) GIS based identification and assessment of groundwater quality potential zones in Puri city, India. J Water Resour Prot 3:440–447. https://doi.org/10.4236/jwarp.2011.36054
10. Krishnaraj S, Kumar S, Elango KP (2015) Spatial analysis of groundwater quality using geographic information system—a case study. IOSR 9:01–06
11. Arulbalaji P, Gurugnanam B (2016) Groundwater quality assessment using geospatial and statistical tools in Salem District, Tamil Nadu, India. Appl Water Sci 7 (6):2737–2751. https://doi.org/10.1007/s13201-016-0501-5
12. Armanuos AM, Negm A, Saavedra Valeriano OC (2016) Groundwater quality investigation using multivariate analysis-case study: Western Nile Delta Aquifer, Egypt. Inter J Environ Sci Devlop 7:1–9. https://doi.org/10.7763/IJESD.2016.V7.732
13. Negm AM, Armanuos AM (2016) GIS-based spatial distribution of groundwater quality in the western Nile delta, Egypt. In: Negm AM (ed) The Nile delta. The handbook of environmental chemistry. Springer. https://doi.org/10.1007/698_2016_66
14. El-Rawy M, Ismail E, Abdalla O (2019) Assessment of groundwater quality using GIS, hydrogeochemistry and factor statistical analysis in Qena governorate, Egypt. Desalin Water Treat 1–16. https://doi.org/10.5004/dwt.2019.24423
15. Fathi H, El-Rawy M (2018) GIS-based evaluation of water quality index for groundwater resources nearby wastewater treatment plants, Egypt. Poll Res 37:105–116
16. Ismail E, El-Rawy M (2018) Assessment of groundwater quality in West Sohag, Egypt. Desalin Water Treat 123:101–108. https://doi.org/10.5004/dwt.2018.22687
17. Liu CW, Lin KH, Kuo YM (2003) Application of factor analysis in the assessment of groundwater quality in a blackfoot disease area in Taiwan. Sci Total Environ 313:77–89
18. Love D, Hallbauer D, Amos A, Hranova R (2004) Factor analysis as a tool in groundwater quality management: two southern African case studies. Phys Chem Earth 29:1135–1143
19. Yu S, Shang J, Zhao J, Guo H (2005) Factor analysis and dynamics of water quality of the Songhua River Northeast China. Water Air Soil Pollut 144:159–169
20. Gao C, Yan J, Yang S, Tan G (2011) Applying factor analysis to water quality assessment: a study case of Wenyu River. In: Li S (ed) Nonlinear mathematics for uncertainty and its applications, AISC 100, pp 541–547
21. Abdalla O, Al-Abri RY (2014) Factors affecting groundwater chemistry in regional arid basins of variable lithology: example of Wadi Umairy, Oman. Arab J Geosci 7:2861–2870. https://doi.org/10.1007/s12517-013-0981-7
22. Kudoda AM, Abdalla OAE (2015) Hydrochemical characterization of the main aquifers in Khartoum, the capital city of Sudan. Environ Earth Sci 74:4771–4786. https://doi.org/10.1007/s12665-015-4464-x
23. Abdelmawgoud A, El-Rawy M, Abu Bakr AM (2021) Evaluating the suitability of groundwater quality for drinking and irrigation purposes in El-Minia governorate, Egypt. J Adv Eng Trends 40(2):21–34. https://journals.ekb.eg/article_128906_dd0e9ad558f7c96ea8a02c0566c153e8.pdf
24. Awad A, Eldeeb H, El-Rawy M (2020) Assessment of surface and groundwater interaction using field measurements: a case study of Dairut City, Assuit, Egypt. J Eng Sci Technol 15(1):406–425
25. Klitzsch E (1986) Plate tectonics and cratonal geology in Northeast Africa (Egypt, Sudan). Geologische Rundschau. Int J Earth Sci 75:755. https://doi.org/10.1007/BF01820645
26. CEDARE (2014) Nubian sandstone aquifer system (NSAS) M&E rapid assessment report. Monitoring & evaluation for water in North Africa (MEWINA) project, water resources management program, CEDARE
27. El-Rawy M, Abdalla F, El Alfy M (2020) Water resources in Egypt. In: Hamimi Z, El-Barkooky A, Martínez Frías J, Fritz H, Abd El-Rahman Y (eds) The geology of Egypt. Regional geology reviews. Springer, Cham. https://doi.org/10.1007/978-3-030-15265-9_18

28. Salama RB (1985) Buried troughs, grabens and rifts in Sudan. J Afr Earth Sci 3:381–390
29. El-Rawy M, De Smedt F (2020) Estimation and mapping of the transmissivity of the Nubian sandstone aquifer in the Kharga Oasis, Egypt. Water 12(2):604. https://doi.org/10.3390/w12020604
30. APHA (American Public Health Association) (2005) Standard methods for the examination of water and wastewater. In: Eaton AD, Clesceri LS, Greenberg AE (eds) 20th edn. APHA, AWWA, WPCF, New York
31. WHO (2011) Guidelines for drinking water quality. World Health Organization Geneva, 4th edn. Recommendations, pp 1–4
32. EHCW (Egyptian higher committee of water) (1995) Egyptian standards for drinking and domestic water according to the act 27/1978 in regulating of the public water supplies, Egypt. Egyptian Governmental Press
33. ArcGIS 10.3 (2014) ArcMap. Esri software. http://desktop.arcgis.com/en/arcmap/
34. Aref F, Roosta R (2016) Assessment of groundwater quality and hydrochemical characteristics in Farashband plain. Iran. Arab J Geosci 9:752. https://doi.org/10.1007/s12517-016-2781-3
35. Hem JD (1985) Study and interpretation of the chemical characteristics of natural water US geological supply paper 2242, p 78
36. Dudeja SS, Singh NP, Sharma P, Gupta SC, Chandra R., Dhar B, Bansal RK, Brahmaprakash GP, Potdukhe SR, Gundappagol RC, Gaikawad BG, Nagaraj KS (2011) Biofertilizer technology and productivity of chickpea in India. In: Singh A, Parmar N, Kuhad RC (eds) Bioaugmentation, biostimulation and biocontrol. Soil biology. Chapter 28, Springer, Berlin, Heidelberg, pp 43–63
37. Sawyer CN, McCarty PL (1967) Chemistry for sanitary engineers. McGraw-Hill, New York
38. Jeevanandam M, Kannan R, Srinivasalu S, Rammohan V (2007) Hydrogeochemistry and groundwater quality assessment of lower part of the Ponnaiyar river basin, Cuddalore district, South India. Environ Monit Assess 132(1–3):263–274
39. Srinivasamoorthy K, Chidambaram S, Prasanna MV et al (2008) Identification of major sources controlling groundwater chemistry from a hard rock terrain—a case study from Mettur taluk, Salem district, Tamilnadu, India. J Earth System Sci 117(1):49–58
40. Ayeni AO, Balogun II, Soneye ASO (2011) Seasonal assessment of physico chemical concentration of polluted urban river: case of Ala river in Southwestern Nigeria. Res J Environ Sci 5(1):22–33
41. Balan N, Liu JY, Otsuak Y, Ram ST, Luhr H (2012) Ionospheric and thermospheric storms at equatorial latitudes observed by CHAMP, ROCSAT and DMSP. J Geophys Res 117: A01313. https://doi.org/10.1029/2011JA016903
42. Abdalla F, Shamrukh M (2016) Quantification of river Nile/Quaternary aquifer exchanges via riverbank filtration by hydrochemical and biological indicators, Assiut, Egypt. J Earth Syst Sci 25(8):1697–1711. https://doi.org/10.1007/s12040-016-0755-1
43. Ramakrishnaiah CR, Sadashivaiah C, Ranganna G (2009) Assessment of water quality index for the groundwater in Tumkur Taluk, Karnataka State, India. E J Chem 6(2):523–530
44. Saeedi M, Abessi O, Sharifi F, Meraji H (2010) Development of groundwater quality index. Environ Monit Assess 163:327–335
45. Yidana SM, Banoeng-Yakubo B, Akabzaa TM (2010) Analysis of groundwater quality using multivariate and spatial analyses in the Keta basin Ghana. J Afr Earth Sci 58(2):220–234
46. Vasanthavigar M, Srinivasamoorthy K, Vijayaragavan K, Ganthi RR, Chidambaram S, Anandhan P, Manivannan R, Vasudevan S (2010) Application of water quality index for groundwater quality assessment: Thirumanimuttar Sub-Basin, Tamil Nadu, India. Environ Monit Assess 171:595–609
47. Tiwari AK, Singh PK, Mahato MK (2014) GIS-based evaluation of water quality index of groundwater resources in West Bokaro coalfield, India. Curr World Environ 9(3):843–850
48. Richards LA (1954) Diagnosis and improvement of saline and alkali soils: agriculture hand book 60. US Department of Agriculture, Washington, DC, p 160
49. Mills B (2003) Interpreting water analysis for crop and pasture, File No. FS0334. DPI'S Agency for food and fiber sciences, Toowoomba

50. Singh V, Khare MC (2008) Groundwater quality evalution for irrigation purpose in some areas of Bhind, Madhya Pradesh (INDIA). J Environ Res Develop 2(3):347–356.
51. Todd DK (1980) Groundwater hydrology, Wiley, New York, p 535
52. Domenico PA, Schwartz FW (1990) Physical and chemical hydrogeology. Wiley, New York
53. Karanth KR (1987) Ground water assessment development and management. McGraw Hill Publishing Company Ltd, New Delhi
54. Raghunath HM (1987) Groundwater, 2nd edn. Wiley Eastern Ltd, New Delhi, India, pp 344–369
55. Doneen LD (1964) Water quality for agriculture. Department of irrigation, University of California, Davis, p 48
56. Kelly WP (1951) Alkali soils—their formation properties and reclamation. Reinhold Pub, New York
57. Szabolcs I, Darab C (1964) The influence of irrigation water of high sodium carbonate content of soils. In: Proceedings of 8th international congress of ISSS, vol 2, pp 803–812
58. Wilkes University Center of Environmental Quality, Geoenvironmental Science and Engineering Department (2002) Corrosion, saturation index, balanced water in drinking water systems—source and cause of corrosion
59. National Academy of Science (1972) Water quality criteria; protection agency, Washington DC, pp 1–594

Groundwater Potential in the Bahariya Oasis, Western Desert, Egypt

Abbas M. Sharaky, Sayed I. Gaber, Kamel Abd El Gelil, and Suad H. Abdoun

Abstract Groundwater is the most important natural source of water in the Egyptian deserts that occupy more than 95% of the country. Bahariya Oasis is one of the Egyptian oases that mainly rely on groundwater for all purposes. Recently, Egypt's plan is to increase the agricultural land by 20% through a mega project called "One and Half Million Fadden" using mostly groundwater. The main purpose of the present chapter is to re-evaluate the hydrogeological conditions of the groundwater aquifers (Nubian Sandstone Aquifer) in Bahariya Oasis through the determination of their potentialities for optimum exploitation, using field measurements, analysis of water samples to investigate the local water exploitation effect on water level and quality. Measurements of the hydrogeological parameters through pumping tests were studied for the groundwater. The piezometric head and groundwater flow were evaluated for 152 wells. Hydrochemical analysis for evaluation of groundwater quality. The transmissivities of the aquifers are 1,134 m^2/d for zone (S1), 471 m^2/d for zone (S2), and 1,023 m^2/d for zone (S3). The annual amounts of water extraction were 18.5, 5 and 84.5 Mm^3/y from the deep zone (S1), middle zone (S2) and shallow zone (S3), respectively. The groundwater is freshwater averaging 286 mg/l, which is suitable for all purposes.

Keywords Bahariya Oasis · Groundwater aquifer · Pumping test · Nubian Sandstone · Hydrochemistry

Introduction

Groundwater is water that occurs underground in a saturated zone beneath the land surface, where it is called the hidden water. According to National Oceanic and

A. M. Sharaky (✉) · S. I. Gaber
Department of Natural Resources, Faculty of African Postgraduate Studies, Cairo University, P.O. Box 12613, Giza, Egypt
e-mail: sharaky@geologist.com

K. A. E. Gelil · S. H. Abdoun
Research Institute for Groundwater (RIGW), Ministry of Water Resources and Irrigation, Kanater El-Kayria, Egypt

A. Negm and A. Elkhouly (eds.), *Groundwater in Egypt's Deserts*, Springer Water,
https://doi.org/10.1007/978-3-030-77622-0_9

Atmospheric Administration (NOAA), groundwater represents about 30% of world's freshwater, where lakes and rivers form only 1%. The rest 69% is locked away in glaciers and ice caps. According to UNESCO's report, the groundwater supplies potable water to about 70% of the world's population, with 51% of countries withdraw more than 100 m^3/capita annually [1]. Groundwater forms in average one third of the freshwater consumed by humans, but in deserts, this percentage can reach up to 100% where annual precipitation is less than 100 mm/yr.

In Africa, the groundwater is the preferred means of supplying water to meet the growing demand of the rural, dispersed communities and the small urban towns [2].

The main advantages of using groundwater as a water source are:

1. the groundwater quality is mostly good,
2. it can be stored for long times without loss (away from evaporation) or pollution,
3. it is often far from sources of pollution,
4. it is widespread in many places,
5. the natural storage capacity is high.
6. it is not affected by short periods of drought,
7. infrastructure is more affordable to poor communities [3], and
8. it can be extracted in places where it is needed.

From several lines of evidence, this water was derived from precipitation during the period of 25,000–5,000 years before the present, when the climate of North Africa was quite different from that of today. The total volume of groundwater in Africa was estimated to be 0.66 million km^3 with a range in uncertainty of between 0.36 and 1.75 million km^3 [4]. It is about 30 times the annual renewable rainfall (22,300 km^3) or 150 times the running freshwater after evaporation on Africa. From several lines of evidence, the groundwater was derived from precipitation during the period of 25,000–5,000 years before the present, when the climate of North Africa was rainy and quite different from that of today.

Groundwater resources in Africa are unevenly distributed, where the largest groundwater volumes are found in the large sedimentary aquifers in the North African countries Libya, Algeria, Egypt and Sudan [4]. The Nubian Sandstone Aquifer System (NSAS) is the world's largest fossil water aquifer system. It is located in the Eastern end of the Sahara.

The NSAS is mainly composed of continental clastics of sandstone with shale and clay intercalations of saturated thickness ranging between 100 and 1,500 m [5]. The NSAS is shared by Egypt, Libya, Sudan, and Chad; covering about 2.2 million km^2 (see Fig. 1). The NSAS extends for 826,000 km^2 in Egypt (38%), 760,000 km^2 (34%) in Libya, 376,000 km^2 in Sudan (17%) and 235,000 km^2 in Chad (11%) [6, 7].

Egypt has planed to reclaim 4 million Fadden in few years from 2016. The present stage is to cultivate 1.5 million Fadden depending mostly (90%) on none-renewable groundwater of the Nubian Sandstone Aquifer, Western Desert. About 20,000 Fadden from the 1.5 Million Feddans Project are located in Bahariya Oasis.

The main purpose of this chapter is to evaluate the hydrogeological conditions of the groundwater in Bahariya Oasis to determine its potentialities for optimum

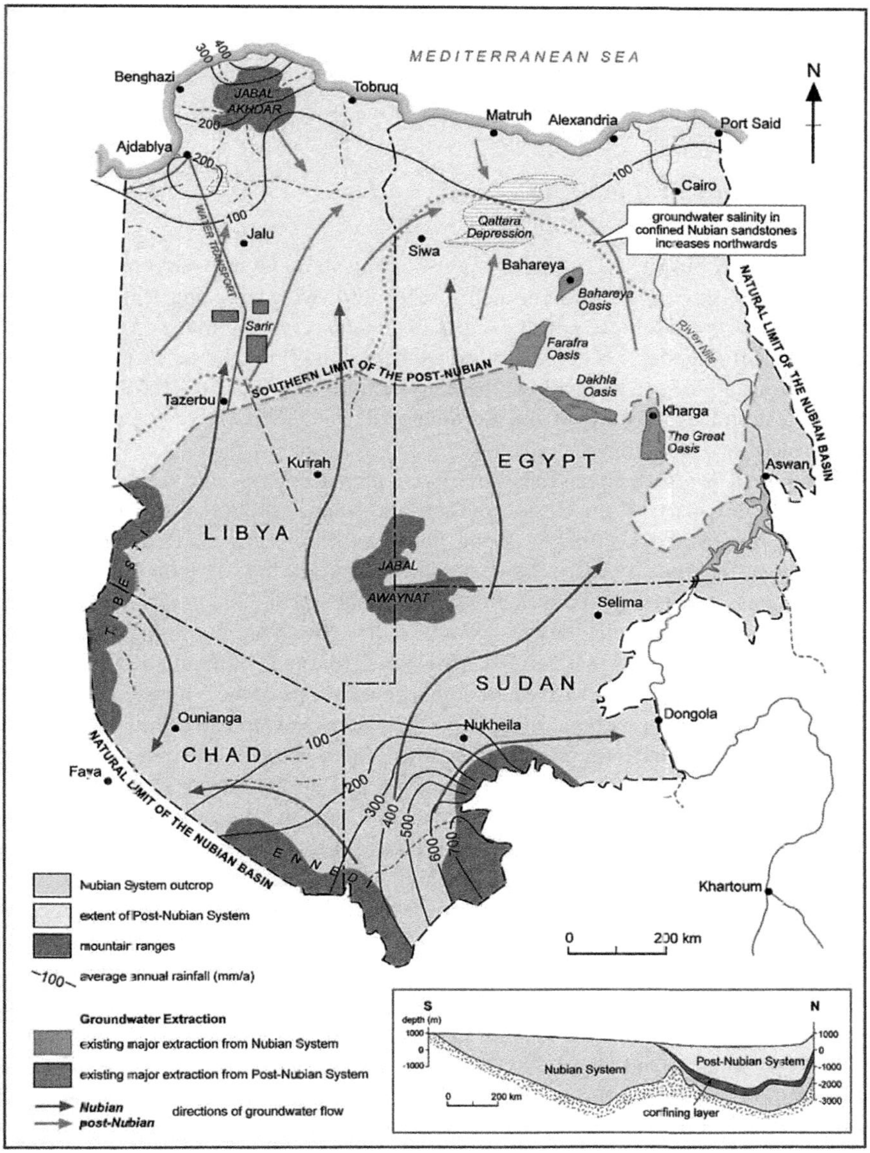

Fig. 1 Hydrogeological sketch map of the Nubian Sandstone Aquifer System (NSAS) covering 2.2 km^2 its extension in Egypt, Libya, Sudan, and Chad; and the direction of the groundwater flow Sudan to Egypt [9]

exploitation of groundwater, using the available data and field measurements. The second target is to investigate the water exploitation effect on water quality of the aquifer.

Hydrogeology of the NSAS

The NSAS is composed of different water-bearing strata laterally or vertically interconnected [8]. It can be differentiated locally into two major aquifer systems [9]. The first one is the oldest and the most extended reservoir that is known as the Nubian Aquifer System (NAS). It underlies the second Post-Nubian Aquifer System (PNAS) located only in the north of the 25th latitude, overlying the NAS in the north of the Western Desert of Egypt and Northeastern Libya [8]. Low permeability layers separate the two reservoir systems [9].

In Egypt, The Cretaceous sandstone covers most of the Western Desert including the New Valley, parts of the Eastern Desert and Sinai. The Precambrian crystalline basement complex is dominated along the coastal zone of the Red Sea, western margins, and the most southern regions of the NSAS (see Fig. 2). It includes granites, granodiorites, diorites, gneisses, schists, and basalts [10]. The complex is overlain by a series of sands, sandstones, clays, and shales which are commonly termed the Nubian Sandstone Series that varies in thickness from tens of meters in the northern regions of Sudan to about 250 m in the southern localities of the Kharga Oasis, 900 m in northern Kharga, and about 1,800 m in the Bahariya Oasis. It attains a thickness of more than 3,500 m in the northern localities of the Dakhla Basin and over 4,500 in the northwestern part [11, 12] of the Kufra Basin of the Libyan Desert [13].

In Egypt, The Cretaceous sandstone covers most of the Western Desert including the New Valley, parts of the Eastern Desert and Sinai. The Precambrian crystalline basement complex is dominated along the coastal zone of the Red Sea, western margins, and the most southern regions of the NSAS (see Fig. 2). It includes granites, granodiorites, diorites, gneisses, schists, and basalts [10]. The complex is overlain by a series of sands, sandstones, clays, and shales which are commonly termed the Nubian Sandstone Series that varies in thickness from tens of meters in the northern regions of Sudan to about 250 m in the southern localities of the Kharga Oasis, 900 m in northern Kharga, and about 1,800 m in the Bahariya Oasis. It attains a thickness of more than 3,500 m in the northern localities of the Dakhla Basin and over 4,500 in the northwestern part [11, 12] of the Kufra Basin of the Libyan Desert [13].

In Sudan, the Basement rocks are unconformably overlain by the Paleozoic sediments which outcrop in SE Libya, NW Sudan, and NE Chad (see Fig. 2). The Precambrian basement forms the base of the NAS. Its surface elevation ranges from Sea level in the southern part to over 5,000 m below mean sea level (bmsl) along the northern boundary. The entire area is divided into Kufra and Dakhla Basins by regional faults and uplifts. The major units of the aquifer system, are predominately filled by 500 to more than 5,000 m thickness of continental classic of Paleozoic and Mesozoic deposits [8]. The groundwater of the Nubian aquifer in Gezirz area is not

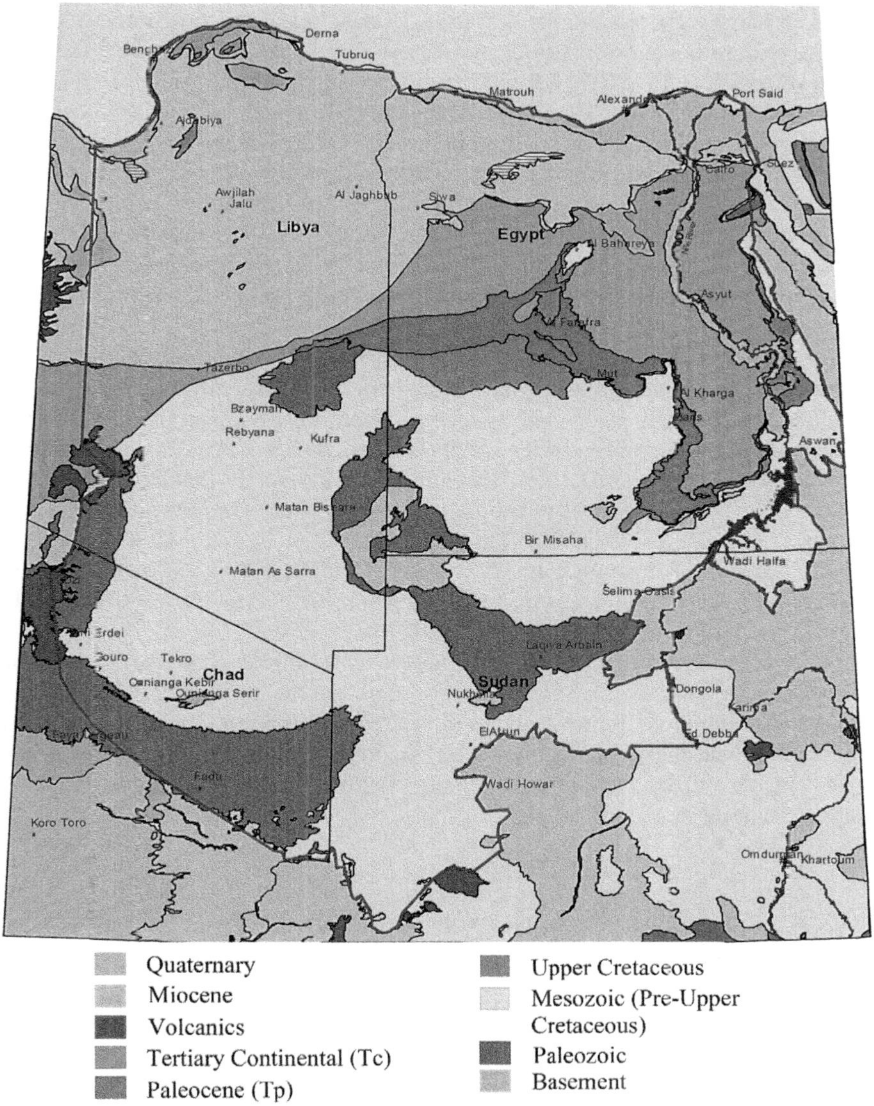

Fig. 2 Regional geology of Nubian Sandstone Aquifer [8]

included in the NSAS. It is recharged from the Blue Nile through the Gezira aquifer [14].

The NSAS has been studied since the beginning of the Twentieth Century. Several authors [15–18] studied the origin of the groundwater in the Western Desert. They concluded that it was derived from the rainfall on western and central Sudan, with partial recharge from the Nile. Others [19, 20] concluded that the groundwater was formed during frequent wet periods in the past. A regional groundwater flow model was developed by CEDARE [8] for the aquifer setting the northern model boundaries to the Mediterranean Sea. The impact of the groundwater extractions on the aquifer characteristics was studied by some authors [13, 21, 22]. The response of the aquifer to the climatic changes during the last 25,000 years was investigated by Gossel et al. [23].

Recent studies [5–7, 9, 24–34] concluded that the NSAS is mainly composed of two water-bearing complexes that are the Nubian facies series and the Post Nubian rocks. There is a partial local recharge of the aquifer in Toshka area from the Aswan High Dam lake [35].

Although the numerous studies of the NSAS, accurate information of NSAS is still vague, such as the total volume of water stored, the rate of groundwater flow, and total annual extraction. The water salinity changes from freshwater in the southern part of the system to low salinity water 1,000–3,000 mg/l (0.1–0.3%), moderate salinity water 3,000–10,000 mg/l (0.3–1%) and high salinity water 10,000–35,000 mg/l (1–3.5%) in the northern part. The estimated volume of the groundwater is not well known. It is estimated as 15,000 km^3 [36]; 60,000 km^3 [37]; 135,000 km^3 [23]; 150,000 km^3 [38] to as much as 372,950 km^3 [8]. The NSAS extends in the four countries: Egypt as 154,716 km^3 (41%), Libya 136,550 km^3 (36%), Chad 47,807 km^3 (13%) and Sudan 33,878 km^3 (9%) [9].

The average volume of water in the NSAS 150,000 km^3 corresponds to annual Nile discharge (84 km^3 at Aswan) of 179–4,440 with an average of 1,786 years or to a water column of 68 mover the aquifer area (2.2 million km^2). The estimated volume is unreasonable and economically impossible to obtain groundwater from great depths over broad areas for economic development. It should be determined in each location for economic usages.

Origin of Groundwater in the NSAS

The origin of the NSAS has been investigated by many authors since the 1920s. Two theories have presented the origin of groundwater in the NSAS [22]. The oldest theory of allochthonous concept that suggested the flow of groundwater from the southern mountainous to the north. According to this theory, the groundwater is partially renewable, and the basin is receiving some recharge of about 1.6 km^3/year. The autochthonous theory supported that the groundwater of the NSAS was formed in situ during the humid pluvial periods of the Holocene [39].

The age dating of the groundwater showed a long wet period ending approximately 20,000 years ago [19]. Between 20,000 and 14,000 b.p. there was a significant recharge representing a semiarid to arid followed by a wet climate period from the beginning of the Holocene (see Fig. 3). The age of the groundwater in the Sahara was measured as one million years old, using krypton-81 and chlorine-36 [40].

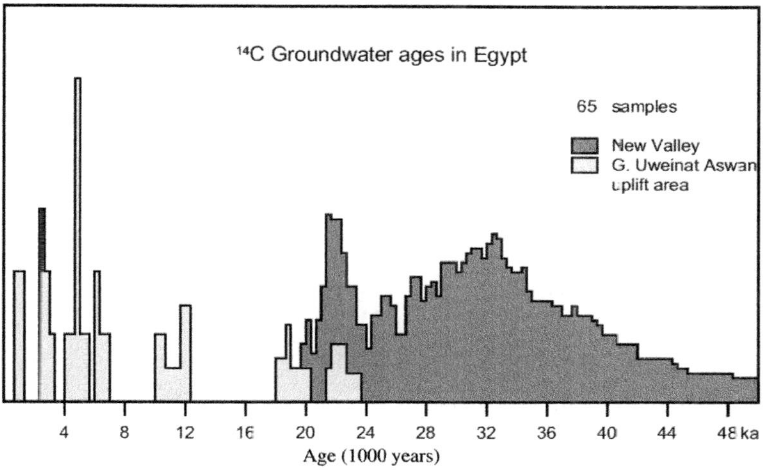

Fig. 3 Frequency distribution of apparent ^{14}C ages of groundwater of the Sahara and in the Egyptian Dakhla Basin of the NSAS, showing the old age (>20,000 years) of the groundwater in the New Valley [19]

In conclusion, the two theories are based on rainfall in the entire NSAS basin or in the southern mountains. During the last millennium, the NSAS basin suffers from the lack of rainfall. Therefore, the NSAS is non-renewable. It is partially renewable in local areas surrounding Aswan High Dam Lake "Nasser Lake" such as Toshka area, Egypt from infiltration of the Nile water [35].

Water Resources in Bahariya Oasis

The groundwater resources in Bahariya Oasis are related to the NSAS that is one of the largest and most important transboundary aquifer systems in Northern Africa. The only source of water resources is the groundwater available for cultivation and domestic uses in the Bahariya Oasis. As a result of increasing population and cultivated areas, deep well drilling was initiated in 1963, declining the piezometer head of the aquifer during the period 1963–1970 to an annual rate of about 1.2 m [5].

Location of Bahariya Oasis

Bahariya Oasis is a pear-like depression in the Western Desert of Egypt, covering about 2,078 km^2. It lies between latitudes 27° 46′ 27″ and 28° 30′ 32″ N and longitudes 28° 29′ 22″ and 29° 08′ 35″ E. The Bahariya Oasis is connected with the Nile Valley by two main roads. The first is 350 km NE–SW asphaltic road extending from Giza town to Bawiti (the oasis capital). The second road is the motor-track that starts from Samalut in the Nile Valley to the north of El-Gedida iron ore mine. A railway for transporting iron ore from El-Gedida mine to the melting factory in Helwan, south Cairo. Bahariya Oasis has connected also with Farafra and Siwa oases via two asphaltic roads about 154 and 373 km, respectively (see Fig. 4).

Sampling and Analysis

The following steps were followed in the present study:

1. Analysis of the climatic elements and meteorological data recorded in the climatic station No. 420 for the period 2000–2010.
2. Landforms and drainage pattern using remote sensing and GIS.
3. Defining the hydrogeological regime in general with special emphasis on the area of study, with respect to the flow directions, hydrologic conditions and hydrologic parameters.
4. Measurements of the hydrogeological parameters (permeability, and transmissivity) through pumping tests that carried out on production wells in the study area by:
    a. using gradual pumping test step,
    b. using the general well equation for producing total drawdown in water level in the well,
    c. drawing a relation between the discharge and the drawdown in the aquifer inside the well,
    d. using continuous pumping test step, and
    e. applying Cooper-Jacob equation.
5. Calculation and analysis of the groundwater extraction.
6. The step-drawdown tests data were used to give an idea about the design and efficiency of the water wells. Computerized Mathematical methods were used for calculating formation loss and well loss.
7. Analysis of 12 pumping tests data (constant discharge tests) was carried out in order to calculate the Nubian aquifer hydraulic parameters (transmissivity, hydraulic conductivity and storage coefficient) using GWW software program with Theis's method.

Climate

The Bahariya depression is characterized by hyper-arid climate conditions receiving an average annual rainfall of 0.6 m and a monthly average maximum temperature of

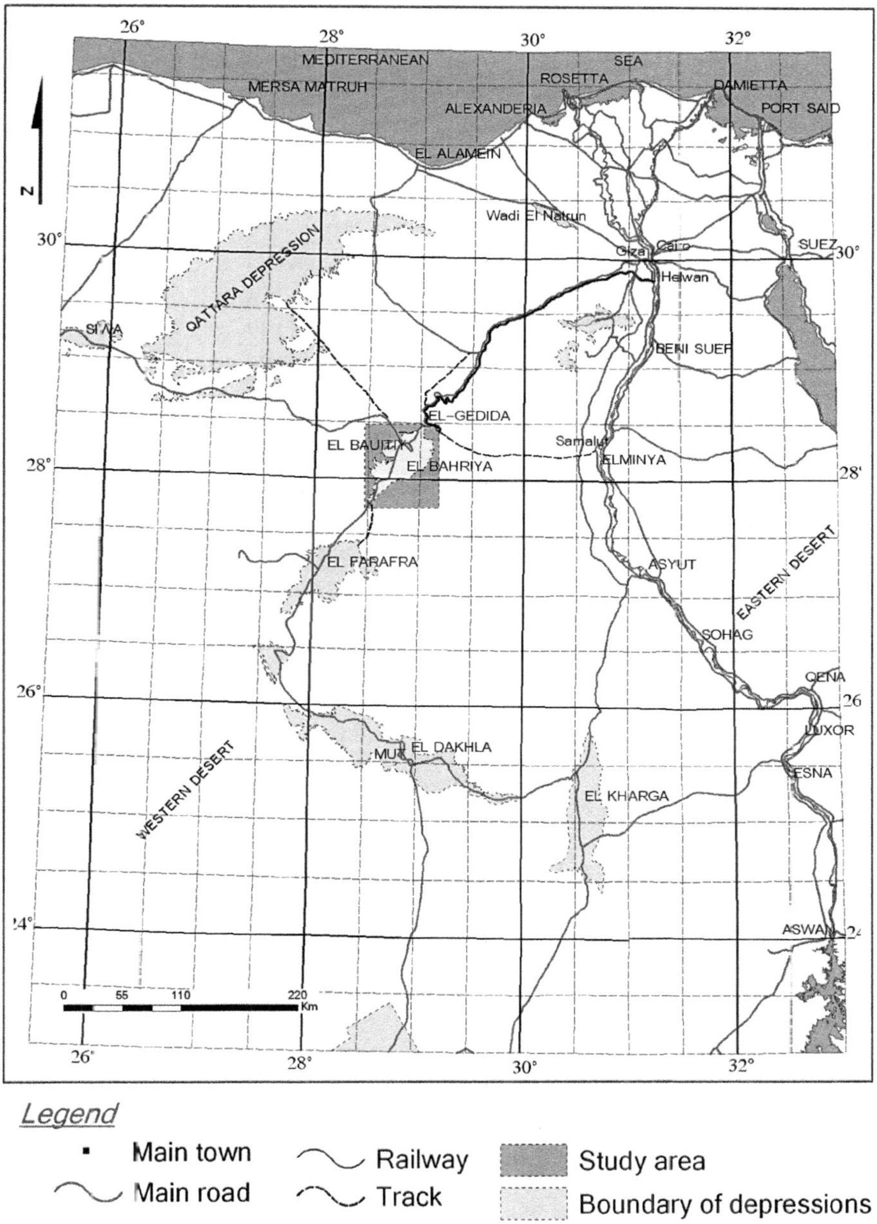

Fig. 4 Location of the Bahariya Oasis and the main roads (Topographic map of Egypt 1:2,000,000 1996)

Table 1 Climatological data of the Bahariya Oasis (Station No. 420) for the period 2000–2010

Months	Temperature (°C)		Average rainfall (mm)	Evaporation (mm)	Relative humidity (%)
	Average maximum	Average minimum			
January	18.50	5.40	1.40	5.29	50.20
February	20.30	6.20	1.30	7.26	40.80
March	24.40	9.30	1.10	9.76	34.70
April	31.50	14.50	0.20	11.03	30.70
May	34.10	17.60	0.00	13.74	28.00
June	37.30	20.40	0.00	14.20	31.00
July	37.30	21.50	0.00	13.21	35.60
August	37.00	21.60	0.00	12.00	38.70
September	34.40	19.30	0.00	10.87	42.10
October	31.10	16.20	0.10	8.91	44.80
November	25.50	10.90	1.30	6.57	52.60
December	20.60	7.00	1.60	4.59	51.30
Average	29.33	14.16	0.59	9.79	40.04

29 °C as recorded from the meteorological station No. 420 at Bahariya depression in the period of 2000–2010 (Table 1).

Air Temperature

The study area is characterized by hot summer with a maximum monthly air temperature of 37 °C from June to August, and cool winter with a minimum monthly air temperature of 5.4 °C in January. The minimum temperature ranges from 5.4 °C in January to 21.6 °C in August (see Fig. 5).

Rainfall and Relative Humidity

The rainfall is extremely rare in Bahariya Oasis all over the year. The average monthly precipitation ranges from 0.0 mm in the summer to 1.6 mm in December with an average of 0.59 mm/y. No rainfall is recorded during May–September (2000–2010) (see Fig. 6). The relative humidity ranges from 28% in May to 52.6% in November with an annual average of 40% (Table 1).

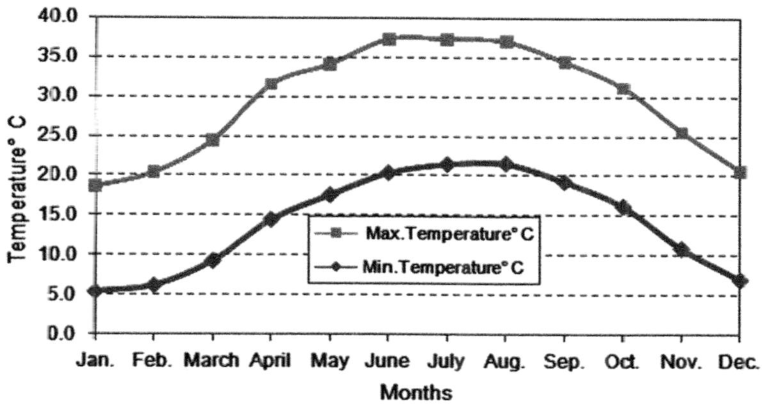

Fig. 5 Average maximum and minimum curve of air temperature (2000–2010)

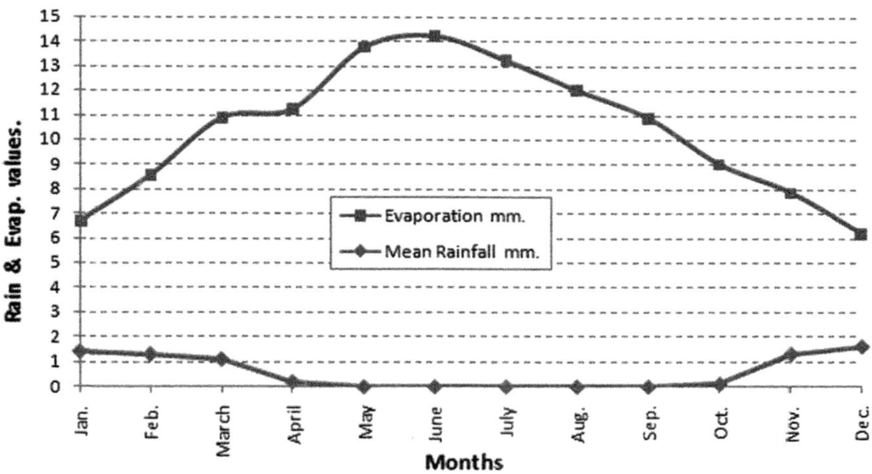

Fig. 6 Average monthly curve of rainfall and of evaporation (2000–2010)

Evaporation

Evaporation in Bahariya Oasis is very high. Its amount is 17 times the precipitation amount. The evaporation ranges from 4.59 mmin December to 14.2 mmin June with an average of 9.79 mm/y (Table 1). In general, the evaporation depends on wind speed, air temperature, moisture profiles, and roughness of the ground surface.

Geology and Geomorphology of the Bahariya Oasis

The Bahariya Oasis has a sedimentary column from Cambrian to Quaternary. The Bahariya floor is mainly composed of Cretaceous sandstone and shale. The Eocene is only represented in the north of the surrounding plateau. The Cretaceous rocks occupy not only the depression itself but also the surrounding plateau on the east and west sides. Tertiary rocks are mainly volcanic basaltic hills and fossiliferous limestone [5]. Quaternary rocks include aeolian sands in the depression and on top of the plateau.

Topography

The Bahariya floor is described as a flat rough stony surface due to the erosion of the surrounding rocks. It varies in elevation from 71 to 370 m above mean see level (amsl). The Bahariya depression is enclosed by a limestone plateau changing in elevation between 200 and 370 m amsl with an average of 230 m amsl (see Fig. 7). The deepest point of the oasis floor is about 80 m amsl at Ain Bagoum area in the north. The highest area is presented in the hilly mountains reaching to 370 m amsl [41]. The depression is classified into northern, center, and southern sectors.

The northern and southern sectors are flat while the central is gagged and higher in elevation ranging between 80 and 140 m. The elevation of the southern sector ranges between 140 and 160 m. The elevation of the central sector ranges between 140 and 180 m. The most important geomorphic features are represented by plateau surface, bounding escarpments, oasis floor, residual hills (questa), volcanic cones and sheets, valleys or wadis, alluvial fans and terraces, dry and wet sabkhas, sand sheets and sand dunes (see Fig. 8) [41, 42].

Geologic Setting

The geologic succession in Bahariya area can be summarized from base to top [28, 43–45] as:

Precambrian Rocks

The Precambrian rocks in the Bahariya Oasis are represented by dense, grey metamorphosed andesite and matrix of fine grains of altered feldspar, chlorite and magnetite.

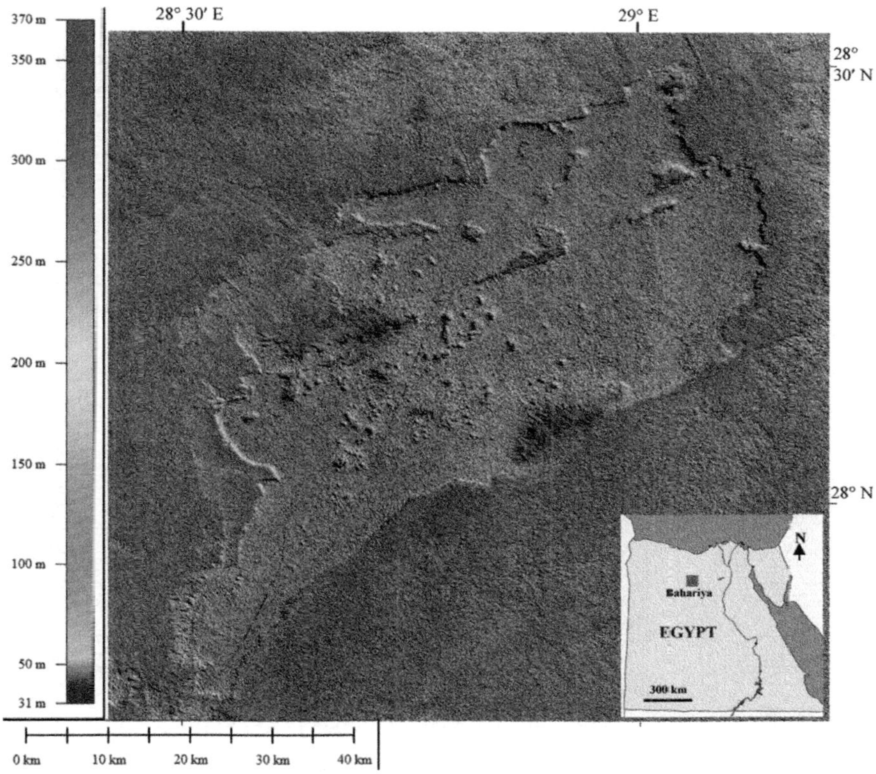

Fig. 7 Digital elevation model of Bahariya depression from Advanced Spaceborne Thermal Emission and Reflection Radiometer (ASTER) resolution of 90 m

Cambrian Rocks

The Cambrian rocks form a section of 458 m thick overlying the metamorphosed igneous basement. It is mainly composed of intercalated siltstone, sandstone, and clay.

Cretaceous Rocks

The Cretaceous rocks overlay unconformably the Cambrian sediments. Their sedimentary thickness is 660 m, and divided into four formations (see Fig. 9) starting from base to top:

a. Bahariya Formation (Kub) of Lower Cenomanian composes the floor of the oasis, the major part of the surrounding scarp and also includes the main mass

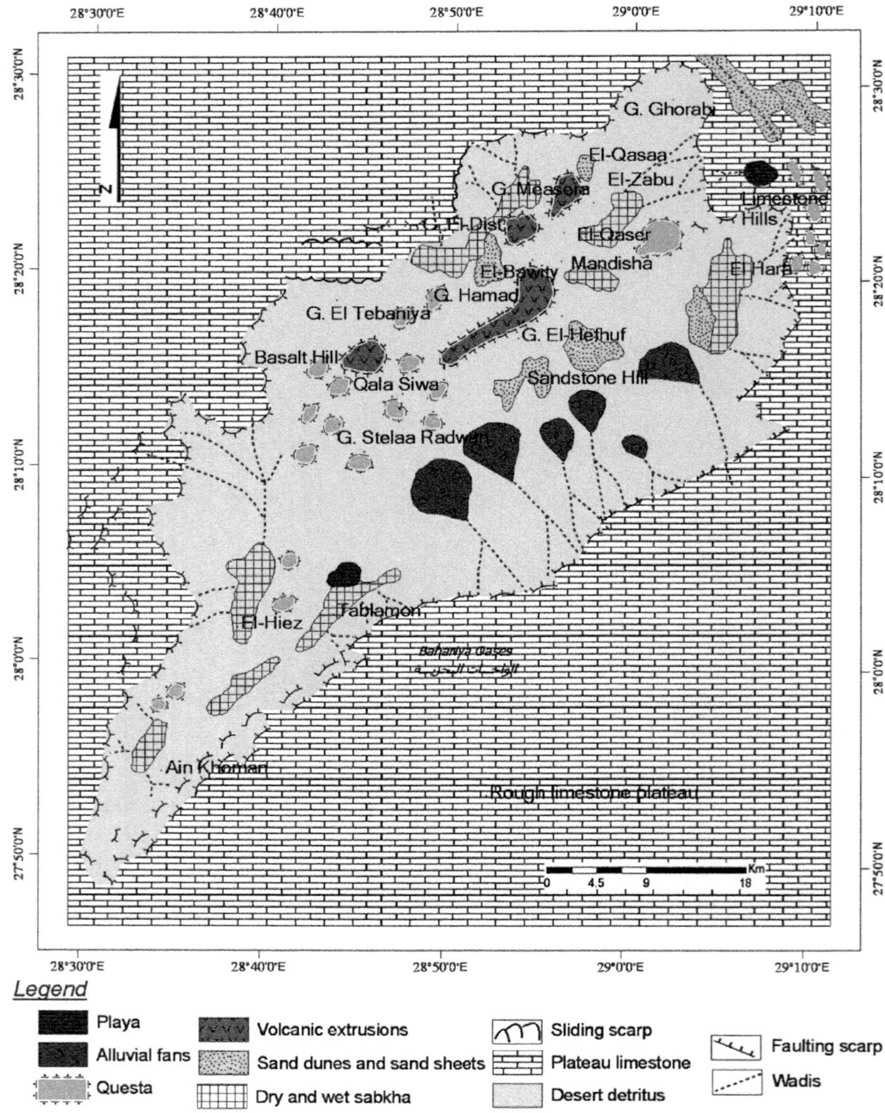

Fig. 8 The Geomorphological features of the Bahariya Oasis [42]

of Gabel Ghorabi. It contains friable, false-bedded, variegated sand, sandstone with harder dark brown ferruginous bands alternated with sandy clay.

b. El-Heiz Fm (Kuz) of Upper Cenomanian consists of shallow marine, brown to grey crystalline limestone and dolomite enclosing a clastic carbonate section.

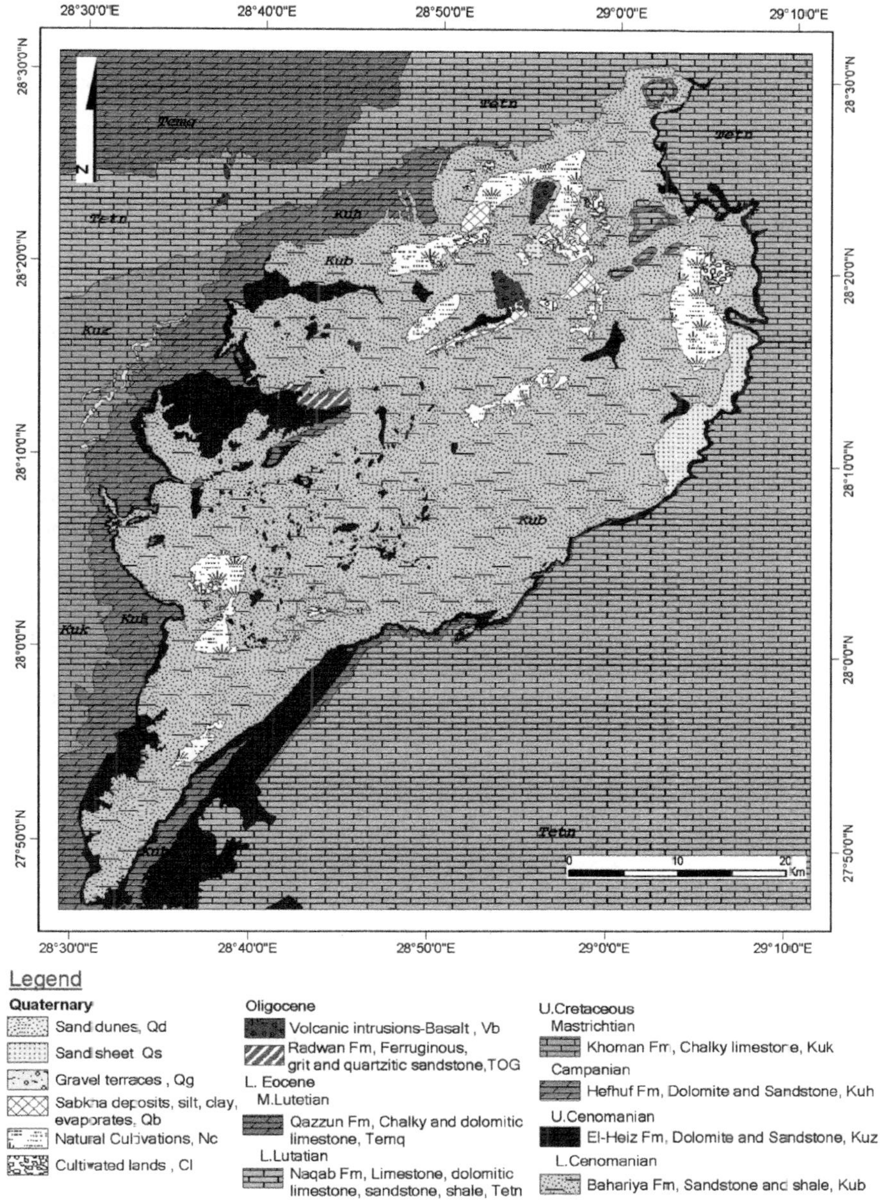

Fig. 9 Exposed geological units in the study area [45]

c. Hefhuf Formation (Kuh) of Campanian-Turonian includes shallow marine multi-colored shale and sandstone locally containing phosphate layers.
d. Khoman Formation (Kuk) of Maastrichtian contains chalk and chalky limestone.

Paleocene Rocks

The Paleocene rocks contain Tarawan Chalk and Esna Shale.

a. Tarawan Chalk (Lower Paleocene) appears at the south of the Bahariya depression (20–30 m thick) and contains chalky limestone and limestone.
b. Esna Shale (Upper Paleocene-Lower Eocene) occurs at the southern escarpment of the plateau separating El-Bahariya and El-Farafra depressions.

Eocene Rocks

The Eocene rocks comprise limestone forming the eroded plateau surface surrounding the Bahariya depression and some of the hills. They overlay unconformably the Upper Cretaceous rocks. The Eocene formations contain:

a. Farafra Formation that constitutes the upper member of the Eocene succession.
b. Naqb Formation (Lower Middle Eocene) that contains pink to violet Nummulitic dolostone and limestone beds.
c. Qazzun Formation (Upper Middle Eocene) that consists of white Nummulitic limestoneand dolostone.
d. El-Hamra Formation (Middle-Upper Eocene) that contains fossiliferous limestone beds overlaying the Qazzun Formation with 63 m in thickness.

Oligocene

The Oligocene rocks are comprise volcanic basaltic rocks and Radwan Formation (TOG) containing ferruginous, grit and quartzitic sandstones.

a. The volcanic rocks are exposed at several localities of the depression (Measera Hill, Mandisha Hill, El-Hefhuf Hill, and Basalt Hill). Some formations are capped with of basaltic sheets or doleritic rocks (Oligocene age).
b. Qatrani Formation (Oligocene) contains weathered sediments from the Cretaceous and Eocene rocks.

Quaternary Rocks

The Quaternary in Bahariya depression is represented by sand dunes (Qd), sand sheets (Qs), gravel terraces (Qg), playa deposits, sabkha deposits, silt, clay, and evaporates (Qb).

Hydrogeology of the Bahariya Oasis

The groundwater in Bahariya Oasis is a part of the major Nubian Sandstone Aquifer System (NSAS). The Precambrian basement complex forms the base of the aquifer. It was tapped at a depth of about 1,824 mbelow mean Sea Level (bmsl) in well BOX1. This area is located at the high structural ridge of the basement rocks, with different depths ranging between 1,400 and 1,700 mbmsl (see Fig. 10). Hydrogeologically, the main groundwater aquifer used in Bahariya Oasis is the sandstone members. The clays and shales form the confining conditions (Table 2).

The Cambrian Rocks

The Cambrian aquifer unit is represented by argillaceous sandstone intercalated with ferruginous shales. It is encountered only in the subsurface as shown in well BOX1 with 467 m in thickness of argillaceous sandstone. It overlays the igneous basement rocks.

The Lower Cretaceous

The Lower Cretaceous (Pre-Cenomanian) aquifer contains sandstone forming zone (S1) capped by a clay unit C1 (see Fig. 11). It is penetrated in the northern part of Bahariya depression by well BOX1 with 451 m in thickness. The Lower Cretaceous unit is partially penetrated in Qubala well-2 and El-Heiz deep well. The sandstone zone (S1) contains also a few clay streaks. It is also characterized by high effective porosity and permeability. The S1 is capped by a clay forming aquiclude (C1) with 225 m in thickness.

The Lower Cenomanian (Bahariya Formation)

The Lower-Cenomanian (Bahariya Formation) unit consists of two water-bearing sandstone zones (S3 and S2) separated by impermeable clay units (C3 and C2), respectively (see Fig. 11).

The Middle Ferruginous Sandstone (S2)

The middle ferruginous sandstone zone (S2) is found out in all drilled water wells. The thickness of S2 zone ranges from 150 mto 300 m. lithologically, it contains

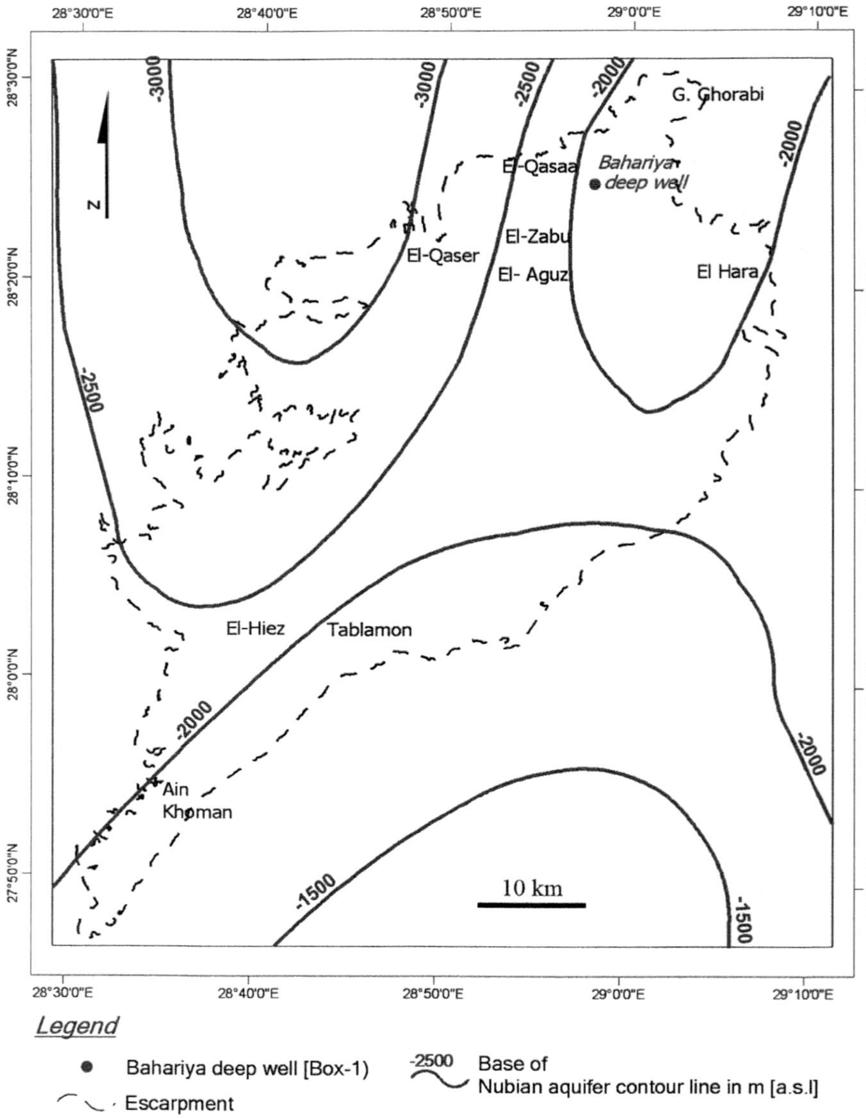

Fig. 10 Base of the groundwater aquifer contour map in Bahariya Oasis [51]

mainly sandstone with clay and shale intercalations. The S2 zone shows harder dark brown ferruginous bands. The middle clay member (C2) forms the basal beds of the ferruginous sandstone rock unit of Bahariya Formation. It is located between two semi-confined clay units C1 and C2. The thickness of C2 ranges from 200 to

Table 2 Simplified stratigraphic succession in the Bahariya Oasis [53]

Era	Age	Stage	Fm	Description	Hydro-geologic unit	Thickness (m)
Cretaceous	U. Cretaceous	Cenomanian	L. Cenomanian (El Bahariya Fm.)	Clay greenish grey, calcareous intercalated with sandstone streeks	C3	50–240
				Sandstone white to grey, fine to coarse with some clay streaks	S3	230–375
				Clay, dark grey, black with sandstone intercalations	C2	80–200
				Sandstone with harder dark brown ferruginous bands	S2	150–300
	L. Cretaceous			Shale, clay, dark, brown, ferruginous with sandstone intercalations	C1	225
				Sand white to yellow, fine to coarse	S1	451
Cambrian				Argillaceous sandstone with dolomitic cement and intercalated with ferruginous shale		467

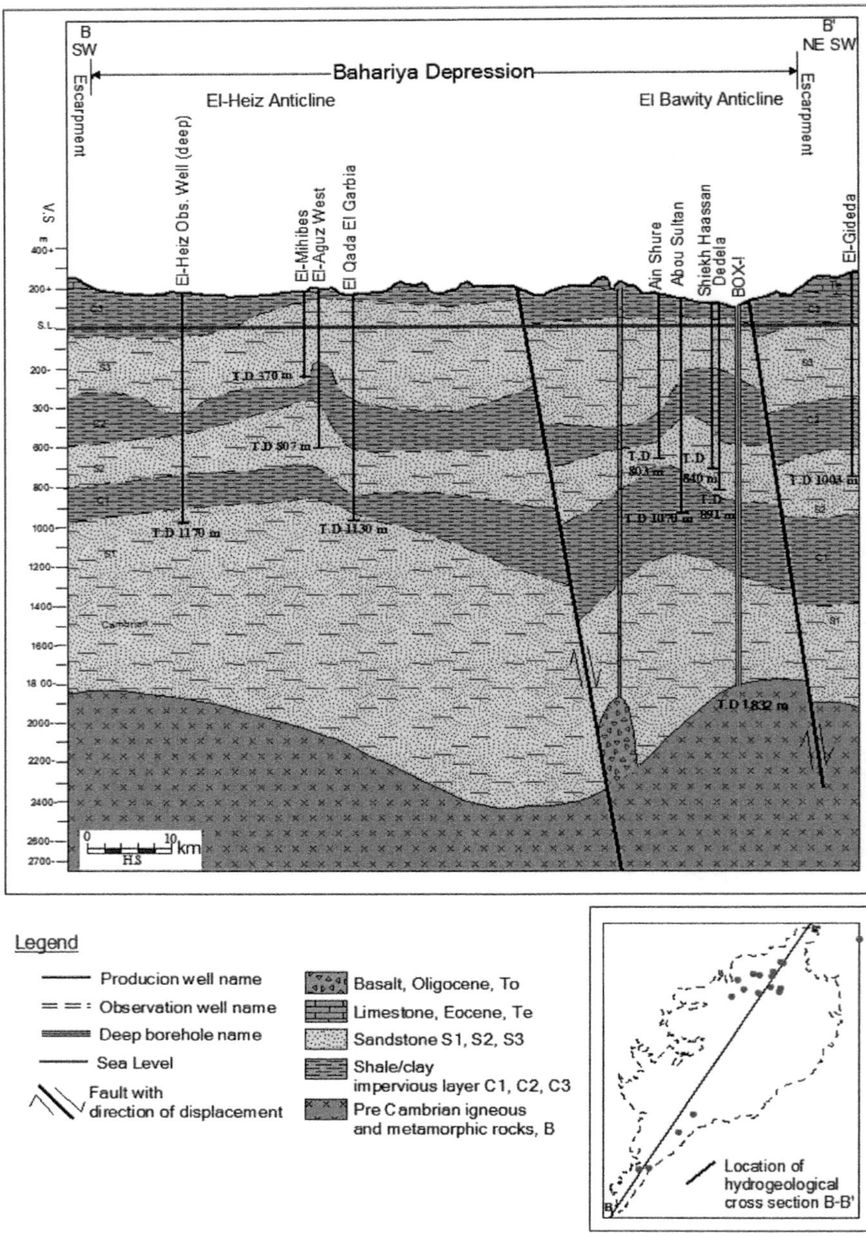

Fig. 11 Hydrogeological cross-section through the Bahariya depression [52]

80 m. Most deep wells penetrating S2 zone lie at the northern part of the Bahariya depression.

The Upper Sandstone Zone (S3)

The upper sandstone zone (S3) shows a wide extension in Bahariya depression. The thickness of S2 reaches to 375 m at El-Bawiti (Rest House well). The minimum thickness is recorded at about 230 m in El-Gedida mining area. The S3 zone contains sand, sandstone, gravels in parts with some clay and shales. It is toped by the confined clay aquiclude (C3) that has a thickness ranging from 50 to 240 m. It is the shallowest zone discharging about 70% of Bahariya water consumptions.

Piezometric Head and Groundwater Flow

The Regional Groundwater Flow

The general pattern flow of the groundwater in the Western Desert is from southwest to northeast [17, 46–49]. The regional annual drop in the piezometric head is 1.4 m/year [50].

The Local Groundwater Flow

The Piezometric Head in S1

The lower sandstone zone (S1) has the highest piezometric head in Bahariya depression, where it is reached 135 mamsl as recorded in El-Heiz and El-Hara areas. The 36 water wells penetrate S1 (see Fig. 12). The water flows to the center of the northern area of (see Fig. 13). The S1 aquifer is capped by semi-permeable clay (C1) with a 451 min thickness as shown in well BOX1. The lowest head (91 m amsl) was recorded at El-Sheikh Hassan and Ain Shwqifor wells (Table 3).

The Piezometric Head in the S2

The middle sandstone zone (S2) ranges from 88 to 157 m amsl in northern Bahariya depression (Table 4). The 32 production wells penetrate the S2 (see Fig. 14). The groundwater flow direction is mainly from the southwest to the northeast (see Fig. 15), with a local flow to the center of the northern area. Korany (1984) concluded A flow from southwest to northeast was recorded by Korany [54], which confirms with the

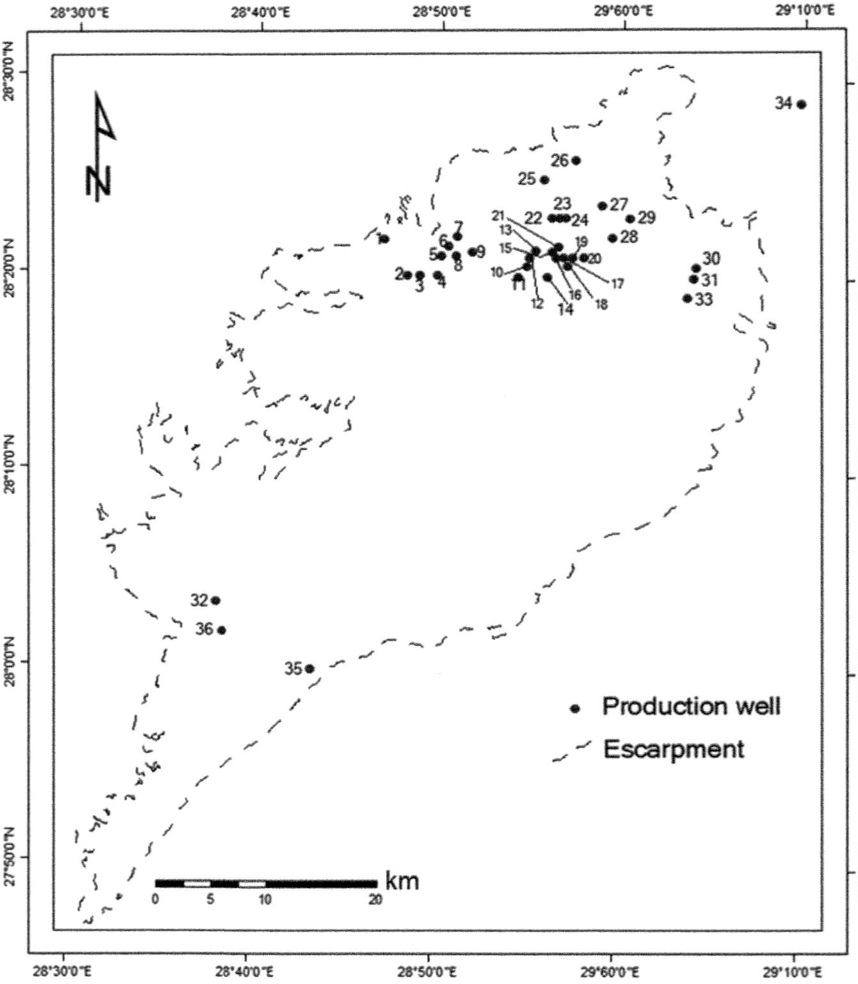

Fig. 12 Location map of the water wells penetrating the water-bearing zone (S1), Bahariya depression

regional general flow. This S2 is capped by the semi-permeable clay unit (C2), with a thickness of about 300 m. The highest piezometric head in depression was 157 mamsl as measured at El-Asila and reached to 88 mamsl at El Bawiti area (Table 4).

The Piezometric Head in the S3

The upper sandstone zone (S3) is shallow, varying in depth between 104 and 152 mamsl. It is penetrated by 84 water productive wells (see Fig. 16). The water flows

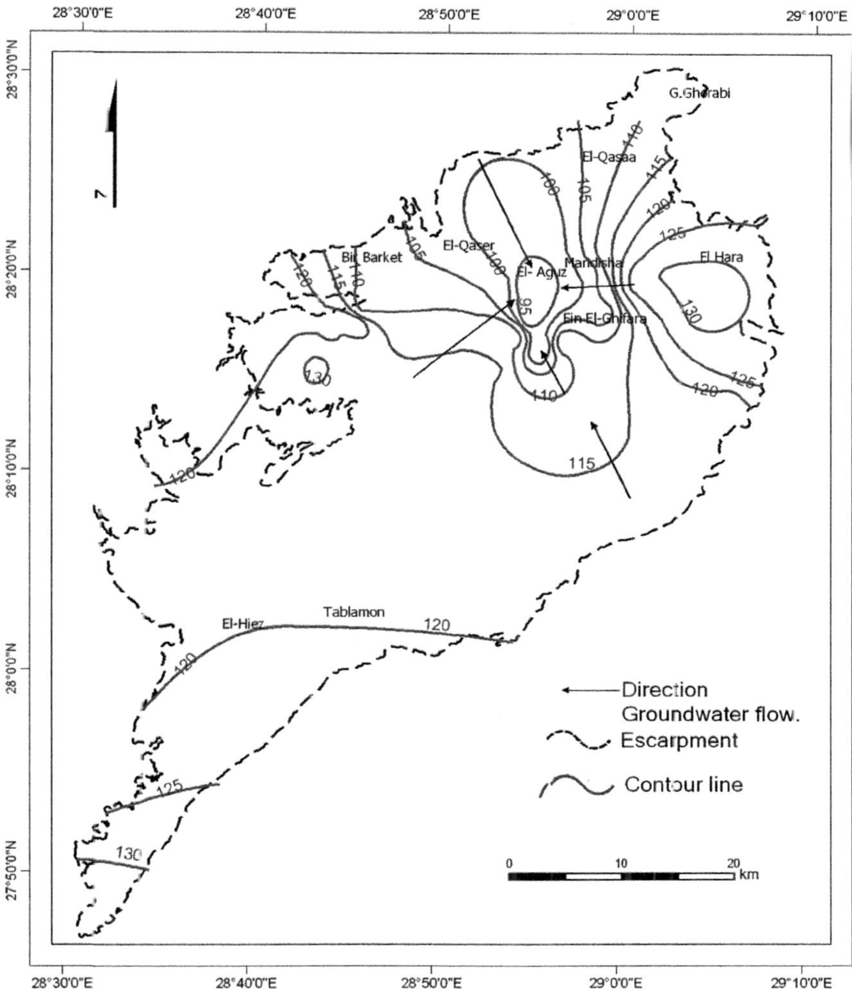

Fig. 13 Contour map of the average piezometric head (m amsl) of the sandstone water-bearing zone (S1) Bahariya depression in 2012 [59]

to the north in the southern sector of the depression. Local down cones are recorded in the northern area (see Fig. 17). The highest piezometric head in the S3 aquifer was 152 mamsl at Um Khalif, while the lowest was 91 mamsl at Ain Youssef (Table 5). The hydraulic gradient of the groundwater was 0.8 m/km. The hydraulic parameters were varied through the successive zones and within the same zone as well due to the variation in lithology and structure.

Table 3 The Water levels in selected wells of the lower sandstone zone (S1) in 2012

No	Well name	District	Total depth (m)	Piezometric head (m amsl)	X-degree	Y-degree
1	Barakat	El-Qaser	834	132	28 46 55	28 21 50
2	El-Dolab El-Gghrbi	El-Qaser	1002	118	28 48 14	28 20 00
3	El-Tabania	El-Qaser	809	111	28 48 55	28 20 01
4	Waled	El-Qaser	1167	118	28 49 55	28 20 01
5	Ain Halfa	El Bawiti	1111	122	28 50 03	28 21 01
6	El-Eza and El-Sheik	El-Heiz	1152	135	28 50 29	28 21 30
7	El-Sarrira	El-Qaser	1015	107	28 50 56	28 22 00
8	Ain Shawer	El-Qaser	803	125	28 50 56	28 21 01
9	El Qlise	El-Bawety	758	120	28 51 50	28 21 13
10	El-Aguz El-Gghrbi	El-Aguz	1045	115	28 54 48	28 20 34
11	El-Aguz El-Sharqi	El-Aguz	794	116	28 54 21	28 20 01
12	El Mehsra	Mandisha	1080	105	28 54 56	28 21 01
13	Abu Ghafer	El Zabu	885	113	28 55 16	28 21 20
14	Abu Singo	El Zabu	1123	115	28 55 56	28 20 01
15	El-Mehbese	El Zabu	771	123	28 56 12	28 21 17
16	Kum Shada	El Zabu	793	123	28 56 22	28 21 01
17	Maroud	El Zabu	1050	97	28 56 46	28 21 00
18	Segam	Mandisha	781	91	28 57 02	28 20 35
19	El-Qabala El-Sharqi	Mandisha	1138	97	28 57 17	28 21 01
20	kum Shroun	Mandisha	902	114	28 57 56	28 21 01
21	Abu Sultan	El Zabu	1076	93	28 56 29	28 21 35
22	Ain El Qaser	El Zabu	1065	101	28 56 09	28 23 01
23	El-Sheikh Hassan	El Zabu	840	90	28 56 33	28 23 00
24	Ain Shwqi	El Zabu	1003	90	28 56 53	28 23 01
25	Qasa 1	El-Qasat	780	97	28 55 40	28 25 01
26	Qasa 3	El-Qasat	954	98	28 57 26	28 26 00
27	Abu Bebi	El Zabu	890	102	28 58 53	28 23 40
28	Qasa 4	El-Qasat	979	101	28 59 29	28 22 01
29	Ain Gemhi	Mandisha	898	130	29 00 26	28 23 00
30	El-Aguz	El-Hara	808	135	29 04 04	28 20 33

(continued)

Table 3 (continued)

No	Well name	District	Total depth (m)	Piezometric head (m amsl)	X-degree	Y-degree
31	Ain Gded	El-Hara	974	129	29 03 57	28 20 01
32	El-Qara El Garbia	El-Heiz	1148	135	28 37 59	28 03 19
33	Ain El Wady-2	El Harra	830	118	29 03 38	28 19 01
34	El-Managem 7	El-Managem	1117	109	29 09 45	28 28 57
35	Tablamun	El-Heiz	1100	133	28 43 16	27 59 56
36	Ain El Eza	El-Heiz	1152	135	28 38 20	28 01 49

Hydraulic Parameters

In Bahariya depression, many pumping and recovery tests were carried out [55]. Three wells were selected to represent the different three zones: the deepest (S1), middle (S2) and the shallow (S3) zones. They are Abou Sultan (1,076 m), Ain El Wadi (790 m), and Ain Haddad (345 m), respectively. Measurements of the hydrogeological parameters through pumping tests which carried out by:

- Gradual pumping test step: It was done on three stages, every stage took 3 h with constant discharge for each stage with measuring the dynamic depth to the water level inside the well through each stage.
- Using the general well equation for producing total drawdown in water level inside the well in m.
- Drawing a relation between the discharge and the drawdown in the aquifer inside the well.
- Continuous pumping test step: A continuous pumping test was done at a constant discharge and the maximum declination in water level is measured.
- By Cooper-Jacob Equation the Transmissivity (T), Permeability (K) were calculated.

Pumping Test of the Deeper Zone (S1) in Abou Sultan Well

Pumping and recovery tests were carried out to determine the hydraulic parameters for the deep zone (S1), using a fixed discharge rate of 160 m^3/h with a duration of 6 consecutive hours. The maximum drawdown in the water surface was about 13 m. The relationship between the times (minutes) with drawdown (meters) is shown in Figs. 18 and 19. The transmissivity (T) and hydraulic conductivity or permeability (K) were calculated by using the Cooper and Jacob equation [56]:

Table 4 The Water levels in selected wells of the middle sandstone zone (S2) in 2012

No	Well name	District	Total depth (m)	Piezometric head (m amsl)	X-degree	Y degree
1	Ahmed El-klhli	El-Asila	380	145	28 54 55	28 17 01
2	Ahmed Arif	El-Asila	357	157	28 55 09	28 16 01
3	El-Habasi	El Zabu	498	97	28 57 02	28 21 01
4	El-Ain El-Bahria	El Harra	750	135	29 03 57	28 21 01
5	El-Ghaba El-Qbli	Mandisha	391	105	28 54 56	28 21 01
6	El-Kber	Mandisha	646	123	28 55 56	28 21 00
7	El-Mehbese El-Zibu	El Heiz	370	135	28 40 54	28 59 01
8	El-Wastani	El Heiz	352	135	28 37 54	28 02 01
9	El-Managem 4	El-Managem	600	115	29 09 48	28 29 58
10	El-Managem 5	El-Managem	615	109	29 11 45	28 28 47
11	Bir Abu Khtal	El Bawiti	370	97	28 52 45	28 22 01
12	Bir Abd El-Said	El Bawiti	370	99	28 50 56	28 23 00
13	El-Managem 8	El Managem	601	112	29 09 57	28 29 25
14	T-Adila	El Bawiti	380	106	28 50 55	28 23 38
15	Ghamel El-Malik	El Bawiti	365	88	28 51 56	28 25 01
16	Hamito	Mandisha	356	136	28 54 55	28 14 00
17	Adel Usif Usif 1	El-Asila	385	156	28 56 56	28 11 00
18	Adel Usif Usif 2	Mandisha	352	146	28 56 56	28 12 00
19	Abed El Raouf 1	El Bawiti	351	101	28 51 55	28 25 01
20	Abed El Raouf 2	El Bawiti	390	91	28 52 36	28 25 00
21	Abd El-Aziz 1	Um Khlifa	370	152	28 32 53	28 05 01
22	Abd El-Aziz 2	Um Khlifa	355	142	28 37 53	28 06 00
23	Aid Ahmed Abd El	El Bawiti	370	144	28 51 56	28 25 19

(continued)

Table 4 (continued)

No	Well name	District	Total depth (m)	Piezometric head (m amsl)	X-degree	Y degree
24	Ain El-Wadi 1	El Harra	700	112	29 02 56	28 19 01
25	Ain Gemha	Mandisha	371	133	28 41 54	28 00 01
26	Fawzi Ibrahim	El-Asila	353	146	28 55 58	28 18 01
27	No name	El-Asila	360	142	28 55 22	28 15 01
28	No name	El-Asila	390	152	28 35 20	28 10 00
29	Mohamed Basuni	El-Asila	385	150	28 54 56	28 13 01
30	Mohamed Farag	El-Asila	355	134	28 55 56	28 17 00
31	Hisham Snousi Ni	El-Asila	355	137	28 57 56	28 14 01
32	Wadi El-Mazare	El Bawiti	598	132	28 46 57	28 21 34

$$T = 2.303\ Q\ /4\pi\ \Delta s$$

where

T Transmissivity of the aquifer (m^2/d).

Q Discharge or pumping rate (m^3/d).

π Constant.

Δs Slope of the fitted line (change in drawdown per log cycle time).

Substituting the value of Q = 3,840 m^3/d and Δs = 0.62 in the Jacob's equation, the T = 1,134 m^2/d. Transmissivity is the rate of flow under a unit hydraulic gradient through a unit width of the aquifer of a given saturated thickness. The hydraulic conductivity (K) was calculated as follows:

$$K = T / H$$

where

K Hydraulic conductivity.

T Transimissivity (m^2/d).

H Saturated aquifer thickness (m).

Substituting the value of T = 1,134 m^2/d and H = 400 m, The K = 1,134 / 400 = 2.9 m/d.

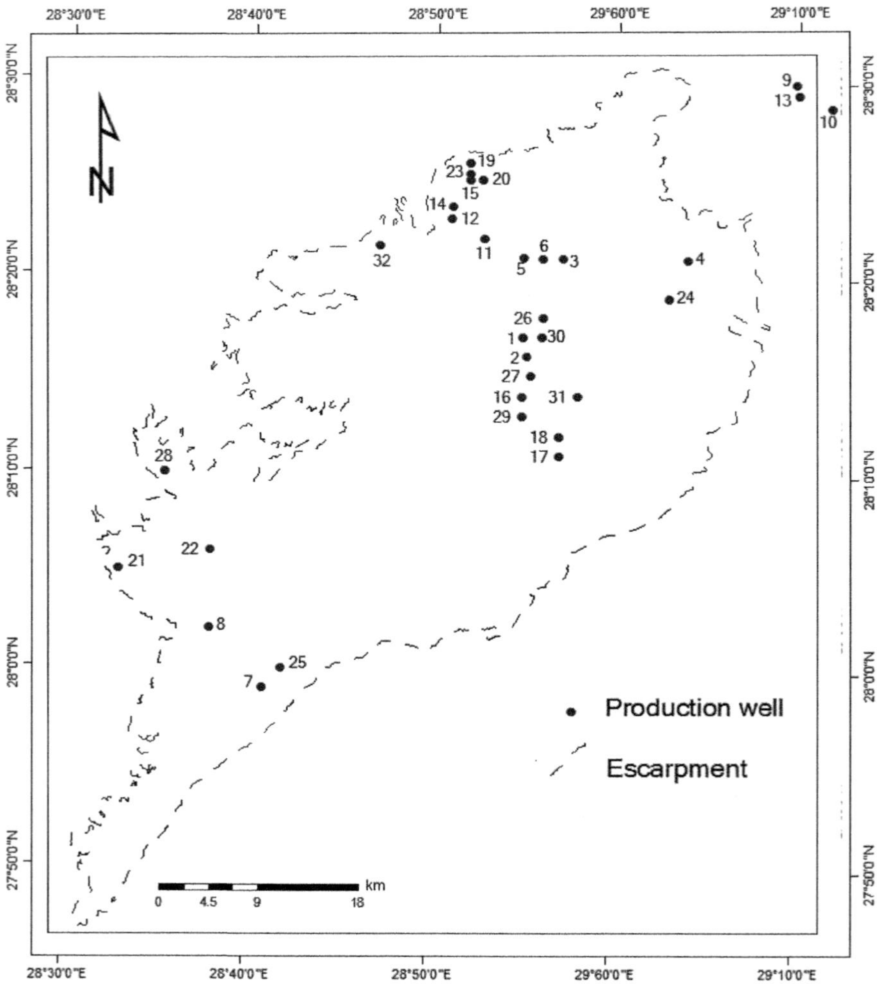

Fig. 14 Location map of the water wells penetrating the water-bearing zone (S2) Bahariya depression

Pumping Test of the Middle Zone (S2) in Ain El Wadi Well

Pumping and recovery tests were carried out to determine the hydraulic parameters for the middle zone (S2), using a fixed discharge rate of 150 m^3/h with a duration of 19 consecutive hours. The maximum drawdown in the water surface was about 16.62 m. Using the value of $Q = 3{,}600\ m^3/d$ and $\Delta s = 1.4$ in Jacob's equation, the transmissivity (T) is 471 m^2/d. Using the value of $H = 100$ m, the hydraulic conductivity (K) is 4.7 m/d.

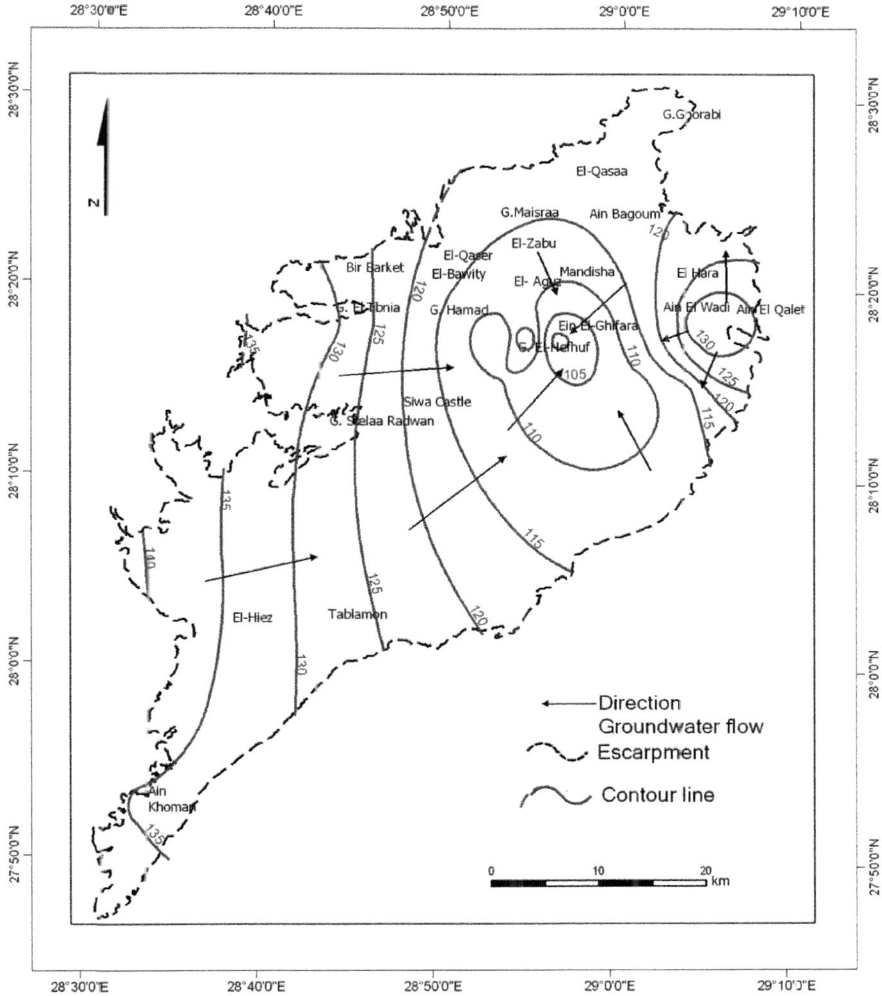

Fig. 15 Contour map of the average piezometric head (m amsl) of the sandstone water-bearing zone (S2) Bahariya depression in 2012 [59]

Pumping Test of the Shallow Zone (S3) in Ain Haddad Well

Pumping and recovery tests were carried out to determine the hydraulic parameters for the shallow zone (S3), using a fixed discharge rate of 135 m3/h with a duration of 17 consecutive hours. The maximum drawdown in the water surface about 8.25 m. Using the value of $Q = 3{,}240\ m^3/d$ and $\Delta s = 0.58$ in the Jacob equation, the $T = 1{,}023\ m^2/d$. Using the value of $H = 200$ m, the hydraulic conductivity (K) is 5.1 m/d.

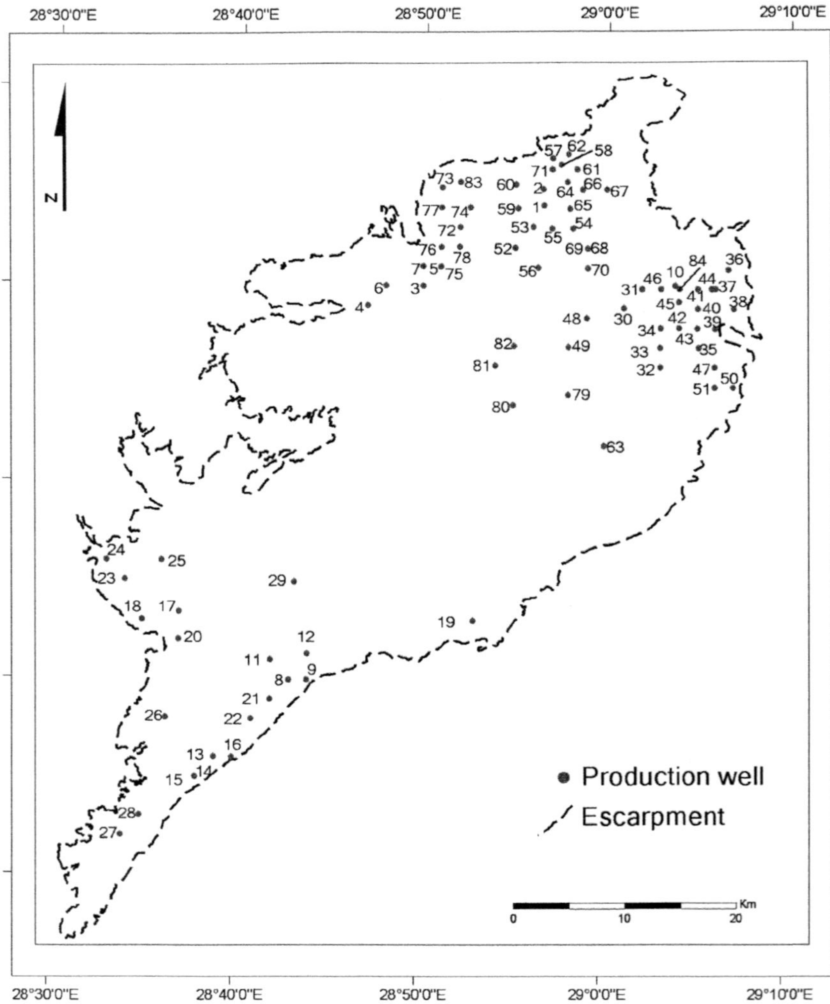

Fig. 16 Location map of the water wells penetrating the water-bearing zone (S3), El Bahariya depression

The Recharge of the Nubian Aquifer System

In the Upper Pleistocene and Holocene, there was a successive series of the pluvial and sub-pluvial interval, with heavy rainfall, which affected directly the Nubian aquifer. The filling process ends when recharge and discharge balance each other. Such state probably existed about 8,000 years ago [8].

In the last 8 thousands, no significant recharge had taken in the aquifer system. The infiltration and the seepage through the modern rain events as well as from

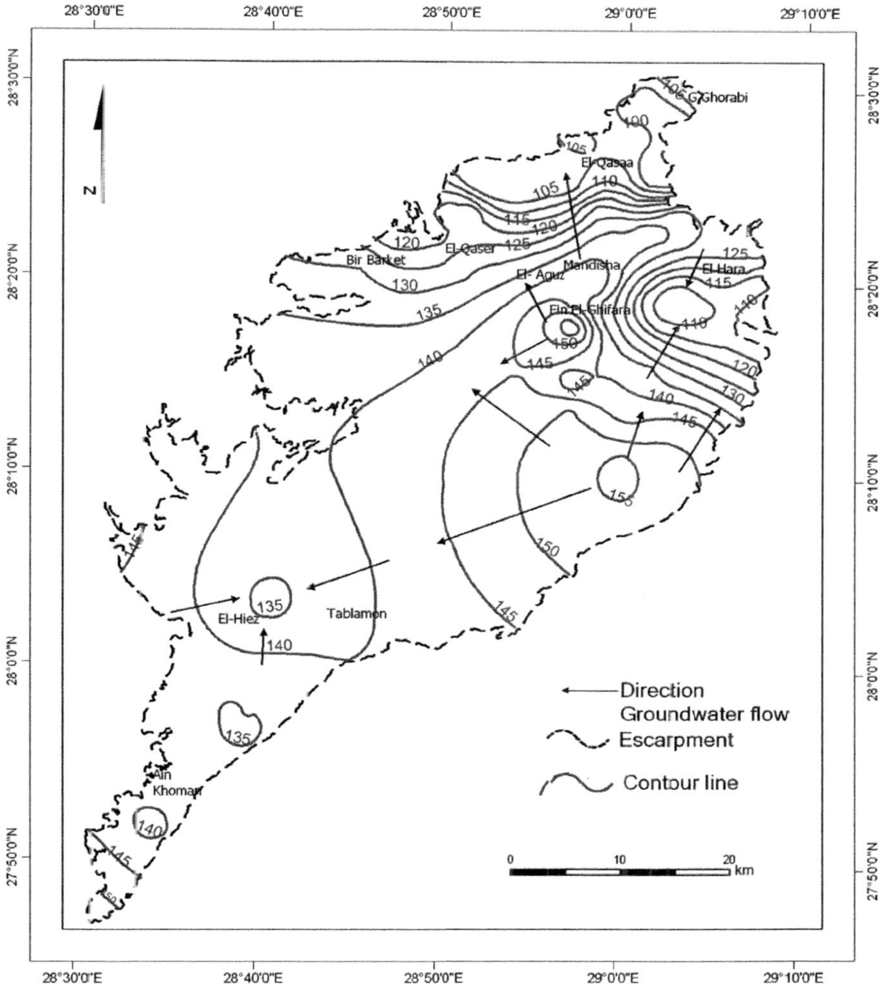

Fig. 17 Contour map of the average piezometric head (m amsl) of the sandstone water-bearing zone (S3), Bahariya depression in 2012

the River Nile and the Lake Aswan Dam show a very limited contribution to the groundwater. In general, the Nubian Sandstone Aquifer System is a non-renewable resource [8, 35, 57].

Table 5 The Water levels in selected wells of the upper sandstone zone (S3) in 2012

No	Well name	District	Total depth (m)	Piezometric head (m amsl)	X-degree	Y-degree
1	EL-Dolab El-Bahri	El Zibu	324	104	29 01 56	28 20 00
2	EL-Ain El-Bahria	El Zibu	301	104	28 33 53	27 52 01
3	Embash	El Qaser	271	118	29 06 57	28 19 01
4	Madi	El Qaser	345	129	28 56 57	28 26 34
5	Estkmal El-Tebnia	El Qaser	305	125	29 45 07	28 19 01
6	Safi and Khawga	El Qaser	306	115	28 00 58	28 22 01
7	El-Bitah and El-Abwar	El Qaser	252	122	29 56 56	28 19 00
8	Sid Ahmed	El Zibu	347	130	28 51 24	28 22 30
9	Tabl Amoun	El Hiez	320	135	28 54 55	28 14 01
10	Ain Hidad	El Harra	195	129	28 52 55	28 03 07
11	Abu Yaser El-Ghrbi 1	Tabl Amoun	115	142	28 42 54	28 00 16
12	Kamid 1	Tabl Amoun	108	144	28 59 56	28 12 01
13	Aelit Mrdi 1	El Rais	107	137	28 39 53	27 56 01
14	Kilani Amin	El Rais	100	143	28 56 56	28 23 01
15	Khiri Ismail	El Rais	100	143	28 58 08	28 23 01
16	Abd Alah Abu Said	El Rais	133	142	28 49 55	28 20 01
17	Kum El-Tel 2	El-Gghrbia	102	135	28 56 13	28 21 00
18	Imam	El-Gghrbia	125	127	28 57 56	28 17 00
19	Abd El-Salam	El-Gghrbia	125	144	28 56 29	28 24 11
20	El-Shibla	El-Gghrbia	127	144	29 06 37	28 21 01
21	El-Bir EL-Sharqi 1	El-Ries	107	137	28 43 07	28 05 01
22	El-Bir EL-Bahri 1	El-Ries	109	141	28 34 53	27 53 01
23	Abd El-Aziz ali 3	Um khalif	155	152	28 50 54	28 21 01
24	Abd El-Aziz ali 4	Um khalif	152	152	28 47 55	28 20 01
25	Abd El-Aziz ali 5	Um khalif	160	147	28 49 55	28 21 01
26	Ghmal Azam	65 km Farafra	170	142	29 03 56	28 19 20

(continued)

Table 5 (continued)

No	Well name	District	Total depth (m)	Piezometric head (m amsl)	X-degree	Y-degree
27	Esam Khlibush	Farafra Road	155	162	29 05 57	23 18 00
28	Abd el Maksud Fawzi	Farafra Road	149	155	28 46 55	23 19 00
29	Nasser Ali 2	El Qasaa 3	199	133	29 05 45	28 20 01
30	Mohmed Abu Zid	Ai El-Wadi	80	125	28 58 18	28 26 01
31	Kabren Walid	Ai El-Wadi	84	130	29 06 57	28 15 01
32	Mahud	El-Khadrwa	110	108	28 57 26	28 2615
33	El-Haj Ali	El-Khadrwa	115	108	29 02 57	28 17 00
34	Mhi Gemha	El-Khadrwa	95	111	28 55 04	28 24 01
35	El-Haj Sliman 1	Hatiet Qebli	65	108	29 05 02	28 17 00
36	Zihum Hassan	El-Ghmiza	85	140	29 05 57	28 20 01
37	Ghmal Khalil 1	El Qasaa3	100	145	28 58 35	28 25 00
38	Ali Snousi	Liqata	120	150	28 36 53	28 02 01
39	Gbril Abd El-Hafiz 1	Liqata	110	150	29 04 56	28 18 00
40	Hamda Mussa	Ain El-banat	70	132	29 02 56	28 20 00
41	Rajab Abd El-Zaher	Bir El-Shariq	100	150	28 57 46	28 25 22
42	Farma 1	Ain Yosif	110	125	29 45 07	28 20 00
43	Abu Nigaa 1	Ain Yosif	141	150	28 37 54	27 55 00
44	Shaban Abd El-Salem	El Mazare El Bahariya	126	150	28 50 55	28 25 01
45	El-Haj Shafia	Ain Abu El Eze	125	137	29 02 56	28 18 00
46	Saleh Mahmod	Ain Hidad	110	135	28 56 56	28 26 01
47	Abd Alah Abd El-Azim	Ain Hidad	115	125	28 56 27	28 25 00
48	Yhia Ibrahim 1	El-Qihaf	75	119	28 51 55	28 25 19
49	Hashim Kahmis	El-Qihaf	85	111	29 05 57	28 16 01
50	Sirkum Company	El Harra	160	130	28 51 55	28 22 00

(continued)

Table 5 (continued)

No	Well name	District	Total depth (m)	Piezometric head (m amsl)	X-degree	Y-degree
51	Sirkum Company	El Harra	165	113	28 57 56	28 14 34
52	Abu Himada 2	Abu Hmidaa	50	101	28 38 53	27 56 01
53	Sheri Mohmed 4	El-Qasafi	99	105	28 50 55	28 21 00
54	Abd El-Hafiz Suktan 1	Shitat 2	150	97	28 43 54	28 00 01
55	Abd El-Hafiz Suktan2	Shitat 3	100	105	29 02 33	28 19 22
56	Abd el-Hakim	El Himed	120	128	28 41 54	28 01 00
57	Emad Mohmed Amir	El Qasaa 3	95	110	28 56 45	28 26 01
58	Mohmed Abd El-Azim	El Qasaa 3	75	104	28 54 56	28 25 12
59	Bir El-Dist 1	El Qasaa 1	125	108	28 33 53	28 05 01
60	Ramdan	El-Mualqa	140	109	28 57 55	28 24 01
61	Saleh El-Khatib	El-Mualqa	120	103	28 51 55	28 23 01
62	Abd El-Hafiz kalif Alah	El-Mualqa	150	104	28 55 55	28 23 04
63	Sheif Mohmed Mursi 1	Ain Yosif	130	91	28 52 29	28 24 01
64	Subhi Radwan himda	Bir Nasser	120	97	28 53 56	28 16 00
65	Abd El-Karim Snousi	El Mazare El Bahariya	125	106	28 43 54	28 01 20
68	Nagi Mansur	El Qasaa 4	150	130	28 57 49	28 26 46
69	Fekri Slim	El Qasaa 4	115	96	29 03 56	28 18 01
70	Aimad Abd El-Bari	El Qasaa 4	140	97	28 34 52	28 03 01
71	Tharwt Mohamed	El-Mualqa	120	88	28 54 56	28 17 01
72	Bir Abu Bahlul	El Bawiti	115	91	28 40 54	27 58 01
73	Bir Abid	El Bawiti	100	99	28 41 54	27 59 01
74	Bir Sid 2	El Bawiti	120	97	28 36 14	27 58 01
75	Bir Saber	El Bawiti	130	97	28 35 53	28 06 01

(continued)

Table 5 (continued)

No	Well name	District	Total depth (m)	Piezometric head (m amsl)	X-degree	Y-degree
76	Bir Ell-Sahil	El Bawiti	140	105	28 32 52	28 06 01
77	El-Frashid 1	El Bawiti	130	106	29 02 57	28 16 01
78	Sid Esa	El Bawiti	120	101	28 50 55	28 24 01
79	Hisham Snousi	El-Aisla	110	136	28 58 56	28 18 28
80	Kalid Ell-Sheikh	El-Aisla	125	157	29 05 57	28 15 00
81	Kalid El-Sudi	El-Aisla	145	142	28 54 56	28 22 01
82	Said fathi Mustfa	El-Aisla	130	150	28 58 56	28 22 01
83	Aid Ahmed	El Bawiti	120	146	28 36 53	28 03 25
84	Saleh Khalil 1	El-Ghmiza	100	140	28 58 56	28 21 01

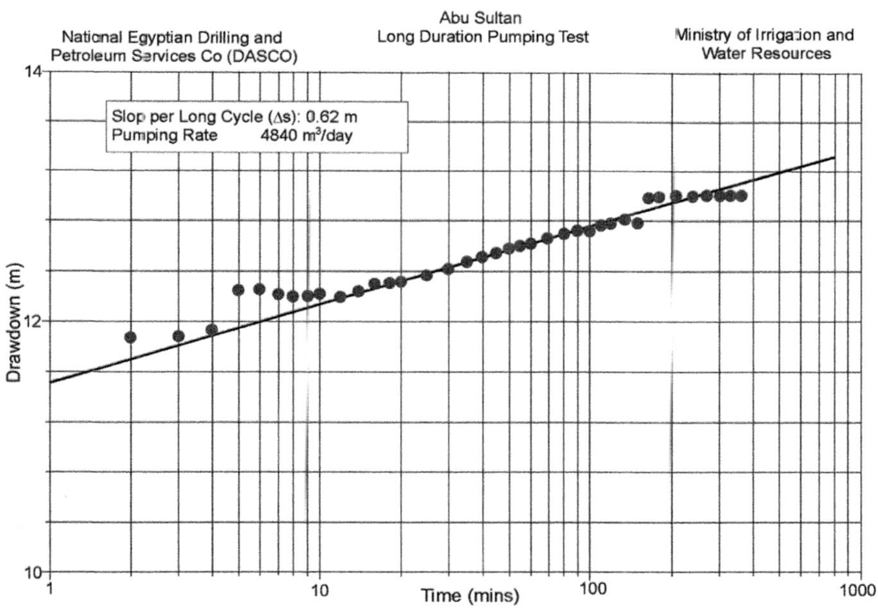

Fig. 18 Relationship between the times and drawdown in pumping test (Abou Sultan well)

The Groundwater Extraction

The groundwater from the Nubian Aquifer has been mostly utilized in Bahariya Oasis through springs and shallow wells. The intensive groundwater usage for different

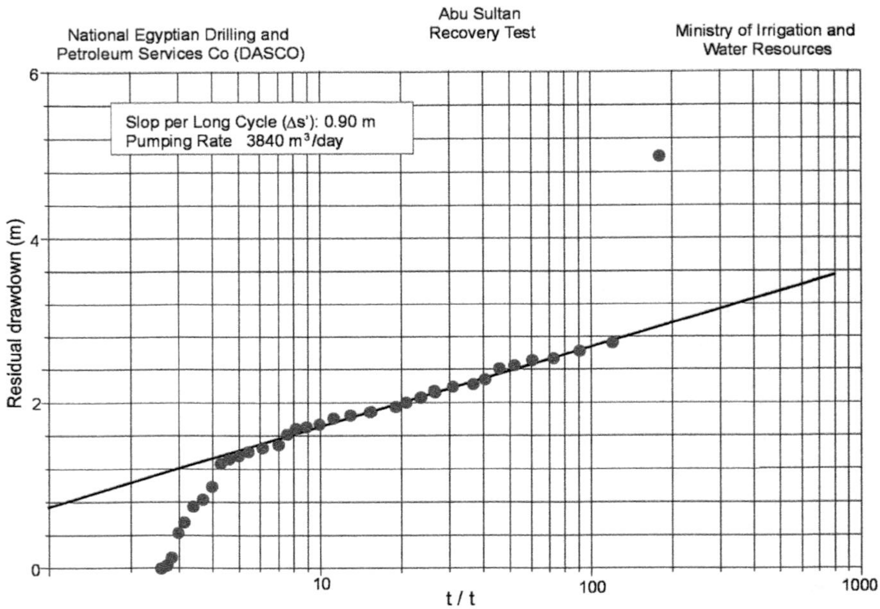

Fig. 19 Relationship between the times and residual Drawdown in recovery test (Abou Sultan well)

Table 6 The amount of water extraction (Q) and the number of wells for each zone [58]

Aquifer	No of wells	Q (m^3/h)	Q (m^3/day) (8 h)	Q (Mm^3/y) (300 d)
S1	66	7,688	61,519	18.5
S2	24	2,084	16,672	5
S3	793	35,192	281,536	84.5
Total	883	44,964	359,727	108

purposes (irrigation, mining, and water supply) may cause severe and dangerous damage on the water aquifer. In 2012, the number of water wells used was 883 produced about 116 million m^3 [58]. They are mostly from the shallow aquifer (S3), where 793 wells were used to discharge 92.3 Mm^3 (Table 6) [58].

Monitoring the Groundwater Flow

Visual Modflow 3.0 model and its required data were collected from the field and other institutions to simulate the groundwater flow in the depression. The data include the

Table 7 Water discharges from the three zones in Bahariya

Aquifer	No. of wells	Area (fed)	Q m^3/h	Q $m^3/h/fed$	Q m^3/d (8 h)	Q $m^3/d/fed$	Q Mm^3/y (300 d)
S1	66	4,780	7,688	1.608	61,519	12.87	18.5
S2	24	1,326	2,084	1.572	16,668	12.57	5.0
S3	793	8,500	35,192	4.14	281,520	33.12	84.5
Total	883	14,606	44,964	7.32	359,707	58.56	108

hydraulic conductivity and specific storage obtained from pumping tests, geological studies, and geophysical logs [58]. The results of the calibration and assuming water allocation per Fadden range from 15 to 20 m^3/d (Table 7), according to the irrigation system and crop type. The drawdowns were estimated for the three groundwater zones S1, S2, and S3 [59].

1. Under the high water allocation per Fadden, the expected drawdown is still in the safe range.
2. The maximum expected drawdown ranges from 3 to 26 m for the minimum water allocation after 25 years, representing less than 5% of the saturated thickness. The drawdown ranges from 4 to 32 m after 50 years, which is consistent with El Hossary [32].
3. The new expected cultivated area within the safe usage is about 50,000 Fadden from the three zones [32].

Water Quality

In general, the groundwater of the Bahariya Oasis is mainly freshwater (less than 1000 mg/l). The salinity of the groundwater in the Bahariya aquifers varies slightly horizontally and vertically from aquifer to another. The total dissolved solids (TDS) calculated from the major ions in the S1 aquifer range from 177 mg/l in the northern east to 318 mg/l with an average of 232 mg/l. The salinity in S2 zone is slightly higher than that of the S1 (Table 8), ranging between 166 and 436 mg/l with an average of 301 mg/l. The average salinity content for the aquifer S3 is 326 mg/l. The three aquifers S1, S2, and S3 have freshwater with an average of salinity of 286 mg/l.

Conclusions and Recommendations

The Bahariya depression is located in a hyper-arid climate zone. The rainfall is scarce and the evaporation is very high. The main water source is non-renewable groundwater aquifer system. The groundwater flows generally from southwest to northeast, following the regional general flow of the Nubian Sandstone aquifer. The

Table 8 Summary of some physio-chemical properties of the Bahariya groundwater [59]

Average	Range		No. of samples	Zone	Physio-chemical parameters
	to	from			
0.36	0.47	0.25	28	S1	EC (dS/m)
0.45	0.68	0.24	12	S2	
0.48	2.24	0.21	44	S3	
7.08	7.84	6.5	28	S1	pH
6.98	7.75	5.39	12	S2	
6.88	7.65	6.03	44	S3	
232	328	177	28	S1	TDS (mg/l)
301	436	166	12	S2	
326	1537	135	44	S3	

hydraulic gradient of the Bahariya groundwater flow is 0.8 m/km. The transmissivity values differ from zone to another. They vary from 134 m^2/d for zone (S1) to 471 m^2/d in zone (S2) and 1,023 m^2/d in zone (S3). The upper zone (S1) and the lower zone (S3) are classified as highly potential aquifers (>500 m^2/d). The middle zone (S2) is a moderate potential aquifer (5–50 m^2/d). The hydraulic conductivity for the Bahariya aquifers ranges from 2.9 m/d in S1 to 5.1 m/d in S3, reflecting very fine sand reservoir rock (1–5 m/d).

The recent annual consumption from the groundwater in Bahariya was 18.5, 5, and 84.5 Mm^3 from the deepest aquifer (S1), middle aquifer (S2) and shallow aquifer (S3), respectively. The expected drawdown will range from 3 to 26 m for the minimum water allocation in 25 years. It represents less than 5% of the saturated thickness. The expected drawdown will range from 4 to 32 m in 50 years. The Bahariya aquifer is good for additional irrigable 50,000 Feddans with a safe usage.

The water salinity in the three aquifer zones averages 286 mg/l. The extension of using the groundwater is recommended to be directed mainly towards industries, truism, and development service projects more than agriculture to achieve sustainable development. It is also strongly recommended to monitor the productive water wells in terms of water level and quality.

References

1. Zektser SI, Everett GL (eds) (2004) Groundwater resources of the world and their use. IHP-VI Series on Groundwater no 6, UNESCO, Paris
2. Pavelic P, Giordano M, Keraita B, Ramesh V, Rao T (eds) (2012) Groundwater availability and use in Sub-Saharan Africa: a review of 15 countries. International Water Management Institute (IWMI), Colombo, Sri Lanka, p 274
3. Adelana SMA, MacDonald AM (2008) Groundwater research issues in Africa. In: Applied groundwater studies in Africa. In: Adelana SMA, MacDonald AM (eds) IAH selected papers on hydrogeology, vol 13. CRC Press/Balkema, Leiden, The Netherlands

4. MacDonald AM, Bonsor HC, Dochartaig BEO, Taylor RG (2012) Quantitative maps of groundwater resources in Africa. Environ Res Lett 7(2):1–7
5. Masoud NH, El Osta MM (2016) Evaluation of groundwater vulnerability in El-Bahariya Oasis, Western Desert, Egypt, using modeling and GIS techniques: a case study. J Earth Syst Sci 12(6):1139–1155
6. Ahmed EHM (2013) Nubian sandstone aquifer system. Merit Res J Environ Sci Toxicol 1(6):114–118
7. Ahmed M, Sultan M, Wahr J, Yan E (2013) Is the Nubian. Geol Soc Am Abs Programs 45(7):681
8. CEDARE (2002) Regional strategy for the utilization of the Nubian sandstone aquifer system, vol II. CEDARE, Heliopolis Bahry, Cairo, Egypt
9. Bakhbakhi M (2006) Nubian sandstone aquifer system, Chapter 5, In: Foster SP, Loucks DP, (eds) Non-renewable groundwater resources UNESCO, pp 75–81
10. Klitzsch E (1994) Geological exploration history of the Eastern Sahara. Geol Rundsc 83:475–483
11. Hermina M (1990) The surroundings of Kharga, Dakhla and Farafra oases. In: Said R (ed) The geology of Egypt. Balkema, Rotterdam, pp 259–292
12. Klitzsch E, Wycisk P (1999) Beckenentwicklung und Sedimentationsprozesse in Kratonalen Bereichen Nordost-Afrikas im Phanerozoikum. In Klitzsch E, Thorweihe U (eds). Nordost-Afrika: Strukturen und Ressourcen, John Wiley & Sons-VCH, Weinheim, pp 61–108
13. Sefelnasr A (2002) Hydrogeological studies on some areas on the New Valley Governorate, Western Desert, Egypt. MSc thesis, Assiut University
14. Sharaky AM, Elewa HH, Kasem AM (2019) Impact of the Grand Ethiopian Renaissance Dam (GERD) on Gezira groundwater, Sudan. In: Negm AM (ed) The handbook of environmental chemistry, vol 79. Springer, Berlin, Heidelberg, pp 519–557
15. Hume WF (1925) Geology of Egypt, vol 1. The surface features of Egypt, their determining causes and relation to geological structure. Egyptian survey Department, Cairo, 408 p
16. Ball J (1927) Problems of the Libyan Desert. Geogr J, London, V. 10
17. Sandford KS (1935) Geological observations on the northwestern frontiers of the Anglo Egyptian Sudan and the adjoining part of the southern Libyan Desert. Quart J Geol Soc Lond 9:323–376
18. Attia MI (1954) Groundwater in Egypt. Desert Inst Bull Cairo 4(1):198–213
19. Thorweihe U (1990) Nubian aquifer system. In: Said R (ed.) The geology of Egypt, 2nd ed. Balkema, Rotterdam
20. Heinl M, Thorweihe U (1993) Groundwater resources and management in SW Egypt. In: Meissner B, Wycisk P (eds) Geopotential and ecology. Catena Suppl., 26:99–121
21. Ebraheem A, Riad S, Wycisk P, Seif El-Nasr A (2002) Simulation of impact of present and future groundwater extraction from the non-replenished Nubian Sandstone Aquifer in southwest Egypt. Environ Geol 43(1–2):188–196
22. Thorweihe U, Heinl M (2002) Groundwater resources of the Nubian aquifer system, NE-Africa. Modified synthesis submitted to: observatoire du Sahara et du Sahel. OSS, Paris, 23 p
23. Gossel W, Ebraheem AM, Wycisk P (2004) A very large scale GIS-based groundwater flow model for the Nubian sandstone aquifer in Eastern Sahara (Egypt, northern Sudan and eastern Libya). Hydrogeol J 12(6):698–713
24. Salem O, Pallas P (2004) The Nubian Sandstone Aquifer System. In Appelgren B (ed) Managing shared aquifer resources in Africa. ISARM-AFRICA. UNESCO, IHP-VI, Ser Groundw 8:19–21
25. Khalifa RM (2006) Study of groundwater resources management in El-Bahariya Oasis. PhD thesis, Fac Sci Alexandria Univ, 226 p
26. Sawires R (2010) Geophysical and hydrogeological studies on El-Bahariya area, Western Desert, Egypt. MSc thesis, Fac. Sci. Assiut University, Egypt
27. Tanner LH, Khalifa MA (2010) Origin of ferricretes in fluvialmarine deposits of the lower cenomanian Bahariya formation, Bahariya Oasis, Western Desert Egypt. J Afr Earth Sci 56:179–189

28. Hamdan M, Sawires F (2011) Hydrogeological studies on the Nubian sandstone aquifer in El-Bahariya Oasis, Western Desert. Egypt, Arab J Geosci 6(5):1333–1347
29. Hamdan M (2012) Multivariate statistical analysis of geochemical data of groundwater in El-Bahariya Oasis, Western Desert. Egypt, Res J Environ Earth Sci 4(6):665–667
30. Hamdan AM, Awad AA, Omran AAA, Rashad F, Sawires RF (2012) Evaluation of hydrogeochemical parameters of the groundwater in El-Bahariya Oasis, Western Desert. Egypt Assiut Univ J Geol 41(1):1–22
31. El Kashouty M, Abdel Aziz A, Soliman M, Mesbah H (2012) Hydrogeophysical investigation of groundwater potential in the El Bawiti. Northern Bahariya Oasis, Western Desert, Egypt. Arab J Geosci 5:953–97032
32. El Hossary MF (2013) Ensuring sustainable development via groundwater management (case study: El Bahariya Oasis). J Am Sci 9(6):6–13
33. Abd El All E, Khalil A, Rabeh T, Osman S (2015) Geophysical contribution to evaluate the subsurface structural setting using magnetic and geothermal data in El-Bahariya Oasis, Western Desert. Egypt. NRIAG J Astron Geophys 4:236–248
34. Rabeh T, Bedair S, Abdel Zaher M (2018) Structural control of hydrogeological aquifers in the Bahariya Oasis. Western Desert, Egypt. Geosci J 22(1):145–154
35. Sharaky AM, El Abd S, Shanab E (2019) Groundwater assessment for agricultural irrigation in Toshka area, Western Desert, Egypt. In: Negm AM (ed) Conventional water resources and agriculture in Egypt, the handbook of environmental chemistry, vol 74. Springer, Berlin, Heidelberg, pp 347–387
36. Ambroggi RP (1966) Water under the Sahara. Sci Am 241:21–29
37. Gischler CE (1976) Present and future trend in water resources development in Arab Countries. UNESCO report
38. Thorweihe U, Heinl M (1996) Groundwater resources of the Nubian aquifer system. Technical University of Berlin, Germany, p 95
39. Sonntag C (1986) A time dependent groundwater model for the Eastern Sahara. Berliner Geowiss. Abh., (A), 72:124–134
40. Sturchio NC, Du X, Purtschert R, Lehmann BE, Sultan M, Patterson LJ, Lu ZT, Müller P, Bigler T, Bailey K, O'Connor TPO, Young L, Lorenzo R, Becker R, El Alfy Z, El Kaliouby B, Dawood Y, Abdallah AMA (2004) One million year old groundwater in the Sahara revealed by krypton-81 and chlorine-36. Geophys Res Lett 31:1–10
41. Parsons RM (1962) Bahariya and Farafra areas, New Valley project, Western Desert of Egypt; Final report, Egyptian Desert Development Organization, Cairo
42. Embabi NS, Mostafa MA (1979) A morphotectonic map of the El Bahariya depression. Ann Geol Surv Egypt 9:179–183
43. El-Akkad S, Issawi B (1963) Geology and iron ore deposits of the Bahariya Oasis: Geological Survey and Mineral Research Department, Egypt, Paper vol 18, p 301
44. Said R, Issawy B (1964) Contributions to the Prehistory of Nubia, No. 1, Preliminary results of a geological expedition to lower Nubia and to Kurkur and Dungul Oases, Egypt, Museum of New Mexico Press Santa Fe, New Mexico
45. CONOCO (1986) Geological Map of Egypt 1, 500,000, Map Sheets NG35NE (Farafra) and NH35SE (El Bahariya). Klitzsch E, List FK, Poehlmann G, Berlin
46. Ezzat MA (1959) Origin of the underground water in the Libyan Desert and preliminary evaluation of General Desert Development Authority, p 862
47. Ayouti MK, Ezzat MA (1961) Hydrogeological observations in the search for underground water in the Western Desert of Egypt. International Association of Scientific Hydrology, Belgium
48. Hemida IH (1964) Artesian water of the oases of the Libyan Desert (U.A.R), Special publication of Moscow Geological Survey Institute (MGRU), Moscow, U.S.S.R. (in Russian Lang)
49. Diab MS (1972) Hydrogeological and hydrochemical studies on the Nubian Sandstone water bearing complex in some localities in U.A.R. PhD Department of Geology, Faculty of Science, Assiut University, p 251

50. Salem MRI (2002) Hydro-geological studies on the Nubian Sandstone Aquifer system in Farafra and El Bahariya Oases, Western Desert, Egypt. PhD thesis, Faculty of Science Ain Shams University
51. General Petroleum Company (1985) (GPC) Bouguer gravity map of Egypt scale 1:100,000
52. RIGW (2010) Save Yield Investigation Project of El Bahariya area (in Arabic)
53. Shata AA (1982) Hydrogeology of the great Nubian sandstone basin. Q J Eng GeolHydrogeol 15:127–133
54. Korany EA (1984) On the demonstration of the hydrogeological control by local geologic structures, El Bahariya Oasis Egypt. E.G.S. Proc of 3rd Ann M 341–355
55. RIGW/ASRT (1998) Hydrogeological map of El Bahariya, scale 1:500.000 project hydrogeological map of Egypt. Int Rep RIGW, Egypt
56. Cooper HH, Jacob CE (1946) A generalized graphical method for evaluating formation constants and summarizing well field history. Am Geophys Union Trans 27:526–534
57. Joint Venture Qattara (1985) Study Qattara depression: topography, regional geology and hydrogeology, vol III. Part I, Lahmeyer International, Salzgitter Consult and Deutsche Project Union, GMBH, German Federal Republic
58. RIGW (2012) Research institute for groundwater and academy of scientific research and technology. Groundwater potentiality maps, internal report submitted to academy scientific research, Cairo, Egypt (2010–2013)
59. Sharaky AM, Abdoun SH (2020) Assessment of groundwater quality in Bahariya Oasis, Western Desert, Egypt. Env Earth Sci 79:145. https://rdcu.be/b3EIZ

Groundwater Quality and Potentiality of Moghra Aquifer, Northwestern Desert, Egypt

Abbas M. Sharaky, Naglaa Saleh, Osama M. Farok, and Elsayed R. Saadany

Abstract The greatest challenge facing Egypt in the last decade is to create new communities in new areas due to the rapid increase in the human population. Egypt has vast areas of desert lands all over the country. Moghra is an area at the northeastern tip of the Qattara Depression that remnants of a larger paleolake. It is a part of the national mega reclamation project called "One and Half Million Feddan". It is planned to reclaim 105,000 ha (250,000 Fadden, 1 Fadden = 4200 km^2) using groundwater in Moghra. Assessment of the Moghra aquifer system has great importance as it is the main source of agriculture in the region. The present chapter aims mainly to investigate the hydrogeological characteristics of the Moghra aquifer using geophysical methods and geochemical analysis of the groundwater. The water samples were collected from 140 productive wells to determine the physicochemical characteristics of the groundwater. Also, it is focused on the analysis and interpretation of the well logging data for 48 deep productive wells. Well logging was used to determine the variations in thickness for sedimentary deposits that affect the quality of groundwater. The physical properties of the Moghra aquifer such as groundwater potentiality, formation water resistivity, formation factor, porosity, and effective porosity, volume of shale, permeability, and formation density are included. The salinity of groundwater ranges between 3,090 and 5,350 ppm with an average of 4,220 ppm. The high salinity is due to the effect of saline lakes as well as the seepage of saltwater from the Mediterranean Sea, low recharge of groundwater and leaching of clay and shale lenses. The pH values of the groundwater range between 7.2 and 8.7. Sodium is the dominant cation compared to others followed by calcium and magnesium, increasing percentage of sodium ion due to interference of seawater.

A. M. Sharaky (✉) · N. Saleh
Faculty of African Postgraduate Studies, Department of Natural Resources, Cairo University, 12613 Giza, Egypt
e-mail: sharaky@geologist.com

N. Saleh
e-mail: naglaasaleh74@cu.edu.eg

O. M. Farok · E. R. Saadany
Ministry of Water Resources and Irrigation, Giza, Egypt

A. Negm and A. Elkhouly (eds.), *Groundwater in Egypt's Deserts*, Springer Water,
https://doi.org/10.1007/978-3-030-77622-0_10

Keywords Groundwater · Moghra · Lower miocene aquifer · Geophysical methods · Well log

Introduction

Water resources in Egypt are confined to the water quota from the Nile River 55.5 km^3/yr according to the Nile water agreement with Sudan in 1959 [1]. Additional resources include a limited amount of rainfall, shallow renewable groundwater in the Nile Valley, and deep non-renewable groundwater in the deserts. The Nile water forms 99% of the annual renewable freshwater. It is guaranteed by the multi-year regulatory capacity provided by the Aswan High Dam (AHD). The total population of Egypt increased from 25 million in 1964 when the AHD inaugurated to 100.7 million in 2019 [2].

The population in Egypt continues to increase and is predicted to exceed 160 million by 2050, which is causing great concern within the water sector. Therefore, Egypt pays great attention to establish food and water security by the installation of new settlements and land reclamation projects depending often on groundwater.

The present chapter deals with the hydrological investigations of the groundwater in Moghra area, Western Desert. It is a part of the recent Egyptian megaproject called "One and Half Million Fadden" (1 Fadden = 0.42 ha) that has been inaugurated in 2014.

Groundwater in Egypt occurs as renewable that is confined to the Nile Valley and the Nile Delta, and non-renewable that extends mainly in the Western Desert. The hydrogeological framework of Egypt includes the following hydrogeological units [3]:

1. Nile aquifer including Nile Valley and Nile Delta aquifers (see Fig. 1)
2. Coastal aquifer
3. Moghra aquifer
4. Aquiclude
5. Carbonate aquifer
6. Nubian aquifer
7. Basement (fissured basement complex aquifer).

The Western Desert covers two-thirds of the land area of Egypt and occupies one of the driest regions of the Sahara. Seven depressions within the Western Desert: Siwa, Qattara, Fayum, Dakhla, Kharga, Baharya, and Farafra. They may represent parts of old drainage systems with extensive erosion and deflection associated with tectonic activity. Geological and geophysical investigations in the Qattara depression indicate the presence of buried fluvial channels with southeast to northwest flow directions from high land areas [4, 5]. A desert lake called Moghra occurs at the northeastern tip of the Qattara depression. It may represent a remnant of larger paleo lake fed by several channels flowed from the southeast during glacial and interglacial stages of post Miocene period, and Moghra aquifer water table with the ground surface [5]. The Egyptian government tries to achieve the greatest project in this area by planning

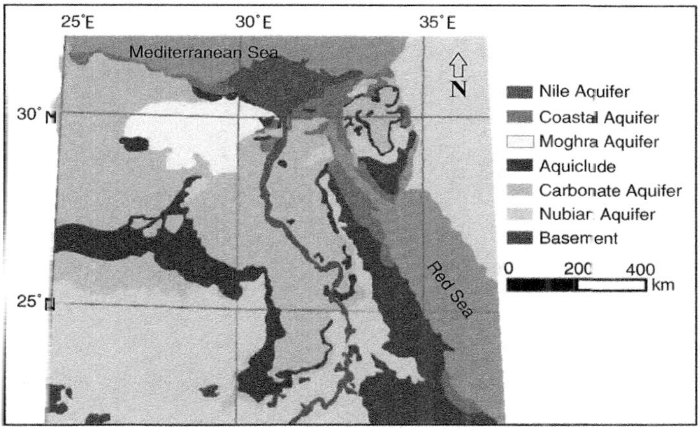

Fig. 1 Distribution map of major groundwater aquifers in Egypt, showing the location of Moghra aquifer in the northern Egypt close to the Mediterranean Sea [3]

a new agricultural and industrial society. The target of the project in Moghra is to cultivate 250 000 fadden and digging 1,352 water wells.

The Moghra area lies at about 100 km south of the Mediterranean coast, 37 km from west of the Qattara depression, 250 km west of Cairo, and 170 km southwest of Alexandria. It is located between latitudes 30° 03′ and 30° 12′ N and longitudes 28° 23′ and 28° 33′ E (see Fig. 2).

The present work deals with the water quality and potentiality of the groundwater of Lower Miocene aquifer in the Moghra area to delineate the subsurface geologic setting and the affecting structural elements (faulting and fractured zones). In addition, we show the occurrence of groundwater resources in the Lower Miocene aquifer. The present study aims mainly to investigate the hydrogeological characteristics of the lower Miocene aquifer by using geochemical and geophysical methods. The main objectives are:

1. Analyzing the properties of Lower Miocene aquifer.
2. Evaluating the subsurface geologic setting, the lithologic succession, and the affecting structural elements (faulting and fractured zones).
3. Examining the chemical characteristics of the extracted water for water quality.
4. Determining the optimum groundwater management system for long term use.

The groundwater in the Western Desert has attracted the attention of many authors since the beginning of the twentieth century. The groundwater is mainly represented by the Nubian Sandstone Aquifer System (NSAS) that was formed from precipitation during the last periods [6, 7]. It was derived from the rainfall on western and central Sudan, with partial recharge from the Nile. Groundwater flow models for the NSAS showed that the groundwater flows under climatic changes for the last 25,000 to 8,000 years [8, 9].

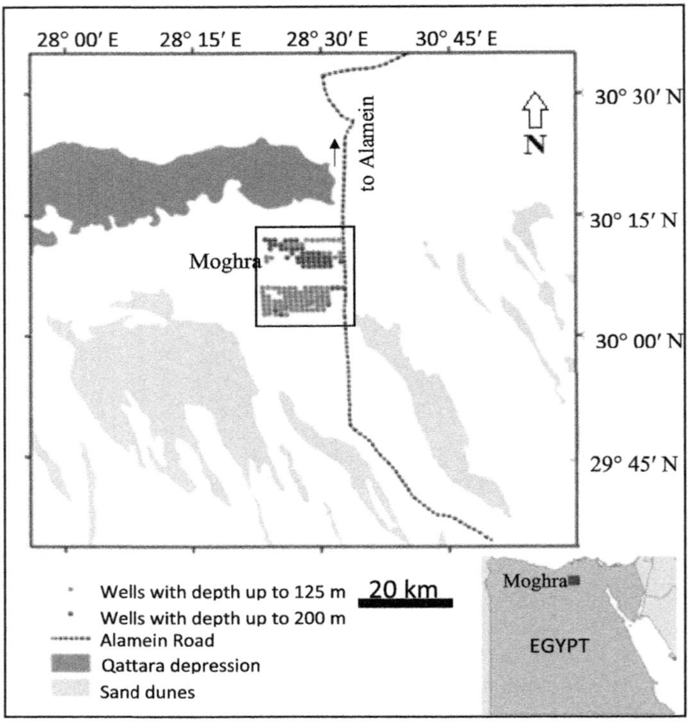

Fig. 2 Location map of the Moghra area, showing the groundwater wells, Qattara Depression, and sand dune belts running to the south-east direction

The previous studies on the hydrogeology of Moghra aquifer are relatively less than those of NSAS. They concluded that the Moghra aquifer consists of Lower Miocene fluviatile and fluviomarine coarse sand and gravel of the Moghra Formation. The water in the Moghra aquifer is a mixture of fossil and renewable water [10–14]. It has an average thickness of 300 m and is generally considered as non-renewable. They concluded that the groundwater of the aquifer is formed from mixing of the modern rainfall, seawater, water from the deeper NSAS, and water from the Quaternary aquifer. The Nile and groundwater in the Western Nile Delta, Egypt, show an interaction. The groundwater quality is unsuitable for drinking and domestic purposes [15].

The northern cliffs of Qattara Depression exhibit distinct outcrops of the Lower Miocene Moghra Formation [16] that occupies most of the floor of the Qattara depression. It extends to the east with dipping beneath younger formations to the north. Both the surface and subsurface boundaries of the Moghra Formation are well described due to the activity of oil exploration in northwestern Egypt as follows [17, 18]:

1. To the north, the Moghra Formation grades sharply into less permeable, clayey facies, especially along the Mediterranean Sea coast. It is overlain by cavernous limestone of Middle Miocene Marmarica Formation.
2. To the south and southwest, the Moghra Formation is limited by Oligocene shale of the Dabaa Formation. The Oligocene basalt flows and a fault escarpment associated with the Bahariya-Abu Roash uplift form the southeastern boundary of the Moghra Formation.
3. To the east, the thickness of the Moghra Formation decreases gradually and its clastic nature changes into a less permeable clayey facies.
4. To the west, Moghra Formation interfingers with less permeable limestone and shale sequence. The top of the underlying shale of Dabaa Formation is higher westward, delineating the western boundary of the Moghra Formation.
5. The top of Moghra Formation is mostly exposed except in the area to the north of Qattara depression, where it is overlain by Middle Miocene limestone of Marmarica Formation.
6. The base of Moghra Formation is marked by Oligocene shale of Dabaa Formation. The thickness of the Moghra Formation varies according to the buried structure on which it was deposited.
7. Hydrologically, the Moghra aquifer is recharged from five different sources:

a. Direct rainfall on the Moghra aquifer outcrops.
b. Groundwater seepage from the overlying Marmarica limestone aquifer.
c. Infiltration of the Mediterranean Seawater.
d. The Nile Delta aquifer.
e. Upward leakage from the NSAS.

Natural discharge from the Moghra aquifer occurs through evapotranspiration from Wadi Natrun in the east and Qattara Depression in the west.

Materials and Methods

The present work was fulfilled through field investigation, data and laboratory analysis.

a. A field survey was carried out to determine the location and level of the 140 drilled wells (see Fig. 2).
b. Identification of the main geomorphic units, structural elements and lithologic units through field investigations.
c. Geomorphologic analyses including the detection of the main landforms using topographic and geologic maps.
d. Critical analysis and interpretation included the well geological data, and well logging data of 48 deep wells. The well logging is widely used as a surveying tool for geologic correlation between wells. It can be used for the determination of petrophysical parameters of the subsurface such as lithology, porosity, effective porosity, permeability, water resistivity, formation factor, and salinity.

e. The wireline logging tools measure the electrical, magnetic, radioactive or acoustic properties of the formation. The measured parameters are recorded continuously as a function of depth. The resistivity logs such as Normal Resistivity 16″, 64″, spontaneous potential (SP) and Gamma logs using in the estimation and quantitative determination of petrophysical characteristics of the lower Miocene aquifer.
f. Hydrogeological measurements and determination of some hydrochemical parameters for the groundwater. They were analyzed chemically in order to determine the major cations and anions. The pH of the examined samples were measured using a laboratory pH meter, Model WTW-Multilane, Germany, with an accuracy of $\pm 1\%$. The electric conductivity was measured by an electric conductivity meter model YSI 32, U.S. with an accuracy of $\pm 1\%$. The total dissolved solids (TDS) were calculated from the sum of all major cations and anions in ppm.
g. The NO_3^- was determined using cadmium reduction method (American Public Health Association. The NH_4^+ was analyzed using the ammonia selective electrode method, PO_4^{2-} using the plasma method.
h. The concentrations of trace elements were performed using the atomic absorption flame spectrometric technique (GBC 908 AA) with lower detection limits ranging from 0.001 to 0.1 ppm.

Climate

The climate of Moghra area is hyper-arid, where the rainfall ranges from 25 to 50 mm/yr with an average evaporation of 1,531 mm/yr [19]. The maximum temperature is 41.4 °C in August and the minimum temperature is 10.7 °C in December. Humidity ranges between 19% in June and 39.5% in December. Wind speed ranges between 2.8 m/s in January and February and 3.7 m/s in September. The rainfall is often rare except sometimes sudden rainstorms. The evaporation varies from 5.4 mm/day in December to 29.2 mm/day in May [14].

Topography and Geomorphology

The studied area occupies a northern portion of the Western Desert of Egypt, which is dominated by arid and semi-arid climatic conditions. It is characterized by low relief and a mild topography with elevations varying from −40 (Qattara Depression) below sea level to about 40 m above sea level, it slopes gently towards the south and west directions. The present landforms are produced as a result of the action of both exogenic and endogenic processes. The landforms reflect the predominant aridity during the recent time, such as sand sheets, sand dunes, salt lakes, and marches.

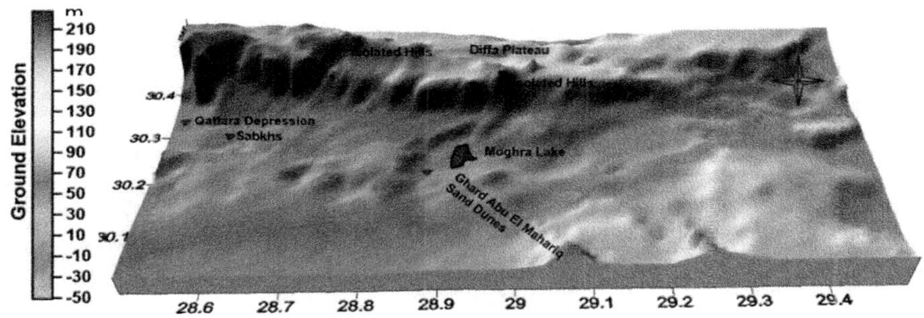

Fig. 3 3-D diagram of the surface area, showing the close relationship between Qattara and Moghra depressions, and the northern high escarpment that is reached to 230 m asl [14]

The Moghra is geomorphologically subdivided into escarpment bounding, sabkha sediments, and sand dunes (see Fig. 3).

Escarpment bounding

The escarpment bounding of tableland called Marmarica that ranges in elevation from 205 to 258 m. It is a steep slope that resulted from faulting. It is widely spread in the northern part of Moghra area.

Sabkha sediments

Composed of silt, sand, and clay intercalated with evaporates present in northern part extend to the west, covers a large area nearest the steep slope of depression. The climatic variation is helping in their formation.

Sand dunes

The recent sediments consist mainly of sabkhas, sand dunes, sand sheets, and gravels. The sand dunes are formed by wind and cover most of the floor of Moghra depression in southern and western parts.

Geologic Setting

The stratigraphic sequence contains most of the sedimentary section from Pre-Cambrian basement rocks to recent deposits in the Western Desert (see Fig. 4). It belongs to the Paleozoic to Miocene ages. It is also affected by deep-seated normal faults, orienting NNE-SSW (Syrian Arc trend) and NNW-SSE (Suez Rift trend) [20, 21].

The Oligocene sediments are represented by horizons of red, violet, and yellow ferruginous sandstone and sands, sometimes with gravel and occasionally quartizite. The sediments are known as Qatrani Formation [22] that was first introduced at Gebel Qatrani, Fayoum. It crops out at the southern part of Moghra area.

The Lower Miocene unit is called Moghra Formation [21] and is composed of fluviomarine sediments that mainly contain coarse sands, sandstone, and clay interbed with vertebrate remains and silicified wood, which becomes gravelly at the

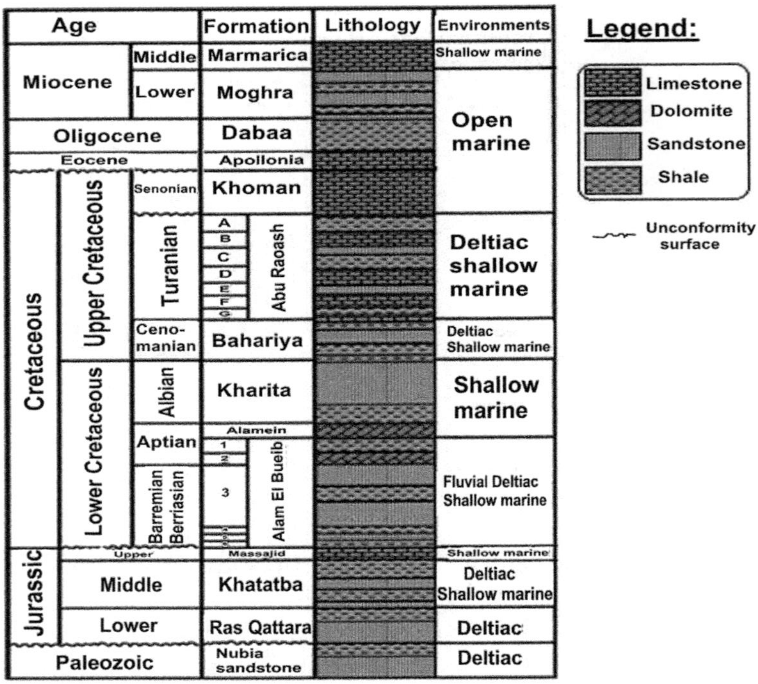

Fig. 4 A generalized stratigraphic section in North Western Desert, Egypt; showing different layers Moghra Formation located between less permeable formations of Dabaa shale and Marmarica limestone [13, 25]

base. The Moghra Formation occurs down at depth more than 200 m. Three members were recognized within the Moghra Formation: El-Raml, Bait Owian, and Monqar El-Dowi [23].

El-Raml Member covers a wide region of the northern part of the Western Desert, between the scarp of G. Qatrani and Moghra depression. It consists of alternating multi-colored sands, sandstone, and gravel of different compositions with scarce intercalating clay bands. The sandstone can be weakly calcareous, gypsiferous, and ferruginous. The maximum thickness of El-Raml member reaches 500 m [24]. It lies unconformably on basaltic flow at different depths. The El-Raml Member is continental to the lagoonal environment of sedimentation due to abundance of well-sorted, rounded and false bedded sands included within most of this member, together with gypseous and salty sandstone [23].

The Bait-Owian Member overlies El-Raml Member, and composed of cross-bedded and ferruginous sandstone interbedded with clay bands. Its thickness ranges from 25 to 30 m.

The Monqar El-Dowi Member contains sandy limestone and calcareous sandstone. It caps Bait Owian Member governed by Lower Miocene ridges. The thickness of this member is structurally controlled and ranges from 2 to 8 m.

The Pliocene- Pleistocene deposits extend along the northern part of the study area, represented by two formations [23]:

a. Issawia Formation is well represented in the northern area. It is mainly dominated by relatively thick white limestone beds and limestone breccia at the upper part intercalated with clay, shale, and silt.
b. Kalakh Formation occurs in the northeastern part, and consists of pink cross-bedded calcarenite with karstified columnar structure at the top.

Well Logging Interpretations

The well logging is the science of recording and analyzing measurements of the physical properties of the rocks in wells. They are commonly used in groundwater and environments investigations. The well logging is well known as a surveying tool for geologic correlation between wells. It can be used for the determination of petrophysical parameters of the subsurface such as lithology, porosity, effective porosity, permeability, water resistivity, formation factor and salinity. The wireline logging tools measure the electrical, magnetic, radioactive or acoustic properties of the formation. The measured parameters are recorded continuously as a function of depth. The resistivity logs such as Normal Resistivity 16″, 64″, Spontaneous Potential (S.P) and Gamma logs were used in the estimation and quantitative determination of petrophysical characteristics of Lower Miocene aquifer and water characteristics. In the investigated area, there are 48 wells penetrated the Lower Miocene aquifer with well depth ±200 m (see Fig. 5).

Electric Logging

Electric log is an excellent correlation tool. This means that the electric log gives a good indication of the general type of material of which bed is composed. Also, it is possible to determine the amount of pore space contained in the formation and the amount and type of fluids contained in the pores space. The porosity and fluid information can be determined depending on some characters such as mud resistivity, temperature, and depth of invasion into the formation by the mud filtrate. They are also affected by how well the interpreter can correct certain inherent errors caused by geometric factors such as sond diameter, borehole diameter, and bed thickness. The conventional electric logs consist of resistivity curves, spontaneous potential (SP), and Gamma Ray.

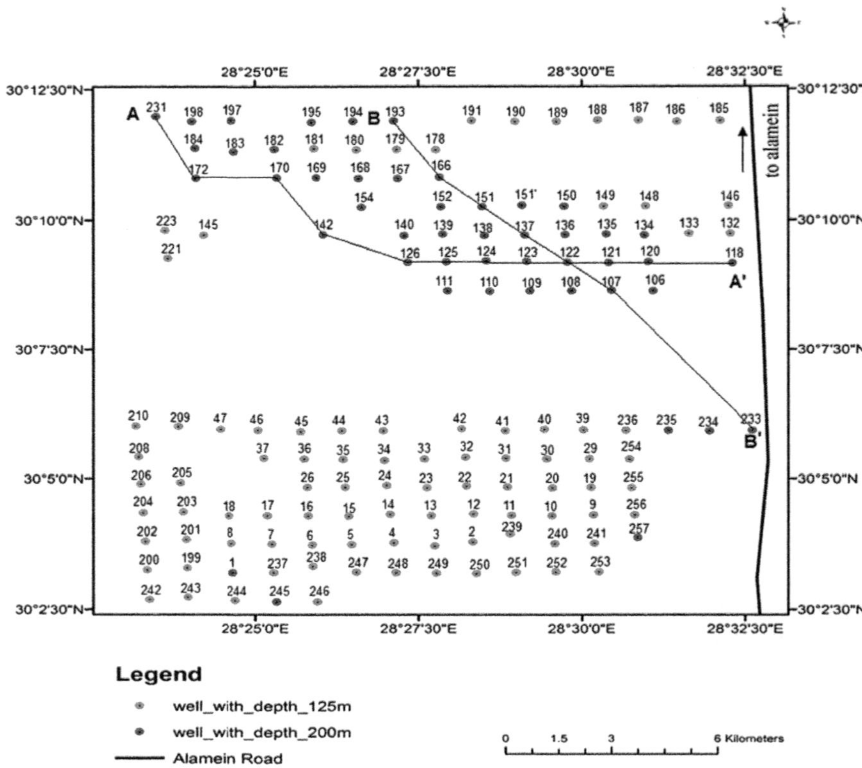

Fig. 5 Location map of the groundwater wells and cross sections shown in Fig. 6

Resistivity Curves

The resistivity curves consist of two curves, short normal resistivity curve (LLS—16″) and long normal resistivity curve (LLD—64″). In these logs the resistivity is proportional to the potential difference between a down whole and surface electrode when a known current is applied to the formation by the sonde according to the theory of log interpretation, long normal has a larger sample volume and, therefore, measures rock properties deeper into the surrounding formation. However, both long- and short-normal logs are affected by borehole fluid and formation, and both are dependent on bed thickness [26]. The resistivity is the inverse of conductivity, where the electrical conductivity is a measure of the ability of rocks to conduct electricity. The resistivity of a formation is a key parameter in determining the water saturation.

Spontaneous Potential (SP)

The spontaneous potential (SP) curve is recorded of the naturally occurring physical phenomena in situ rocks. The SP is naturally occurring electrical potential (voltage) that results from chemical and physical changes at the contact between different types of subsurface geological materials. The SP curve records the electrical potential (voltage) produced by the interaction of the formation contain water, conductive drilling fluid, and certain ion-selective rock (shale). In a borehole, potential occur between the drilling fluid and the filter cake (26).

Gamma-Ray Log

Gamma logging measures the naturally occurring radiation coming from the materials encountered in the borehole. As well as the natural radioactivity of the formation. Gamma radiation is emitted from certain elements in geologic materials that are unstable and decay spontaneously into other more stable elements. Gamma rays have a great ability to penetrate other materials, where gamma-ray has a short wavelength. Certain radioactive elements occur naturally in igneous and metamorphic rocks and as depositional particles in sedimentary rocks. Clay and shale contain high concentrations of radioactive isotopes, usually potassium. Sand and gravels, on the other hand, contain primarily silica, which is a stable substance and therefore emit only very low level of radiation. Limestone and dolomite also emit little radiation. The materials normally found in sedimentary materials such as clay, limestone, and sandstone contains decay products of Uranium and Thorium. Potassium is an important constituent of clay, mica, feldspar, and shale which emits gamma ray [27].

Results and Discussion

Lithology

The electric wireline logs (LLS, LLD, SP, Gamma-ray) curves provide a continuous recording of formation parameters versus depth that can be very useful for lithology determination. The qualitative interpretations of electric logs are useful in differentiation between shale, shaley sand, and clean sand intervals. The interpretation of the 48 logs (electric log) shows that the Moghra aquifer consists of different lithologic units that are mainly composed of sand, clay, and sand with shale intercalation.

The electric logs are also used in correlation of equivalent strata from one well to others. This correlation permits accurate subsurface mapping which determines the elevation of the formation and the presence or absence of faults. By using the SP curves which are recorded in all wells, it is easy to determine the upper bed boundary of Moghra Formation. The combination of SP curve, the resistivity curve, and Gamma curve shows that the sand percentage increases towards the eastwards and sand shale ratio increases gradually northwards and to the center of Moghra area. The good logs which most frequently used for correlation are the resistivity, SP and Gamma-ray, in the investigated area. The logs assist in the construction of

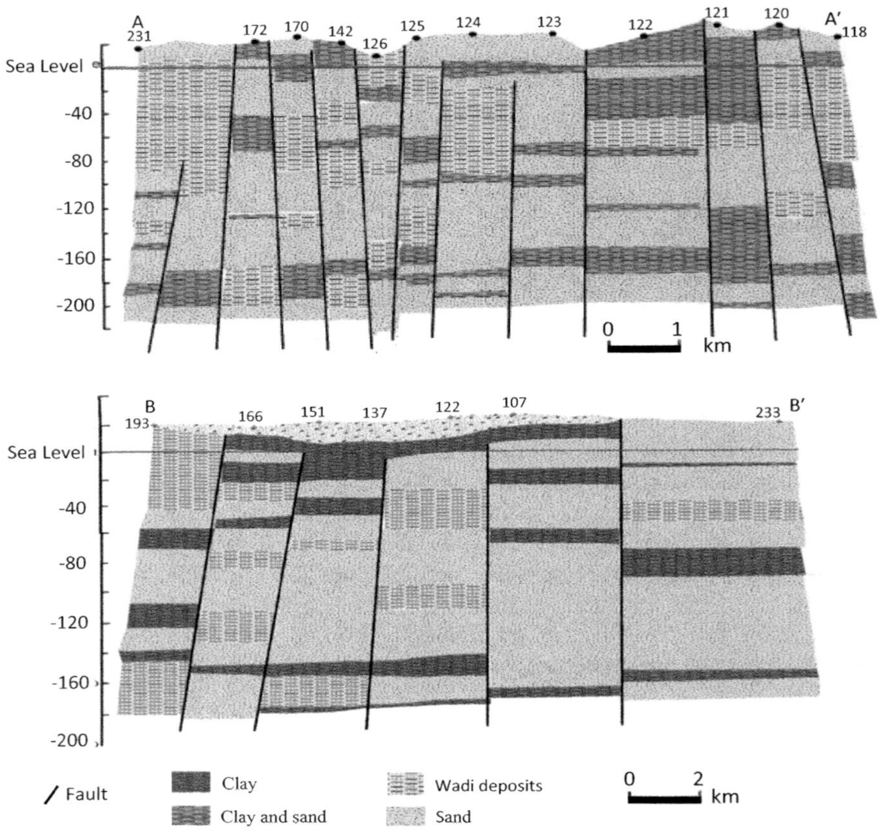

Fig. 6 Cross-sections A-A′ and B-B′ in Moghra area (from map Fig. 5), showing the dominant sand and clay lithology of Moghra Aquifer crossed by a series of normal faults

correlation cross-sections demonstrating the present thickening and thinning of the lithologic sections and lateral changes of sedimentations (see Fig. 6).

Groundwater Potentiality

Groundwater potentiality represents a supply of water that can be drawn for use. It can be readily transported to achieve a better balance between the locations of supply and demand. The main factor affecting groundwater potential is the continuity of the source, both in terms of quantity and quality. Quantity refers to the availability for the source, while quality refers to its suitability for a specific use. In a saturated rock or soil, the resistivity is largely dependent on the density and porosity of the material and on the salinity of the saturating fluid. So, the groundwater potentiality within a certain layer can be determined by the deviation of the electrical resistivity log of a layer along a profile of an area. The apparent or true resistivity that is calculated from the electric log (especially long normal resistivity LLD and short normal resistivity LLS) can be used to define aquifer limits or variations in groundwater potentiality.

Dry formations are poor in electrical conductors and show very high resistivity. Saturation of a formation reduces its resistivity, the reduction in resistivity is partially controlled by the porosity. This occurs because water is an electrical conductor, so, its presence in the interconnected pores reduces the overall formation resistivity. Silt, clay, and shale have the lowest resistivity because the inertial water contains dissolved salt. Sand and gravel with freshwater have moderate to high resistivity. The highest resistivity values are found in sandstone and limestone saturated with freshwater and in dense igneous and metamorphic rocks such as granite and dry shale [27].

Good interpretation of borehole resistivity data involves minimizing the effect of drilling fluid resistivity. Ideally, the borehole diameters should be kept as small as possible, which are less or equal to 20 cm. So, the short normal resistivity curve gives the resistivity closed to the borehole which called flushed zone resistivity (Ro). Generally, the short normal resistivity curve R16″ equals the resistivity of the flushed zone (Ro). The effect of the drilling fluid can be reduced if the electrode spacing is relatively large in comparison with the borehole diameter, so the long normal resistivity curve gives the resistivity of the noninvaded zone after reduced the drilling fluid effect.

Generally, the long normal resistivity curve (R64″) must be corrected to obtain the resistivity of non invaded zone. The correct resistivity of non invaded zone is the true resistivity (R_t) which can compute from long normal resistivity (LLD) and short normal resistivity (LLS) by using Schlumberger equations:

$$\begin{aligned} &1.\text{ If LLD} > \text{LLS} \\ &R_t = 1.7\,\text{LLD} - 0.7\,\text{LLS} \end{aligned} \tag{1}$$

$$\begin{aligned} &2.\text{ If LLD} < \text{LLS} \\ &R_t = 2.4\,\text{LLD} - 1.4\,\text{LLS} \end{aligned} \tag{2}$$

where

LLS are the averages of the short normal resistivity reading from the resistivity logs along the L. Miocene aquifer.

LLD are the averages of the long normal resistivity reading from the resistivity logs along the L. Miocene aquifer

R_t true resistivity which is computed along the Moghra aquifer.

By applying the Schlumberger equations on the logging data of the Moghra Formation for the 48 wells the result is the true resistivity shown in Fig. 7. The groundwater

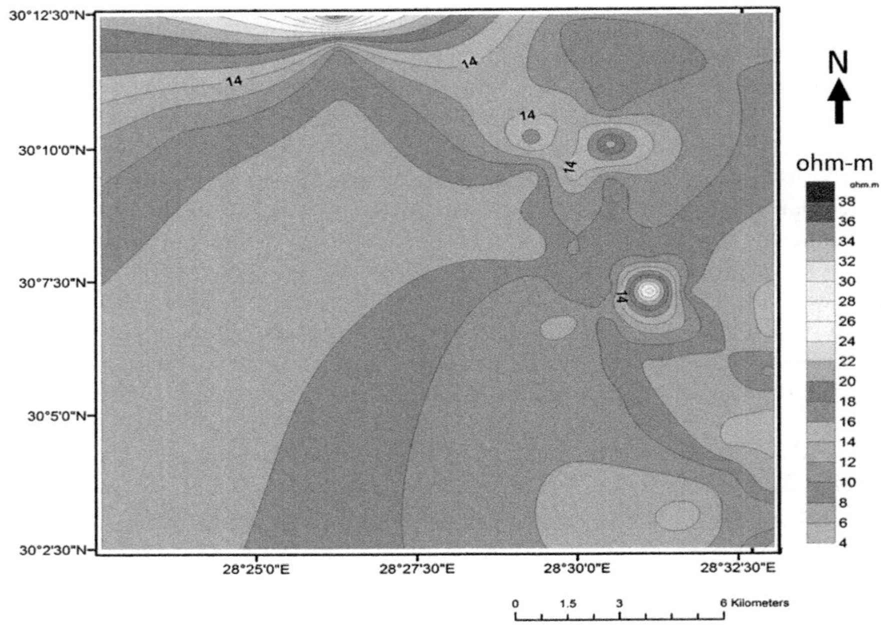

Fig. 7 Formation true resistivity (R_t) contour map of Moghra aquifer, showing the gradual increase of the groundwater potentiality to the north and eastwards

potentiality within this aquifer has been determined by iso-resistivity contour map showing that the groundwater potentiality increases gradually northwards in wells 193, 194, and 195, and to the eastwards in wells 107, 106, 235 and 234. In conclusion, the quality and quantity of groundwater in the northern and eastern areas of the study area are considered good. Generally, the gentle slope in iso-resistivity contour lines may indicate a gradual change in lithology.

Formation Water Resistivity (R_w)

Formation water resistivity sometimes called connate water or interstitial water, the resistivity of the formation water (R_w) is an important interpretation parameter. It is required to give information about the quality of water and groundwater potentiality. There are different methods to determine the value of (R_w) from SP and from resistivity (LLS an LLD) logs with the known salinity of all wells distributed in the investigated area using TDS.

Based on the chemical analysis of the water sampling of all wells by can be calculated (R_w) using the Guyod equation [28]:

$$TDS = K/R_w \quad (3)$$

where

TDS Total dissolved salt (ppm)
R_w Formation water resistivity (Ω-m)
K Constant factor which depends on the water solution as the following:

K = 5300 for NaCl solution
= 4200 for $MgCl_2$ solution
= 6700 for $MgSO_4$ solution
= 10000 for $NaHCO_3$ solution
= 12000 for Ca (HCO_3) solution.

In the majority of formation water, there is enough NaCl, that the K value for sodium chloride can be used. However, in very freshwaters, when certain salts are predominant, the K value may be affected. By applying the Eq. (3), the Rw is obtained for the Moghra aquifer (see Fig. 8). The Rw values in Moghra Formation show a wide variation representing the sand and shaley sands. The formation water resistivity of Moghra aquifer increases northwards in wells 197,198 and 231. It also increases

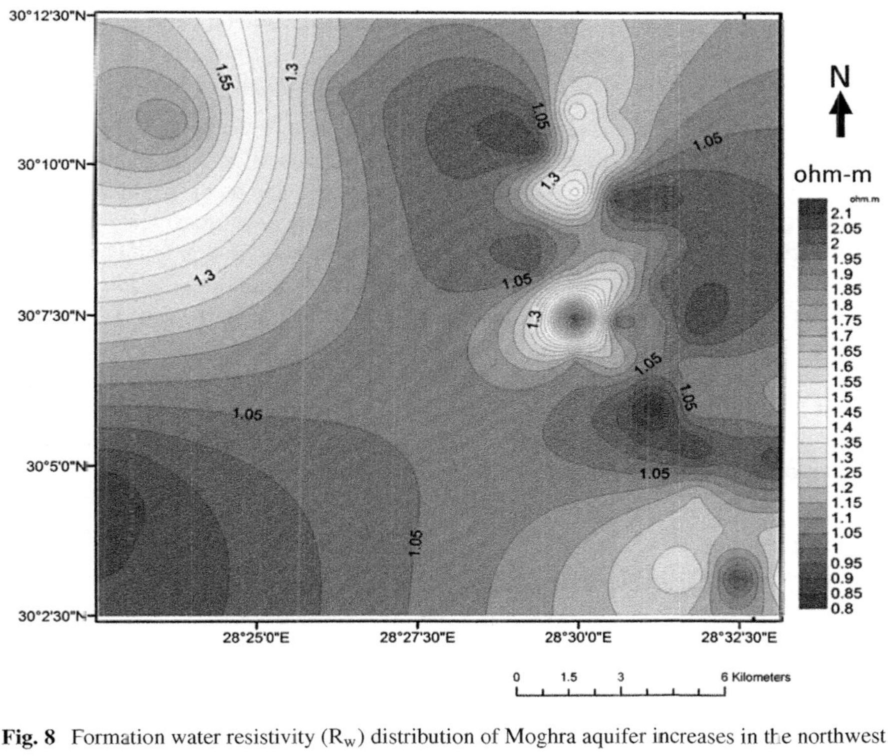

Fig. 8 Formation water resistivity (R_w) distribution of Moghra aquifer increases in the northwest

eastwards in wells 108,107 and 109, while it decreases in southwards. The decreasing of formation water resistivity may be attributed to the decreasing in sand content besides the existence of fine sand intercalated by shale and silt minerals.

Formation Factor (F)

Experimentally, the resistivity of clean formation is proportional to the resistivity of the brine with which it is fully saturated. The constant of proportionality is called formation factor that is the ratio of resistivity of a formation to the resistivity of water with which it is saturated. Generally, the formation factor, originally (F) is defined as the quotient from the resistivity of 100% water-saturated rock and the resistivity of the pore water or is defined as the ratio of rock resistivity to water resistivity of a fully saturated rock (Archie equation):

$$F = R_t/R_w \tag{4}$$

where F = Formation resistivity factor

- R_t Formation resistivity (Ω-m) that can be computed from long normal resistivity LLD by using schlumberger formulas (1) and (2)
- R_w Formation water resistivity which can be computed using a formula (3)

Most rock grains have a very high resistance relative to water, so the formation factor is roughly greater than 1. However, experiments show that in high water resistivity the value of (F) is reduced as (Rw) increases, and as the grain size of the sand decreases. This phenomenon is attributed to a greater proportionate influence of the surface conductance of the grains in more freshwaters. By applying the Archie Eq. (4) on the Moghra Formation, the formation factor can be obviously computed using the electric logs recorded through the 48 wells. Generally, the formation factor (F) increases northwards in wells 193, 194 and 195 (see Fig. 9), while it decreases south- and westwards. The formation factor (F) decreases with the decreasing grain size of the sand and increasing of (Rw) which is attributed to the freshwater. The values of formation factor of Moghra Formation indicated that this formation has a smaller grain size and a higher percentage of the shaley content.

Porosity and Effective Porosity

The porosity is defined as the ratio of the pore space to the total volume of the material. It is expressed as a percent or fraction of the total bulk volume of the rock [29]. Porosities are classified according to the distribution and shape of pores into intergranular, sucrosic or matrix porosity. It is existed in the formations since the

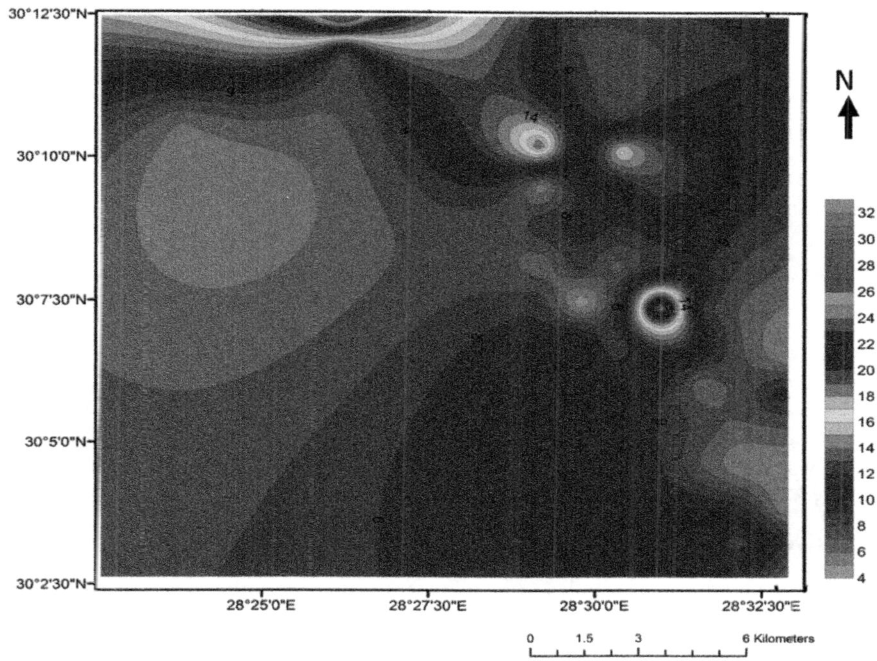

Fig. 9 Formation resistivity factor distribution of Moghra aquifer increases northwards where smaller grain size and a higher percentage of the shaley content, and decreases westwards

time they were deposited, for this reason; it is also referred to as primary porosity. Secondary porosity is caused by the action of the formation of water or tectonic forces on the rock matrix after deposition. Also stresses in the formation may also occur and cause networks of cracks, fissures or fracture which add to the pore volume. There are two different types of porosity, absolute (total) porosity measuring of the total pore spaces in a rock as a function of its bulk volume and effective porosity measuring of the interconnected pore spaces [29].

Total Porosity

The porosity is the ratio of voids to the total volume of soil or rock. So, for a given porosity, the ratio (R_t/R_w) remains nearly constant for all values of (R_w) below one Ω-m. Then, the formation factor is a function of porosity and also of pore structure and pore-size distribution. Archie proposed based on observations, a formula relating porosity and formation factor, the relationship is:

$$F = a/(\Phi)^n$$

where: F = Formation factor

n Cementation exponent
a Constant
Φ Formation porosity

General acceptance of the following formation factor, porosity relationship depending on lithology or pore structure.

$$F = 0.62/(\Phi)^{2.15} \quad \text{for sand} \tag{5}$$

$$F = 1/(\Phi)^{2} \quad \text{for compacted formation} \tag{6}$$

By applying the Archie formula (5) on the sand of Moghra Formation after obtaining the average values of formation factor (F), the average total porosity (Φ) of the water aquifer can be obviously obtained. The maximum average total porosity attains 37%, while the minimum average total porosity attains 12% (see Fig. 10).

Effective Porosity

The effective porosity includes only interconnected pores that sponsor fluid flow. Even in pure sand, not all of the total pore space participates in the groundwater flow whereas on the other hand, in pure shale not all of the total pore space is excluded from the groundwater flow. Then, the effective porosity is the share of the pore space through which the groundwater flows under normal pressure conditions. That means, the effective porosity (Φ_{eff}) is a function with formation water factor (F)

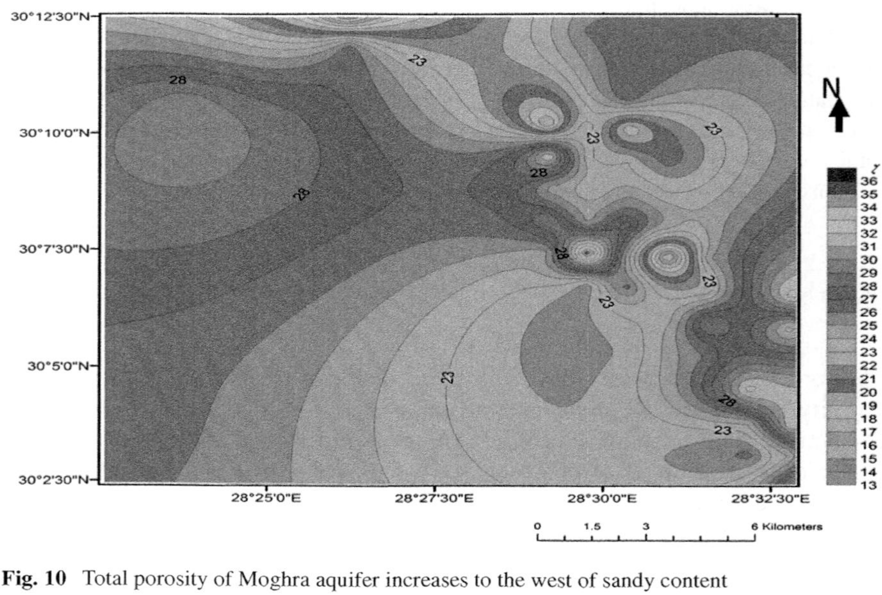

Fig. 10 Total porosity of Moghra aquifer increases to the west of sandy content

and a function with (TDS) of water flow through pore space under normal pressure condition. Due to the majority of formation waters includes sufficient NaCl, then, the effective porosity can be computed from Guyod equation [28]:

$$\Phi_{eff} = [(R_w * 5300)/R_t]^{1/2} \quad (7)$$

where

Φ_{eff} Effective porosity (%)
R_w Formation water resistivity (Ω-m)
R_t True resistivity which is obtained from LLD (long normal resistivity) (Ω-m)

By applying Eq. (7) on Moghra Formation after obtaining (R_w) and (R_t), the average values of effective porosity can be calculated. The effective porosity distribution in average ranges from 4 to 13.5% (see Fig. 11). It increases westwards in wells (142, 169, 170 and 172) then decreases eastwards at a certain part then increases again towards 233, 234 and 235 wells. It also decreases gradually in northwards at 194 and 195 wells. In general, the average porosity and effective porosity of Moghra Formation increase as the thickness, sand percentage, and sand/shale ratio increase.

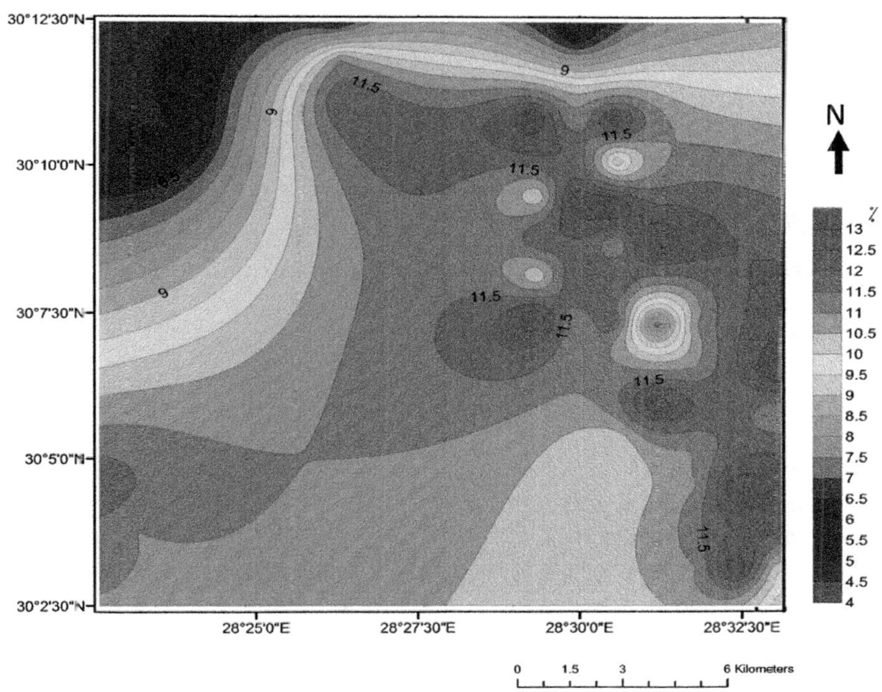

Fig. 11 Effective porosity distribution of Moghra aquifer ranges from 4 to 13.5%, indicating moderately good porous aquifer

The distribution of porosity and the effective porosity shows that the Moghra is a moderately good porous aquifer.

Volume of Shale

The presence of shale in any formation is one of the problems in the determination of porosity and contained fluid. The most significant effect of shale in the formation is the reduction of resistivity specially those containing water. Shale content can be determined in a variety of ways from borehole measurements. The simplest way is the calculation from the gamma ray log and from porosity and effective porosity. The amount of shale and its physical properties depend on the way of shale distribution in the formation.

Two analytical methods are used for determination of shale volume [30]:

a. **From Gamma ray (GR)**:

$$V_{sh} = \left[G_{log} - G_{min}\right] / [G_{max} - G_{min}] \tag{8}$$

where

V_{sh} shale content or volume of shale (%)
G_{log} value of GR in the log observed
G_{min} minimum value of GR in clean sand
G_{max} maximum value of GR in shale

b. **Form porosity and effective porosity**:

$$\Phi_{eff} = \Phi_{total} * (1 - V_{sh}) \tag{9}$$

where

Φ_{eff} effective porosity (%)
Φ_{total} absolute or total porosity (%)
V_{sh} shale content or volume of shale (%).

The obtained shale content is corrected as in the gamma-ray method using the porosity and effective porosity equation. By applying the Eq. (8) and corrected the result value in Eq. (9) into the Moghra Formation. The average V_{sh} of this formation can obviously be computed for different wells (see Fig. 12). In general, the distribution of shale volume across the Moghra area shows that the volume of shale increases north-westwards gradually in 172, 184, 198 and 231 wells, while it is

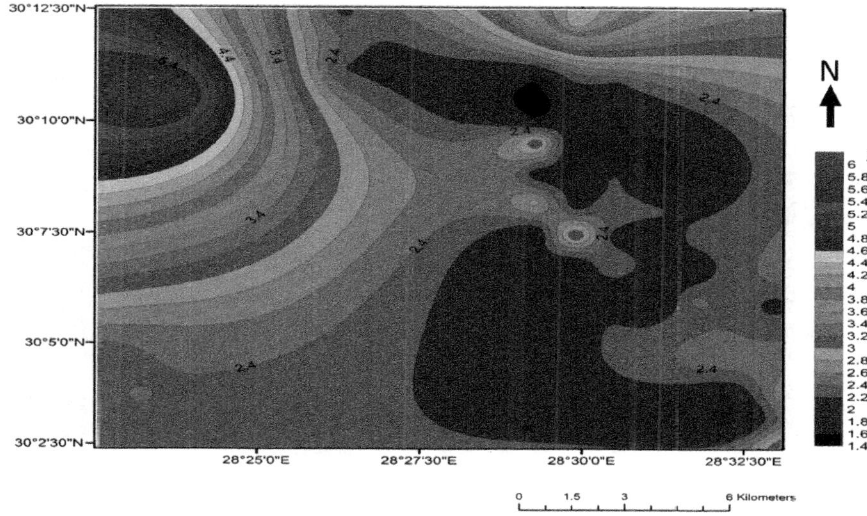

Fig. 12 Distribution of shale in the Moghra aquifer shows increasing shale to the north-west

decreased towards the middle and southern of the studied area in 257, 106, 107, 118, 120, 121, 122, 150, 151, 168 and 169 wells.

Permeability

Permeability is defined as the volume of water of unit viscosity passing through a unit cross-section of the material under unit pressure gradient in unit time. The unit of permeability is the Darcy that is defined as the permeability allowing the flow of one cubic centimeter per second of a fluid of one centipoises viscosity through a cross-sectional area of one square centimeter under pressure gradient of one atmosphere per centimeter [26].

Darcy is a very large unit so in practice. The millidarcy (md) is a unit commonly used, which equals 0.001 Darcy. Some rough relationships between effective porosity and permeability have existed, greater permeability, in general, corresponds to greater effective porosity, but this is far from being an absolute rule. Shale and some sand have high porosity, but the grains are so small that the path available for the movement of fluid are quite restricted and tortuous. Thus their permeability may be very low.

Methods and procedures for estimating permeability for borehole by geophysical logs have been an active subject of research [31, 32].

Recently, the permeability estimating procedures require data only on the cementation exponent factor (n) and effective porosity (Φ_{eff}). Calculation of the permeability (k), in millidarcies is determined using equation (Jorgensen et al. 1980) [33]:

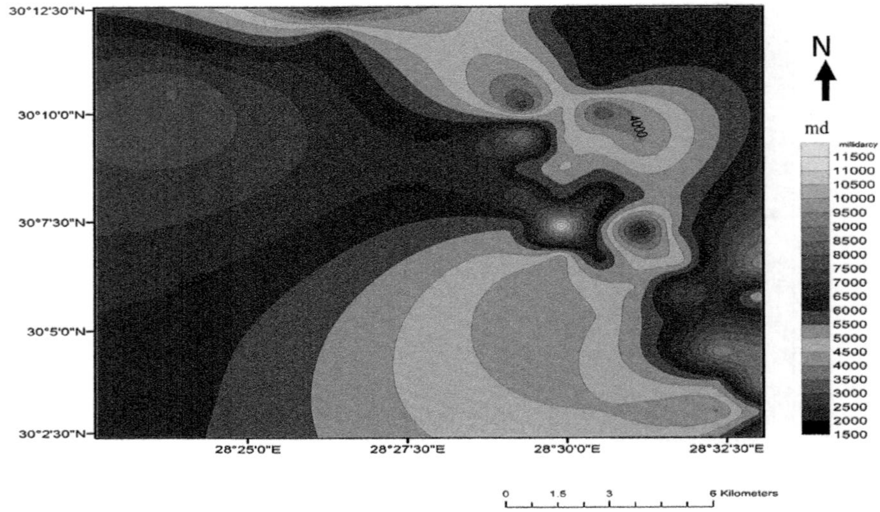

Fig. 13 Permeability (md) distribution of Moghra aquifer indicates medium permeability aquifer based on the lithology distribution, where sand content and effective porosity are high

$$K = 84000\,[\Phi_{eff}^{n+2}/(1 - \Phi_{eff})^2] \tag{10}$$

where

K	Permeability (md)
Φ_{eff}	Effective porosity (%)
n	Cementation factor (2.15 for sand).

By applying the Jorgensen Eq. (10) on the Moghra aquifer using the effective porosity, the average permeability of the Moghra Formation, was calculated for each well. The average permeability distribution is high in the eastern and western parts as recorded from 106, 107, 233, 234, 235,172, 184 wells (see Fig. 13) where sand content and effective porosity are high. On the contrary, the formation permeability is decreased at the central part and towards the south in 257, 247, 1, 233, 120, 136, 135, 134, 150, 194 and 195 wells.

In general, the formation permeability in the Moghra is described as a medium permeability aquifer based on the lithology distribution.

Hydrochemistry

Water quality plays an important role in the identification of uses of water in different ways (drinking, domestic, industrial and in agriculture uses). Nowadays, Egypt depends mainly on groundwater as an important source of water in great projects to get the benefit of that water. It is important to carry out chemical analysis for that water to determine EC, TDS, and pH, and salinity from the concentrations of major (Na^+, K^+, $Mg^{2+,}$ and Ca^{2+}) and anions (Cl^-, $SO_4{}^{2-}$, $HCO_3{}^-$, $CO_3{}^{2-}$ and $NO_3{}^-$). It is important to compare the concentrations of dissolved elements to the permissible standard limit set by the World Health Organization (WHO) to test the suitability of water to be used for drinking, domestic, and irrigation.

Major Cations and Anions

Sodium is the most dominant cations followed by calcium, magnesium, and potassium. The average values of the major cations in the 148 water wells are 1108, 195, 101 and 21 ppm, respectively (Table 1). The high sodium content is due to Mediterranean seawater intrusion. Increasing of calcium content is related to the type of water-bearing strata in which calcite, dolomite, gypsum, and anhydrite are responsible for enriching the groundwater with calcium ions. The leaching process of clay originated from the lagoon and marine environments add more magnesium [34]. Low potassium content is related to less dissolving of its salts in the groundwater.

The chloride is the dominant anion in the groundwater of Moghra aquifer followed by bicarbonate and sulfate. It varies from 1415 to 2642 ppm with an average of 2107 ppm. Bicarbonate ranges from 284 ppm in well (197) to 531 ppm in well (132) with an average of 392 ppm. Sulfate is recorded due to intrusion of seawater and dissolution of gypsum included in the water-bearing strata. It ranges from 30 to 45 ppm with an average of 46 ppm (Table 1).

Total dissolved solids (TDS)

The groundwater of Moghra aquifer is brackish water. It ranges between 3,090 and 5,350 ppm with an average of 4,217 ppm. The TDS distribution contour map of Moghra aquifer (see Fig. 14) shows that there is no general trend of salinity. The high salinity is due to the effect of saline lakes located to the east, seepage of seawater from the Mediterranean Sea, low recharge of groundwater and leaching of clay and shale lenses.

The pH values

The groundwater of the Moghra aquifer is slightly alkaline due to the chemical composition of the aquifer lithology and seawater intrusion. The pH values range between 7.2 and 8.7 with an average of 8.0.

Table 1 Maximum, minimum and average of major ions and salinity (ppm) of the148 water wells in Moghra aquifer

	Na^+	Ca^{2+}	Mg^{2+}	K^+	Cl^-	$HCO_3{}^-$	$SO_4{}^{2-}$	TDS
Max	1405	245	127	40	2642	531	57	5350
Min	752	131	68	14	1415	284	30	3090
Average	1108	195	101	21	2107	392	46	4217

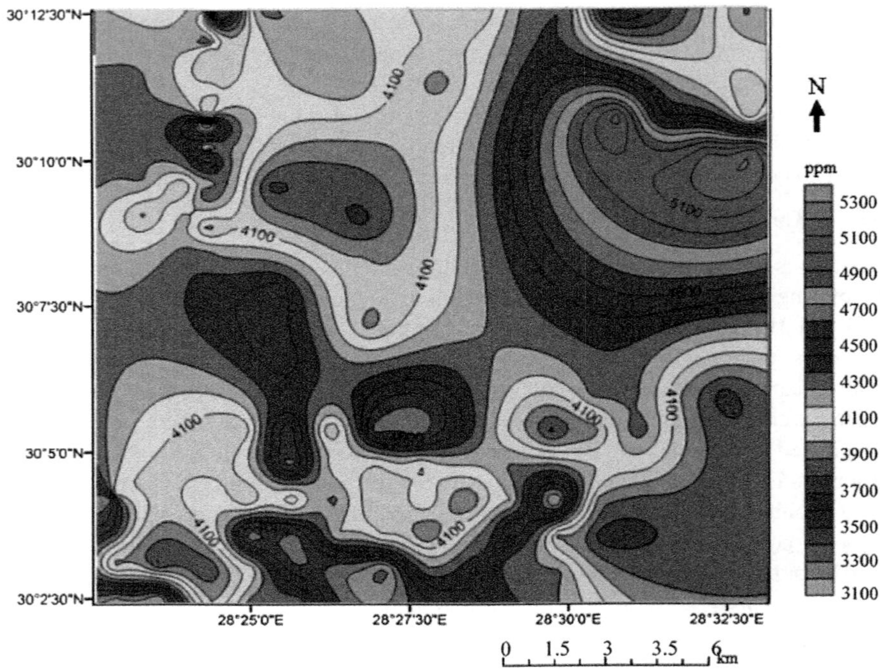

Fig. 14 Distribution of salinity (TDS) in Moghra aquifer shows that there is no general trend of salinity that ranges between 3,090 and 5,350 ppm with an average of 4,217 ppm

Groundwater Type

There are different classifications of groundwater depending on its chemical constituent. They depend on the percentage of the major cations and anions prevailing in groundwater in addition to the relationship between some cations and anions. Piper diagram [35] indicates different water types. It consists of three separate fields for plotting chemical data (see Fig. 15). The major cations and anions are expressed in epm percent. The calcium, magnesium, and sodium plus potassium are plotted in the lower left triangle, while bicarbonate plus carbonate, chloride plus sulfate are plotted in the lower right triangle. The intersections of the two points on the diamond-shaped field are projected, representing the water sample. The diamond-shaped field of this diagram consists of two equal triangles.

A Piper diagram (see Fig. 15) was created for the Moghra aquifer using the analytical data obtained from the hydrochemical analysis. The groundwater is classified in the piper diagram into 6 fields. They are type 1 (Ca-HCO_3), type 2 (Na–Cl), type 3 (Ca–Mg-Cl), type 4 (Ca–Na–HCO_3), type 5 (Ca–Cl), and type 6 (Na- HCO_3). The groundwater types of Moghra aquifer were confined to (Na–Cl) and (Na-HCO_3) types, indicating marine during the reduction process and meteoric water of shallow aquifers and leaching carbonate rocks.

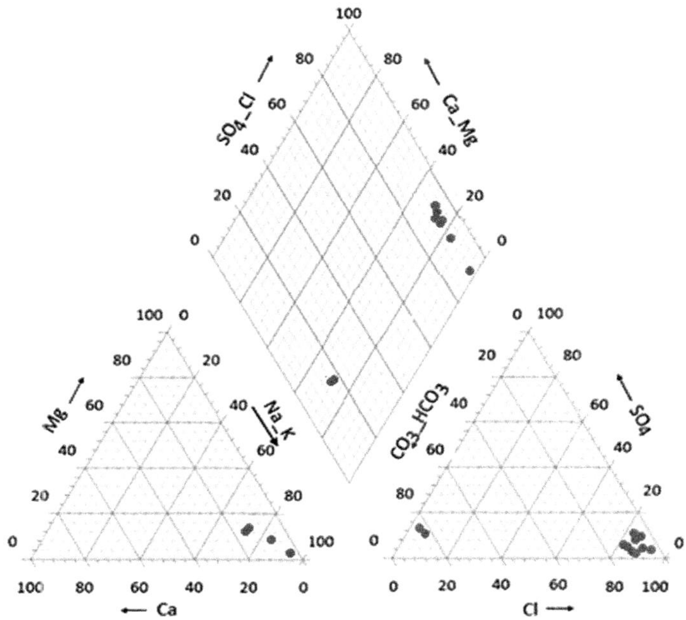

Fig. 15 Piper classification of groundwater in Moghra aquifer shows (Na–Cl) and (Na–HCO_3) types

Pumping Tests

In the study area, many pumping and recovery tests were carried out for many wells. Two of them were selected to represent the different zones. The first well (141) with 200 m, the second one (39) with 125 m, measurements of hydrogeological parameters through pumping tests that were carried out by:

a. Gradual pumping test step was done in four stages, each one took 240 min with constant discharge Q (m^3/h) for each one with measuring the dynamic depth to the water level inside the well.
b. Using the general well equation for producing total drawdown in water level inside the well.
c. Drawing the relationship between the discharge (Q) and the drawdown in the aquifer inside the well.
d. Continuous pumping test was done at constant (Q) and the maximum declination in water level is measured.
e. Cooper-Jacoub equation for the transmissivity (T) and Permeability (K) calculation.

Pumping test of well 140

Pumping and recovery tests were carried out in four stages at different discharges of 45, 62, 87 and 100 m^3/h with drawdown 1.5, 3, 3.8, and 4.5 m, respectively. Each stage in 240 min. and under a fixed discharge of 95 m^3/h during 2880 min. with maximum drawdown 4.5 m. The relationship between time in minutes and drawdown in meters is illustrated in Figs. 16 and 17). The time elapsed since the start of the recovery phase (in minutes) is denoted by t′. For all the residual drawdowns, the time elapsed since the very start of the pumping phase of the test (t) was calculated (in minutes).

Jacob equation [36] was used to determine the transmissivity of the aquifer as follows:

$$T \text{ (transmissivity)} = 2.3\, Q/4\pi\, \Delta S$$

where

Q	discharge (m^3/d)
Π	constant
ΔS	the slope of the straight line of residual drawdown.
Permeability (K)	T/H

where: T = transmissivity

H = saturated thickness

T = 1595 m^2/day. K = 1595/54 = 29.5 m/day.

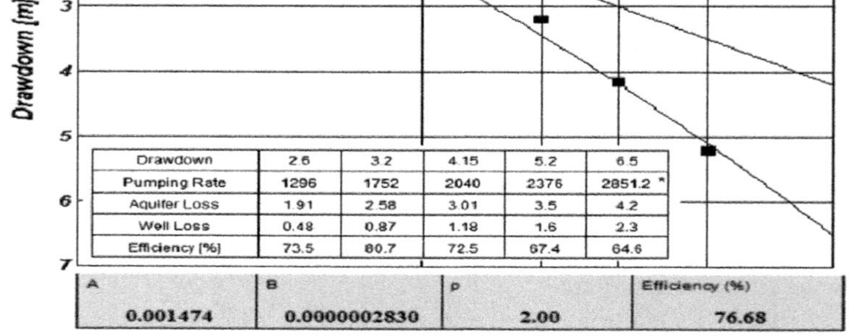

Fig. 16 Stages for steps of pumping test of well 140. Four stages at different discharges of 45, 62, 87 and 100 m^3/h with drawdown 1.5, 3, 3.8, and 4.5 m, respectively

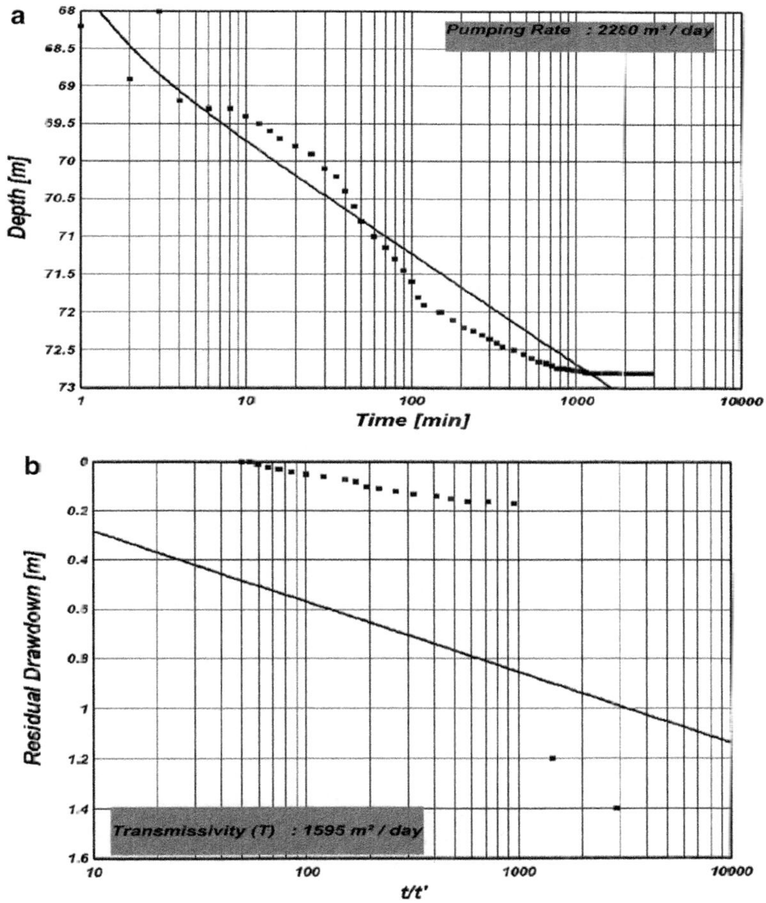

Fig. 17 Long duration pumping test of well 140, the slope of the straight line of residual drawdown is used in the calculation of transmissivity (T) of 1595 m^2/day and permeability (K) of 29.5 m/day: **a** long duration pumping test. **b** The recovery tests

Pumping Test of Well 39

Pumping test and recovery tests carried out to determine hydraulic parameters used four stages at different discharges (Q), 56, 68, 77, and 98 m^3/h with drawdowns of 1.3, 2.04, 2.64, and 3.02 m). Each stage in 240 min. and under a fixed discharge of 95 m^3/h during 2880 min with maximum drawdown 3.16 m. The relationship between time and drawdown was constructed and Jacob equation [36] was used to determine transmissivity of the aquifer as follows: T = 836 m^2/day and K = 836/36 = 23.2 m/day.

Hydraulic Properties

The transmissivity distribution in the study area has been mapped for wells 140 and 39 with different depths ±125 and ±200 m. The groundwater potential briefly in the study area is shown in Fig. 18. The maximum transmissivity value was 3980 m^2/day in well 219 and the minimum one was 569 m^2/day in well 133 in all wells with

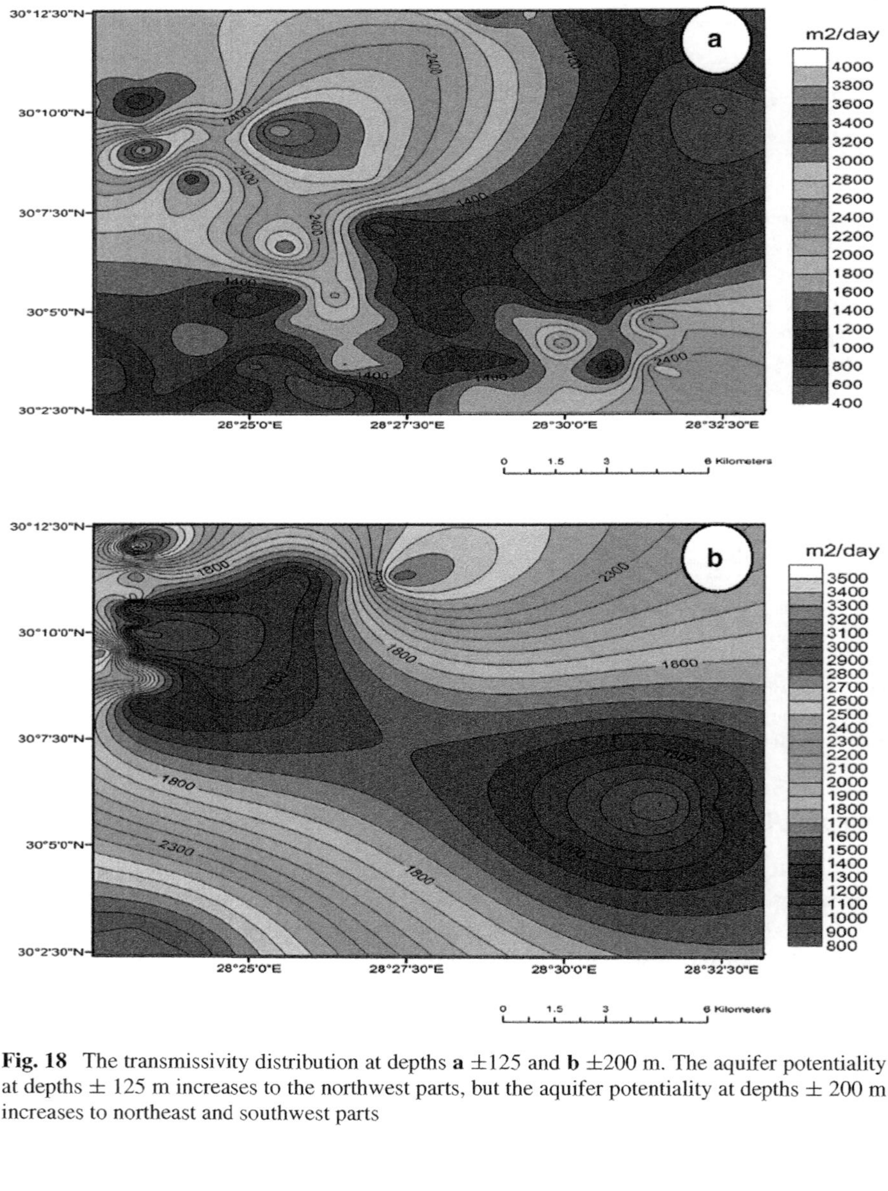

Fig. 18 The transmissivity distribution at depths **a** ±125 and **b** ±200 m. The aquifer potentiality at depths ± 125 m increases to the northwest parts, but the aquifer potentiality at depths ± 200 m increases to northeast and southwest parts

depth ±125 m. At depths ±200 m, the maximum transmissivity was 3707 m^2/day in well 184, and the minimum one was 735 m^2/day in well 134. The estimated average transmissivity value for the selected 48 wells (±200 m in depth) was 1985 m^2/day, while it was 1385 m^2/day for the 98 wells (±125 m in depth).

The hydraulic conductivity was also calculated, and its maximum value was 110 m/day, and the minimum value was 18.9 m/day in wells with depths ±125 m. It is also calculated in wells with depths ±200 m. The maximum value was 61.7 m/day and the minimum value was 11.13 m/day. The transmissivity distribution maps (see Fig. 18) indicate that the aquifer potentiality at depths ±125 m increases to the northwest parts, while the low values are in the east and south parts. But the aquifer potentiality at depths ±200 m increases to northeast and southwest parts, while the low values are mainly in the center.

The expected increase of hydraulic conductivity is associated with the increasing of the sand content and the grain size. As the percentage of shale is 1.4 and 2.4% in southwest and northeast parts, respectively with increasing in the amount of fine sand at shallow depths ±125 m, the aquifer potentiality decreases at depths ±125 m.

Water Evaluation and Suitability to Use

Quality of water is very important to determine uses of water; quality standards for domestic, industrial and agricultural use are variable and have wide limitations. The evaluation of quality of water for different purposes determined by parameters like salinity and chemical composition. In general, water for drinking and domestic uses should be colorless, odorless, clear, and free from excessive dissolved solids as well as harmful organisms. Table 2 shows the standard concentration given by the World Health Organization [37]. It would be used for judging water quality for drinking

Table 2 Permissible limit for drinking water [37] and water quality in Moghra

Chemical	Maximum acceptable limit (ppm)	Average in Moghra (ppm)
TDS	500	4220
pH	6.5–8.5	8
Ca	75	373
Mg	50	194
Na	200	1675
Cl	200	1400
B[a]	0.5	0.99
Fe[a]	0.3	4.56
Mn[a]	0.05	0.88
Zn[a]	5	1.18

[a]Data from [31]

purposes. The salinity, Cl, and Na contents of groundwater in Moghra indicate that Moghra groundwater is unsuitable for drinking because it has higher concentrations of TDS, B, Ca, Mg, Na, and Cl than the maximum permissible levels recommended by WHO (Table 2).

Suitability of water for drinking of livestock and poultry

The maximum concentration limit of total dissolved salts and specific ions in drinking water for livstock and poultry are recommended by the national academy of science and national academy of engineering. Water should not have dissolved salts level higher than 3000 ppm to be suitable for animals and poultry. Young animals and pregnant or locating animals are less resistant to high salt levels than old animals. Water with TDS of more than 7000 ppm becomes increasingly risks for watering farm animals. According to data, it is not allowed to depend on water of Moghra aquifer for drinking by livestock and poultry.

Suitability of groundwater for irrigation purposes

The agriculture expansion largely depends on the groundwater resources and its quality. The present section deals with the suitability of the examined groundwater wells for irrigation purposes besides the soil texture and salt tolerance of crops to avoid any harmful effect. The water of high saline more than 3000 ppm is not suitable for irrigation purposes. More than 1000 ppm not suitable for public water supply and it used only for irrigation of plants of high tolerance [38]. Therefore, the Moghra aquifer is not suitable for most of the irrigation purposes.

The suitability of water for irrigation is usually determined based on salinity, sodium adsorption ratio (SAR) and boron content. The concentration of boron in the Moghra aquifer ranges between 0.5244 and 1.804 ppm with an average of 0.99 ppm. The Moghra groundwater is suitable for semi tolerant and tolerant crops [13]. El-Sayed and Morsy concluded that some wells belong to S3-C4 zone (see Fig. 19) and have high sodium hazard and very high salinity hazard. It can not be used for irrigation under normal conditions [13]. They recommended using very salt tolerant plants with good drainage system, high leaching, and organic additions. Permeable soils, adequate drainages and adding gypsum and others amendments probably make the use of this water feasible for salt tolerant plants [13].

Conclusions

The Moghra area is a remnant of a larger paleolake including the mouth of a paleo-river. It is characterized by low relief and a mild topography with elevations −10 m below sea level to about 40 m above sea level. The geophysical study for 48 wells logs indicates great variations in lithology that consists of sand, sandstone, and clay. The Moghra aquifer is under unconfined conditions. The formation water resistivity of the aquifer generally increases towards the north. The decreasing of formation water resistivity in the south may be attributed to the decreasing in sand content

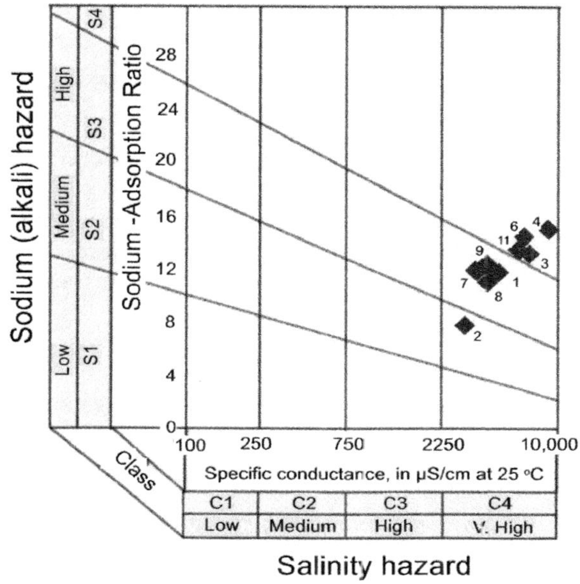

Fig. 19 Classification of groundwater of the Moghra aquifer for irrigation. Most of the wells belong to S3-C4 and S3-C4 zones with high sodium very high salinity hazards [13]

and the existence of fine sand intercalated by shale and silt minerals. The formation factor increases towards the north. The effective porosity averages from 12 to 33%, and increases towards the west. The volume of shale is increased gradually towards the northwest. The origin of the Moghra aquifer is a mixture of water from different modern rainfalls, water of post Moghra aquifer, seawater and groundwater of Nubian Sandstone aquifer.

The Moghra groundwater is brackish with TDS ranging from 3090 to 5350 ppm with an average of 4220 ppm, due to saline lakes towards the east, seepage of saltwater from the Mediterranean Sea, low recharge of groundwater and leaching of clay and shale lenses. The contents of major ions and TDS increase to the northwest. The Moghra aquifer is slightly alkaline with pH values range between 7.2 and 8.7 with an average of 8.0 due to the alkaline chemical composition of the aquifer rocks and the effect of the seawater.

The Moghra groundwater is unsuitable for drinking and domestic purposes due to the high content of salinity Na, Cl, Ca, Mg, B, Fe, and Mn. It is also unsuitable for livestock and poultry. Under certain conditions of permeable soil, good drainage system, and using agricultural fertilizers, it may be suitable for the irrigation of salt-tolerant and semi-tolerant crops.

Recommendations

In the light of the above analysis and presented results, the following recommendations are highlighted:

1. Monitoring the water quality to discover the changes in water types by continuous uses.
2. Converting two water wells with different depths (125 m and 200 m) to be test wells for monitoring the groundwater level.
3. Carrying out more chemical analysis and trace elements to archive the optimum uses of groundwater in the study area.
4. Reviewing the development plan of water uses according to it is chemical properties.
5. Using modern irrigation systems for irrigation of salt-tolerant and semi-tolerant crops.

References

1. MWRI: Ministry of Water Resources and Irrigation (2014) Water scarcity in Egypt: the urgent need for regional cooperation among the Nile Basin Countries, 5 pp
2. WPR: World Population Review (2019) Egypt population (2019-08-13). http://worldpopulationreview.com/countries/egypt/. Accessed 19 Aug 2019
3. RIGW: Research Institute for Groundwater (1998) Hydrogeological map of Egypt, 1:2,000,000 with explanatory note. Academy of Scientific Research and Technology, Cairo
4. Albritton CC, Brooks JE, Issawi B, Swedan A (1990) Origin of the Qattara Depression, Egypt. Geol Soc Am Bull 102:952–960
5. Khan SD, Fathy MS, Abdelazeem M (2014) Remote sensing and geophysical investigations of Moghra Lake in the Qattara Depression, Western Desert, Egypt 207:10–22
6. Shata AA (1955) An introductory note of the geology of the North portion of the Western Desert. Bull Desert Institute of Egypt V.71–81
7. Said R (1962) The geology of Egypt. Elsevier Pupl. Co., Amsterdam, p 377
8. Thorweihe U, Heinl M (2002) Groundwater resources of the Nubian Aquifer System, NE Africa. Modified synthesis submitted to Observatiore du Sahara et du Sahel (OSS, Paris) 1998, Aquifers of major basins non-renewable water resource
9. Gossel W, Ebraheem AM, Wycisk P (2004) A very large scale GIS-based groundwater flow model for the Nubian sandstone aquifer in Eastern Sahara (Egypt, northern Sudan and eastern Libya). Hydrogeol J 12(6):698–713
10. Ezzat MA (1982) Impact of a future Qattara salt-water lake on the Nubian Sandstone aquifer system in the Western Desert, Egypt. In: Proceedings of the exeter symposium, July 1982. IAHS Publ. No. 136, pp 297–314
11. Abd-Alla A (2001) Sedimentology, mineralogy and depositional environments of Moghra Formation (Lower Miocene) at Minqar Abu Duweis, Qattara Depression, Egypt
12. El Tahlawi MR, Farrag AA, Ahmed SS (2008) Groundwater of Egypt: "an environmental overview". Environ Geol 55:639–652
13. El-Sayed SA, Morsy SM (2018) Hydrogeological assessment of Moghra aquifer, north western desert, Egypt. Annals Geol Surv Egypt V XXXV:110–130
14. Yousf AF, El Fakharany MA, Abu Risha UA, Afifi MM, Al Sayyad MA (2018) Contributions to the geology of Moghra-Qattara area, North Western Desert, Egypt. J Basic Environ Sci 5:1–19

15. Sharaky AM, El Hasanein AS, Atta SA, Khallaf KM (2016) Nile and groundwater interaction in the Western Nile Delta, Egypt. In: Negm A (ed) The Nile Delta. The handbook of environmental chemistry, vol 55. Springer, Cham, pp 33–62
16. Hassan SM (2013) Sequence stratigraphy of lower Miocene Moghra Formation, in the Qattara Depression. North Western Desert, Egypt
17. Joint Venture Qattara (1985) Study Qattara Depression Vol. III. Part I: topography, regional geology and hydrogeology. Lahmeyer International, Salzgitter Consult and Deutsche Project Union. GMBH, German Federal Republic
18. Rizk ZS, Davis AD (1991) Impact of the proposed Qattara Reservoir on the Moghra aquifer of Northwestern Egypt
19. NWRP: National Water Resources Plan (2005) Facing the challenge, Ministry of Water Resources and Irrigation Integrated Water Resources Management Plan for 2017
20. Shata A (1953) New light on the structural development of the Western Desert of Egypt. Bull Desert Inst 5:63–77. Cairo
21. Said R (1962) The geology of Egypt. Elsevier, Amsterdam, p 377
22. Beadnell HL (1905) The relation of the Eocene and Cretaceous systems in the Esna-Aswan reach of the Nile Valley. Quat J Geol Soc 61:667–678
23. Omara S, Sanad S (1975) Rock stratigraphy and structural feature of the area between Wadi El Natrun and the Moghra Depression (Western Desert), Egypt. Geologisches Jahrbuch 16:45–73
24. Marzouk I (1970): Rock stratigraphy and oil potentialities of the Oligocene and Miocene in the Western Desert of Egypt. In: 7th Arab petroleum congress, vol 54, pp 1–37
25. Shalaby MR, Hakimi MH, Abdullah WH, Islam MDA (2016) Implications of controlling factors in evolving reservoir quality of the Khatatba Formation, Western Desert, Egypt. Scientia Bruneiana Special Issue. Geology
26. Paillt FL, Zaghlol E, and El Daftar T (1990) Applications of geophysical well log analysis to characterization of aquifers in the Sinai Region, Republic of Egypt. U.S. Geological Survey, Report 90-4194
27. Sharma PV (1997) Geophysical methods in geology, 428 pp
28. Guyod H (1965) Interpretation of electric and Gamma ray log in water well, the well log analysis. American Geophysical Union Technical Paper, Mandrel Industries Inc., Houston, TX
29. Martin FD, Robert M, Colpitts PG (1996) Reservoir engineering. In: Lyons WC (ed) Standard handbook of petroleum and natural gas engineering, vol 2
30. Schlumberger (1989) Log interpretation principles/application: Houston, Schlumberger educational service. 13 chapter
31. Timur A (1968) An investigation of permeability, porosity, and residual water saturation relationship for sandstone reservoirs. Log Anal 9(4):8
32. Coats GR, Dumanoir JL (1974) A new approach to improved log-derived permeability. The Log Analyst (January-February) 17
33. Jorgensen DG (1980) Relationships between basic soils-engineering equations and basic ground-water flow equations. U.S. Geological Survey Water-Supply Paper 2064, 40 pp
34. El Abd EA (2005) the geological impact on the water bearing formations in the area southwest Nile Delta, Egypt. Unpublished PhD thesis, Fac. Sci., Menufiya Univ., Egypt, 319 pp
35. Piper A (1944) A graphic procedure in geochemical interpretation of water analysis. Trans. Amer
36. Jacob CE (1946) Drawdown test to determine effective radius of artesian well, vol 112. Trans. Amer. Society Civil engrs, pp 1047–1070
37. World Health Organization (WHO) (2011) Guidelines for drinking-water quality, 4th edn, 541 pp
38. Taylor GG, Oza MM (1954) Geological survey of India. Bull Series B 45, 29 pp

Transboundary Groundwater Management Issues in the Nubian Sandstone Aquifer System (NSAS)

Caroline King-Okumu and Ahmed Abdelkhalek

Abstract Transboundary resource management refers to the principle of reasonable utilisation of land and water. This requires consideration of the social and ecological dimensions of groundwater use, in addition to hydrological monitoring and modelling systems. Egypt is a world leader in transboundary groundwater management and regional cooperation on water management issues more broadly. This chapter discusses Egypt's experiences in leading knowledge generation and sharing with other riparian countries to further the sustainable utilization of the major transboundary aquifer in Africa: the Nubian Sandstone Aquifer System (NSAS). To date, Egyptian transboundary groundwater management initiatives have been state-led, and have focused on the generation of studies with the intent to share data between countries on the hydrological conditions and volumes of water flowing through the aquifer. This is intended to enhance management and coordination. However, this data sharing is slow and sometimes sensitive. Furthermore, it has been difficult for the transboundary groundwater management initiatives to get to grips in any practical way with the challenges of sustainable groundwater use. Practices to conserve the health of the land that stores the groundwater reserves, and enable their replenishment also receive little attention in transboundary cooperation for groundwater management. To monitor, manage and sustain the hydrological, ecological and socio-economic aspects of the transboundary system requires the engagement of local institutions to conduct or facilitate the monitoring and to implement the management practices. For them, it is important not only to pursue the sustainable management of the aquifer over the longer term, but also to consider how to improve the viability of local resource management in the short term. In light of this, the main recommendation is to build and share knowledge about local capacities to manage ecosystem

C. King-Okumu (✉)
The Borders Institute (TBI), Africa, Nairobi, Kenya
e-mail: caroking@yahoo.com

GeoData Institute, University of Southampton, Southampton, UK

A. Abdelkhalek
Technical Office of Groundwater Sector, Ministry of Water Resources and Irrigation, Giza, Egypt
e-mail: ahmedimail@yahoo.com

A. Negm and A. Elkhouly (eds.), *Groundwater in Egypt's Deserts*, Springer Water,
https://doi.org/10.1007/978-3-030-77622-0_11

service production and recharge patterns in the surface layers, as well as to monitor and jointly manage the reserves stored in the deeper layers.

Keywords Transboundary · Nubian sandstone aquifer system · Sustainable management · Egypt

Introduction

Transboundary resource management refers to the principle of reasonable utilisation of land and water [1]. This requires consideration of the social and ecological dimensions of groundwater use [explored in 2], in addition to hydrological monitoring and modelling systems. Egypt is a world leader in transboundary groundwater management and regional cooperation on water management issues more broadly. This chapter discusses Egypt's experiences in leading knowledge generation and sharing with other riparian countries to further the sustainable utilization of the major transboundary aquifer in Africa: the Nubian Sandstone Aquifer System (NSAS).

The main of objective of the chapter is to explore how Egypt can lead the way forward toward the sustainable utilization of the NSAS. Egypt is doing this by sharing insights with the other riparian countries concerning available management technologies and innovations as it pushes the boundaries of science, engineering and ecosystem management to respond to pressing shared water management challenges.

The remainder of the introduction to this chapter provides an overview of the NSAS and the growing sustainability challenges as they are currently experienced in Egypt. Section 2 then discusses the principles of sustainable utilization of groundwater in transboundary systems. In the third part of the chapter, a brief overview of transboundary cooperation in the NSAS is provided. The fourth part is devoted to discussion of the challenges and way forward. Finally, some conclusions and recommendations are offered.

Overview of the Nubian Sandstone Aquifer System (NSAS) and Its Sustainability Challenge

Egypt receives transboundary flows of groundwater through the Nile Basin and also through the Nubian Sandstone Aquifer System (NSAS). The NSAS has been described by Nijsten, Christelis [3] as the 'heavyweight' amongst all of the transboundary aquifer systems in Africa and the largest known fossil aquifer in the world. This is due to its extent—underlying some 2,500,000 km^2 of Egypt, Libya, Chad and Sudan, and to the estimated volume of non-recharging water reserves that it contains [4] (see Fig. 1). The countries sharing the aquifer system include: Sudan, with approximately (17%) (373,000 km^2 located in the north-western part of the country; Chad, with about 11% (around 233,000 km^2 located in the north-eastern part of the country;

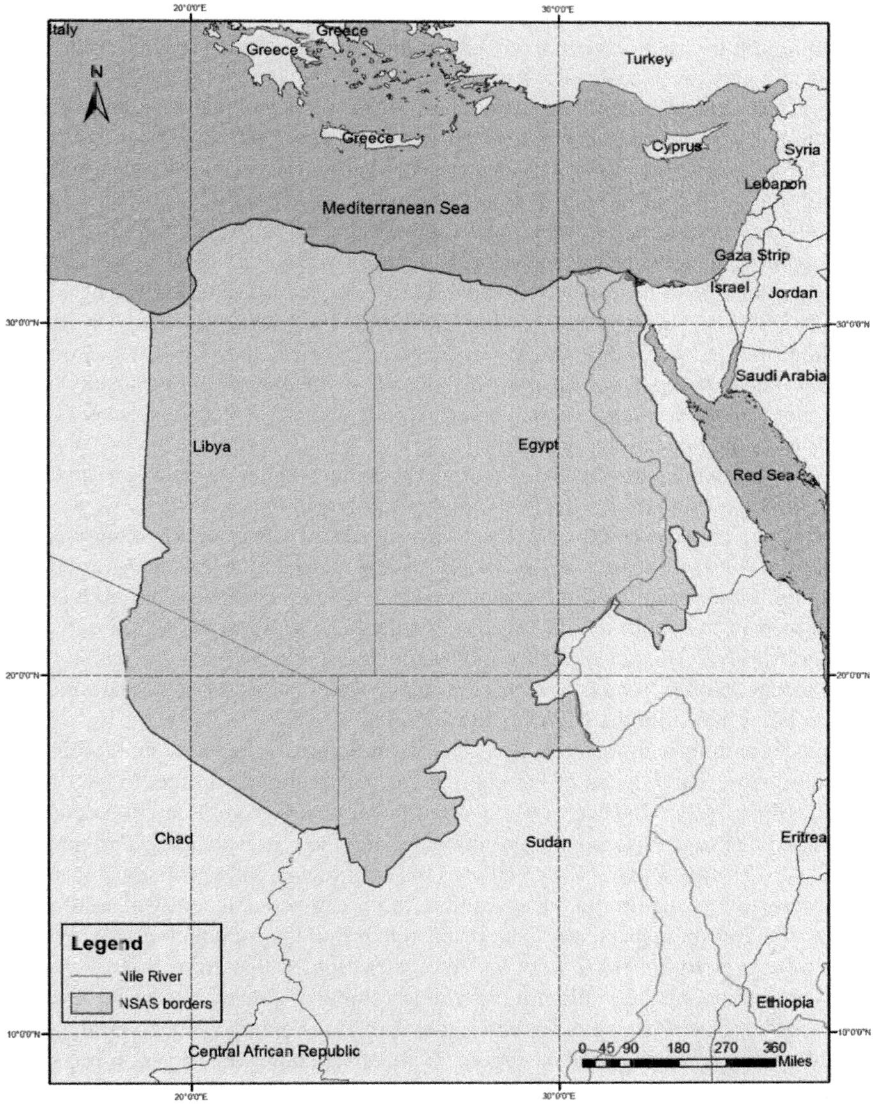

Fig. 1 Extent of the Nubian Sandstone Aquifer System according to the Joint Authority for the Study and Development of the Nubian Sandstone Aquifer System (JASD-NSAS). *Source* Ahmed Abdelkhalek, based on: [4] and https://www.nsasja.org/domain_en.php

Libya, where about 37.5% of the aquifer covers around 754,000 km^2 of south-eastern Libya; Egypt, which receives the remainder, underlying around 816,000 km^2 [5]. The results of a very large-scale GIS-based groundwater flow model for the NSAS in the Eastern Sahara (Egypt, north Sudan and east Libya) indicated that the groundwater in this aquifer was formed by infiltration during the wet periods 20,000 and 5,000 years b.p. The recharge of groundwater due to regional groundwater flow from more humid areas in the south was excluded. It also indicates that the NSAS is a fossil aquifer, which had been in unsteady state conditions for the last 3,000 years [5–8].

According to the latest estimates [9], the total volume of fresh groundwater in storage in the NSAS is around 372, 950 billion cubic metres (BCM) of which 41.5% (154,715 BCM) is in Egypt, 36.6% (136,550 BCM) in Libya, 12.8% (47,807 BCM) in Chad, and 9.1% (33,878 BCM) in Sudan. However, the bulk of this water is either too deep to reach and abstract with the present techniques, or too salty to use, particularly in the northern areas. Therefore, only about 3.9% of the reserve (14,459 BCM) is recoverable at the present.

NSAS has two aquifers; the most important is the Nubian Aquifer System (NAS) which is located underneath the Post Nubian Aquifer (PNA) (see Fig. 2). Low permeability layers are located between these two aquifers. PNA is an unconfined aquifer located only in the northern region of NSAS and is used by Egypt and Libya. NAS covers the whole area of NSAS and is used by all four countries. Although NAS is unconfined in the south of NSAS, it is confined in the northern region due to the presence of PNA [6]. Across Egypt's Western Desert, the NSAS is overlain by other more recent aquifers, which do receive recharge water vertically and horizontally, as well as being fed from the NSAS below [10, 11].

NSAS contains a large amount of groundwater amounting to about 475,753 km^3 assuming storativity values of the confined and unconfined aquifers to be 104 and 7,102, respectively. However, only a small portion of this volume can actually be developed due to deep depths to groundwater and the corresponding high pumping costs [12, 13]. Bakhbakhi [12] calculated the total recoverable groundwater in each country assuming maximum water declines in unconfined and confined aquifers are 100 m and 200 m, respectively. The result found that the total recoverable groundwater in Egypt to be 5,367 km^3 and the extraction at that time was only 0.506 km^3/year, indicating the availability of a large volume of unused water [12]. Contemporary management strategies for sustainable use of the multi-layered aquifer system should consider the integrated management of both the renewable and non-renewable aspects in different locations and contexts. For ongoing and future development potential in Libya, Salem and Pallas [14] have observed that the Post Nubian reservoir corresponding to the Post Eocene deposits in Egypt and Libya are also more important than the older Palaeozoic and Mesozoic deposits that extend over the whole Nubian Basin. This is partly because the Nubian System itself becomes very saline in the northern part in Egypt and Libya.

In Libya, the "Great River" the Man-made River began extracting substantial amounts of water from this aquifer By the end of the twentieth century [15]. The Libyan "Great River" is the Manmade piping system in the Sahara Desert of Libya as huge network of pipes to supply water from the Nubian Sandstone Aquifer. It is

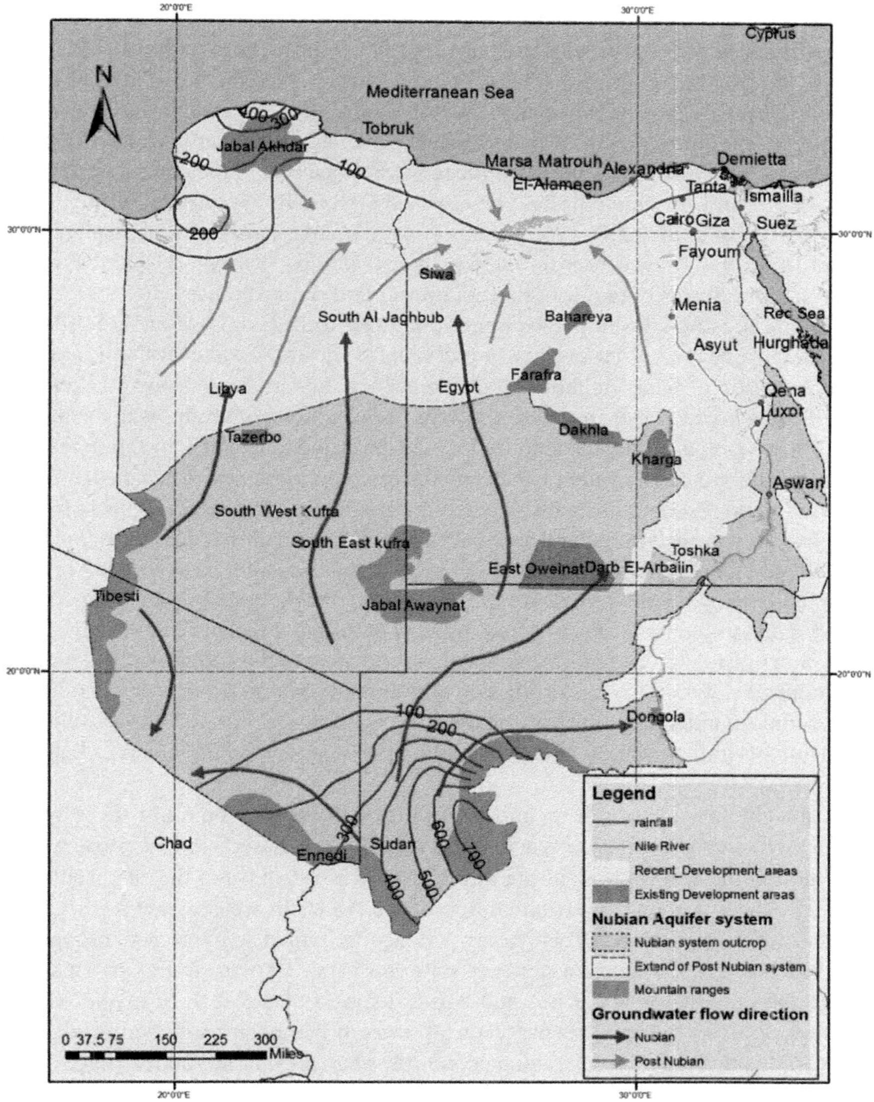

Fig. 2 Map of major flows and features in the Nubian Sandstone Aquifer. *Source* Ahmed Abdelkhalek. based on [4] and https://www.internationalwaterlaw.org/blog/2013/10/20/adoption-of-regional-strategic-action-plan-on-the-nubian-sandstone-aquifer/

the world's largest irrigation project. The extraction is estimated 2.37 km^3 per year. The extracted water is primarily used to supply the Kufra Oasis with water [16].

In Egypt, groundwater of the NSAS is the backbone for the development in the Western Desert, where it represents the sole source for water supply [16]. Interestingly, as the development of water resource management and use is evolving in Egyptian parts of the NSAS, the dynamic renewable aspects of the aquifer system are transforming and growing rapidly. The Use of NSAS in the Western Desert region of Egypt started in in the early 1960s in an effort to increase agricultural productivity. Groundwater withdrawal was initiated at five locations; Kharga, Dakhla, Farafra, Bahariya, and Siwa Oases for both agricultural and domestic uses. In 1990, a new agricultural area was developed using groundwater from East Oweinat [15, 17].

In 2015, a new national project to reclaim 1.5 million Feddan was initiated for the reclamation of lands in about 15 different locations in Egypt, most of them are located in the Western Desert region due to the large volume of groundwater available from NSAS (see the Chapter authored by Abdelkhalek and King in this book). In addition to this, a canal system has been constructed along the northern edge of the Western Desert (Al Hamam canal extension), bringing in additional supplies of water to the former marginal rangeland areas. A series of desalination plants have also been constructed along the Northern coast to supply drinking water to expanding coastal settlements (personal communication, Cairo Water Week, http://www.cairowaterweek.eg/). Human interventions are creating new patterns of recharge at the surface of the system due to increasing extractions of groundwater and return flows of irrigation and drainage waters. This has resulted in additional problems affecting groundwater quality and quantity in almost all of the five locations mentioned earlier (e.g. the expansion of drainage lakes and waterlogging problems in the Siwa depression [18, 19]).

At present, these new inflows and return flows of water to the multi-layer hydrological system do not appear to reach far into the Western desert, and the waste waters from the Northern coastal areas are still mostly flushed out into the sea. There are growing concerns about the sustainability of the NSAS in areas where it is increasingly used for irrigation [20]. However, a concerted effort to explore the scope for the productive reuse of wastewaters and nutrients is taking place in Egypt. This may change the patterns of water use and waste disposal at the surface of the NSAS. Although to date, the total recharge to the system has been estimated at only 3–4% [21], this does not yet take into account the changes that are taking place at the surface. Nor does it consider the full significance of human powers for innovation that are already transforming desert landscapes across Egypt and lining the coasts with desalination systems. The balance of recharge sources to the multi-layered system in Egypt's Western Desert is therefore could change rapidly over the coming years.

Periodically, ambitious proposals for water transfers to transform the Western Desert have been made. These have included many ambitious schemes that have not been implemented, e.g. to flood the Qattara depression (see Fig. 2) with seawater, or to pump drainage water out of waterlogged areas around the lakes in the Siwa depression. But other such schemes have been realized across Egypt—leading to the creation of artificial lakes and rich new biodiverse ecosystems such as Wadi Rayan.

Another visionary plan for a "Desert Development Corridor" in Egypt, researched and created by Boston University geologist Dr. Farouk El-Baz. El-Baz's idea has two components: first, an axis composed of a north-south running eight-lane highway, a high-speed train, an electricity line, and a water pipeline from the Toshka Canal to supply freshwater for human consumption along the 1,200 km strip of desert [17].[1] In a world where such feats of innovation and environmental engineering are possible, the population is expanding, and real-estate values are continuously rising, the further spread of developments relying on engineered flows of water across the Western desert is probably inevitable.

The very important questions concern the nature of the new systems of engineered hydrological and ecological processes that will emerge, and the future balance of flows from desalination of seawater, wastewater and brackish groundwaters versus the conservation and use of the precious fresh groundwater reserves from the deeper layers of the NSAS. These are human decisions that the current generation in Egypt is making for the future generations. A large number of wells have been installed across the Western Desert, and many of them have begun pumping out non-renewable sources of water (see the Chapter authored by Abdelkhalek and King in this book). The Egyptian government expects to be able to control how rapidly these wells will be pumped and to manage the sustainability of the new system that it is creating.

Extraction rates in Libya, Western Sudan and Chad may be less predictable. However, at the local level, communities across all four countries aspire to conserve and enhance the basic functioning of the systems that will enable them and their children to subsist and prosper. This reality is a sufficient basis for cooperation and mutual benefit (Fig. 3).

Background to the Sustainable Utilization of Groundwater in Transboundary Systems

Two major international instruments are available to guide the sustainable utilization of groundwater in transboundary systems, such as the NSAS. These provide a series of norms and options but require the States within each transboundary system to negotiate and shape their own context-specific agreements, according to their own particular capabilities and needs.

First, the UN General Assembly adopted Resolution A/RES/63/124 on the law of transboundary aquifers in 2009.[2] This Resolution provided draft articles in an annex, including provisions on defining opportunities for cooperation among aquifer States including the regular exchange of data, monitoring (jointly or not), and joint management. These articles did not fully engage with the significance of human innovation in land and water management processes as a driving force in the sustainable or

[1] http://blogs.bu.edu/professorvoices/2011/03/01/development-corridor/.

[2] https://www.internationalwaterlaw.org/documents/intldocs/UNGA_Resolution_on_Law_of_Transboundary_Aquifers.pdf.

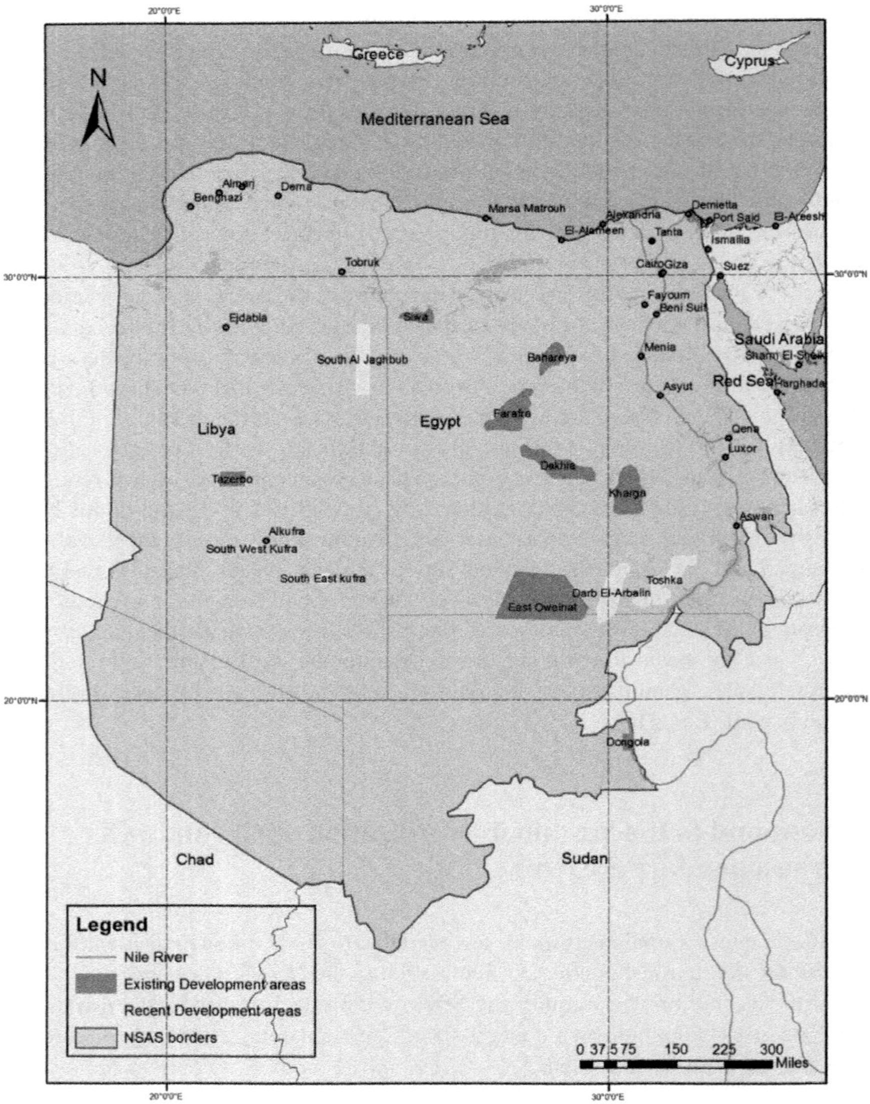

Fig. 3 Extraction zones in the Nubian Sandstone Aquifer System. *Source* Ahmed Abdelkhalek based on: [4] and https://www.nsasja.org/domain_en.php

unsustainable management of aquifer systems as is currently encountered in the NSAS. Nor do they refer in details to the full scale of the opportunity for States to peer review, backstop, encourage and independently validate one another's' efforts to understand and maximise these opportunities.

The general principles in the Draft Articles on the Law of Transboundary Aquifers adopted by the United Nations International Law Commission in 2008 (ILC Draft Articles, appended to UNGA Resolution No.63/124 of 11 December 2008) offer basic norms on equitable and reasonable use, the duty not to cause significant harm, and procedural and environmental protection norms.

Secondly, a Convention on the Protection and Use of Transboundary Watercourses and International Lakes (Water Convention) further aims to enable the sustainable use of transboundary water resources by facilitating cooperation. Initially negotiated as a regional instrument for use between States within Europe, this Convention has been amended to become universally available. As of March 2016, all Member States can accede to this Convention. Nonetheless, they must each still define their own needs and constraints and highlight the vision and capabilities that they are each bringing to the cooperative arrangement for sustainable utilization.

Overview of Transboundary Cooperation in the NSAS

The following summary of cooperation in the NSAS is partially based on previous accounts [9] as discussed previously by Quadri [22]:

An agreement for a Joint Authority for the Study and Development of the NSAS (JASD-NSAS) was first signed in 1992 (see https://www.nsasja.org/about_en.php and description at: https://programme.worldwaterweek.org/Content/ProposalResources/allfile/the_joint_authority_for_the_study_and_development_of_the_nubian_sandstone_aquifer_system.pdf). An "internal regulation" of the Authority, sets out the internal structure, functions, decision-making process, and funding of the Authority. The agreement included no provisions regarding the management of the NSAS or the groundwater stored in it. The mandate of the JASD-NSAS is to collaborate and develop co-operative activities for the sustainable mutual development of the NSAS, including monitoring the status of utilization of the aquifer and evaluation of the progress and activities enacted on the regional and national levels. As of 2019, the JASD-NSAS has held 20 meetings.

Two agreements made in 2000 through a Global Environment Facility (GEF)-funded project implemented by CEDARE provided the basis for a "Programme for the Development of a Regional Strategy for the Utilization of the NSAS". These agreements required that regular monitoring, updating and sharing of data and information from the NSAS should inform the sustainable use of the aquifer's groundwater resources. Regarding monitoring and information exchange, the four NSAS countries agreed to share data collected and analysed. For information on the Formulation of an Action Programme for the Integrated Management of the Shared Nubian Aquifer (2006–2011) see: https://iwlearn.net/iw-projects/2020/.

A further step in the process of cooperation between the NSAS countries was the "Regional Action Programme for the Integrated NSAS Management", funded by the GEF and implemented by UNDP, IAEA, and UNESCO-IHP [21]. This project supported the development of a regional strategy for the integrated NSAS management, aimed at the equitable long-term exploitation of the aquifer. The project fostered a better understanding of aquifer issues and responses, while laying the basis for a regional Strategic Action Plan (SAP).

The SAP agreement, signed by the NSAS countries and the Joint Authority in 2013, committed Parties to agree on actions for the sustainable management of the aquifer. In 2018, support for a project on Enabling implementation of the Regional SAP for the rational and equitable management of the NSAS was approved by the GEF [9].

Challenges and Way Forward

To date, Egyptian transboundary groundwater management initiatives have been state-led, and have focused on the generation of studies with the intent to share hydrological data between countries on the hydrological conditions of the aquifer. This is intended to enhance management and coordination. However, this hydrological data sharing is slow and sometimes sensitive. Furthermore, it has been difficult for the transboundary groundwater management initiatives to get to grips in any practical way with the challenge of sustainable groundwater use. Also, the political instability which still continues to this day in Libya (following revolutions in Egypt, Sudan and Libya) is also another big challenge affecting initiatives for cooperation between the four neighbouring countries who share the NSAS water resource.

Egypt can share its expertise in well-drilling technologies, and try to encourage other states to establish more effective archives to store and analyse the hydrogeological profiles and information that is generated during the drilling process. This will help to improve groundwater management knowledge, across the shared aquifer, including in Egypt.

Ecological management practices to conserve the health of the land that stores the groundwater reserves, and enable their replenishment (Table 1) have so far received little attention in transboundary groundwater management programmes. Yet these practices can make a major difference to water resource availability and its productive uses. They can conserve the quality of groundwater reserves from contamination threats, enhance the conditions for recharge (enlarging the volume of recharge), and establish virtuous cycles of use and reuse to make more uses of less water. Effective drainage and treatment technologies, including desalination can also further add to water resource availability. All of these affect the water balance calculations and the sustainability of the groundwater resource on a local scale.

Also the use of non-conventional sources of energy such as solar and wind for the pumping of wells' water can help in the conservation of the ecology and groundwater quality. In Egypt, there is an increasing expansion in the use of solar energy for

Table 1 National targets for land degradation neutrality (sustainable development goal 15.3) in NSAS

Egypt 2018 (2015–2030)	Sudan 2018 (2010–2030)	Chad 2015 (2040)	Libya (NA)
LDN at the sub-national scale: • LDN is achieved in the land degradation hotspots: Kafr El Sheikh Governorate, Demiata Gov., Rasheed area, El Minia Gov., Sohag Gov., Al Fayoum, Mersa Matrouh Gov. (Fuka – El Sallum), El Khattara area, El Tina Plain area, El Farafra oasis and North Sinai in Egypt by 2030 as compared to 2015 and an additional 10% of the degraded hotspot areas has improved (net gain) Specific targets to avoid, minimize and reverse land degradation: • Improve productivity and carbon stocks of 3,342 km^2 (802,080 feddan) of cultivated areas by 2030 • Restore and increase the productivity of 11,666 km^2 (2,800,000 feddan) of cropland using modern agricultural techniques and SLM practices in the northern areas, western and eastern fringes of reclaimed lands of Nile Delta and El Tina Plain area by 2030 • Rehabilitate and increase the productivity of 8,000 km^2 (1,920,000 feddan) of rangeland and rainfed areas using SLM practices in the north coastal areas (rangelands and rain-fed farming areas) by 2030 • Rehabilitate and increase the productivity of 7,500 km^2 (1,800,000 feddan) of cropland using SLM practices in the reclaimed areas in western desert fringes of middle and upper Egypt Governorates by 2030	• Reservation of 15% of the country to be registered by the government as renewable natural resource area • LDN is achieved (no net loss); Sub-national targets: • Determine the productivity of pastoral land in each state and increase it to 2.5 tons/ha; • Forest conservation and reforestation of 66 square kilometers of degraded forests • Improving the quality of pastures' soil and SOC; • Improve production in rainfed agricultural areas (clay) to mean 10 Shawwal per feddan (and sand) 5 Shawwal per feddan; • Increase the carbon stock in the soil by 30,5742 tons • Cultivation of trees and shrubs of high nutritional value in the pasture lands in the semi-desert and savannah-poor areas, especially in the area of 12,563 km^2; • Raising the productivity of declining agricultural areas (15,496 km^2) and cropland with early signs of decline (59,719 km^2) and stable but stressed (32,467 km^2) (Sudan also has set specific targets to improve land use inventory, sustainability planning, implementation and evaluation in state agencies)	• By 2040, 1 738.8 km^2 of forest will be restored • By 2040, 17.95 km^2 of wetlands will be restored By 2040, 29 000 km^2 of degraded land (bare soils and other) will be restored	na

(continued)

Table 1 (continued)

Egypt 2018 (2015–2030)	Sudan 2018 (2010–2030)	Chad 2015 (2040)	Libya (NA)
• Reclamation and cultivation of 6,300 km^2 (1.5 million feddan) of virgin land in reclaimed desert soils at different locations in the western desert of Egypt by 2030 • Gain in land productivity and SOC stocks in about 8,333 km^2 of cropland in reclaimed desert lands at different location (cultivated areas) by 2030 as compared to 2015 • Halt the conversion of cropland to other land cover classes by 2030 • Increase by 25% forest cover/tree cover through agroforestry and SLM in existing forests by 2030 as compared to 2015 • Halt the occurrence of soil erosion by rain water, creating dams for water harvesting to be utilized for agricultural purposes for an area of 2,500 km^2 in dry valleys of elevated areas of the inland Sinai and Eastern Desert by 2030 Rationalize water consumption by growing crops of low water requirements and adopting modern irrigation systems for around 1,000 km^2 in some oases in the western desert of Egypt by 2030			

(continued)

Table 1 (continued)

Egypt 2018 (2015–2030)	Sudan 2018 (2010–2030)	Chad 2015 (2040)	Libya (NA)
Link: https://knowledge.unccd.int/home/country-information/countries-having-set-voluntary-ldn-targets/egypt	https://knowledge.unccd.int/home/country-information/countries-having-set-voluntary-ldn-targets/sudan	https://knowledge.unccd.int/home/country-information/countries-having-set-voluntary-ldn-targets/chad	na
Date of Voluntary National Review of Progress report on SDG: 2018 Link: https://sustainabledevelopment.un.org/content/documents/20269EGY_VNR_2018_final_with_Hyperlink_9720185b45d.pdf	2018 https://sustainabledevelopment.un.org/content/documents/21741VNR_Sudan.pdf	2020 https://sustainabledevelopment.un.org/content/documents/23405RAPPORT_NATIONAL_VOLONTAIRE_FINAL_TCHAD.pdf	2019 https://sustainabledevelopment.un.org/content/documents/26753Libya_VNRLIBYA2020part1.pdf

running wells in the Western desert and some other areas in the recent years. The number of wells running with solar energy is still not too big compared to the total number of wells all over Egypt but it's increasing. Moreover, the applications of decentralized solar water pumping solutions in the agro-food sector have a strong potential for creating jobs in both small and medium enterprises [23].

Egypt has a major wealth of emerging experience to share concerning land and water management practices that can be integrated with and contribute to the sustainable utilization of groundwater, as described earlier in this chapter (see CWWW proceedings: http://www.cairowaterweek.eg/).

Notable technological innovations include growing insights from the management of the new systems of engineered hydrological and ecological processes [e.g. 24, 25]. Practices to conserve the health of the land that stores the groundwater reserves, including reductions in the use of agrochemicals and the safe disposal of wastes at the surface are particularly important [26–30]. The study of runoff and recharge process that enable the local replenishment of aquifers is also critical. There are also new management options appearing for the future balance of flows from desalination of seawater, wastewater and brackish groundwaters versus the conservation and use of the precious fresh groundwater reserves from the deeper layers of the NSAS.

Groundwater users in Egypt can draw on a range of technologies that could help to support and sustain productive activities in groundwater dependent environments. These include technologies to deliver, regulate and monitor the flows of water for different uses, including uses of saline water either with or without desalination. All of these can be of direct practical use and interest to groundwater dependent communities in the other three riparian countries.

To monitor and sustain the hydrological, ecological and socio-economic aspects of groundwater management requires the engagement of local institutions to conduct or facilitate the monitoring and to implement the management practices [31]. For them, it is important not only to pursue the sustainable management of the transboundary system over the longer term, but also to consider how the monitoring systems and information can be used to improve the viability of local resource management in the short term.

Several promising areas for potential cooperation and knowledge exchange can be identified. States could exchange points of view to articulate specific good practices and intentions in relation to the following:

(1) There is an opportunity for new monitoring systems to be incorporated into new groundwater developments, such as those in the Western Desert of Egypt. These should support sustainable integration of ecological and hydrological management, as well as enabling and contributing to monitoring of broader changes in the hydrological systems.
(2) Increasing understanding of the functions and value of the desert ecosystems is inspiring new approaches to their conservation and use.

(3) Innovative approaches to the management of wastewater, drainage and nutrients and plant assemblages at the surface of the systems can reshape the availability and demands for water to be extracted from the deeper layers and affect localized recharge processes
(4) In some cases, innovations in the use of solar-powered energy systems are decreasing the operating costs of groundwater management technologies.
(5) Database creation and management at the local and national levels remains a significant challenge area that is holding back effective knowledge accumulation, analysis and exchange across the transboundary systems. There is scope for identification of good practices and lessons learned e.g. in the compilation and analysis of hydrogeological profiles gained from drilling activities.
(6) The over-arching challenge of institution-building for groundwater management requires continuous attention. This includes the need to build, resource and sustain functioning institutions at the local level.

Conclusions

This chapter has highlighted some of Egypt's experiences in monitoring and managing the groundwater in the NSAS. The NSAS is the heavyweight of aquifer systems in Africa and worldwide. Egypt is already a world leader in transboundary groundwater management and regional cooperation on water management issues more broadly. Egypt has growing capability and options to reshape and manage the balance between extraction and recharge of the NSAS over the coming years. There is no physical or technological barrier nor any other reason why Egypt and its neighbours should not achieve the sustainable utilization of their aquifer system. However, this will require a major concerted scientific and institutional effort. To achieve this, Egypt and its partners must look well beyond the current international norms and conventions for transboundary groundwater management cooperation, and lead the way forward to a more sustainable future.

Recommendations

- The main recommendation is to build local capacities to manage land quality and recharge patterns in the surface layers, as well as to conserve and protect the reserves stored in the deeper layers.
- Egypt and its partners must look well beyond the current international norms and conventions for transboundary groundwater management cooperation since these do not give sufficient consideration to best practices in land and water management
- States could more fully engage with the significance of human innovation in land and water management processes as a driving force in the sustainable or

unsustainable management of aquifer systems as is currently encountered in the NSAS.

- Egypt should develop and evaluate possible strategies to reshape and manage the balance between extraction and recharge of the NSAS and its associated multi-layered aquifer systems over the coming years, taking into consideration local ecological effects and return to the aquifer, as well as flows of water through the aquifer.
- States should peer review, backstop, encourage and independently validate one another's' efforts to understand and maximise the sustainable management of the aquifer system.
- Egypt should demonstrate leadership by leading the way forward to achieve the sustainable utilization and management of the shared aquifer system, taking into consideration ecological management processes and options as well as quantifying hydrological flows and deepening understanding of hydrogeological conditions.
- This will require the strengthening of management institutions at both local and national levels

References

1. Loucks SFDP (ed) (2006) Non-renewable groundwater resources: a guidebook on socially-sustainable management for water-policy makers. UNESCO, Paris
2. King C, Salem B (2012) A socio-ecological investigation of options to manage groundwater degradation in the western desert, Egypt. Ambio 41(5):490–503
3. Nijsten G-J et al. (2018) Transboundary aquifers of Africa: review of the current state of knowledge and progress towards sustainable development and management. J Hydrol Reg Stud
4. CEDARE (2001) Regional strategy for the utilization of the Nubian Sandstone Aquifer System, vol II–Hydrogeology. Centre for Environment and Development for the Arab Region and Europe
5. Gossel W, Ebraheem AM, Wycisk P (2004) A very large scale GIS-based groundwater flow model for the Nubian sandstone aquifer in Eastern Sahara (Egypt, northern Sudan and eastern Libya). Hydrogeol J 12(6):698–713
6. Moghazy NH, Kaluarachchi JJ (2020) Assessment of groundwater resources in Siwa Oasis, Western Desert, Egypt. Alexandria Eng J 59(1):149–163
7. Heinl M, Thorweihe M (1993) Groundwater resources and management in SW Egypt. Catena Suppl 26:99–121
8. Thorweihe U (1990) The Nubian aquifer system. In: Said R (ed) The geology of Egypt. Balkema, Lisse, The Netherlands, pp 601–614
9. GEF (2018) Enabling implementation of the Regional SAP for the rational and equitable management of the Nubian Sandstone Aquifer System (NSAS), in Project Document. Global Environment Facility, Washington DC, p 94
10. WPRP (1998) Hydrogeology of deep aquifers in the Western desert and Sinai. Minist. Water Resour. Irrig. Water Policy Reform Program: Cairo, Egypt
11. CEDARE (2001) Regional strategy for the utilization of the Nubian sandstone aquifer system, vol I. Centre for Environment and Development for the Arab Region and Europe
12. Bakhbakhi M (2006) Nubian sandstone aquifer system. In: Loucks SFDP (ed) Non-renewable groundwater resources: a guidebook on socially-sustainable management for water-policy makers. UNESCO, Paris, pp 75–81

13. Moghazy NH, Kaluarachchi JJ (2020) Sustainable agriculture development in the western desert of Egypt: a case study on crop production, profit, and uncertainty in the Siwa region. Sustainability (Switzerland) 12(16)
14. Salem O, Pallas P (2002) The Nubian Sandstone Aquifer System (NSAS), in managing shared aquifer resources in Africa: regional and basin organization. In: Appelgren B (ed) IHP-VI, series on Groundwater no. 8. UNESCO, Paris, p 19–21
15. Abdel-Shafy HI, Kamel AH (2016) Groundwater in Egypt issue: resources, location, amount, contamination, protection, renewal, future overview. Egypt J Chem 59(3):321–362
16. El Nahry AH, Elewa HH, Qaddah AA (2010) Soil and groundwater capability of East Oweinat area, Western Desert, Egypt using GIS spatial modeling techniques. Nat Sci 8(8):1–17
17. Sheweka SM (2012) New Egypt with a new hybrid skin "A New Hybrid Architecture Vision for the Egyptians'Development Corridor". Energy Procedia 18:449–457
18. Aly AA (2019) Soil and groundwater salinization in Siwa Oasis and management opportunities: twenty year change detection and assessment. Arid Land Res Manage
19. Abdulaziz AM, Faid AM (2015) Evaluation of the groundwater resources potential of Siwa Oasis using three-dimensional multilayer groundwater flow model, Mersa Matruh Governorate, Egypt. Arab J Geosci 8(2):659–675
20. Ibrahem SMM (2019) Effects of groundwater over-pumping on the sustainability of the Nubian Sandstone Aquifer in East-Oweinat area, Egypt. NRIAG J Astron Geophys 8(1):117–130
21. IAEA, UNDP, and GEF (2013) Regional strategic action programme for the Nubian Aquifer System
22. Quadri E (2019) The evolving framework for transboundary cooperation in the Nubian Sandstone Aquifer System. Water Int 44(3):363–377
23. Abou-Khodier I, Mahmoud DM (2017) Market assessment study of socio-economic impacts for solar pumping systems in terms of local job and value creation in Egypt. Regional Center for Renewable Energy and Energy Efficiency "RCREEE", Egypt, p 57
24. Rashed AA, El-Sayed EA (2014) Simulating agricultural drainage water reuse using QUAL2K model: Case study of the Ismailia Canal catchment area, Egypt. J Irrig Drain Eng 140(5)
25. Rashed AA (2016) Vegetable production on rice bales using Brackish treated drainage water. Irrig Drain 65(5):664–672
26. Gad M, El-Hattab M (2019) Integration of water pollution indices and DRASTIC model for assessment of groundwater quality in El Fayoum depression, western desert, Egypt. J Afr Earth Sci 158
27. El Osta M, Hussein H, Tomas K (2018) Numerical simulation of groundwater flow and vulnerability in Wadi El-Natrun depression and vicinities, West Nile Delta, Egypt. J Geol Soc India 92(2):235–247
28. Masoud MH, Schneider M, El Osta MM (2013) Recharge flux to the Nubian Sandstone aquifer and its impact on the present development in southwest Egypt. J Afr Earth Sc 85:115–124
29. Salem B et al (2017) Beyond DRASTIC in Wadi Natrun. In: Renaud F, Schuster-Wallace C (eds) Groundwater and human security. UNU, Hamilton, Ontario, pp 25–96
30. Ahmed AA (2009) Using generic and pesticide DRASTIC GIS-based models for vulnerability assessment of the quaternary aquifer at Sohag, Egypt. Hydrogeol J 17(5):1203–1217
31. King C et al. (2010) Management challenges and opportunities in the Endorheic basins of the Northern Saharan Transboundary aquifers. In: Proceedings of international conference "Transboundary Aquifers: Challenges and New Directions" (ISARM2010). UNESCO IHP, Paris

Groundwater Quality for Irrigation as an Aspect of Sustainable Development Approaches: A Case Study of Semi-Arid Area Around Ismailia Canal, Eastern Nile Delta, Egypt

Zenhom El-Said Salem, Mohamed Sobhy Fathy, Abdel-Fattah Ibrahim Helal, Sofia Yahia Afifi, and Abdullah Muhammad Attiah

Abstract Irrigation water is a basic factor in maintaining productivity and conserving agricultural lands, in addition, irrigation is the most important water-consuming sector. Approximately, forty percent of the world's food is produced by irrigation that highly depends on the resources of groundwater. Seventy-six groundwater samples were collected from the Quaternary aquifer, Eastern Nile Delta region, Egypt. Irrigation water quality parameters: pH, TDS, EC, Sodium Adsorption Ratio (SAR), Sodium to Calcium Activity Ratio (SCAR), Adjusted-SAR (adj. R_{Na}), Adjusted Sodium to Calcium Activity Ratio (adj. SCAR), Soluble Sodium Percentage (SSP), Kelley's Ratio (KR), Residual Sodium Carbonate (RSC), Residual Sodium Bicarbonate (RSBC), Corrosively Ratio Index (CRI), Magnesium Adsorption Ratio (MAR), Permeability Index (PI), Potential Salinity (PS), Heavy metal Pollution Index (HPI) and Irrigation Water Quality Index (IWQI) were determined to assess the groundwater quality of the Quaternary aquifer in the study area for irrigation purpose. Generally, the groundwater in the study area has a moderate quality for irrigation purposes, and as indicated by the obtained results of the determined evaluation parameters, management changes and practices should be considered for maintaining the soil properties and crop productivity.

Z. E.-S. Salem (✉) · M. S. Fathy
Geology Department, Faculty of Science, Tanta University, Comprehensive Composite/Tanta University Building—Central Delta, 31527 Tanta, Egypt

M. S. Fathy
e-mail: msfathy@science.tanta.edu.eg

A.-F. I. Helal · A. M. Attiah
Central Laboratory for Elemental and Isotopic Analysis (CLEIA), Nuclear Research Center (NRC), Atomic Energy Authority—Inshas, P.O. Box 13759, Abu Zabal, Egypt
e-mail: abdullah.attiah@eaea.org.eg

S. Y. Afifi
Nuclear Materials Authority, El Maadi, P.O. Box 530, 11381 Cairo, Egypt

A. Negm and A. Elkhouly (eds.), *Groundwater in Egypt's Deserts*, Springer Water,
https://doi.org/10.1007/978-3-030-77622-0_12

Keywords Groundwater · Quaternary Aquifer · Irrigation Water Quality Parameters · Eastern Nile Delta · Ismailia Canal · Egypt

Introduction

Water is the ever most important vital source of life for the survival of all living organisms, as well as human survival itself depends essentially on water: water is a rights issue [1, 2]. Water is the very core of any sustainable development and the key factor for all aspects of socio-economic development and any future development. The second goal "Zero Hunger" of the Sustainable Development Goals (SDGs) set by the United Nations General Assembly for the year 2030, aims to eliminate hunger, achieve food security, improve nutrition and health and encourage sustainable agriculture in a way that ensures the protection of natural resources of lands and water [2]. Improving the global food and agriculture system is an essential requirement to help end hunger by increasing the capacity for agricultural productivity, maintaining the productivity of agricultural lands and sustainable food production [2]. The linkage between agriculture and food security, nutrition and health, rural development and growth, and environmental protection, makes the agriculture to be the main driver of achieving these targets [1].

Increasing demand for food requires increasing crop production and ensure sustainability for a long time to come. Agriculture has to increase the production of safe and nutritious food to meet the increasing demand driven by population growth [3, 4]. In addition, agriculture plays the main role in the sustainable management of natural resources, especially land and water resources [1]. Improving agricultural productivity, enhancing and conserving natural resources, such as soil and water, are essential requirements to increase global food supplies on a sustainable basis [1]. Water resources management and agricultural food production are increasingly becoming global issues, which are inseparable [5]. Wastewater from irrigation as well as other sectors would increase with increasing of water usage. In order to avoid pollution of land and water resources, safe use of these waters in agriculture is called for [3, 4].

Irrigation using groundwater has become widespread in semi-arid and arid regions, where groundwater is considered the ultimate valuable source of water when surface-water resources are depleted. In addition, groundwater aquifers are accessible and flexible sources for quick and easy access to water for individual users than the traditional surface-water irrigation systems that can hardly match. Furthermore, groundwater is less susceptible to contamination and droughts than surface water [6–12].

The availability of water by itself is not the key factor for sustainable development, but it should fit a specific purpose with respect to both quantity and quality. The concept of quality must deduce how well a water resource achieves the requirements of the concerned user and should be assessed based on its suitability for the intended use. In a serious attempt to achieve the 2nd goal "Zero Hunger" of the Sustainable

Development Goals, it is important to initially identify the quality of the available water resources for irrigation use. Water used for irrigation should achieve the crop productivity, maintain the soil quality and productivity, protect the environment, keep soil from degradation, and maintain both physical and mechanical properties of the soil, e.g. soil permeability and soil structure. Irrigation water can affect the fertility requirements of the crops, the crop yield and quantity, performance and longevity of irrigation system, and methods of water application. Understanding irrigation-water quality has major importance in determining the appropriate crop and soil management practices that are necessary for long-term productivity [1, 2, 13–21].

The suitability of water for irrigation is evaluated by its potential to cause problems for both soil and plant. The common problems that related to irrigation water quality are salinity hazard that related to the amounts and kind of soluble salts in the water, sodicity hazard that related to the relative proportion of Na to other cations, alkalinity hazard, i.e., bicarbonate concentration as related to the concentration of calcium plus magnesium, or calcium alone, permeability and irrigation water infiltration, i.e., how well the water is able to seep into the soil, toxicity of specific ions and metals and hazards related to excessive nitrogen, high bicarbonate water and an unusual pH of the water [2, 3, 12, 17–20, 22, 24–29, 33].

Several studies were carried out for evaluating the groundwater quality in the Nile Delta, such as Armanuos el al [30]; El Ramly [31]; Eltarabily and Negm [32]; Negm and Armanuos [33]; Salem et al. [17–20]. Taha et al. [34] evaluated the groundwater quality of the Quaternary aquifer, in the Eastern Nile Delta region, based on TDS and the major anions and cations and concluded that the groundwater of Quaternary aquifer is suitable for drinking in its western part, while its quality is unsuitable for drinking or domestic uses in the northern and eastern parts. El-Fakharany et al. [35] studied the quality of the Quaternary aquifer groundwater, at the southwestern part of the Eastern Nile Delta region, using multivariate statistical techniques and concluded that the hydrochemical characteristics of the groundwater were influenced mainly by the anthropogenic activities that included intensive use of fertilizers, septic tanks, industrial and domestic effluents and landfill waste sites. Eltarabily and Negm [36] concluded that the Quaternary aquifer in the Eastern Nile Delta is recharged primarily from Ismailia Canal and Damietta Branch in addition the infiltration from the agricultural fields during irrigation. Hegazy et al. [37] evaluated the main factors affecting the quality of the groundwater around Abu Zaabal, Eastern Nile Delta region, and suggested that the industrial effluent, domestic sewages, agricultural wastes and the excessive use of chemical fertilizers and pesticides are the main factors affecting the groundwater quality through complex hydrochemical processes with mutable sources of recharge. Abu El Ella [38] studied the suitability of groundwater in the Eastern Nile Delta region for different purposes and concluded that the groundwater of Quaternary aquifer is generally suitable for drinking and irrigation purposes in the south of the region, while it is affected by seawater intrusion in the north. Embaby et al. [39] assessed the quality of groundwater, in El-Salhia Plain, Eastern Nile Delta region, for drinking and irrigation purposes, and concluded that the groundwater in the upper and lower portions of the plain is good for drinking, while in the middle portion it had a poor quality for drinking. In addition, for irrigation purpose special management should be concerned for controlling salinity hazard. Eltarabily et al. [40] studied the groundwater quality for irrigation in the southwest of the Eastern Nile Delta region, and indicated the validity of groundwater for irrigation based on pH, TDS, RSC and SAR.

This study is conducted to evaluate the groundwater quality of the Quaternary aquifer in the semi-arid Eastern Nile Delta region around Ismailia Canal, Egypt, for irrigation purpose as a basic evaluation study for understanding what management changes and practices that should be considered for conserving the soil from degradation, maintaining the soil productivity, restoring the maximum production capability and maintaining the full crop and long-term productivity under the given set of conditions.

Study Area

Egypt is located in a semi-arid to arid zone as a part of Sahara of North Africa, and facing great challenges due to its limited water resources represented mainly by its fixed share of the Nile water and its aridity as a general characteristic [41]. Eastern Nile Delta region is an important area among the newly developed regions in Egypt, where there are several land reclamation projects and many industrial activities, and it is considered an essential development axis for the heavily populated regions of the Nile Valley and Delta. Eastern Nile Delta region constitutes a portion of a semi-arid to the arid belt of north Egypt, while it is affected by the Mediterranean climate along the coastline. It is characterized by a long dry summer and short temperate winter with low precipitation with a high evaporation rate [42].

The study area is a part of the Eastern Nile Delta region, extending along and around the course of Ismailia Canal that streams from the River Nile in Cairo governorate and passes throughout the Quaternary sediments until it ends into El Temsah Lake in Ismailia governorate (Fig. 1). It is placed between latitudes 30° 03′ and 30° 42′ N and longitudes 31° 03′ and 32° 24′ E, bounded by the River Nile and Damietta branch from west, the Suez Canal from east and from south by the desert rolling plains and foot-hills of Cretaceous to Tertiary structural exposures [43].

Geological Setting

Generally, the Eastern Nile Delta region has a low topographic relief with a gently sloping surface towards the north and it takes a rolled shape towards the south, where the land rises as a moderately elevated plateau. The Eastern Nile Delta region is mainly occupied by rock units (Fig. 2), which belong to Quaternary and Tertiary periods. Quaternary of the Nile Delta was subdivided on the basis of their lithologic composition into two formations, which from oldest to youngest are Mit Ghamr and Bilqas formations. Mit Ghamr Formation (Late Pliocene–Pleistocene) is composed

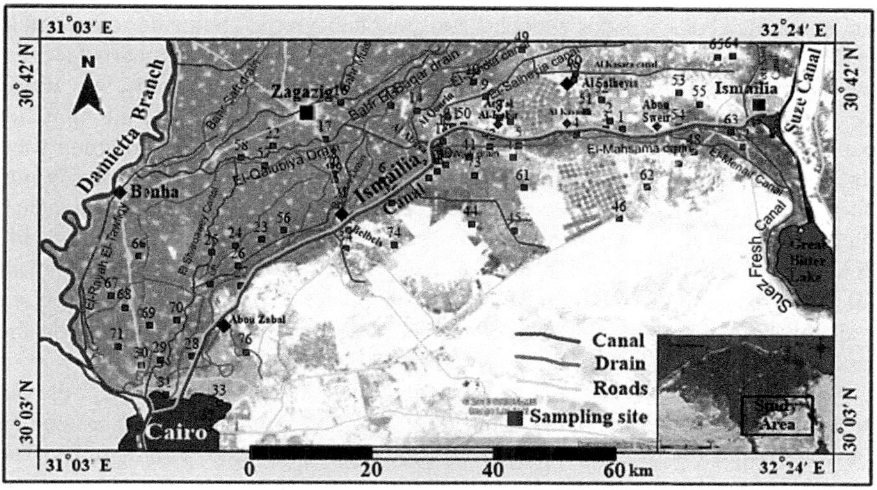

Fig. 1 Map of the study area showing the sampling sites and the surface water system (The satellite image was cropped from Google Earth Pro—Data SIO, NOAA, U.S. Navy, NGA, GEBCO—Image Landsat/Copernicus—Image IBCAO)

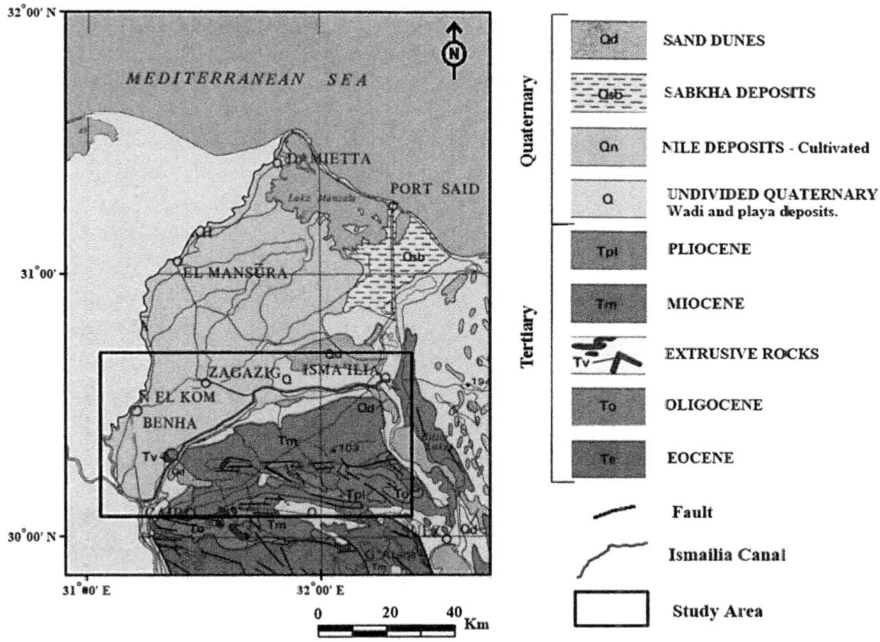

Fig. 2 Geological map of Eastern Nile Delta (*Source* Geologic Map of Egypt 1981)

mainly of thick beds of sands with clay and gravel interbeds. The sand is quartz-rich, and medium to course-grained. Pebbles are found and consisted of quartzite, chert and dolomite grains with shell fragments of gastropods and pelecypods, in addition the coquina layers are found intermittently, whereas the gravels are composed of flint and dolomites. It was suggested that the sediments of Mit Ghamr Formation were probably deposited under deltaic and/or coastal marine to fluvial conditions. While, Bilqas Formation (Late Pleistocene to Holocene) is dominated by sand, silt and clay interbeds, however, the sediments are more calcareous in the northern part of the Nile Delta and are described as Nile mud of flood basin deposits in the eastern part of Nile Delta. The sand is fine-grained, in addition, plant remains, and peat deposits are frequent. Continental, fluviatile, lagoonal, and beach environments are the probable depositional environments of these sediments [43–52].

North of Ismailia canal, the flood deposits of the Nile silt that represent the old cultivated lands are the main landscape of the area. The desert lands south to Ismailia Canal have a low to moderate relief and are gradually elevated towards south. The western part of the study area is covered mainly by the Eastern Nile Delta flood plain deposits that constitute the highly fertile lands. The eastern part of the study area, in which Ismailia Canal runs through Wadi El-Tumilat that extends from El-Abdassa in the west to El-Temsah Lake in the east, where the area south to Wadi El-Tumilat that bounded by the cultivated lands in the west and Suez Canal in the east is occupied by sand and gravels that are developed into a typical desert pavement. Wadi El-Tumilat depression separates between two higher elevated geomorphologic units that are Um-Gedam gravely slopes toward the south and El-Salheyia gravely plain which slopes toward the north [45, 50, 51, 53, 54].

Surface Water System

The Eastern Nile Delta region is dissected by intersected sets of canals and drains; e.g., Ismailia canal, Kashmir El-Yumna canal, El-Sharqawiyah canal, El-Rayah El-Tawfiqy, Kashmir El-Yusra canal, El-Salheyia canal, El-Kasara canal, El-Saidia canal, Port-Said canal, Suez fresh canal, El-Menaif canal, Bahr Muis, Bahr El-Baqar drain, Bahr Saft drain, El-Qalubiya drain, Belbies drain, El-Mahsama drain, Gabal Al Asfer drain, Shibin El Qanater drain, and El-Wadi drain (Fig. 1). Generally, the surface water levels of canals vary seasonally in response to quantities of the recharging water. These canals and drains are cutting through both the Holocene Nile silt and clay deposits and the Pleistocene sediments, passing through large population densities and activated industrial areas, and receiving the untreated liquid wastes and effluents. As the thickness of the top clay layer varies, contaminated water may infiltrate into the soil towards the groundwater aquifer. Ismailia Canal passes through the Quaternary sand and gravels, in the eastern part of the study area, without any lining, where such conditions largely permit the seepage from Ismailia canal to the surrounding groundwater aquifer. Hence, Ismailia Canal may present a recharging source for the groundwater aquifer especially in the eastern part while to the western part its effect is limited [45, 55–59].

Groundwater System

The Quaternary aquifer in the Eastern Nile Delta region is a part of the Nile Delta Aquifer system that was classified as a phreatic aquifer near the western and eastern borders and a leaky aquifer in the southern, northern and middle parts. The Quaternary aquifer was divided into two hydrogeological units; the upper unit is the Holocene aquitard that is represented by Bilqas Formation, while the lower unit is the Pleistocene aquifer that is represented by Mit Ghamr Formation. In the floodplain region, the Quaternary aquifer is underlain by Pliocene plastic clay, which acts as an aquiclude. Towards the eastern and southern parts, the Quaternary aquifer directly overlies the fissured carbonate aquifer of Miocene and Oligocene [45, 60–70].

Methodology

Samples Collection and Analyses

Groundwater samples were collected from 76 groundwater wells, with depths ranging from 10 to 146 m, from the Quaternary aquifer in the study area. Geographical locations of the sampling sites were determined by a handheld global positioning system (GPS) (Explorist 200, Megellan), and a map of sampling sites was constructed as shown in Fig. 1. Groundwater samples were collected in new pre-cleaned, acid rinsed and high-density polyethylene bottles to prevent unpredictable changes in characteristics. From each location, two samples were collected; one sample for elemental analysis which was acidified by adding 2 mL of 10% HNO_3 to pH less than 2 to keep the metal ions remain in solution and reduce precipitation, microbial activity and sorption losses to bottles walls. While the other sample was collected for major and minor ions analyses and was not acidified. All samples bottles were labelled, kept in cooler box, moved to the laboratory, and then kept cool at a temperature less than 5 °C until conducting the chemical analysis. Collection, storing and analysis of samples were made according to the standard methods [71].

Electrical conductivity (EC), TDS and pH measurements were performed in situ using Hach's portable EC/TDS meter and portable pH meter (Model P 314). Concentrations of Na, K, Ca, Mg, Cl, SO_4, CO_3, HCO_3 and NO_3 were determined using Ion Chromatography (GBC, TIMERLINE INSTRUMENT, Boulder, Colorado). The elemental concentrations of Al, As, Cd, Cr, Co, Cu, Fe, Pb, Li, Mn, Ni, Se, V and Zn were determined using Inductively Coupled Plasma-Optical Emission Spectrometer

ICP-OES [Prodigy HDICP, TELEDYNE, Leeman Labs]. Double distilled deionized water was used to prepare all solutions. Preservation, preparation and chemical analyses of the groundwater samples were performed in the Central Laboratory for Elemental and Isotopic Analysis (CLEIA), Nuclear Research Center, Atomic Energy Authority (Inshas), and the Geochemical Laboratory, Nuclear Material Authority, (Inshas) Egypt. The obtained results were tabulated, statistically studied and graphically presented. Surfer program (version 11.1.719—Surface Mapping System—Golden Software, Inc.) was used for mapping purposes.

Groundwater Quality for Irrigation

Quality of water is determined based on the purpose for which water will be used. For irrigation purposes, the irrigation water quality depends on several criteria such as salinity hazard, sodicity hazard, alkalinity hazard, permeability hazard and specific ion toxicity hazard [4, 12, 22–24, 26, 28, 29]. In the present study several parameters are considered in evaluating the quality of the Quaternary aquifer groundwater in Eastern Nile Delta region for irrigation, which are pH, TDS, EC, Sodium Adsorption Ratio (SAR), Sodium to Calcium Activity Ratio (SCAR), Adjusted Sodium Adsorption Ratio (Adj. R_{Na}), Adjusted Sodium to Calcium Activity Ratio (Adj. SCAR), Soluble Sodium Percentage (SSP), Kelley's Ratio (KR), Residual Sodium Carbonate (RSC), Residual Sodium Bicarbonate (RSBC), Corrosively Ratio Index (CRI), Magnesium Adsorption Ratio (MAR), Permeability Index (PI), Potential Salinity (PS), Heavy metal Pollution Index (HPI) and Irrigation Water Quality Index (IWQI). These parameters are determined as presented in Table 1, which presents the classification of water quality for irrigation based on the concerned parameters, where all ions concentrations are in meq/L.

Results and Discussion

Seventy-six groundwater samples were collected from the Eastern Nile Delta region, Egypt, to assess the groundwater quality of the Quaternary aquifer for irrigation. Twenty-six physio-chemical parameters: pH, TDS, EC, K, Na, Ca, Mg, Cl, SO_4, HCO_3, CO_3, NO_3, Al, As, Cd, Cr, Co, Cu, Fe, Pb, Li, Mn, Ni, Se, V and Zn were determined for the collected groundwater samples. Table 2 shows the descriptive statistics of the obtained hydrochemical results and Table 3 presents the descriptive statistics of the calculated irrigation water quality parameter.

Table 1 Irrigation water quality parameters and classification of water quality for irrigation

Parameter		Range	Water classification	Reference
pH		6.5–8.4	Normal pH range of irrigation water	[12, 72, 73]
		6.5 > pH > 8.4	Abnormal pH value and water needs further evaluation	
TDS (mg/l)		> 3000	Suitable for irrigation	[12, 74, 75]
		>3000	Unsuitable for irrigation	
EC (dS/m)		<0.2	Sever restriction as water has no fertility value and may create severe permeability problems in the soil	[3, 4, 73]
		0.2–0.7	No restriction	
		0.7–3.0	Moderate restriction	
		>3.0	Sever restriction	
$SAR = \frac{Na^+}{\sqrt{\frac{(Ca^{2+}+Mg^{2+})}{2}}}$	Sodicity parameters	<10	Low sodicity hazard	[3, 4, 25, 73, 76–79]
$SCAR = \frac{Na^+}{\sqrt{Ca^{2+}}}$		10–18	Medium sodicity hazard	
$*Adj.R_{Na} = \frac{Na^-}{\sqrt{\frac{(Ca_x{}^{2+}-Mg^{2+})}{2}}}$		18–26	High sodicity hazard	
$^a Adj.SCAR = \frac{Na^+}{\sqrt{Ca_x^{2+}}}$		>26	Very high sodicity hazard	
$SSP = \frac{Na^+}{Ca^{2+}+Mg^{2+}+Na^+} x100$		<60	Safe	[26, 80]
		>60	Unsafe	
$KR = \frac{Na^+}{Ca^{2+}-Mg^{2+}}$		<1	Suitable	[24, 81, 82]

(continued)

Table 1 (continued)

Parameter	Range	Water classification	Reference
	>1	Unsuitable	
$RSC = \left(HCO_3^- + CO_3^{2-}\right) - (Ca^{2+} + Mg^{2+})$	<1.25	Suitable and safe	[3, 4, 23, 29, 73, 83–86]
	1.25–2.5	Suitable with certain management	
	>2.5	Unsuitable	
$RSBC = HCO_3^- - Ca^{2+}$	<2.5	Low alkalinity	
	2.5–5.0	Medium alkalinity	
	5.0–10.0	High alkalinity	
	>10.0	Very high alkalinity	
$CRI = \frac{\left(\frac{Cl}{35.5}+2\left(\frac{SO_4}{96}\right)\right)}{\left(2\left(\frac{HCO_3+CO_3}{100}\right)\right)}$	>1	Corrosive water	[87]
	<1	Safe and non-corrosive water	
$MAR = \frac{Mg}{Ca+Mg} * 100$	<50	Suitable	[12, 23, 24, 73]
	>50	Unsuitable with a harmful effect on soils	
$PI = \frac{Na^+ + \sqrt{HCO_3}}{Ca^{2+}+Mg^{2+}+Na^+} * 100$	>75%	Excellent	[88–91]
	25–75%	Good	
	<25%	Unsuitable	
$PS = Cl^- + \left(\frac{SO_4^{2-}}{2}\right)$	<3.0	Excellent to good	[28, 92, 93]
	3.0–5.0	Good to injurious	
	>5.0	Injurious to unsatisfactory	
[b] $HPI = \frac{\sum_{i=1}^{n} W_i Q_i}{\sum_{i-1}^{n} W_i}$	<100	Not polluted by trace elements	[94–102]
	>100	Polluted by trace elements	
[c] $IWQI = \sum_{i=1}^{n} q_i w_i$	85–100	No restriction for use	[103–105]

(continued)

Table 1 (continued)

Parameter	Range	Water classification	Reference
	70–85	Low restriction for use	
	55–70	Moderate restriction for use	
	40–55	High restriction for use	
	0–40	Sever restriction for use	

[a] Ca_x^{2+} is the adjusted calcium concentration is calculated based on the procedures adapted from Suarez [79]; Ayers and Westcot [73]; Gupta and Gupta [3]; Gupta and Gupta [4]
[b] HPI is calculated by the procedures adapted from Mohan et al. [99]; Prasad and Bose [101]; Trivedy [106]; Yusuf et al. [102]
[c] IWQI is calculated by the procedures adapted from Meireles et al. [105]; Besser et al. [23]; Abbasnia et al. [104]

Hydrogen Ion Activity (pH)

The obtained results of pH measurements indicated that the pH of the groundwater samples ranged from 7.13 to 9.63 with a mean value of 7.78 (Table 2). For irrigation suitability, the normal pH range of irrigation water is from pH 6.5 to pH 8.4 [12, 73]. Approximately, 96% of the groundwater samples are within the normal range of pH for irrigation, while only 3 groundwater samples (about 4%) have pH values more than 8.4, which are not suitable for irrigation [12, 73].

Salinity Hazard

Water salinity is usually measured by TDS or EC. TDS of the groundwater samples, expressed in mg/l, ranged from 205 to 4965 with a mean value of 1058 (Table 2). Approximately 92% of the groundwater samples have TDS values less than 3000 mg/l and are suitable for irrigation, while the remaining 8% of the groundwater samples having TDS values more than 3000 mg/l that are not suitable for irrigation as presented in Table 4 [12, 74, 75]. Electrical conductivity (EC), measured in dS/m, of the groundwater samples ranged from 0.3 to 7.8 with a mean value of 1.7 (Table 2). Table 5 shows the classification of the groundwater samples based on salinity level according to EC, where, about 70% of the groundwater samples have low salinity,

Table 2 Summary of the descriptive statistics of the obtained results of the determined physio-chemical parameters

Parameter	Min	Max	Mean	Median	Stand. dev
pH	7.13	9.36	7.78	7.80	0.40
TDS (mg/l)	205	4965	1058	651	968
EC (dS/m)	0.32	7.76	1.65	1.02	1.51
K (meq/l)	0.00	1.77	0.34	0.28	0.29
Na (meq/l)	1.00	48.30	9.42	4.28	10.74
Ca (meq/l)	0.50	18.04	3.71	2.34	3.71
Mg (meq/l)	0.50	16.00	3.28	2.00	3.45
Cl (meq/l)	0.57	52.70	7.90	2.72	10.35
SO_4 (meq/l)	0.10	19.71	4.94	3.86	4.73
HCO_3 (meq/l)	0.50	10.51	2.40	1.00	2.56
CO_3 (meq/l)	0.00	10.33	0.92	0.52	1.54
NO_3 (meq/l)	0.11	0.85	0.36	0.27	0.20
Al (mg/l)	0.001	0.826	0.208	0.188	0.138
As (mg/l)	0.000	0.081	0.007	0.005	0.011
Cd (mg/l)	0.001	0.017	0.004	0.004	0.002
Cr (mg/l)	0.000	0.013	0.001	0.000	0.002
Co (mg/l)	0.001	0.132	0.026	0.026	0.021
Cu (mg/l)	0.001	0.033	0.009	0.009	0.004
Fe (mg/l)	0.000	3.091	0.496	0.094	0.712
Pb (mg/l)	0.000	0.091	0.008	0.004	0.012
Li (mg/l)	0.005	0.104	0.021	0.016	0.016
Mn (mg/l)	0.000	1.591	0.272	0.123	0.369
Ni (mg/l)	0.000	0.034	0.007	0.007	0.005
Se (mg/l)	0.000	0.079	0.017	0.018	0.013
V (mg/l)	0.000	0.240	0.051	0.052	0.035
Zn (mg/l)	0.000	5.305	0.104	0.000	0.631

13% of the groundwater samples have medium salinity, 12% of the samples have high salinity, and the remaining 5% of the groundwater samples have very high salinity.

Based on salinity hazard of the groundwater samples according to EC (dS/cm), approximately 24% of the groundwater samples have no restriction on their use for irrigation, while the majority of samples (about 59%) have slight to moderate restriction on their uses for irrigation, and the remaining 17% of the groundwater samples have severe restriction on their uses as presented in Table 6 [3, 4, 73]. Figure 3 shows the spatial distribution of groundwater quality for irrigation according to restriction on the use and based on salinity hazard throughout the study area. The restriction on the use of the groundwater for irrigation increases due north and southeast of the study area.

Table 3 Summary of the descriptive statistics of the determined irrigation quality parameters

Parameter	Min	Max	Mean	Median	Stand. dev
SAR	0.71	33.62	5.25	2.88	5.30
SCAR	0.68	36.67	5.27	3.02	5.58
Adj. R_{Na}	0.73	38.53	5.49	2.95	5.92
Adj. SCAR	0.71	55.77	6.50	3.39	8.61
Sodicity hazard	0.72	55.77	6.35	3.16	8.09
SSP (%)	14	93	53	52	19
KR	0.17	12.38	1.78	1.10	2.07
RSC (meq/l)	−24.93	8.33	−3.67	−1.73	5.66
RSBC (meq/l)	−14.03	7.14	−1.32	−0.50	3.85
CRI	0.39	24.21	6.20	3.63	6.20
MAR (%)	25	78	48	50	10
PI (%)	21.16	103.57	65.57	68.19	19.66
PS (meq/l)	0.89	61.25	10.37	5.30	11.96
HPI	5.92	222.47	50.90	49.31	25.29
IWQI	13.67	96.06	62.61	69.43	19.90

Table 4 Classification of groundwater samples for irrigation based on TDS in mg/l

TDS	Water class	Samples	%
<3000	Suitable for irrigation	1, 3–7, 9–15, 17–45, 48–63, 65–76	92
>3000	Unsuitable for irrigation	2, 8, 16, 46, 47, 64	8

Table 5 Classification of groundwater samples based on salinity hazard according to EC

EC (dS/m)	Salinity	Samples	%
0.2–1.5	Low	1, 3, 4, 6, 11–15, 17–21, 24–30, 32, 33, 35, 37, 39–42, 44, 48, 50, 52–59, 61, 63, 65–70, 72–76	70
1.5–3.0	Medium	5, 22, 23, 31, 34, 36, 43, 49, 51, 60	13
3.0–5.0	High	7–10, 16, 38, 45, 62, 71	12
5.0–10.0	Very high	2, 46, 47, 64	5

Table 6 Classification of groundwater samples according to restriction on use for irrigation and based on salinity hazard

EC (dS/m)	Restriction on use	Samples	%
< 0.2	Sever restriction	–	0
0.2–0.7	No restriction	3, 11–15, 21, 25, 28, 40, 50, 56, 59, 61, 67, 73, 75, 76	24
0.7–3.0	Slight to moderate restriction	1, 4–6, 17–20, 22–24, 26, 27, 29–37, 39, 41–44, 48, 49, 51–55, 57, 58, 60, 63, 65, 66, 68–70, 72, 74	59
>3.0	Severe restriction	2, 7–10, 16, 38, 45–47, 62, 64, 71	17

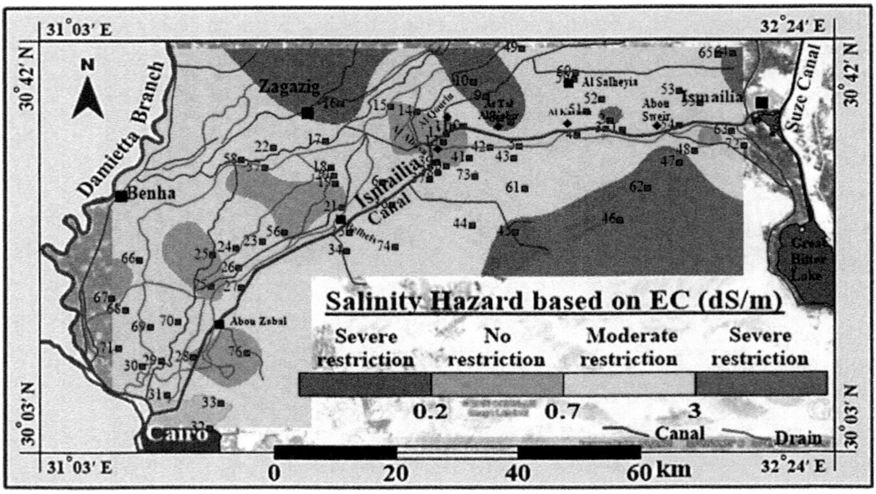

Fig. 3 Spatial distribution map of the salinity hazard based on EC (dS/m) presenting the restriction on use of the groundwater for irrigation in the study area (The satellite image was cropped from Google Earth Pro—Data SIO, NOAA, U.S. Navy, NGA, GEBCO—Image Landsat/Copernicus—Image IBCAO)

Sodicity Hazard

The Sodicity hazard, which is the potential sodium hazard, of the groundwater samples is evaluated depending on Ca/Mg and Ca/HCO_3 ratios, considering the integrated effect of Na, Ca, Mg, HCO_3 and CO_3, and determined based on SAR, SCAR, Adj. R_{Na} and Adj. SCAR as presented in Table 7 [4, 77]. The sodicity hazard of the groundwater samples ranged from 0.72 to 55.77 with a mean value of 6.35 (Table 3). About 80% of the groundwater samples have low sodicity hazard, about 15% of the samples have medium sodicity hazard, only one sample has high sodicity hazard, while about 4% of the samples have very high sodicity hazard. Figure 4 shows the spatial distribution of the sodicity hazard, where the hazard of sodium appears in the northwestern part of the study area, and in some locale areas such are

Table 7 Criteria of sodium hazard calculation according to Gupta and Gupta [4]; Gupta [77]

Ca/Mg	Ca/HCO_3	Sodicity parameter used	Samples
>1	>1	SAR	1, 3–6, 9–15, 17, 18, 24, 26–28, 30–32, 34–36, 39, 40, 45–47, 56–71, 73–75
<1	>1	SCAR	20, 23, 43, 72
>1	<1	Adj. R_{Na}	37, 38, 41, 44, 48–55, 76
<1	<1	Adj. SCAR	2, 7, 8, 16, 19, 21, 22, 25, 29, 33, 42

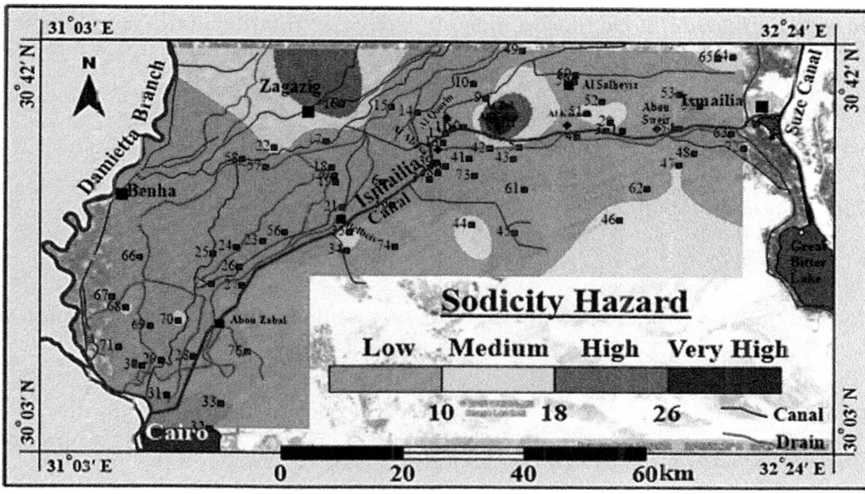

Fig. 4 Spatial distribution map of the sodicity hazard presenting the groundwater quality for irrigation in the study area (The satellite image was cropped from Google Earth Pro—Data SIO, NOAA, U.S. Navy, NGA, GEBCO—Image Landsat/Copernicus—Image IBCAO)

At Tal Al-Kabir, Al Kassasin, and Belbeis, while the groundwater in the majority of the study area has low sodium hazard.

SAR and EC

According to the guidelines set by Ayers and Westcot [73]; Miller and Gardiner [107] for the degree of restriction on use of water for irrigation that based on SAR and EC (Table 8), it was indicated that approximately 50% of the groundwater samples can be used safely for irrigation with no restrictions on their use, while the remaining

Table 8 Guidelines for irrigation water assessment based on SAR and EC [73, 107]

Permeability problem	Degree of restriction on use		
	Severe	Slight to moderate	None
	EC (dS/m)		
SAR = 0–3 and EC =	<0.2	0.2–0.7	>0.7
SAR = 3–6 and EC =	<0.3	0.3–1.2	>1.2
SAR = 6–12 and EC =	<0.5	0.5–1.9	>1.9
SAR = 12–20 and EC =	<1.3	1.3–2.9	>2.9
SAR = 20–40 and EC =	<2.9	2.9–5.0	>5.0

Table 9 Classification of groundwater samples based on restriction on their use for irrigation according to the guidelines set by Ayers and Westcot [73]; Miller and Gardiner [107] that based on SAR and EC

Restriction on use	Samples	%
No restriction	1, 2, 4–6, 9, 10, 17, 18, 23, 26, 27, 30–34, 36–39, 41, 43, 45–47, 53–55, 58, 60, 62–64, 66, 69, 71, 74	50
Slight to moderate Restriction	3, 7, 8, 11–16, 19–22, 24, 25, 28, 29, 35, 40, 42, 44, 48–52, 56, 57, 59, 61, 65, 67, 68, 70, 72, 73, 75, 76	50

50% of the samples have slight to moderate restriction on their uses for irrigation as they may negatively affect the soil permeability (Table 9).

Soluble Sodium Percentage (SSP)

The obtained results of SSP of the groundwater samples ranged from 14 to 93% with a mean value of 46% (Table 3). According to SSP classification of water suitability for irrigation depending on the hazard of sodium (Table 1), about 64% of the groundwater samples are safe for irrigation with SSP values less than 60%, and the remaining 36% of the groundwater samples have SSP more than 60% that are unsuitable for irrigation, where such water may result in sodium accumulations that will negatively affect the physical properties of the soil. Figure 5 shows the spatial distribution of

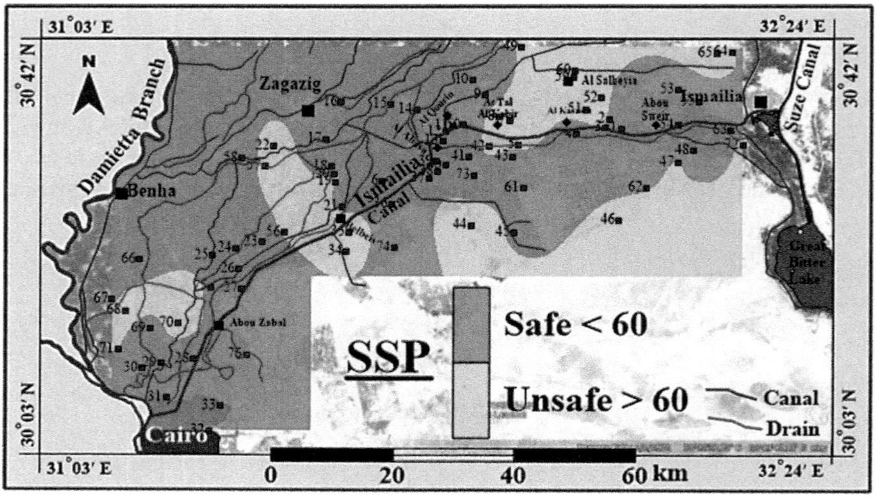

Fig. 5 Spatial distribution map of the SSP presenting the groundwater quality for irrigation in the study area (The satellite image was cropped from Google Earth Pro—Data SIO, NOAA, U.S. Navy, NGA, GEBCO—Image Landsat/Copernicus—Image IBCAO)

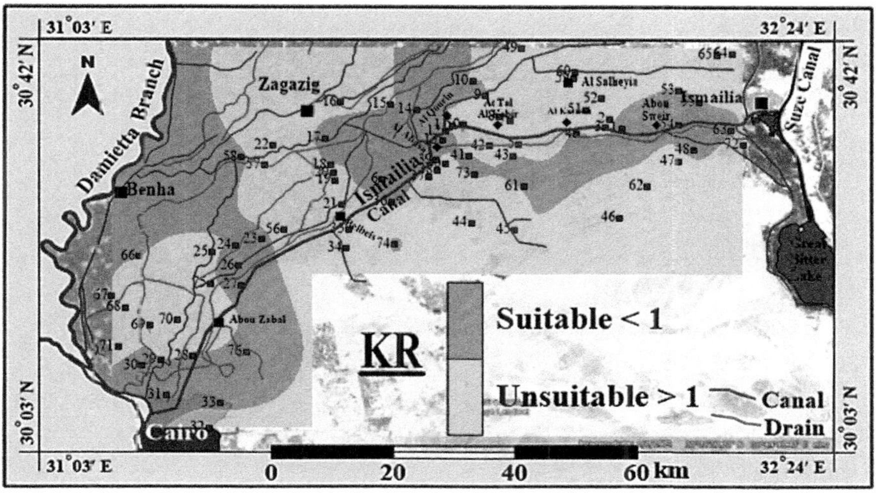

Fig. 6 Spatial distribution map of the KR presenting the groundwater quality for irrigation in the study area (The satellite image was cropped from Google Earth Pro—Data SIO, NOAA, U.S. Navy, NGA, GEBCO—Image Landsat/Copernicus—Image IBCAO)

the different quality types of the groundwater samples for irrigation based on SSP, where the quality of the groundwater for irrigation decreases greatly in the eastern part of the study area, and in some local areas in the western part of the study area.

Kelley's Ratio (KR)

Kelly's ratio relates the sodium amount to calcium and magnesium amount in irrigation water. The KR of the groundwater samples ranged from 0.17 to 12.38 with a mean value of 1.78 (Table 3). The obtained results of KR indicated that approximately, 47% of the groundwater samples came with KR values less than one and therefore they are safe and suitable for irrigation, while the remaining 53% of the groundwater samples had values of KR more than one, hence they are unsafe and unsuitable for irrigation as shown in Fig. 6 [81, 82, 108–110].

Alkalinity Hazard (RSC and RSBC)

The RSC of the groundwater samples ranged from −24.9 to 8.3 with a mean value of −3.7 (Table 3). The obtained results of RSC revealed that approximately, 91% of the groundwater samples came with RSC values less than 1.25 indicating a low alkalinity hazard and therefore they could be used safely for irrigation with little or no

Table 10 Classification of groundwater for irrigation use based on RSC [83]

RSC	Water quality	Samples	%
<1.25	Water can be used safely	1–6, 9–21, 23–41, 43, 45–48, 50–70, 72–76	90.8
1.25–2.5	Water can be used with certain management	7, 22, 42, 44, 49	6.6
>2.5	Unsuitable for irrigation purposes	8, 71	2.6

Table 11 Classification of groundwater for irrigation use based on RSBC [3, 4]

RSBC	Alkalinity Hazard	Samples	%
<2.5	Low alkalinity	1, 3–6, 9–15, 17–20, 23–28, 30–41, 43, 45–48, 50–76	86.9
2.5–5.0	Medium alkalinity	2, 7, 16, 21, 22, 29, 42, 44, 49	11.8
5.0–10.0	High alkalinity	8	1.3
>10.0	Very high alkalinity	–	–

development of the alkalinity hazard, while about 7% of the groundwater samples had RSC vales more than 1.25 and less than 2.5 indicating a medium alkalinity hazard, hence they could not be used directly for irrigation without special management practices, in addition there were about 3% of the samples had RSC values more than 2.5 indicating a high alkalinity hazard and therefore they are not suitable for irrigation as presented in Table 10 [83].

The RSBC of the groundwater samples ranged from −14.0 to 7.1 with a mean value of −1.3 (Table 3). According to RSBC classification of irrigation water (Table 1), there are about 87% of the groundwater samples have low alkalinity hazard, while about 12% of the groundwater samples have a medium alkalinity hazard, in addition only one groundwater sample has a high alkalinity hazard (Table 11). According to the obtained results of RSC and RSBC of the groundwater samples, it is indicated that the majority of the groundwater samples have low alkalinity hazard and could be used safely for irrigation. Based on Gupta and Gupta [3]; Gupta and Gupta [4], the groundwater samples: 2, 8, 9, 10, 16 and 62 that have CO_3 ions concentration ranges from 2 to 5 meq/l, should be considered as critical, because the presence of CO_3 ions in that amounts have the tendency for eliminating the ions of Ca from the irrigation water causing strong alkalisation in the irrigated soils and also cause a permeability problem in soil.

Corrosivity Ratio Index (CRI)

The CRI values of the groundwater samples ranged from 0.39 to 24.21 with a mean value of 6.20 (Table 3). The obtained results of the CRI indicated that the majority of

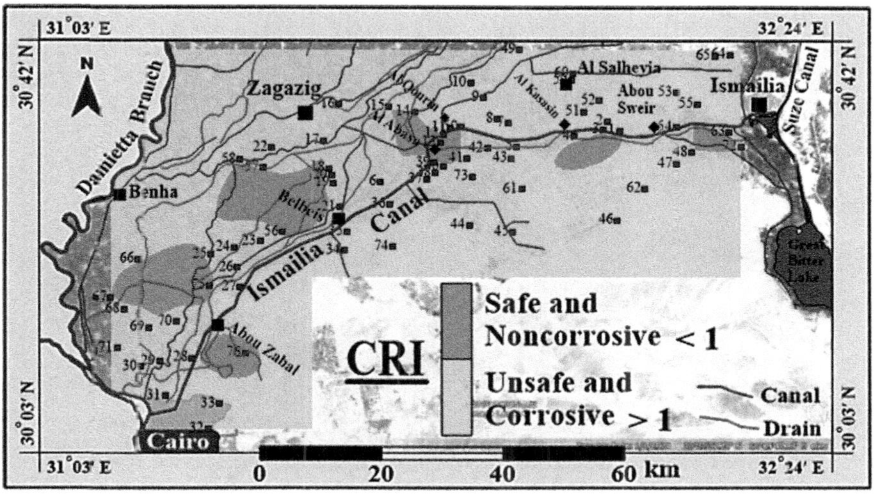

Fig. 7 Spatial distribution map of the CRI presenting the groundwater impact on the irrigation equipment in the study area (The satellite image was cropped from Google Earth Pro—Data SIO, NOAA, U.S. Navy, NGA, GEBCO—Image Landsat/Copernicus—Image IBCAO)

the groundwater samples (about 82%) have CRI values more than 1, they are not safe for irrigation because of their tendency to be corrosive for the irrigation equipment, while about 18% of the groundwater is safe for irrigation uses that were distributed in some local areas adjacent to the Ismailia canal as shown in Fig. 7.

Magnesium Adsorption Ratio (MAR)

The MAR of the groundwater samples ranged from 25 to 78% with a mean value of 48% (Table 3). The majority of the groundwater samples (about 80%) are suitable for irrigation with little or no magnesium hazard, while about 20% of the groundwater samples are unsuitable for irrigation, which is mainly found in the western part of the study area as shown in Fig. 8.

Permeability Hazard

Permeability Hazard is evaluated using the permeability index (PI), where the PI values of the groundwater samples ranged from 21.16 to 103.57 with a mean value of 65.47 (Table 3). Based on the PI classification of irrigation water quality, about 29% of the groundwater samples are suitable for irrigation, while the majority of the groundwater samples (about 68%) have marginal quality, and only two groundwater samples are unsuitable for irrigation, as distributed in Fig. 9.

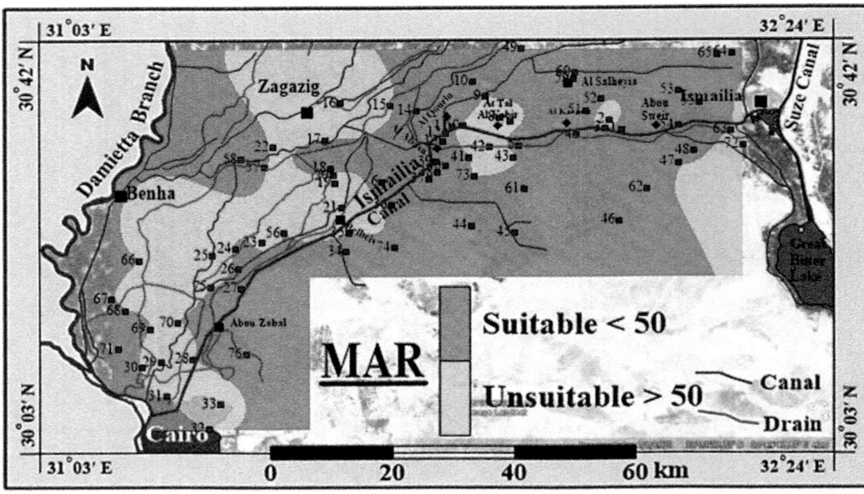

Fig. 8 Spatial distribution map of the MAR presenting the groundwater quality for irrigation in the study area (The satellite image was cropped from Google Earth Pro—Data SIO, NOAA, U.S. Navy, NGA, GEBCO—Image Landsat/Copernicus—Image IBCAO)

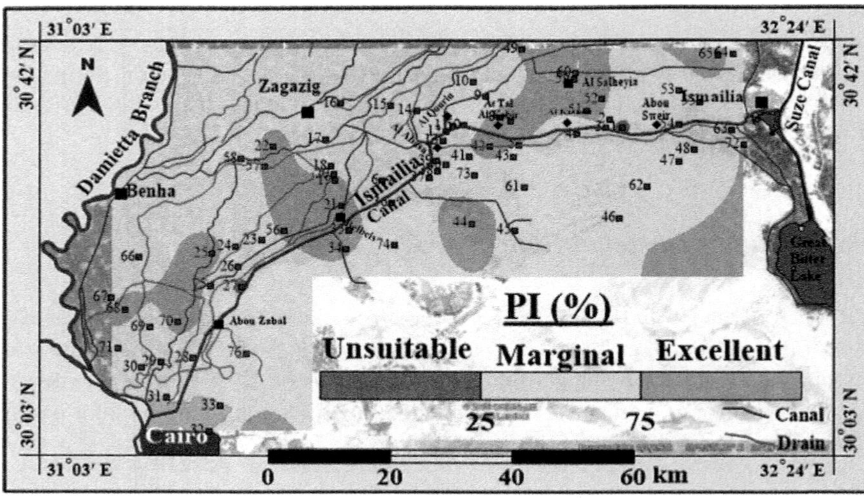

Fig. 9 Spatial distribution map of the PI presenting the groundwater quality for irrigation in the study area (The satellite image was cropped from Google Earth Pro—Data SIO, NOAA, U.S. Navy, NGA, GEBCO—Image Landsat/Copernicus—Image IBCAO)

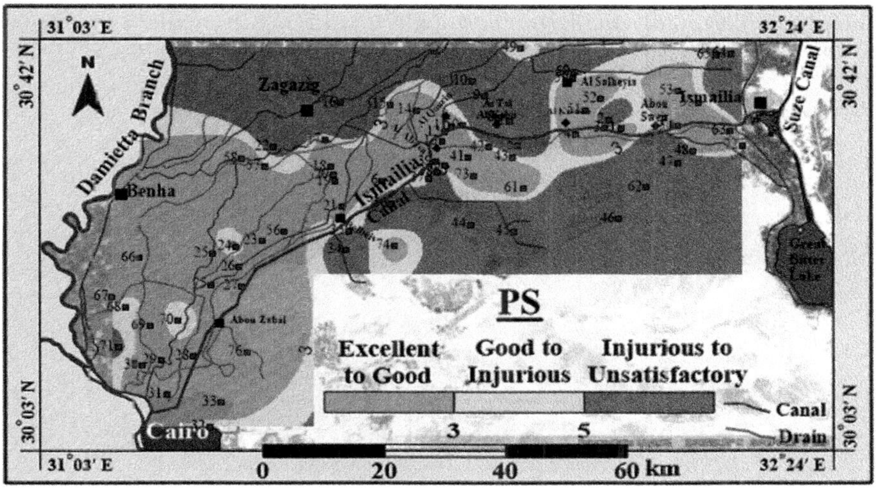

Fig. 10 Spatial distribution map of the PS presenting the groundwater quality for irrigation in the study area (The satellite image was cropped from Google Earth Pro—Data SIO, NOAA, U.S. Navy, NGA, GEBCO—Image Landsat/Copernicus—Image IBCAO)

Potential Salinity (PS)

The values of PS of the groundwater samples ranged from 0.59 to 61.25 with a mean value of 10.37 (Table 3). Approximately 29% of the groundwater samples were classified as excellent to good, 18% of the groundwater samples were classified as good to injurious, while 53% of the groundwater samples were classified as injurious to unsatisfactory for irrigation. Figure 10 shows the spatial distribution of the groundwater quality for irrigation based on PS, where the potential salinity increases due north and southeast of the study area.

Heavy Metal Pollution Index (HPI)

Heavy metal pollution index (HPI) is a rating method that reflects the integrated influence of different dissolved metals and trace elements on the overall groundwater quality for human consumption according to a set of recommended limits for drinking. Calculating the HPI includes the identification of the pollution parameters on which the index-based and the establishment of a rating scale for each selected parameter giving weightage. The critical pollution value of HPI is 100 [94, 96, 101, 102, 111–118].

In the present study, HPI is used to evaluate the groundwater quality for long-term irrigation use to estimate the pollution and toxicity of plant by the trace elements: Al, As, Cd, Cr, Co, Cu, Fe, Pb, Li, Mn, Ni, Se, V and Zn, according to the recommended maximum concentrations of trace elements set by the National Academy of Science

Table 12 Recommended maximum concentrations of trace elements in irrigation water that were set by the National Academy of Science and the National Academy of Engineering (1972) [3, 4, 26, 119]

Element	Recommended maximum concentration (mg/l)	
	For long-term use	For short-term use
Al	5.0	20.0
As	0.1	2.0
Cd	0.01	0.05
Cr	0.10	1.0
Co	0.05	5.0
Cu	0.2	5.0
Fe	5.0	20.0
Pb	5.0	10.0
Li	2.5	2.5
Mn	0.2	10.0
Ni	0.2	2.0
Se	0.02	0.02
V	0.1	1.0
Zn	2.0	10.0

and the National Academy of Engineering (1972) for the long and short-term use of irrigation water based on the protection of soils for plant production as presented Table 12 [3, 4, 26, 119].

The obtained results revealed that, values of HPI of the groundwater samples ranged from 5.9 to 222.5 with a mean value of 50.9 (Table 3). Approximately, 98.7% of the groundwater samples had values of HPI less than the critical HPI value (100), and they are safe for irrigation, while only 1.3% of the groundwater samples had HPI value more than the critical pollution value. The quality of the groundwater in the long-term use for irrigation is suitable with respect to metals and trace elements pollution in all the study area except in the northwestern part as presented in Fig. 11.

Irrigation Water Quality Index (IWQI)

The obtained results of IWQI of the groundwater samples indicated that the IWQI values ranged from 13.67 to 96.06 with a mean value of 62.61 (Table 3). According to the IWQI classification defined by Meireles et al. [105], approximately 8% and 42% of the groundwater samples have no and low restriction on their use for irrigation with no toxicity risk for most plants, respectively. In addition, there are about 21% of the groundwater samples have moderate restrictions on their use for irrigation, while 9% and 20% of the groundwater samples have high and severe restrictions on their direct application for irrigation without management, respectively. The spatial distribution

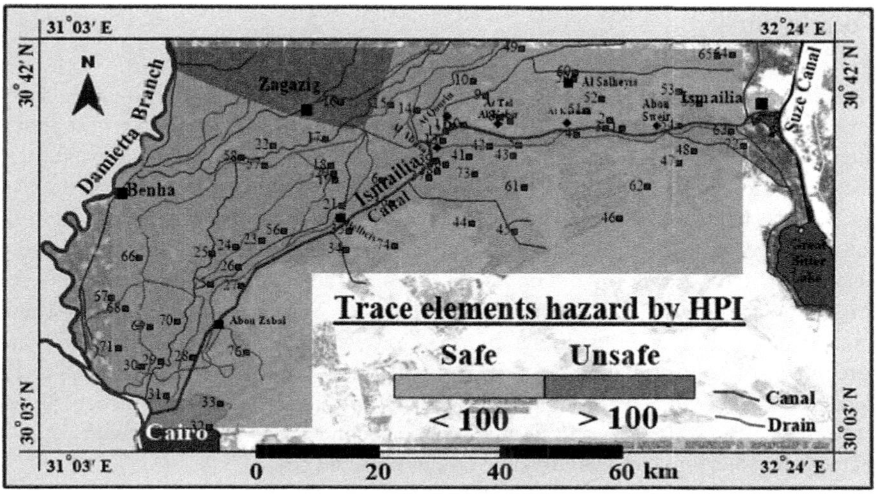

Fig. 11 Spatial distribution map of the trace elements hazard based on HPI presenting the groundwater quality for irrigation in the study area (The satellite image was cropped from Google Earth Pro—Data SIO, NOAA, U.S. Navy, NGA, GEBCO—Image Landsat/Copernicus—Image IBCAO)

of the groundwater quality for irrigation throughout the study area according to the IWQI is shown in Fig. 12, where, the restrictions on their uses for irrigation increase in northern part of the study area and south of the Ismailia canal.

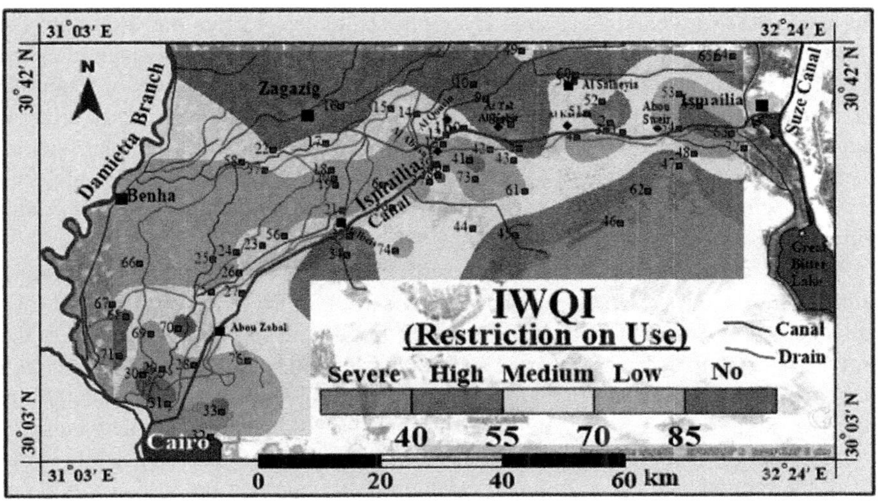

Fig. 12 Spatial distribution map of the restriction on use of the groundwater for irrigation based on IWQI in the study area (The satellite image was cropped from Google Earth Pro—Data SIO, NOAA, U.S. Navy, NGA, GEBCO—Image Landsat/Copernicus—Image IBCAO)

Conclusion

The quality of irrigation water is evaluated according to its potential to cause problems for both soil and plant [23]. Irrigation water should achieve the crop productivity, keep soil from degradation, and maintain both physical and mechanical properties of the soil to ensure the agricultural productivity, maintaining the productivity of agricultural lands and sustainable food production as main requirements for eliminating hunger, achieving food security, improving nutrition and health, and ensuring sustainability for a long time to come [1–3, 14, 15, 45]. The suitability of water for irrigation depends on content, amounts and types of soluble salts in irrigation water, impacts of the mineral constituents in irrigation water on both plants and soils, and percentage of solute accumulation in soils after irrigation [23]. Problems of these salts are of two types; the first one is associated with total salinity that affects plants and can lead to saline soil condition, while the second one is associated with sodium that affects soils and can lead to sodic soil condition [26].

Based on the obtained results of the quality parameters that were considered, it was found that the groundwater in nearly all the study area has normal pH values for irrigation without causing a nutritional imbalance or containing toxic ions. While, there are about 4% of the groundwater samples have pH values outside the normal pH range (6.5–8.4), and this abnormal pH may need to be corrected by introducing an amendment into the water or try to correct the soil pH problems that could be developed rather than try to treat the water itself by adding sulfur to correct the high pH or lime to correct a low pH. In addition, irrigation equipment should be chosen carefully to resist potential corrosion [73].

TDS results showed that only about 8% of the groundwater samples have TDS more than 3000 mg/l and therefore they are not suitable for irrigation. However, the estimated salinity hazard according to the EC values of the groundwater samples, in dS/cm, indicates that the potential for salinity hazard and restriction on water use for irrigation increase toward north, northeast and southeast of the study area, as indicated by Fig. 3, where, there are about 17% of the groundwater samples that are severely restricted for irrigation use and about 59% of the groundwater samples that are slight to moderately restricted, while only 24% of the samples can be used safely without restrictions for irrigation.

According to the results of sodicity hazard (Fig. 4), about 80% of the groundwater samples have a low sodium hazard and could be used for irrigation on nearly all soils and for all crops except those are very sensitive to sodium. Approximately, 15% of the groundwater samples have a medium sodium hazard and could be used for irrigation on almost all soils for crops that are semi-tolerant or tolerant to sodium. In addition, only one sample has a high sodium hazard and it could be used for irrigation for crops that are tolerant to sodium on soils provided with good drainage, while about 4% of the groundwater samples have very high sodium hazard and so they could not be used directly for irrigation without special treatment.

The indexes of permeability hazard (Fig. 9) and potential salinity (Fig. 10) showed that about 29% of the groundwater samples are suitable and excellent to good water for irrigation use. While the remaining 71% of the groundwater samples have quality varies from marginal to unsuitable for irrigation. As suggested by Doneen [120], the recommended limits of potential salinity (PS) for soils of good, medium and low permeability are 5–20, 3–15 and 3–7, respectively, the obtained results of PS indicate that about 43% of the groundwater samples are suitable for the good permeable soil, about 62% of the groundwater samples are suitable for soils of medium permeability, while about 41% of the groundwater samples are suitable for soils of low permeability [3, 4]. Indexes of Potential salinity (PS) (Fig. 10) and Kelley's ratio (KR) (Fig. 6) showed that about 53% of the groundwater samples are unsuitable for irrigation uses.

Based on the RSC results, about 91% of the groundwater samples can be used safely for irrigation, about 7% of the samples need certain management before using for irrigation, while only two samples are unsuitable for irrigation [83]. While, according to Gupta and Gupta [4], it was suggested that the alkalinity hazard should be determined by the index of RSBC because CO_3 ions do not occur very considerably in perceivable amounts and because HCO_3 ions do not precipitate Mg ions. Accordingly, about 87% of the groundwater samples have low alkalinity and can be used for irrigation on nearly all soils for all crops without permeability problems, and about 12% of the samples have medium alkalinity that can be used on nearly all soils and all crops except those are sensitive for CO_3 and HCO_3, in addition, this water with medium alkalinity can give optimum yields with several alkali tolerant crops. While only one sample has a high alkalinity that requires good drainage to be used for irrigation.

There is an agreement between the results of IWQI and irrigation water quality assessment based on SAR and EC, where the obtained results showed that 50% of the groundwater samples have no or little restriction on their use for irrigation while the other 50% of the samples have more restrictions and so a special managements are required before their direct use for irrigation.

Generally, the groundwater in the study area has a moderate quality for irrigation purposes, and as indicated by the obtained results of the used evaluation parameters, therefore, management changes and practices should be considered for maintaining the soil properties and crop productivity. The similarity between the distribution maps of the determined parameters (Figs. 3, 4, 5, 6, 7, 8, 9, 10, 11, 12) with respect to the distribution of zones with the better water quality indicating the effect of Ismailia canal on enhancing the quality of groundwater, where the groundwater receive seepage of the canal water, especially at the area where lithology can permit the infiltration of the canal water through the soil recharging the groundwater.

For maintaining the soil properties, conserving water and decreasing the possible pollution of the groundwater by the irrigation return flow, enhancement of the drainage network and using modern irrigation methods are important. In addition, using irrigation equipment that can resist corrosion is recommended because of the high tendency of the groundwater in the majority of the study area for corrosion as indicated by the obtained results (Fig. 7). Certain management is required in parts of the study area that having groundwater with high levels of salinity and sodium hazards, because using such water may result in the accumulation of high concentrations of salts in the soil, which in turn may cause a physiological drought condition.

Although, the field appears to have plenty of moisture, the plants wilts because their roots are unable to absorb the water. In addition, continuous use of water having a high sodium hazard may lead to a breakdown in the physical structure of the soil affecting its permeability and the soil becomes hard and compact when dry and increasingly impervious to water penetration [26].

Recommendations

Minimization and prevention of adverse impacts of agricultural activities on water quality for other socio-economic activities and purposes, in addition, establishment and operation of cost-effective water quality monitoring systems for agricultural water uses. Proper disposal of sewage from human settlements and of manure produced by livestock breeding, in addition, improving agricultural drainage systems. Minimization of the adverse impacts from agricultural chemicals using integrated pest management is of urgent importance, in addition, heightening of the societal awareness and educating the agricultural communities about the pollution impacts of using chemicals and fertilizers on the water quality and food safety and the proper pesticide handling and application. Reducing and controlling the drilling operations of the private production wells and the extensive and unplanned groundwater withdrawal. Application of artificial recharging techniques for the groundwater aquifer for salinization control. Treatment of the industrial effluents and wastes before discharging into the drains. Enhancement of the sanitary drainage network in the urban and rural areas, in addition to lining the surface water canals and drains. Prohibition and conviction of using sewage and drainage water for irrigation without adequate treatment according to the soil conditions and the types of plants. Periodical monitoring for both surface water and groundwater quality. Establishing biological, physical and chemical water quality standards for agricultural water users. Development and implementation of water-resources monitoring systems with predefined indicators, parameters, frequency and sampling points, in combining with quantity data to avoid the cumulative effect of poor decisions of water-resources management because of inadequate water quality data [121–124].

References

1. FAO (2017) Water for sustainable food and agriculture: a report produced for the G20 presidency of Germany. Food and Agriculture Organization of the United Nations, Rome, Italy
2. UNGA (2015) United Nations general assembly: resolution adopted by the general assembly—transforming our world: the 2030 agenda for sustainable development. https://www.un.org/ga/search/view_doc.asp?symbol=A/RES/70/1&Lang=E
3. Gupta IC, Gupta SK (2003) Use of saline water in agriculture. A study of arid and semi-arid zones of India. Revised 3rd edn. Scientific Publishers, Jodhpur, India

4. Gupta SK, Gupta IC (2015) Management of saline & waste water in agriculture. Scientific Publishers, Jodhpur, India
5. Bayoumi A (2003) Overview on decentralization and participatory irrigation management. Water demand management forum on decentralization and participatory irrigation management, 2–4 Feb 2003, Cairo, Egypt
6. Döll P (2009) Vulnerability to the impact of climate change on renewable groundwater resources: a global-scale assessment. Environ Res Lett 4:035006. https://doi.org/10.1088/1748-9326/4/3/035006
7. FAO (2010) AQUASTAT—FAO's global information system on water and agriculture. FAO, http://www.fao.org/nr/aquastat. Last access: 16 March 2010, Rome, Italy
8. Garrido A, Martínez-Santos P, Llamas MR (2006) Groundwater irrigation and its implications for water policy in semiarid countries: the Spanish experience. Hydrogeol J 14(3):340–349. https://doi.org/10.1007/s10040-005-0006-z
9. Nickson R, McArthur J, Shrestha B, Kyaw-Myint T, Lowry D (2005) Arsenic and other drinking water quality issues, Muzaffargarh District, Pakistan. Appl Geochem 20:55–68. https://doi.org/10.1016/j.apgeochem.2004.06.004
10. Shiklomanov IA (2000) Appraisal and assessment of world water resources. Water Int 25(1):11–32
11. Siebert S, Burke J, Faures JM, Frenken K, Hoogeveen J, Döll P, Portmann FT (2010) Groundwater use for irrigation—a global inventory. Hydrol Earth Syst Sci 14(10):1863–1880. https://doi.org/10.5194/hess-14-1863-2010
12. Yasmin G, Islam D, Islam MT, Shariot-Ullah M, Adham AKM (2019) Evaluation of groundwater quality for irrigation and drinking purposes in Barishal district of Bangladesh. Fundamental Appl Agric 4(1):632–641. https://doi.org/10.5455/faa.301258
13. Bohn HL, Brain LM, George A, O'Connor (1985) Soil chemistry, 2nd edn. Wiley, New York, pp 234–248
14. Brady NC (2002) The nature and properties of soil, Upper Addle River, 13th edn. Macmillan, New Jersey, pp 413–436
15. Brhane GK (2016) Irrigation water quality index and GIS approach based groundwater quality assessment and evaluation for irrigation purpose in Ganta Afshum selected Kebeles, Northern Ethiopia. Int J Emerg Trends Sci Technol 3(9):4624–4636. https://doi.org/10.18535/ijetst/v3i09.10
16. Khan F, Husain T, Lumb A (2003) Water quality evaluation and trend analysis in selected watersheds of the Atlantic Region of Canada, Environmental Monitoring and Assessment, 88, pp 221–242. Scheme Water SA 33(1):101–106
17. Salem ZE, Elsaiedy G, ElNahrawy A (2017a) Hydrogeochemistry and quality assessment of groundwater under some central Nile delta villages, Egypt. In: Negm AM (ed) Groundwater in the Nile Delta, Hdb Env Chem. https://doi.org/10.1007/698_2017_111
18. Salem ZE, Elsaiedy G, and ElNahrawy A (2017b) Assessment of the groundwater quality for drinking and irrigation purposes in the Central Nile Delta Region, Egypt. In: Negm AM (ed) Groundwater in the Nile Delta, Hdb Env Chem. https://doi.org/10.1007/698_2017_137
19. Salem ZE, Temamy AMA, Salah MK, Kassab M (2018) Evaluation of water resources qualities for agriculture irrigation in Abu Madi area. Northern Middle Nile Delta 74:277–316. https://doi.org/10.1007/698_2018_273
20. Salem ZE, El Nahrawy A, Ghobara M (2018) Spatiotemporal hydrochemical evaluation and quality assessment of drainage water compared to canal surface water in the Middle Nile Delta, Egypt 75:123–161. https://doi.org/10.1007/698_2018_296
21. Sargaonkar A, Deshpande (2003) development of an overall index of pollution for surface water based on a general classification scheme in Indian context. Environ Monit Assess 89:43–67
22. Ayers RS, Westcot DW (1994) Water quality for agriculture: FAO irrigation and drainage paper 29. Revision 1:1–130

23. Besser H, Mokadem N, Redhouania B, Rhimi N, Khlifi F, Ayadi Y, Omar Z, Bouajila A, Hamed Y (2017) GIS-based evaluation of groundwater quality and estimation of soil salinization and land degradation risks in an arid Mediterranean site (SW Tunisia). Arab J Geosci 10:350–370. https://doi.org/10.1007/s12517-017-3148-0
24. Bhat MA, Wani SA, Singh VK, Sahoo J, Tomar D, Sanswal R (2018) An overview of the assessment of groundwater quality for irrigation. J Agri Sci Food Res 9(1):209
25. Collins R, Jenkins A (1996) The impact of agricultural land use on stream chemistry in the Middle Hills of the Himalayas. Nepal. J Hydrol 185:71–86
26. Fipps G (2015) Irrigation water quality standards and salinity management strategies. Agri Life Extension Texas A&M Syst B-1667:4–3
27. Mass (1990). In: Tanji KK (ed) Crop salt tolerance. Agricultural salinity assessment and management manual. ASCE, New York, pp 262–304
28. Tahmasebi P, Mahmudy-Gharaie MH, Ghassemzadeh F, Karouyeh AK (2018) Assessment of groundwater suitability for irrigation in a gold mine surrounding area, NE Iran. Environ Earth Sci 77:766. https://doi.org/10.1007/s12665-018-7941-1
29. Wadie AST, Abduljalil GADS (2010) Assessment of hydrochemical quality of groundwater under some urban areas within Sana'a secretariat. Ecletica Quimica 35(1):77–84. www.SCIELO.BR/EQ
30. Armanuos AM, Negm A, Valeriano OCS (2016) Groundwater quality investigation using multivariate analysis—case study: western Nile Delta Aquifer. Egypt. Int J Environ Sci Develop 7(1):1–9. https://doi.org/10.7763/IJESD.2016.V7.732
31. El Ramly IM (1997) Hydrogeological and water quality characteristics of the saturated zone beneath the various land uses in the Nile Delta region, Egypt. Freshwater Contamination in Proceedings of Rabat Symposium S4. IAHS Publ. no. 243. pp 255–261
32. Eltarabily MGA, Negm AM (2018) Groundwater management for sustainable development plans for the Western Nile Delta. In: Negm AM (ed) Groundwater in the Nile Delta, Hdb Env Chem. https://doi.org/10.1007/698_2018_247
33. Negm AM, Armanuos AM (2016) GIS-based spatial distribution of groundwater quality in the Western Nile Delta, Egypt. In: Negm AM (ed) The Nile Delta, Hdb Env Chem. https://doi.org/10.1007/698_2016_66
34. Taha AA, El Mahmoudi AS, El-Haddad IM (2003) Evaluation of the water quality in new communities south east the Nile Delta, Egypt. Emirates J Eng Res 8(2):51–67
35. El-Fakharany MA, Mansour NM, Yehia MM, Monem M (2017) Evaluation of groundwater quality of the Quaternary aquifer through multivariate statistical techniques at the southeastern part of the Nile Delta, Egypt. Sustain Water Resour Manag. https://doi.org/10.1007/s40899-017-0087-6
36. Eltarabily MGA, Negm AM (2017) Groundwater management for sustainable development east of the Nile Delta aquifer. In: Negm AM (ed) Groundwater in the Nile Delta, Hdb Env Chem. https://doi.org/10.1007/698_2017_102
37. Hegazy D, El-Bastaweesy M, Garamoon H, Melegy A, Elsaid MA (2018) The impacts of human and agricultural activities on the groundwater quality in Mostorod Abu Zabal Area, East Nile Delta, Egypt. Al Azhar Bullet Sci 29(2):59–71
38. Abu EL, Ella EM (2018) Environmental management of groundwater in the eastern Nile Delta, Egypt. Int J Water Res Arid Environ 7(1):130–143
39. Embaby AAA, Beheary MS, Rizk SM (2017) Groundwater quality assessment for drinking and irrigation purposes in El-Salhia plain East Nile Delta Egypt. International Journal of Innovations in Engineering and Technology—Special Issue on EGE, pp 51–73
40. Eltarabily MG, Negm AM, Yoshimura C, Abdel-Fattah S, Saavedra OC (2018) Quality assessment of southeast Nile Delta groundwater for irrigation. Water Resour 45(6):975–991. https://doi.org/10.1134/S0097807818060118
41. ENCID (1996) Egyptian national committee on irrigation and drainage. https://www.icid.org/v_egypt.pdf
42. Arnous MO, El-Rayes AE, Green DR (2015) Hydrosalinity and environmental land degradation assessment of the East Nile Delta region, Egypt. J Coastal Conserv 19:491–513. https://doi.org/10.1007/s11852-015-0402-z

43. Abd El-Gawad AM (1997) Shallow geophysical exploration for defining the water occurrences in the area east of the Nile Delta, Egypt. Ph.D. Thesis Fac Sci Ain Shams Univ., p 410
44. El-Dairy MD (1980) Hydrogeological studies on the eastern part of Nile Delta using isotope techniques. M.Sc. thesis, Fac of Sci Zagazig Univ., Egypt
45. El-Haddad IM (2002) hydrogeological studies and their environmental impact on future management and sustainable development of the new communities and their surroundings, East of the Nile Delta, Egypt. Ph.D. Thesis, Fac Sci Mansoura Univ, Egypt, p 435
46. Ismail AIM, Ryden N (2012) Engineering geological characteristics of soil materials, East Nile Delta, Egypt. World academy of science, engineering and technology international journal of geological and environmental engineering. Int Schol Sci Res Innov 6:282–287
47. Said R (1981) The geological evolution of the Nile River. Springer, New Work, p 151
48. Salem ZE (2009) Natural and human impacts on the groundwater under an Egyptian village, Central Nile Delta—a case study of Mehallet Menouf. In: 13th international water technology conference (IWTC, 13), Hurghada, Egypt, 3, pp 1397–1414
49. Sallouma MKM (1983) Hydrogeological and hydrochemical assessment of the Quaternary aquifer in the eastern Nile Delta, Egypt. Ph.D. thesis. Faculty of Science, Ain Shams University, Egypt, p 166
50. Sandford KS, Arkell WJ (1939) Paleolithic man and the Nile Faiyum divide. Chicago Univ Oriental Inst I:1–77
51. Shata AA, Abd El-Salam AA, Hargu AA (1979) Project soil map of Egypt. Soil map of eastern Delta region (El-Tumilat—Suez). 4th report. Academy of Scientific Research Technology, Cairo, pp 89–105
52. Stanley DJ, Warne AG, Davis HR, Bernascone MP, Chen Z (1992) Nile delta: the late quaternary evolution of the north-central Nile delta between Manzalla and Burullus lagoons, Egypt. Natl Geogr Soc Res Explor 8:22–51
53. Gad MI (1995) Hydrogeological studies for groundwater reservoirs, east of Tenth of Ramadan City and vicinities. M.Sc. Thesis, Faculty of Sci. Ain Shams University Egypt
54. Kaiser MF, El Rayes A, Ghodeif K, Geriesh B (2013) GIS data integration to manage waterlogging problem on the eastern Nile Delta of Egypt. Int J Geosci 4:680–687. https://doi.org/10.4236/ijg.2013.44063
55. Abdel-Fattah MK, Helmy AM (2015) Assessment of water quality of wastewaters of Bahr El-Baqar, Bilbies and El-Qalyubia Drains in East Delta, Egypt for irrigation purposes. Egypt J Soil Sci 55:287–302
56. Ali OA, Tarek SJ (2012) Removal of polycyclic aromatic hydrocarbons from Ismailia Canal water by chlorine, chlorine dioxide and ozone. Desalin Water Treat 1:289–298. https://doi.org/10.5004/dwt.2009.131
57. El-Fakharany MA, Mansour NM (2009) Assessment of water resources quality at the southeastern part of the Nile Delta, Egypt. The International Conference on Water Conservation in Arid Regions
58. Ghobara M, Salem ZE (2017) Spatiotemporal fluctuations in phytoplankton communities and their potential indications for the pollution status of the irrigation and drainage water in the Middle Nile Delta Area. Egypt 74:317–345. https://doi.org/10.1007/698_2017_185
59. Goher ME, Hassan AM, Abdel-Moniem IA, Fahmy AH, El-sayed SM (2014) Evaluation of surface water quality and heavy metal indices of Ismailia Canal, Nile River, Egypt. Egyptian J Aquatic Res 40:225–233. https://doi.org/10.1016/j.ejar.2014.09.001
60. Elewa HH, Shohaib RE, Qaddah AA, Nousir AM (2013) Determining groundwater protection zones for the quaternary aquifer of northeastern Nile Delta using GIS-based vulnerability mapping. Environ Earth Sci 68:313–331
61. El-Kashouty M, Sabbagh AAE (2005) Hydrogeology and hydro-geochemistry of the quaternary aquifer in the middle Nile Delta Area, Egypt. In: Ninth international water technology conference, IWTC9, Sharm El-Sheikh, Egypt, pp 395–4114
62. El-Sayed SA (2018) Contribution to the ground water hydrology of the quaternary aquifer in West Ismailia Area, Egypt. J Geosci Environ Protect 6:134–158. https://doi.org/10.4236/gep.2018.67010

63. Geriesh MH (1994) Hydrogeological and hydrogeochemical evaluation for the groundwater resources in the Suez Canal region, Egypt. Ph.D. Thesis, Suez Canal University, Faculty of Science, Ismailia, Egypt
64. Geriesh MH, EL-Rayes AE (2001) Municipal contamination of shallow groundwater beneath South Ismailia villages. In: 5th international conference on geochemistry, Alex. Univ. Egypt, pp 241–253
65. Geriesh MH, Balke KD, El-Rayes AE, Mansour BM (2015) Implications of climate change on the groundwater flow regime and geochemistry of the Nile Delta, Egypt. J Coastal Conserv 19:589–608. https://doi.org/10.1007/s11852-015-0409-5
66. Mabrouk B, Ramadan F, Nagaty M, Azeem YE (2016) Sedimentological and hydrogeochemical studies of the quaternary groundwater aquifer in El Salhyia area, Sharkia governorate, Egypt. Middle East J Appl Sci 6:120–138
67. Morsy WS (2009) Environmental management to groundwater resources for Nile Delta region. Ph.D. thesis, Faculty of Engineering, Cairo University, Egypt
68. Salem ZE, Al Temamy AM, Salah MK, Kassab M (2016) Origin and characteristics of brackish groundwater in Abu Madi coastal area, Northern Nile Delta, Egypt. Estuarine, Coastal Shelf Sci 178:21–35
69. Sefelnasr A, Sherif M (2013) Impacts of seawater intrusion in the Nile Delta aquifer, Egypt. Natl Groundwater Assoc 52:264–276
70. Yehia MM (2000) Environmental impacts of sewage irrigation water on groundwater quality of northeast Cairo, Egypt. Helwan Univ. Fac Eng Res J 72:176–193
71. APHA (2005) Standard methods for the examination of water and wastewater, 21st edn. American Public Health Association, Washington, DC, USA
72. Ayers RS (1977) Quality of water for irrigation. J Irrigation Drainage IV ASCE 103:135–154
73. Ayers RS, Westcot DW (1985) Water quality for agriculture. FAO Irrigation Drainage Paper 29(1):1–144
74. Davis S, Dewiest R (1966) Hydrogeology. Wiley, New York, USA
75. Puntamkar S, Kant K, Mathur S (1988) The effect of saline water irrigation on soil properties. Trans Indian Soc Dexter Technol Krishi Bhawan, Bikaner, Rajasthan, India 15:69–72
76. Wilcox LV (1955) Classification and use of irrigation waters. U.S. Dept. Agric. Circular 969, Washington DC, USA
77. Gupta IC (1984) Reassessment of irrigation water quality criteria and standards. Curr Agric 8:113–126
78. Calvache ML, Pulido-Bosch A (1994) Modelling the effects of salt-water intrusion dynamics for a coastal karstified block connected to a detrital aquifer. Groundwater 32:767–777
79. Suarez DL (1981) Relationship between pH and SAR and an alternative method for estimating SAR of soil or drainage water. Soil Sci Soc Am J 45:469–475
80. Todd DK (1980) Groundwater hydrology, 2nd edn. Wiley , New York, USA, p 535
81. Kelly WP (1963) Use of saline irrigation water. Soil Sci 95(4):355–391. https://doi.org/10.1097/00010694-196306000-00003
82. Ramesh K, Elango L (2012) Groundwater quality and its suitability for domestic and agricultural use in Tondiar river basin, Tamil Nadu, India. Environ Monit Assess 184:3887–3899
83. Wilcox LV, Blair GY, Bower CA (1954) Effect of bicarbonates on suitability of waters for irrigation. Soil Sci 77:259–266
84. Eaton FM (1950) Significance of carbonate in irrigation waters. Soil Sci 67(3):128–133
85. Domenico PA, Schwartz FW (1990) Physical and chemical hydrogeology. Wiley, New York, p p824
86. Aghazadeh N, Mogaddam AA (2010) Assessment of groundwater quality and its suitability for drinking and agricultural uses in the Oshnavieh area, Northwest of Iran. J Environ Prot 1:30–40. https://doi.org/10.4236/jep.2010.11005
87. Aher K, Gaikwad S (2017) Irrigation groundwater quality based on hydrochemical analysis of Nandgaon block, Nashik district in Maharashtra. Int J Adv Geosci 5(1):1
88. Patel RL, Dhiman SD (2017) Groundwater quality assessment for irrigation water use in Mahi right bank command area, Gujarat, India. Kalpa Publ Civil Eng 1:343–351

89. Sheriff KMM, Hussain AZ (2017) Evaluation of groundwater quality and its suitability for irrigation use on the bank of Noyyal River at Tiruppur, Tamilnadu, India. Environ Pollut Protect 2(3). https://doi.org/10.22606/epp.2017.23003
90. Hasan M, Shang Y, Akhter G, Jin W (2017) Evaluation of groundwater suitability for drinking and irrigation purposes in Toba Tek Singh District, Pakistan. Irrigation Drain Syst Eng 06(01). https://doi.org/10.4172/2168-9768.1000185
91. Doneen LD (1964) Notes on water quality in agriculture. Published as a water science and engineering, Paper 4001, Department of Water Sciences and Engineering, University of California, Davis
92. Kumarasamy P, Dahms H, Jeon H, Rajendran A, James RA (2014) Irrigation water quality assessment—an example from the Tamiraparani river, Southern India. Arab J Geosci 7:5209–5220. https://doi.org/10.1007/s12517-013-1146-4
93. Doneen LD (1962) The influence of crop and soil on percolating water. In: Proceedings of the Biennial conference on ground water recharge, pp 156–163
94. Amadi AN (2012) Quality assessment of Aba River using heavy metal pollution index. Am J Environ Eng 2:45–49
95. Balakrishnan A, Ramu A (2016) Evaluation of heavy metal pollution index (HPI) of groundwaterin and around the coastal area of Gulf of Mannar Biosphere and Palk Strait. J Adv Chem Sci 2:331–333
96. Sirajudeen J, Arul MS, Manivel V (2014) Heavy metal pollution index of groundwater of Fathima Nagar area near Uyyakondan Channel, Tiruchirapalli District, Tamilnadu, India. World J Pharm Pharmaceut Sci 4:967–975
97. Reza R, Singh G (2010) Heavy metal contamination and its indexing approach for river water. Int J Environ Sci Tech 7:785–792
98. Lou S, Liu S, Dai C, Tao A, Tan B, Ma G, Chalov RS, Chalov SR (2017) Heavy metal distribution and groundwater quality assessment for a Coastal Area on a Chinese Island. Pol J Environ Stud 26:733–745
99. Mohan SV, Nithila P, Reddy SJ (1996) Estimation of heavy metal in drinking water and development of heavy metal pollution index. J Environ Sci Health 31:283–289
100. Rama P, Dubey RK, Dubey SK, Singh AK (2017) Assessment of heavy metal pollution through index analysis for Yamuna Water in Agra Region, India. Int J Curr Microbiol App Sci 6:1491–1498
101. Prasad B, Bose JM (2001) Evaluation of heavy metal pollution index for surface and spring water near a limestone mining area of the lower Himalayas. Environ Geol 41:183–188
102. Yusuf A, Olasehinde A, Mboringong MN, Tabale RP, Daniel EP (2018) Evaluation of heavy metals concentration in groundwater around Kashere and its environs, upper Benue trough, Northeastern Nigeria. Global J Geol Sci 16:25
103. Chukwuma C, Chukwuma C, Uba I, Orakwe C, Ogbu N (2016) Irrigation water quality index assessment of Ele River in parts of Anambra State of Nigeria. Arch Current Res Int 4(3):1–6. https://doi.org/10.9734/acri/2016/25885
104. Abbasnia A, Radfard M, Mahvi AH, Nabizadeh R, Yousefi M, Soleimani H, Alimohammadi M (2018) Groundwater quality assessment for irrigation purposes based on irrigation water quality index and its zoning with GIS in the villages of Chabahar, Sistan and Baluchistan. Iran. Data Brief 19:623–631
105. Meireles ACM, Andrade EMd, Chaves LCG, Frischkorn H, Crisostomo LA (2010) A new proposal of the classification of irrigation water. Revista Ciência Agronômica 41:349–357
106. Trivedy RK (1995) Encyclopedia of environmental pollution and control Karad, India. Environ Media 2:342
107. Miller RW, Gardiner DT (2007) Soils in our environment, 11th edn. Prentice Hall, Upper Saddle River
108. Dhar I, Walia D (2017) Hydrochemical characterization of groundwater of West Jaintia Hills, Meghalaya for irrigation purpose. IOSR J Environ Sci Toxicol Food Technol 11(8):83–89. https://doi.org/10.9790/2402-1108018389

109. Patel Y, Vadodaria GP (2015) Hydro-chemical analysis of groundwater quality for irrigation of Mehsana District, Gujarat State, India. Int J Sci Res 4(2):1841–1846
110. Shah SM, Mistry NJ (2013) Groundwater quality assessment for irrigation use in Vadodara District, Gujarat, India. World Acad Sci Eng Technol 23:7–23
111. Abdullah E (2013) Quality assessment for Shatt Al-Arab river using heavy metal pollution index and metal index. J Environ Earth Sci 3:114–120
112. Edet AE, Offiong OE (2002) Evaluation of water quality pollution indices for heavy metal contamination monitoring. A study case from Akpabuyo-Odukpani area, Lower Cross River Basin (southeastern Nigeria). Geo J 57:295–304
113. Enaam AJ (2013) Quality assessment for Shatt Al–Arab River using heavy metal pollution index and metal index. J Environ Earth Sci 3:114–120
114. Milivojevic J, Krstic D, Smit B, Djekic V (2016) Assessment of heavy metal contamination and calculation of its pollution index for Ugljesnica River, Serbia. Bull Environ Contam Toxicol 97:737–742
115. Rama P, Dubey RK, Dubey SK, Singh AK, Sharma TC (2017) Assessment of heavy metal pollution of Yamuna water in Mathura region through index analysis approach. Int J Chem Stud 5:1286–1289
116. Rizwan R, Gurdeep S, Manish KJ (2011) Application of heavy metal pollution index for groundwater quality assessment in Angul District of Orissa, India. Int J Res Chem Environ 1:118–122
117. Sobhanardakani S, Taghavi L, Shahmoradi B, Jahangard A (2017) Groundwater quality assessment using the water quality pollution indices in Toyserkan Plain. Environ Health Eng Manage 4:21–27
118. Yankey RK, Fianko JR, Osae S, Ahialey EK, Duncan AE, Essuman DK, Bentum JK (2013) Evaluation of heavy metal pollution index of groundwater in the Tarkwa mining area, Ghana. Elixir Pollut 54:12663–12667
119. Rowe DR, Abdel-Magid IM (1995) Handbook of wastewater reclamation and reuse. CRC Press, Inc., p 550
120. Doneen LD (1963) Water quality for agriculture. Department of irrigation, University of California, Davis., p 48
121. Ahmed MA, Abdel Samie SG, Badawy HA (2012) Factors controlling mechanisms of groundwater salinization and hydrogeochemical processes in the Quaternary aquifer of the Eastern Nile Delta, Egypt. Environ Earth Sci. https://doi.org/10.1007/s12665-012-1744-6
122. Andreoli CV (1993) The influence of agriculture on water quality. In: Prevention of water pollution by agriculture and related activities. In: Proceedings of the FAO expert consultation, Santiago, Chile, 20–23 Oct. 1992. Water Report 1. FAO, Rome, pp 53–65
123. Mahmoud MA (2017) Groundwater and agricultural in the Nile Delta. In: Negm AM (ed) Groundwater in the Nile Delta, Hdb Env Chem. https://doi.org/10.1007/698_2017_94
124. Sagardoy JA (1993) An overview of pollution of water by agriculture. In: Prevention of water pollution by agriculture and related activities, proceedings of the FAO expert consultation, Santiago, Chile, 20–23 Oct. 1992. Water Report 1. FAO, Rome, pp 19–26

Assessment of Groundwater Resources in Egypt's Deserts

Mohamed Elkholy

Abstract Egypt needs to re-evaluate the role of groundwater to reallocate them as freshwater sources to meet the projected water shortage in the near future due to the high population growth of 2.05% annually, the increasing demand and the expected negative impacts after the operation of the Ethiopian Renaissance dam. It is necessary to plan to increase water resources and encourage investments aimed at treating and desalinating semi-saline groundwater in non-coastal provinces. Hydrogeological mapping of groundwater resources becomes one of the key tools for the development of groundwater. Integrating remote sensing and GIS have been successfully utilized in groundwater mapping. With the advance in satellite imagery, locating regions with high potentiality for groundwater abstraction is becoming more cost-effective, and convenient than many other invasive alternatives. Monitoring the groundwater quality is important to quantify the quality changes in the long term from pollution and seawater intrusion and to describe the overall quality status on a countrywide scale in the main aquifers. Different studies are presented to give a full view of the status of the groundwater in different parts in Egypt.

Keywords Deep groundwater · GIS · Groundwater quality · Hydrogeological maps · Remote sensing · Shallow groundwater

Introduction

Most of the water used by Egyptians comes from the Nile River, which is fueled by heavy rains that hit the Ethiopian highlands. However, the situation is different in areas far from the river, where water is scarce, and there is a steady increase in demand due to high rates of population growth and human activities. According to a study conducted by the US Geological Society, these factors, in addition to the near completion of the construction and construction of the Ethiopian Renaissance Dam, will cause a crisis in water resources in Egypt.

M. Elkholy (✉)
Faculty of Engineering, Irrigation Engineering and Hydraulics Department, Alexandria University, Alexandria, Egypt
e-mail: mohamed.elkholy@alexu.edu.eg

A. Negm and A. Elkhouly (eds.), *Groundwater in Egypt's Deserts*, Springer Water,
https://doi.org/10.1007/978-3-030-77622-0_13

Groundwater is therefore expected to play an important role in providing Egypt with part of its water needs, along with desalination of seawater. Therefore, knowing the quantity of water available in aquifers and how quickly they are replenished is vital to provide the population with water for various purposes, especially for drinking and irrigation. The determination of the life of water sources greatly helps in these calculations.

Groundwater in the Nile Basin

Groundwater in the Nile Basin mainly occurs in different compositions of rock systems with confined and unconfined conditions shown in Fig. 1. The main aquifers are:

(1) *Victoria artesian aquifer* which is extremely abundant in surface water presented as numerous swamps, rivers, and lakes;
(2) *Congo artesian aquifer* occupies more than 3.2 million km^2 of Equatorial Africa with plentiful surface water;
(3) *Upper Nile artesian aquifer* which is located southern of Bahr El-Ghazaal, White Nile, and the Sobaat plain with water at depths from 25 to 100 m. Rocks of the Nubian aquifer underlain the northern parts of the basin;
(4) *Volcanic rock aquifers are* occurring mainly in the Ethiopian Highlands and covering large parts of the Gaembela plains and the Lake Tana area. Its yields vary according to the weathering conditions;
(5) *Nubian sandstone aquifer* which covers an area of approximately 2 million km^2 with a thickness from 200 to 600 m. It spans many parts of Egypt, Sudan, Chad, and Libya and is characterized by high porosity and high hydraulic conductivity up to about 4000 m^3/day. Other notable consolidated alluvial aquifers include the Um Rouwaba, Geziraa, and Al Atshan aquifers in Sudan; karsified and fissured carbonate aquifers in the Wadi Aaraba areas in the Eastern Desert in Egypt; and the Moghra Aquifer found between the Rosetta branch and the El-Qattara Depression in the Western Desert in Egypt.
(6) *Nile Valley aquifer* that consists of reworked and fluvial sands, silts, and clays ranging in thickness from a few meters to more than 300 m. This high transmissivity combined with high storage capacity and active refill from the Nile River and the surrounding irrigation canals make the aquifer a highly valued resource; and
(7) *Nile Delta aquifer*, which is like the Nile Valley aquifer, consists of sand and gravel with intercalated clay layers having a thickness of 1000 m in some regions and high hydraulic conductivity of about 25,000 m^3/day, makes is also a valuable groundwater resource.

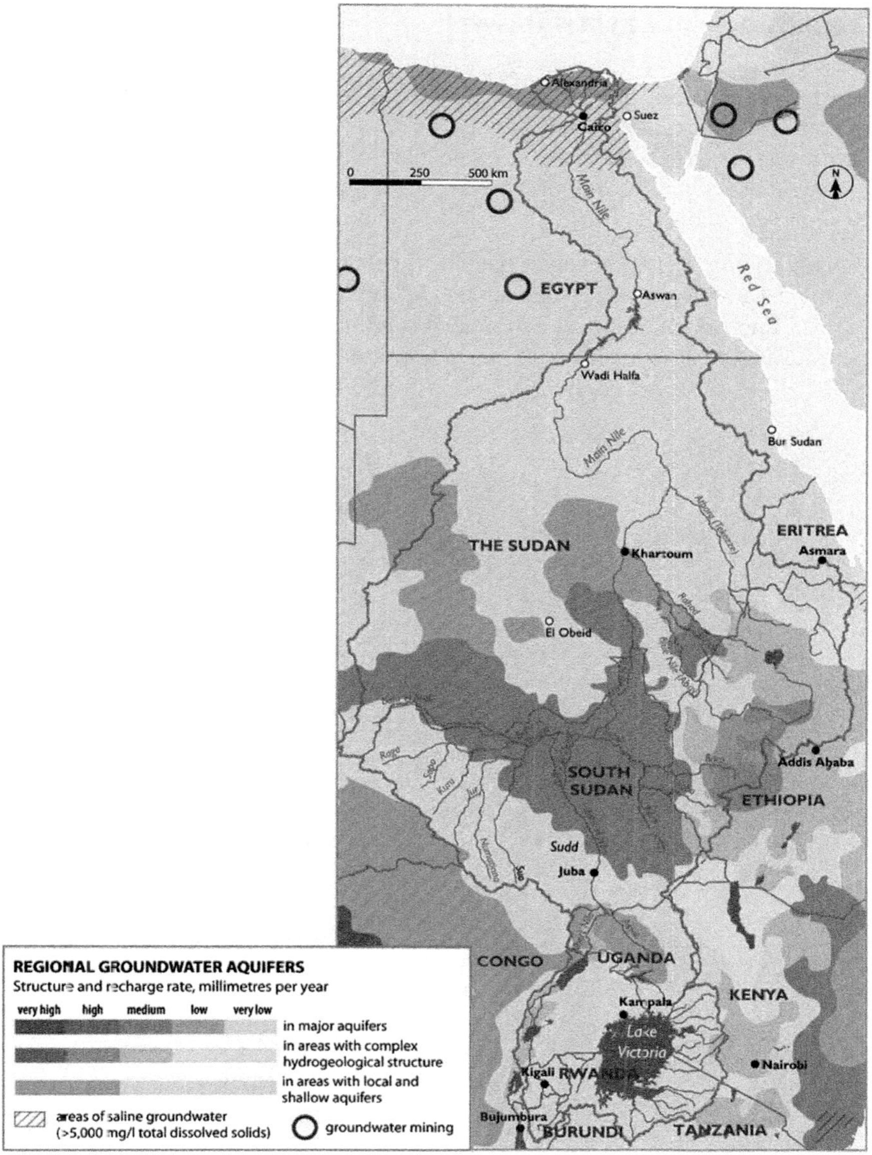

Fig. 1 Groundwater aquifers within the Nile basin [1]

Groundwater in Egypt's Desert

There has been a steady increase in the abstraction of groundwater for the last 20 years in Egypt [2]. Although it is very limited in withdrawal quantity compared to the total water resources, groundwater becomes the only source of water for people living in the remote areas and the deserts. The six main groundwater systems in Egypt, as shown in Fig. 2 are:

1. Nile aquifer: mostly occupies the Nile flood plain region (including Cairo) and the desert fringes and covers almost 4% of the country total area. It consists of a thick layer of gravel and graded sand topped by a layer of clay with a thickness that varies from 300 m at Sohag city to few meters to the north of

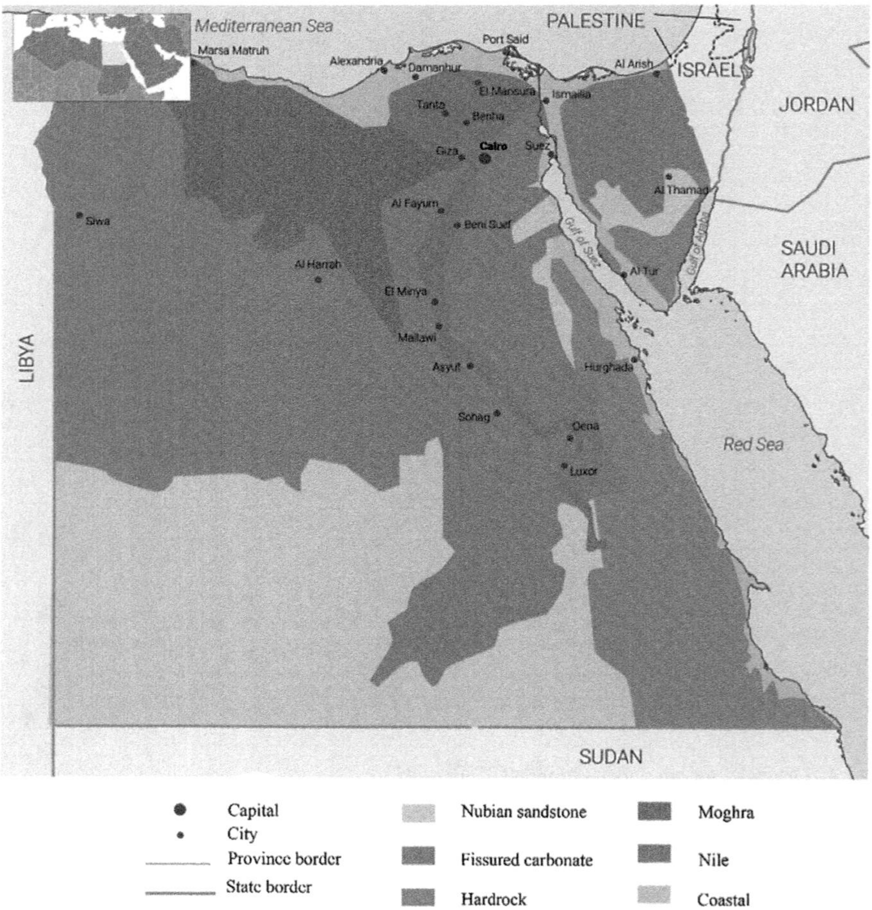

Fig. 2 Major aquifer systems in Egypt [5, 6]

Cairo. The groundwater of the Nile Valley and the Delta is not considered as a water source to rely on since it depends on seasonal infiltration and seepage losses from the Nile water, the irrigation and drains networks and the percolation losses from irrigation lands. Hence, their yield must not be added to Egypt's total water resources. Their storage capacity in the Nile Valley and Delta system are estimated to be 200 km^3 (with a salinity of 800 ppm) and 300 km^3, respectively (Hamza and Mason, 2004). Although, the shallow groundwater water table allows the abstraction of large quantities of water (100–300 m^3/h) at relatively low pumping cost using relatively shallow wells, it is very susceptible to pollution. In term of abstractions, it provides about 85% of the total groundwater abstractions in the country [3].

2. Nubian sandstone aquifer: is fossil groundwater located in the southwest part of the country shared with Sudan, Chad, and Libya where the latter made use of it through its great river project. This aquifer contains about 150,000 billion cubic meters of fossilized water at depths reaching up to 2000 m. The Nubian Aquifer is unconfined in the south-western parts of the country and confined beneath the Eastern Desert. Recent studies and hydrogeological maps show that the shallow aquifers of the Nile delta and Moghra aquifers are connected to this deep aquifer through faults and fissures, thus providing good potential for groundwater development projects [2]. Also, it has been found that the aquifer is in a state of unsteady conditions for the last 3000 years [4].
3. Moghra aquifer: is located mainly in the western edge of the Delta and outcrops on the surface of Wadi El-Farigh and Wadi El-Natroun. The groundwater flow is in general directed towards the El-Qattara Depression. Rainfall and lateral direct inflow from the irrigation canals and leakages from the Nubian aquifer recharge the aquifer. Due to the excessive abstractions rates for groundwater-based reclamation projects in the western desert and growth in industrial and municipal demands, particularly in the western peripheries of the Nile Delta, the sustainability and water quality of this resource is at risk.
4. Fissured carbonate aquifer: widely spreads over more than 50% of the country's area, topping the Nubian aquifer. It is the least exploited and explored aquifer in Egypt. It is recharged mainly from leakages from the underlying Nubian aquifer and local rainfall in some areas.
5. Coastal aquifer: is a renewable aquifer extended on northern and Red sea coasts, that is recharged mainly by rainfall. However, the presence of salty water beneath it limits the abstracted quantities.
6. Hardrock aquifer: mostly in southern Sinai and the eastern desert. This aquifer is influenced by several factors, namely lithologic, tectonic, and morpho-tectonic. The filled fractures and dykes present in the tectonic factors hinder the movement o the groundwater.

There is, however, a major determinant and logical question of the sustainability of the deep groundwater described by the public and some groundwater hydrologists as non-renewable. To answer this question, it is necessary to provide a brief historical overview of the development of the exploitation of underground water in the desert of

Egypt and the consequences of both quantitative and qualitative deterioration or the positive results achieved in some areas. This will help to get the maximum benefits from the development of comprehensive development plans away from the valley and the Delta.

History has witnessed the existence of Pharaonic and Roman civilizations in the oases scattered in the Western Desert in Egypt (El-Kharga–El-Dakhla–El-Bahariya–Siwa and some other extinct oases). Without the presence of underground water flowing from springs, civilizations would not have existed. Some of the oases ceased not as a result of depletion of the reservoir, but due to sand dunes that led to the rise of the level of the surface of the earth and then stopped the springs from flowing by itself. In the early 1960s, with the start of the agricultural revolution since 1952, the country turned to the development of hand-dug wells as well as self-flowing springs. As a result of the increase in agricultural land and the increase in demands for water, the government started through the institution of desert reclamation to dig surface as well as deep wells in the deserts. Moreover, since the water at that time in surface aquifers was available (depth of 150 m or less), the economic cost of drilling wells was acceptable, in addition modern irrigation systems (sprinkler and drip irrigation) were not yet widespreaded, which led to wasteful and inefficient use of water represented in excessive planting of water crops such as rice, sugarcane, and others. Besides, farmers transferred from the valley, and the delta to the areas of agricultural development projects in the western desert continued the same methods of irrigation and the cultivation of seasonal crops, which led to the waste of large quantities of water due to the needs of sandy soil. All of these contributed to the over-extraction and reduced water pressure within the aquifers and ceasing of the springs or decreasing its flows.

Therefore, the Ministry of Irrigation set up a rationalization plan for the most efficient use of groundwater. Many meetings were held between the Ministry of Irrigation, universities, research institutes, oil companies and public drilling companies. These meetings aim to determine the potential of groundwater reservoirs, complete the required data, and follow up the latest developments and studies available in various aspects. Hopefully, to have an integrated study in hand to help decision-makers and political leaderships to optimize the utilization of our underground stock.

Deep Groundwater (up to 1500 m in some areas) are mostly located in the western desert and Sinai and are non-renewable which make it hard to utilize since the cost of pumping will be high besides water quality will be deteriorated at these depths [7]. These deep storages in the western desert are estimated to be 40,000 km^3 with a salinity of 400–700 ppm, as illustrated by [8].

Results of previous investigations about the long-term extraction rate of groundwater from the non-replenished Nubian Sandstone aquifer in the southern part of the western desert indicate that the aquifer is still in an unsteady depletion process [9].

Moreover, according to a study on the groundwater potentials in East Oweinat area (southern of the western desert) [10], it was stated that the quality of the groundwater is fresh and has Total Dissolved Solids (TDS) that ranges from 200 to 700 mg/l. The residual sodium carbonate ranges from 0.69 to 2.45 epm while the sodium adsorption ratio varies from 0.5 up to 1.7 epm which highlights the suitability of this groundwater for irrigation purposes with a safe extraction rate of about 4.7×10^6 m^3/day. However, the withdrawal rate to sustain a long-term development project has to be assessed carefully. The area has high porosity and hydraulic conductivity that ranges from 20 to 30% and from 10 to 20 m/day, respectively [11]. Ibrahem [12] investigated the safe withdrawal rates of groundwater from the Nubian aquifer in East Oweinat area and found that although the groundwater depletion levels range from 0.35 m/year in the eastern part to 2.34 m/year in the western part of the area in 2018 due to over-pumping, the sustainable withdrawal rates may reach 1.89×10^9 m^3/year (≅220,000 feddan) and may last for 174 years under good management. However, the water quality and its appropriateness for use in irrigation has to be checked since the rate of dropping of groundwater levels is high.

El Nahry et al. [13] combined the use of Geographic Information system (GIS) with a weighted spatial capability model (WSCM) to evaluate the different soil/groundwater capabilities of a pilot area in East Oweinat that depends on the Nubian Sandstone aquifer as the only groundwater resources used for domestic and agricultural purposes in this area.

From more than 380 production wells in this area, he found that the depth to groundwater ranges from 28 to 34 m below ground levels (Fig. 3a), and the safe yield is in the range of 200 to 250 m^3/h (Fig. 3b). Also, he found that the sodium absorption ratio (SAR) (where high concentration of sodium in the soil significantly affects the salinity of the soil and its infiltration rate) is below 10 which indicates its safe use for irrigation purposes. However, irrigation should be under certain precautions of selecting crops with high tolerance to salinity and with continuous leaching requirements. Figure 3 shows the depth to groundwater, the well safe yield, the TDS and the SAR values. The classification of irrigation water according to the SAR values can be found in College of Agriculture Sciences [14].

Another study was conducted on the El-Dakhla depression [15], where they developed an optimization model using both the genetic algorithm and MODFLOW simulation to maximize the pumping rate of groundwater from the Nubian aquifer during the period from 2008 to 2050. They found that the transmissivity of the Nubian aquifer in El-Dakhla depression varies widely from as low as 494 to 2350 m^2/day. Moreover, there is a significant decline in the measured piezometric surface of the groundwater estimated at 0.1 to 6 m in the period from 2005 to 2008 due to intensive irrigation strategies. Moreover, from their results, it is found that the present abstraction rate of groundwater estimated at 511,783 m^3/day will cause a significant decline of the piezometric head of the aquifer by about 26 m in some wells in the northern east pf El-Dakhla area by the year 2050.

Sharaky et al. [16] assessed the relationship between the recharging of the Nubian sandstone aquifer in Toshka area from Aswan high dam reservoir and the drop in the groundwater levels for the irrigation of the 25,000 feddan using 102 wells as part of the mega project entitled "1.5 Million Feddan Project". They developed a groundwater model using MODFLOW to estimate the hydraulic parameters of the aquifer. They found that the groundwater level in 2006 drops to almost 14 m as a result of high withdrawal rate compared to the recharging rate. And this drop continued to

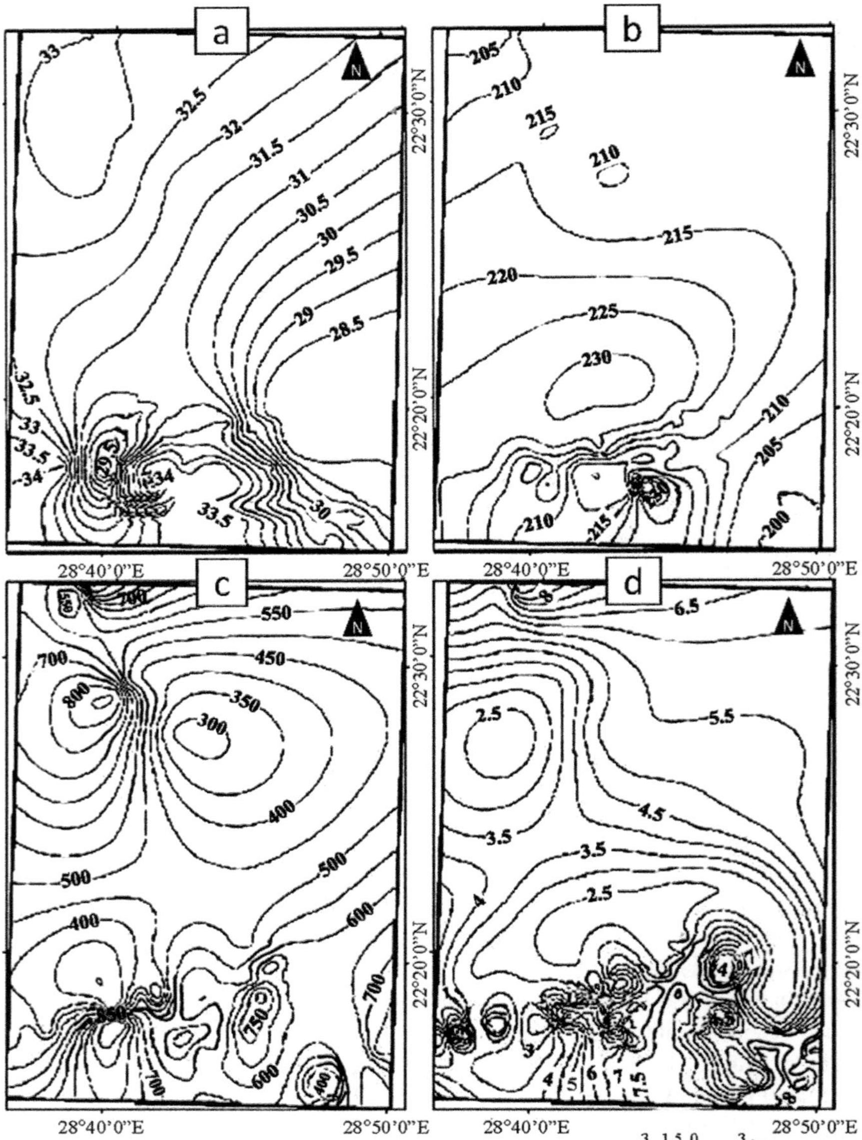

Fig. 3 East Oweniat area, **a** depth to groundwater (mbgl), **b** well safe yield (m^3/h), **c** TDS (mg/l), **d** SAR [13]

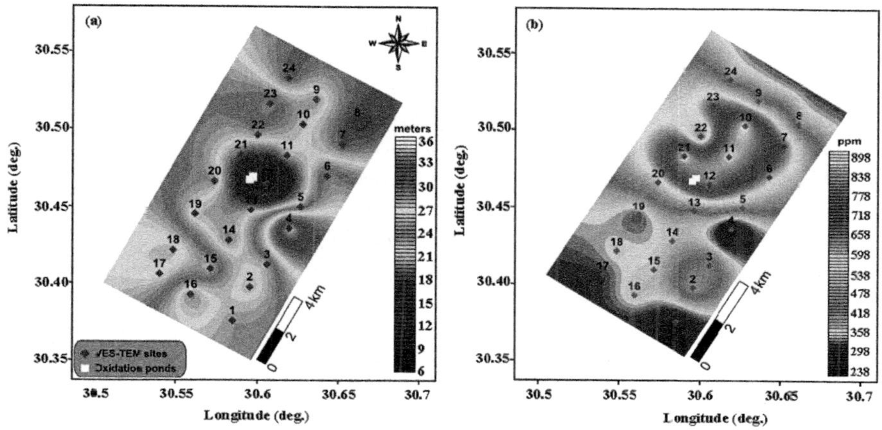

Fig. 4 Contour maps for the **a** depth to aquifer top surface (mgbl), and **b** salinity levels (ppm) [20]

increase in 2016 because of the new wells. They estimated that the safe extraction rate from the 102 wells is about 1007 m^3/day for the coming 100 years to irrigate the 25,000 feddan.

Many studies addressed the geology, and the geomorphology of the region of the western Nile Delta, among them, are [17, 18] and Massoud et al. [20] used the Transient Electromagnetic.

[TEM] and the Vertical Electrical Sounding (VES) surveys to characterize the groundwater aquifers at El-Sadat industrial area. The Pleistocene aquifer in this area is the main source of water with high water content and quality as it is recharged mainly from the Nile Delta fresh aquifer and seepage from El-Beheiry, El-Nasery and El-Nubariya canals. Figure 4 shows that the top surface of the aquifer is located at depths ranging from 6 to 36 mgbl, and the soil salinity ranges from 217 to 925 ppm (i.e., 0.33 to 1.42 ds/m which can be classified as areas of medium to high salinity levels according to the US Salinity Laboratory).

Abouelmagd et al. [19] used also 46 TEM devices to investigate the extension of the groundwater aquifer in west El-Minia governorate. And found that the aquifer mainly composed of fractured limestone with water content covering almost 60% of the whole area.

In another study by Salem and El-Bayumy [21], they studied an area that lies west of the Nile Delta between the Rosetta branch and Wadi El-Natroun including Sadat city. Electric logs were used to estimate the petrochemical properties (i.e., porosity, water resistivity, permeability, and formation factor) of the aquifer while geochemical investigation provided the salinity levels in the region. Sobeih et al. [22] employed MODFLOW and applied different scenarios to simulate the groundwater budget in El-Sadat city and its surrounding areas. The results showed that about 54% of the inflow to the aquifer is from infiltration from the surrounding canals and about 33% from excess irrigation water. On the other hand, about 78% is discharged from the wells, and 18% is lost through open drains. Ibrahem [23] studied the area of the Northwest of Cairo at 34 km on Cairo-Alexandria Desert road using data from 119 wells tapping the aquifers layers. This area is fed from Tertiary and Quatenary aquifers. The results show that the average specific capacity is about 7.85 $m^3/h/m$ and the salinity ranges from as low as 288 to as high as 13,120 mg/l.

Different studies successfully integrated modern techniques to identify prospective groundwater potential zones. These techniques are Geographic Information system (GIS), Enhanced Thematic Mapper Plus (ETM+), Weighted Spatial Probability Modeling (WSPM), Watershed Modeling System (WMS), and Remote Sensing (RS). Omran [24] investigated the groundwater potential areas in Sohag region using GIS and remote sensing. He assigned different weights for each relevant layer like slopes, lineaments, drainage attributes, aquifer depth and thickness, water resistivity to generate a groundwater potential map, as shown in Fig. 5 and verified it with yield data to validate his model. Higher values indicate a very good potential for groundwater abstraction while lower values less than 5 indicate poor abstraction potentials.

Elewa and Qaddah [25] successfully integrated ETM+ , GIS, WMS, and WSPM to identify groundwater potential areas (Fig. 6) in Sinai Peninsula using eight different

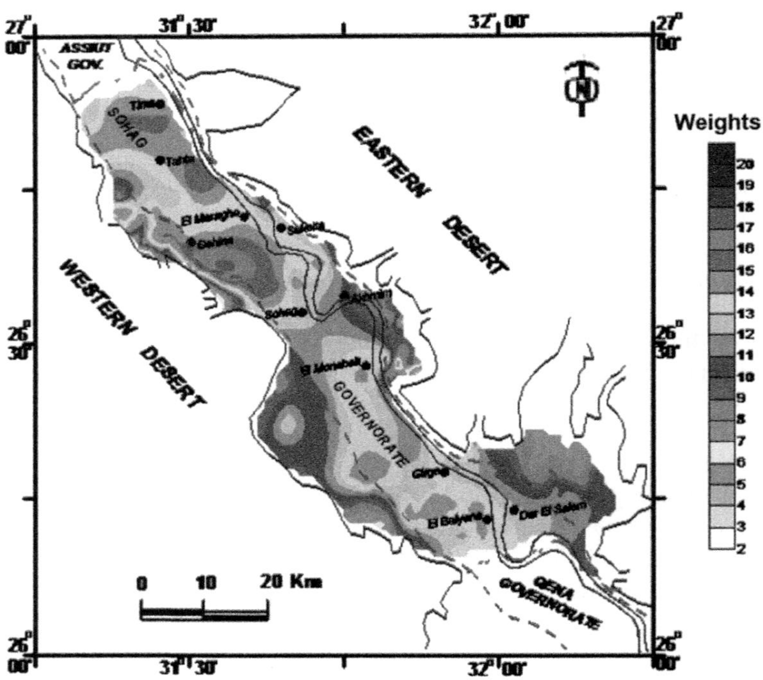

Fig. 5 Groundwater potentiality map, Sohag Governorate area [24]

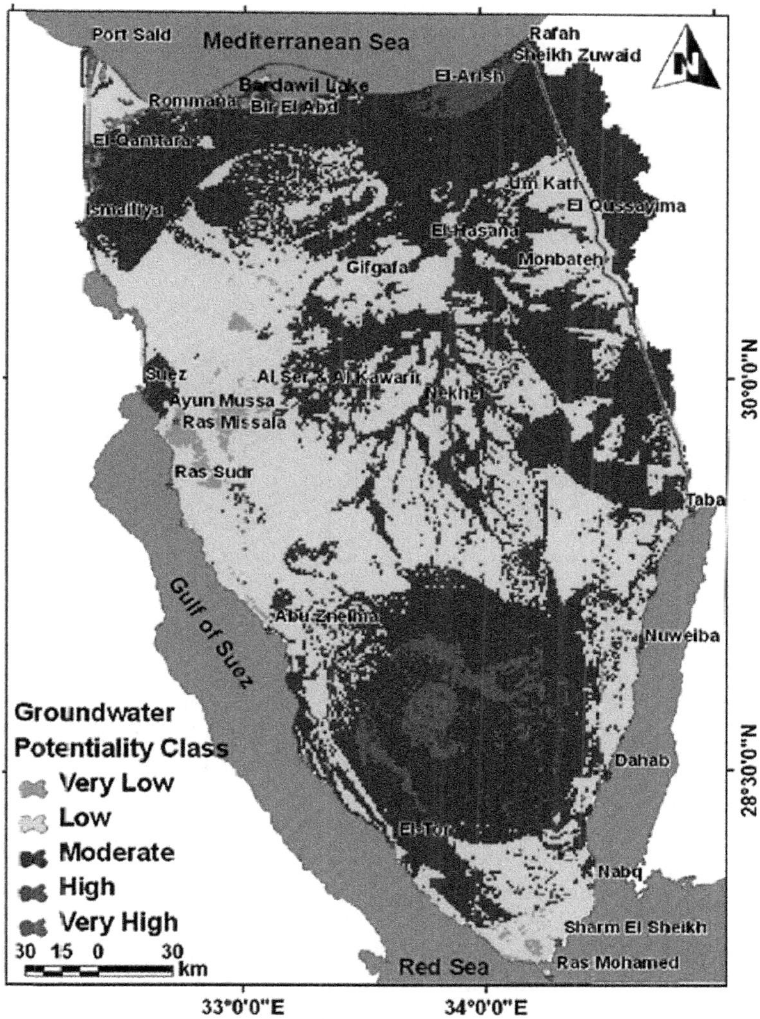

Fig. 6 Groundwater potentiality map of Sinai [25]

weighed thematic layers: rainfall, infiltration, net groundwater recharge, slope, lineament density, drainage density, water quality and depth to groundwater. The validity of their GIS model was verified by comparing it to published hydrogeological maps and actual yield data from boreholes. Five different regions were identified from very low to very high potentiality. Elevated areas like in El-Arish, Sheikh Zuwaid, Rafah, Wadi Feiran wells, the alluvial fan of El-Qaa plain, and St. Catherine area are identified as areas with high potentiality of groundwater and represent about 0.001% of the total Sinai area. However, the overall potential for Sinai is for a moderate class.

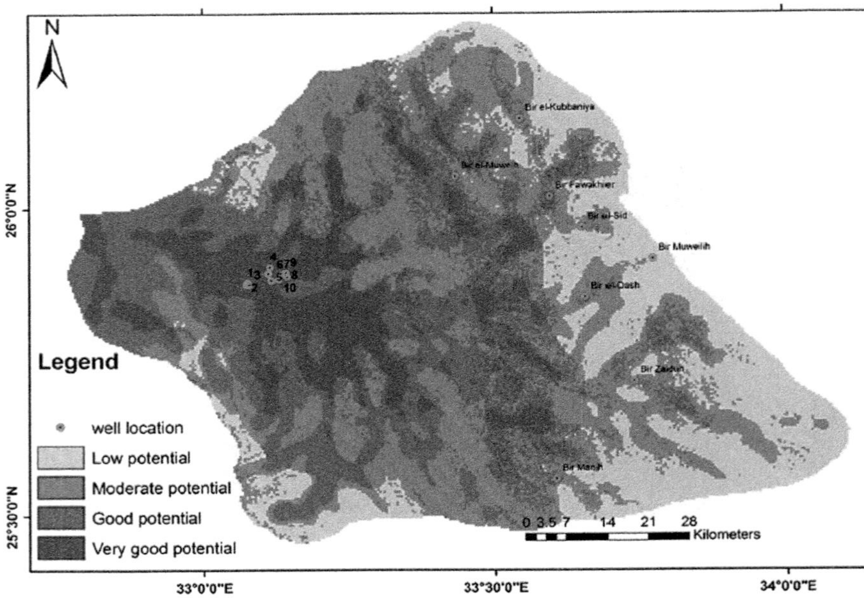

Fig. 7 Groundwater prospect map of Wadi El-Laqeita in central-eastern desert [26]

Abdallah [26] also prepared thematic maps such as lithology, drainage, lineaments, topography and slopes maps and integrated it through a GIS module to delineate and identify prospective groundwater zones (as shown in Fig. 7) in Wadi El-Laqeita in the central-eastern desert and its surrounding areas. This area is considered one of the most promising areas for agricultural development in Egypt.

Aly et al. [27] raised awareness of the deteriorating water quality in Siwa oasis which makes it unsuitable for irrigation as a result of excessive withdrawals of groundwater from the deep sandstone aquifer. All samples showed sodicity and salinity problems besides some iron toxicity. They recommended mixing poor water quality with high water quality, leaching, selection of crops with high tolerance to salinity, changing irrigation methods, and controlling point sources of pollution. Moghazy and Kaluarachchi [28] found that the salinity levels in Siwa has increased from 4.5 to 10.5 ds/m (very high salinity) during the period from 1996 to 2013 due to overuse of groundwater withdrawals, poor drainage system and poor management of wells in the area. In another study by Moghazy and Kaluarachchi [29], they found that the area to be cultivated is limited to about 16,400 feddan in Siwa to ensure that government policies are achieved. This area actually represents almost half of the planned area by the government. Abdulaziz and Fiad [30] developed a 3D multilayer groundwater model to understand the aquifer system and predict the future pumping rates and the potential for groundwater abstraction over 50 years. They found that there is a local recharge from the Nubian sandstone aquifer to the carbonate aquifer topping it and the optimum pumping rate should be regulated not to exceed about 520,000 m^3/day.

It should be noted also that climate change will have a negative impact on increasing sea levels. It is certain that sea water level has increased over the past 20 years. It is said that if the mean global temperature increases by 4°, the seawater will correspondingly increase by 0.71 m above the present levels [31]. This will increase the seawater intrusion into the coastal aquifer which will consequently increase the salinity levels of pumped groundwater, and movement of the shoreline inland reducing the Nile delta. However, this can be controlled on a small scale by injection of freshwater, subsurface barriers, etc. Mahmoud [32] investigated in detail the effect of climate change on the agricultural sector in Egypt. Climate changes will not affect Egypt only but all the Nile basin countries that will then put measures trying to adapt to this change which will ultimately put more pressures on our water resources. However, regarding the future impact of the climate change on the Nile flows, there was no complete agreement between models whether the Blue Nile (the main contributor of the Nile flows) will get drier or wetter, which make the future projections of the climate change uncertain [33].

The Egyptian government has to focus on creating, innovating, and improving the utilization of groundwater resources. This includes the expansion of the use of groundwater in the valley and the delta, noting that the water tables in these reservoirs are close to the surface and therefore the initial cost and operating cost of its wells will not be too much. Wells have to be distributed in the best way to avoid a sudden change in the water table, and the monitoring of the quality of water withdrawn must be kept under control. The exploitation of the shallow groundwater in the valley and the Delta will supply the necessary water at the end of the canals and mesqas where Nile water does not reach it. This would solve the problems of canal ends and increase the economic value of groundwater and protect the agricultural land from the continuous rise in the level of groundwater in some places.

Sustainability of Groundwater in Egypt's Deserts

The Nubian aquifer is the main source of water for the irrigation of much of the lands in the Western Desert in Egypt. It is theoretically non-renewable, but some evidence confirms its renewability. Among these indications is the self-flow of water springs that have been scattered for thousands of years in the oases of Al-Bahr, Al-Kharja, Dakhla and Al-Farafra. Recent study by elmansy et al. [34] in Farafra oasis showed that a safe extraction rate of 3000 m^3/d can be achieved under strict sustainability criteria set by the ministry of Water Resources and Irrigation of Egypt.

This underground reservoir feeds the areas of the ancient Farafra (190,000 feddans), the area of new Farafra (40,000 feddans), El-Dakhla (50,000 feddans) in the New Valley governorate, in the south-east of El-Qattara depression (50,000 feddans) and its extension in Giza (90,000 feddans), in Siwa (30,000 feddans), west of Minya governorate (400,000 feddans) and west-west of Minya (220,000 feddans).

While the reservoir that feeds the western areas of Al-Marashda (18,000 feddans), Toshka wells (25,000 feddans) and West Kom Ombo (25,000 feddans) is a renewable one due to its connection to the Nile and Lake Nasser. The safe withdrawal rates from these wells ensure the stability of groundwater levels proportionate with the feeding rates of the Nile. In addition, the city of Al-Tur (20,000 feddans) is dependent on the sedimentations in Sinai valleys, which is renewable, because it is fed from rainwater and floods that fall on the mountains of St. Catherine.

Moreover, the area of Moghra in Marsa Matrouh with 170,000 feddans is treated as a non-renewable reservoir. The rate of feeding the underground reservoir is very low besides the extracted water from it is relatively saline.

The life span for the existing projects in this area is estimated to be more than 100 years due to the availability of groundwater quantities greater than the required withdrawal rates during this period. Provided that the project is managed according to some conditions of which are: (a) restrict controls set by the government, (b) operation of wells by using solar energy, which ensures that the process of withdrawal from the reservoir will be limited on the daytime hours, and allows the reservoir to replenish itself during the night hours, (c) design of the wells fields in an optimal manner to ensure the stability of water levels, (d) use of modern irrigation methods as there is no room for surface irrigation to be used anymore, (e) criminalization of the cultivation of water-consuming crops, (f) strict monitoring of the operation of wells, monitoring of changes in groundwater levels and salinity levels, and taking precautionary measures to prevent any risks to the underground reservoir, (g) conducting educational seminars for wells users before handing it to them to inform them of the importance of periodic maintenance and allow them to know the principles of management and preservation of aquifers to preserve its life-time and to be partners in taking responsibility.

Conclusions

Groundwater is ranked second after the Nile River in the list of water resources in Egypt. There are aquifers that vary from shallow renewable groundwater aquifers to deep groundwater aquifers. Groundwater in Egypt is found in six main water basins, namely: the Nile aquifer, which is located under the Nile Valley and the Delta and is a renewable and vulnerable groundwater reservoir. The deep Nubian sandstone aquifer covering the Western Desert and is also located below the rest of the aquifers in other areas. It is non-renewable and characterized by water that is free of contaminants. Moreover, Fissured carbonate aquifer that spans about 50% of the area of Egypt in the eastern and western deserts and the Sinai Peninsula. Al Moghra and Hardrock aquifers, although water is poor in quantity and quality, it can be used in some saline-tolerant crops. Lastly, the coastal aquifer along the coastal strip of the Red Sea and the Mediterranean Sea.

From recent studies, it is certain that Egypt desert is rich in groundwater and is enough to satisfy its demands. However, the government must develop good plans for its exploitation and management and to prepare long- and short-term plans with periodic review to monitor the behavior of the underground reservoir to follow the changes that may occur in terms of quantity and type and the need to use the latest techniques to ensure sustainable future projects that rely on groundwater.

Recommendations

The government has to carefully set plans to monitor the water quality of the different aquifers. Besides, a monitoring network for the groundwater is needed to track the piezometric levels, salinity levels, rate of change in heads with time, and abstraction rates within the aquifers. Data should be shared between different institutional levels, authorities, research institutions, universities to get a good picture of the overall status of the groundwater. Recent advances in satellite imagery and the integration of using GIS, Remote Sensing and other invasive techniques should be applied. More attention should be directed to the use of the Nubian sandstone aquifer as it has the potential for satisfying the demands of the new groundwater-based reclamation projects in the western desert. More studies and investigations are required to make sure of its sustainability in the long term.

References

1. Nile Basin initiative, NBI (2015) The water resources of the Nile Basin. http://nilebasin.org/index.php/about-nile-basin/42-the-water-resources-of-the-nile-basin
2. Ministry of Water Resources and Irrigation Egypt (MWRI) (2005) Water for the future—National Water Resources Plan 2017. Technical report
3. AfDB (2015) Environmental and social management framework (ESMF). National Drainage Programme Egypt. African Development Bank
4. Gossel W, Ebraheem AM, Wycisk P (2004) A very large scale GIS-based groundwater flow model for the Nubian sandstone aquifer in Eastern Sahara (Egypt, northern Sudan and eastern Libya). Hydrogeol J 12:698–713
5. RIGW/IWACO (1999) Contributions to environmental management of Egypt's groundwater resources. Final report, EMGR Project, Cairo
6. RWSN (Rural Water Supply Network) (2009) Triple S briefing: providing reliable rural water services that last. http://reliefweb.int/report/uganda/triple-s-briefing-providing-reliable-rural-water-services-last
7. MWRI, Ministry of Water Resources and Irrigation (Egypt) (2014) Water scarcity in Egypt: the urgent need for regional cooperation among the Nile Basin countries. Report No. 6p
8. Hamza W, Mason S (2004) Water availability and food security challenges in Egypt. In: Proceedings international forum on food security under water scarcity in the middle east: problems and solutions. Como, Italy, pp 24–27
9. Ebraheem AM, Riad S, Wycisk P, Seif El-Nasr AM (2002) Simulation of impact of present and future groundwater extraction from the non-replenished Nubian Sandstone aquifer in Southwest Egypt. Environ Geol 43:188–196
10. Nour S (1996) Groundwater potential for irrigation in the East Oweinat area, Western Desert Egypt. Environ Geol 27:143–154
11. El Tahlawi MR, Farrag AA, Ahmed SS (2008) Groundwater of Egypt: '"an environmental overview."' Environ Geol 55:639–652. https://doi.org/10.1007/s00254-007-1014-1

12. Mohamed Ibrahem SM (2019) Effects of groundwater over-pumping on the sustainability of the Nubian sandstone aquifer in East-Oweinat area, Egypt. NRIAG J Astron Geophys 8(1):117–130
13. El Nahry AH, Elewa HH, Qaddah AA, Gedida N (2010) Soil and groundwater capability of East Oweinat area, Western Desert, Egypt using GIS spatial modeling techniques. Nat Sci 8(8):1–17
14. College of Agricultural Sciences. Irrigation Water Quality (2002) The Pennsylvania State University, USA. http://www.cas.psu.edu/docs/casdept/Turf/Education/Turgeon/CaseStudy/OldRanch/IrrWatQual.html
15. Gad MI, El Sheikh AE, El Osta MM (2011) Optimal management for groundwater of Nubian aquifer in El Dakhla depression, Western Desert, Egypt. J Water Resour Environ Eng 3(14):393–409
16. Sharaky AM, El Sayed A, Shanab EF (2018) Groundwater assessment for agricultural irrigation in Toshka area, Western Desert, Egypt. In: Conventional water resources and agriculture in Egypt. Springer, Cham, pp 347–387
17. Dawoud MA, Darwish MM, El-Kady MM (2005) GIS-based groundwater management model for western Nile Delta. Water Resour Manage 19:585–605
18. El Kashouty M, Sabbagh A (2011) Distribution and immobilization of heavy metals in Pliocene aquifer sediments in Wadi El Natrun depression, Western Desert. Arab J Geosci 4(5–6):1019–1039
19. Abouelmagd R, Barseem M, Metwally S, Farag M, Mousa S (2020) Groundwater exploration in fratures rocks using transient electromagnetic technique (TEM), west El-Minia, Egypt. Egypt Geophys Soc J 18(1):145–154
20. Massoud U, Kenawy AA, Ragab ESA, Abbas AM, El-Kosery HM (2014) Characterization of the groundwater aquifers at El Sadat city by joint inversion of VES and TEM data. NRIAG J Astron Geophys 3(2):137–149
21. Salem ZE, El-Bayumy DA (2016) Hydrogeological, petrophysical and hydrogeochemical characteristics of the groundwater aquifers east of Wadi El-Natrun, Egypt. NRIAG J Astron Geophys 5(1):124–146
22. Sobeih MM, El-Arabi NE, Helal EEDY, Awad BS (2017) Management of water resources to control groundwater levels in the southern area of the western Nile delta, Egypt. Water Sci 31(2):137–150
23. Mohamed Ibrahem SM (2020) Groundwater hydrology and characteristics of the tertiary aquifers, Northwest Cairo, Egypt. NRIAG J Astron Geophys 9(1):420–432
24. Omran AA (2008) Integration of remote sensing, geophysics and GIS to evaluate groundwater potentiality—a case study in Sohag Region, Egypt. In: The 3rd international conference on water resources and arid environments and the 1st Arab Water Forum
25. Elewa HH, Qaddah AA (2011) Groundwater potentiality mapping in the Sinai Peninsula, Egypt, using remote sensing and GIS watershed-based modeling. Hydrogeol J 19:613–628
26. Abdalla F (2012) Mapping of groundwater prospective zones using remote sensing and GIS techniques: a case study from the Central Eastern Desert, Egypt. J Afr Earth Sci 70:8–17
27. Aly AA, Kishk FM, Gaber HM, Al-Omran AM (2016) Long-term detection and hydrochemistry of groundwater resources in Egypt: case study of Siwa Oasis. J Saudi Soc Agric Sci 15(1):67–74
28. Moghazy NH, Kaluarachchi JJ (2020) Assessment of groundwater resources in Siwa Oasis, Western Desert, Egypt. Alex Eng J 59(1):149–163
29. Moghazy NH, Kaluarachchi JJ (2020) Sustainable agriculture development in the Western Desert of Egypt: a case study on crop production, profit, and uncertainty in the Siwa region. Sustainability 12(16):6568
30. Abdulaziz AM, Faid AM (2015) Evaluation of the groundwater resources potential of Siwa Oasis using three-dimensional multilayer groundwater flow model, Mersa Matruh Governorate, Egypt. Arab J Geosci 8(2):659–675
31. Van Vuuren DP, Isaac M, Kundzewicz ZW, Arnell N, Barker T, Criqui P, Berkhout F, Hilderink H, Hinkel J, Hof A et al (2011) The use of scenarios as the basis for combined assessment of climate change mitigation and adaptation. Glob Environ Change 21:575–591

32. Mahmoud MA (2017) Impact of climate change on the agricultural sector in Egypt. In: Conventional water resources and agriculture in Egypt. Springer, Cham, pp 213–227
33. Nour El-Din M (2013) Climate change risk management in Egypt. Report on the proposed climate change adaptation strategy for the ministry of water resources & irrigation in Egypt
34. Elmansy Y, Mohamed MS, Hassan AA, El-Gindy AM, Riad P (2020) Sustainable agricultural groundwater management for new reclaimed areas in Farafra Oasis, Western Desert of Egypt. Int J Eng Adv Technol (IJEAT) 9(4):118–127

Potential Use of Groundwater and Future Expansion

Groundwater Exploitation in Mega Projects: Egypt's 1.5 Million Feddan Project

Ahmed Abdelkhalek and Caroline King-Okumu

Abstract Grand feats of engineering and environmental management are a well-established tradition in Egypt. Following the political changes in 2013, and the gradual stabilization later on, the Government of Egypt embarked on applying the policy of horizontal expansion through a comprehensive development plan which included the reclamation of about 1.5 million feddan. This is the first stage of an ambitious plan to reclaim 4 million feddans from the desert. The aims are not only to increase the agricultural farmlands but also to create an integrated society in the new lands and to achieve a comprehensive agro-industrial development. Moreover, the project also aims to reduce the need for food imports. This chapter provides a brief description of the project, including an overview of the general vision and strategy adopted by the MWRI to guarantee its sustainability and success. It highlights the MWRI's plans to avoid the mistakes that have been observed in past development projects that are dependent on groundwater management and to establish an effective system for predicting and monitoring economic and environmental impacts. To help with this, the chapter observes that techniques for community engagement, data-logging, remote sensing and geographic information systems are available. The chapter concludes that the establishment of the environmental monitoring system should be a high priority both for the groundwater resource users and for the government. This should be integrated with participatory planning for long-term sustainability, environmental risk assessment and disaster preparedness.

Keywords Egypt · Megaproject · Desert reclamation · Groundwater · 1.5 million feddan project · Sustainable development

---

A. Abdelkhalek (✉)
Technical Office of Groundwater Sector, Ministry of Water Resources and Irrigation, Giza, Egypt
e-mail: ahmedimail@yahoo.com

C. King-Okumu
The Borders Institute (TBI), Nairobi, Kenya
e-mail: carocking@yahoo.com

C. King-Okumu
GeoData Institute, University of Southampton, Southampton, UK

A. Negm and A. Elkhouly (eds.), *Groundwater in Egyptâ s Deserts*, Springer Water,
https://doi.org/10.1007/978-3-030-77622-0_14

Abbreviations

MWRI Ministry of water resources and irrigation
ECDC Egyptian countryside development company

Introduction

Mega construction projects represent a strategic option for achieving sustainable development objectives in many countries [1]. Government actions to construct infrastructural, industrial, educational, cultural, transportational, medical, and residential projects can aim to provide societies with their needs and fulfill their requirements [1]. Environmental impact assessment, monitoring and periodic evaluation is another important element of project management. This is highly required to enable the project managers and intended beneficiaries to ensure and enhance the sustainability of the project.

The early announcement of the 1.5 million feddan project occurred as a part of the presidential election campaign of Abdel Fatah El-Sisi in 2014, during which the proposed project was publicized to target the reclamation of 4 million feddan. The 1.5 million feddan project is intended to be the first stage of action toward the 4 million feddan target. In its early planning, the area of the project was initially estimated to be one million feddans. However later on it was extended to 1.5 million feddan and hence the name of the project was modified. Since the early announcement of the project, much skepticism and criticism emerged both socially and politically regarding the possibility and feasibility of such a project. This lively and open public debate has later extended to the mechanisms and implementation steps of the project.

This Chapter provides a brief description of the project, an overview of the general vision and strategy adopted by the Ministry of Water Resources and Irrigation (MWRI) to guarantee the sustainability and success of the project. It highlights their plans to avoid the past mistakes with the development of groundwater. The best available scientific capacities and expertise must be applied to enable the management of the groundwater resources, conserve the strategic national reserves of non-renewable fossil water and ensure national security for the long- and short-term. The chapter concludes that the establishment of the environmental monitoring system should be a high priority both for the groundwater resource users and for the government. This should be integrated with participatory planning for long-term sustainability, environmental risk assessment and disaster preparedness.

Background

Egypt is a highly arid country and receives very little annual precipitation. It has a high degree of risk to natural hazards and is highly vulnerable to climate change impacts [2]. Egypt has four main agro-ecological zones: (1) the northern coastal belts, (2) the Nile valley, (3) the inland Sinai and Eastern desert, and (4) the western desert each of which has specific attributes of resources base, climatic features, terrain and geographic characteristics, land use patterns and socio-economic implications [3]. The country is facing great challenges due to its limited water resources, where the dependency ratio of 96.9% of the total renewable water resources flowing into the country from neighbouring countries is one of the world's highest [4]. The Nile River provides about 96% of Egypt's water [5].

While the country's total land area is 995,450 square kilometers, only 3.6% of that land is arable. The remaining 96.4% of the country's area is dominated by a vast desert plateau [6]. About 9 million feddans (3.8 million hectares) of land is available for cultivation in Egypt [7]. The latest population census in Egypt shows that there are more people now living along the narrow strip of land by the Nile than ever before (about 95 million) [8, 11]. Only sparse settlements exist in desert oases sustained solely by groundwater. Successive governments have observed that with the continuous increase in population, demand for land and shortages in the Nile Valley, desert reclamation could offer a solution [9]. Increasingly, this reclamation has relied on the use of non-renewable groundwater.

Since ancient times, the agriculture sector has been a cornerstone in Egyptian economy [3]. Agriculture and agribusiness sectors contribute about 35% of economy's domestic product. Agriculture provides jobs for about 22 million Egyptians (which is above a fifth of the total population [10].

The value of agricultural export revenues has been increasing over the past decade. Noticeably, the value of agricultural exports compared to the value of agriculture imports shows an upward trend, which implies an increase in export revenues relative to the costs of imported agriculture products [11] (see Fig. 1).

Expanding arable land by reclaiming it from desert areas has long been used as a method to extend agricultural production in Egypt [12]. Successive governments have sought solutions to the increasing problems caused by the overcrowding around the River Nile through the encouragement of development outside of the Delta and away from the river Nile. Since the 1950s the Egyptian government and international investors have made continuous efforts to reclaim land from the desert. The North Sinai Agricultural Development Project (NSADP) and the Toshka project (also known as the New Valley Project) resulted in an increase of over 80% in reclaimed agricultural land in Egypt in 2006/07 and 2007/08 [13] (see Fig. 2).

The Nile River is a transboundary water source shared by ten other countries. Egypt is especially affected by upstream Nile projects, an ever increasing population, and climate change, all of which intensify water scarcity [12]. Since these trends are expected to continue, the need for alternatives to continuing dependence on the Nile will grow. Other alternatives besides increasing use of groundwater

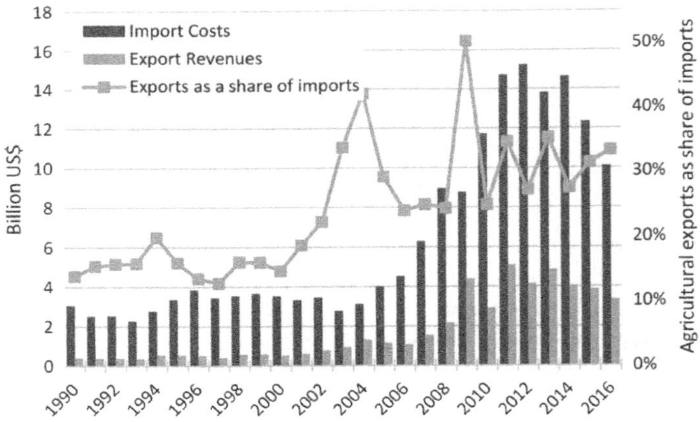

Fig. 1 Trends in value of Egypt's agricultural exports and imports, 1990 to 2016 [11]

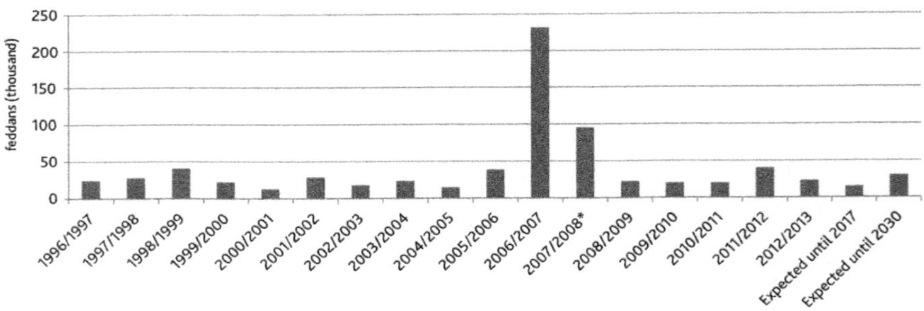

Fig. 2 Reclaimed agricultural land in Egypt 1996/1997 [13]

such as desalination of seawater or increased treatment and reuse of wastewater are believed to be more expensive.

History of Groundwater Exploitation in the Egyptian Desert [14]:

- In ancient times, water used to emerge naturally from springs in many desert areas (like Kharga–Dakhla–Baheria–Siwa and other ancient valleys) and this helped the expansion of the Pharaonic and Roman civilization in those desert areas.
- As time passed, springs got rarer and Egypt's population began to be highly dependent on the Nile Valley and Delta for agriculture and urban development.
- In the early 60 s the government began to develop the man dug and artesian wells in order to increase the cultivated areas. The Authority for Reconstruction of Desert also dug many shallow and deep wells in many areas.

- Due to the availability of water at relatively small depths (not exceeding 150 m), the economic feasibility of digging those wells was acceptable. However, due to the lack of knowledge about low flow irrigation methods (like drip and sprinkle irrigation) the beneficiaries of those wells were not using water in a conservative way and used to cultivate water-loving crops (like rice, corn and Alfalfa... etc.). The farmers who were used to irrigation methods in the Delta and the Nile valley used the same methods in their new desert lands and this rapid extraction of groundwater led to significant decreases in artesiansim. The number of artesian wells fell, and drainage problems caused the formation of expanding wastewater lakes in many areas.
- The lack of vision of consecutive administrations regarding the sustainability of the aquifers allowed the farmers to use the same water quotas they were used to in the Nile valley and Delta which led to a significant drop in water levels in a vast number of wells and this condition mostly requires the replacement of the well.
- In order to maintain the already existing development in those reclaimed desert areas the government and private investors were then forced to dig new deep wells. They aimed at meeting the water needs of the irrigated areas. However, the high financial cost of digging and running those wells did exceed the annual economic revenues generated in the cultivated lands. In some cases, it may also cancel out their present real-estate value. This situation led to an apparent imbalance between the financial, social, environmental and political sustainability of the developing land reclamation areas.

The MWRI has gained experience in the management of groundwater extraction. The total groundwater volume has been estimated at about 40,000 billion m^3. However, abstraction is estimated to be about 2.0 billion m^3 per year [3]. During the last few decades, the MWRI has observed a rapid development and increasing stress on the available groundwater sources and lands in the Western Nile Delta area. New communities have been initiated, accompanied by industrial and agricultural activities [15].

Authorities' enforcement of regulations on use of pumps for groundwater extraction is low and the bureaucratic complexity of getting a license for new wells leads to less effective regulation in practice. Illegal pumping is widespread [5]. Other obstacles in utilizing groundwater resources are the great depths of these aquifers (up to 1500 m in some areas) and the deteriorating water quality at the increasing depths [3]

The Political Dimension of the Project

Throughout the past few decades, the agricultural sector In Egypt was subject to major policy changes that had substantial impact on this sector [3]. During the 1960s, agriculture was pivotal to Egypt's development policies, the rural citizen and the factory worker were at the center of post-independence national Egyptian identity [11]. Nasser's regime adopted a socialist policy characterized by central planning and price controls. The government imposed crop rotation schedules, crop area allocations, and compulsory procurement for most major crops, usually at prices lower than the market price [16]. The intention of this policy was to transfer agricultural surpluses to the government in order to finance industrial growth [11].

Starting in 1986 and through the 1990s, a reversal of the above mentioned interventionist policies began with a sweeping wave of liberalizing economic reforms, giving the private sector a larger role in the agriculture sector and reducing that of the government [11].

Efforts to expand outwards from the Nile Delta and Valley have been in place since the early 1900s. However, the period beginning in the 1970s, following the completion of the Aswan High Dam, is notable for a series of government desert development megaprojects [11]. The aim is to expand arable land into the desert and the presence of groundwater made this a possibility.

Following the political changes in 2013 and the gradual stabilization in the country conditions, the Government of Egypt embarked on applying the policy of horizontal expansion through a comprehensive development plan that included the reclamation of about 1.5 million [3]. The project is part of a national strategy to attract renewed foreign direct investment following the social and political unrest that the country has witnessed since the 25th of January revolution in 2011 [17]. The government has realized the necessity for immediate reform of the economic sector to revive the economy. Due to political instability since 2011, GDP growth has averaged 2%. The Egyptian government embarked on an ambitious economic reform plan after the 2014/2015 fiscal year. The intention was to demonstrate to the IMF and to Egypt's regional and international partners as well as to the Egyptian people that the government had effective mechanisms in place to reform and recalibrate the economy. In light of this, commentators have observed some reasons for optimism after several years of sluggish growth [17].

President Abdel Fattah al-Sisi refers to mega national projects as a locomotive of development [18]. And in the government's statement before the House of Representatives on March 3, 2016, many such projects were mentioned, including the development of the Suez Canal Corridor, establishment of new cities, reclamation of one million and half feddans, the development of the western north coast by establishing a set of horizontal and longitudinal corridors as well as industrial, tourist, urban, agricultural and power generation projects [18].

There is a huge debate both socially and politically about the feasibility, mechanisms of implementation and timing of some of these projects. Their impact on ordinary citizens and their ability to attract national and foreign investments are still under the test. Their environmental impacts also require continuous monitoring, evaluation and adaptive management by the responsible authorities and communities [19]. Everyone is interested to ensure that these projects should have a positive impact on the economy at least over the long term.

MWRI's Vision of the Project

Management of the groundwater reserve is the most important element in the project. Therefore, the exploitation of this strategic reserve should take into consideration the concept of sustainability with all its economic, social, ecological and political dimensions. This requires continuous study, monitoring and evaluation by the responsible authorities and user groups.

The vision, goals and scope of work for the 1.5 million feddan project have been published by the MWRI (on their official site) and the following general goal and objective of development using groundwater are stated:

> *Expanding the use of our strategic groundwater reserve to overcome challenges imposed by any possible decrease of water availability from the Nile due to dry seasons (short or long term) or the construction of storage dams on the Nile's tributaries.* [20]

> *Increasing the exploitation of groundwater resources can also maximize the economic and social value of the resource. This goal must be achieved using the most sustainable methods of management to avoid any overexploitation of the resource and to provide welfare for the next generations.* [20]

The Strategy of Work for Development Using Groundwater

Sustainable development of groundwater (either renewable or non-renewable) completely depends on studying the feasibility of using this groundwater reserve to achieve an acceptable economic and social outcome without exhausting the aquifer or causing unacceptable negative impacts on the environment. Therefore, the biggest determinants of sustainable development of groundwater are not only the baseline understanding of the groundwater conditions and the management plan, but also the beneficiaries and day to day on the ground managers of their source.

To manage the shared groundwater resource at the strategic level, it is important to work with the individual water users who manage it on the ground. Their culture and practices determine the ways in which they handle the water. In light of this, engaging them in the management and monitoring of the resource conditions is essential (see many previous discussions in: Attia [21], King and Salem [22], Villholth et al. [23]). In light of the significant costs of establishing comprehensive environmental monitoring systems, the regular voluntary participation of water users, beneficiaries and younger generations in the environmental management system is the key to management efficiency regardless of the aim of the development project whether it is agricultural, industrial or touristic or other.

The MWRI intends to ensure that the concepts of sustainability and economic use of water must be well established in the new forming communities to guarantee the continuity of the project. This will require environmental education and public awareness programs to be put in place. The MWRI is determined not to repeat previous mistakes of past development of groundwater (see Box 1).

Box 1: Lessons learned from past development of groundwater in the Egyptian deserts [14]

The MWRI have assessed and summarized the main problems with previous groundwater development projects as follows:

- Over-extraction of water from wells that largely exceeds the safe yield of wells.
- Permitting the use of big water pumps with high flow rates was a condition which allowed water users to exhaust the groundwater reserve leading to a significant drawdown in water levels in the wells and expanding the cone of depression of each well which has negative interference effect on neighboring wells. These pumps also allow over-consumption of the energy sources subsidized by the government and this increases the severity of the economic loss during the running of wells.
- The running of wells was trusted to workers from the same beneficiary families that use the water and this condition allowed users to increase the working hours of wells based on their needs and regardless of the safe yield of wells.
- The continuous extraction of water from wells throughout the year with no or little breaks has led to the continuous drawdown of water levels because water is not given enough time for recovery.
- The political support of governors to farmers' demands to increase their water quotas has led to exhaustion of groundwater resources in many areas.
- Using flood irrigation in the old lands and also some of the newly reclaimed lands while abiding by the same water quota stated on the old contracts issued from the government during the 60 s (which allows for 22 cubic meters/feddan/day) was also a major cause leading to the exhaustion of aquifers.
- Formation of agricultural wastewater lakes due to excess drainage water caused by expanding use of flood irrigation and the expanding areas of these lakes created a threat to the existing cultivated land and the environment.
- Assigning the sole responsibility for running and maintenance of wells to the MWRI and the Ministry of Agriculture with no or little liability and participation of water users has led to the misuse of wells by the users while putting the blame only on the government for any negative impact of this misuse.
- The low economic outcome of the traditional crops being cultivated in the irrigated areas (mostly for crops such as corn, rice and—alfalfa).
- Lack of marketing and manufacturing of the crops produced led to low economic income of the cultivated crops.

Description of the Project

The 1.5 million feddan project is the first stage within an ambitious plan to reclaim 4 million feddans. The project aims to increase the agricultural land by 20% in addition to creating promising investment opportunities in various fields, including reclamation of agricultural lands, establishment of projects targeting food industries and establishment of logistics areas in addition to developing these urban areas to create an integrated and sustainable environment according to the 2030 strategy [24]. Moreover, the project aims to narrow the gap between supply and demand of the foodstuffs and reduce imports [18].

These objectives of the project are relatively similar to previous land reclamation projects in Egypt. What is different about the new project is that it has the benefit of all of the lessons and experiences of the past. Moreover, the project is pursuing the comprehensive goal of creating new integrated agro-industrial societies in the new lands. To ensure the realization of this goal a new governmental entity was established for the administration of the project under the name of the Egyptian Countryside Development Company (ECDC) (the role of the company is discussed in the next section).

Groundwater is the major water source for the project. However, it is also important to remember that there are other water resources that are periodically available in desert areas. These include occasional rainwater and also surface drainage water. Table 1 shows the intended contribution percentages of both surface and groundwater to the total annual irrigation water budget [25].

The following tables (Table 2, 3a–c) along with (Fig. 3) give an overview of all of the officially published information about the locations of the project, intervention areas and size proposed for each location, number of wells proposed for each location, number of feddans proposed for reclamation for each location, well depth proposed in each location and water source. These tables were officially published by the MWRI and the ECDC.

The Egyptian Countryside Development Company (ECDC)

The main vision of the 1.5 million feddans project is to build communities in the newly reclaimed lands of Egypt, presenting an internationally appreciated model for sustainable economic development that is aligned with Egypt's vision 2030 [26]. This requires the cooperation and coordination of a number of Ministries and institutions. To unify the entity responsible for the marketing and allocation of the land to beneficiaries, the idea of forming the ECDC has come into existence.

Table 1 Surface and groundwater as % intended contribution to the 1.5 million feddan project (*source* MWRI) [25]

Water source	Area (thousand feddan)	Percentage (%)
Surface water	172	11.50
Groundwater	1328	88.50
Total area	1500	100

Table 2 Proposed areas for the 1.5 million feddan project (*source* MWRI) [25]

1.5 Million feddan project							
No.	Location name	Water source	Governorate	Area (Thousand feddan)	Total area/governorate (Thousand feddan)	No. of wells	Well depth (meter)
1	Old Al-Farafra	Groundwater	New Valley	190	280	800	900–1000
2	New Al-Farafra			40		170	800–1000
3	Extension of El-Dakhla			50		210	450–750
4	South East of Qatara Depression		Al-Giza	90	90	380	1000±
5	Al-Moghra		Matrouh	170	170	720	125–200
6	Al-Amal Village	Surface water	Al-Ismailia	3.5	3.5	Surface water	
7	West of Kom-Ombo	Groundwater	Aswan	25	193	100	250–300
8	Toshka	Surface water		143		Surface water	
9	Toshka	Surface water		25		102	250
10	West of Al-Marashda	Surface water	Qena	25.5	43.5	Surface water	
	West of Al-Marashda	Groundwater		18		70	200–300
11	West west of El-Minya (1)		El-Menia	220	220	755	250–650
12	Extension of South East of Qatara depression		New Valley	50	50	210	1000±
13	East of Siwa		Matrouh	30	30	125	1000±
14	El-Tor		South of Sinai	20	20	144	250±
15	West of El-Minya (Wind Farm)		El-Menia	250	400	700	250–450
16	West west of El-Minya (2)			150		625	250–450
Total				1500	1500	5111	

Table 3 a–c Showing the three phases of the 1.5 million feddan project (*source* MWRI) [25]

No.	Location name	Water source	Area (thousand feddan)
a First phase			
1	Old Al-Farafra	Groundwater	30
2	New Al-Farafra		20
3	Extension of Eldakhla		20
4	Al-Moghra		135
5	Al-Amal Village	Surface water	3.5
6	Toshka		143
7	Toshka	Groundwater	25
8	West of Al-Marashda	Surface water	25.5
	West of Al-Marashda	Groundwater	18
9	West west of El-Minya (1)		80
Total area			500
b Second phase			
1	Old Al-Farafra	Groundwater	120
2	New Al-Farafra		20
3	Extension of El-Dakhla		30
4	West of Kom-ombo		25
5	El-Moghra		35
6	West west of El-Minya (1)		140
7	South East of Qatara Depression		40
8	Extension of South East of Qatara depression		50
9	East of Siwa		30
Total area			490
c Third phase			
1	Old Al-Farafra	Groundwater	40
2	South East of Qatara Depression		50
3	El-Tor		20
5	West of El-Minya (Wind Farm)		250
6	West west of El-Minya (2)		150
Total area			510

ECDC is a shareholding company instituted by the provisions of the investment guarantees and investment law no. 8 the year 1997, its executive board and amendments by law 17, the year 2015. It is a 100% Egyptian company founded in 2015 for the purpose of reclamation and development of the 1.5 million feddan project areas [26].

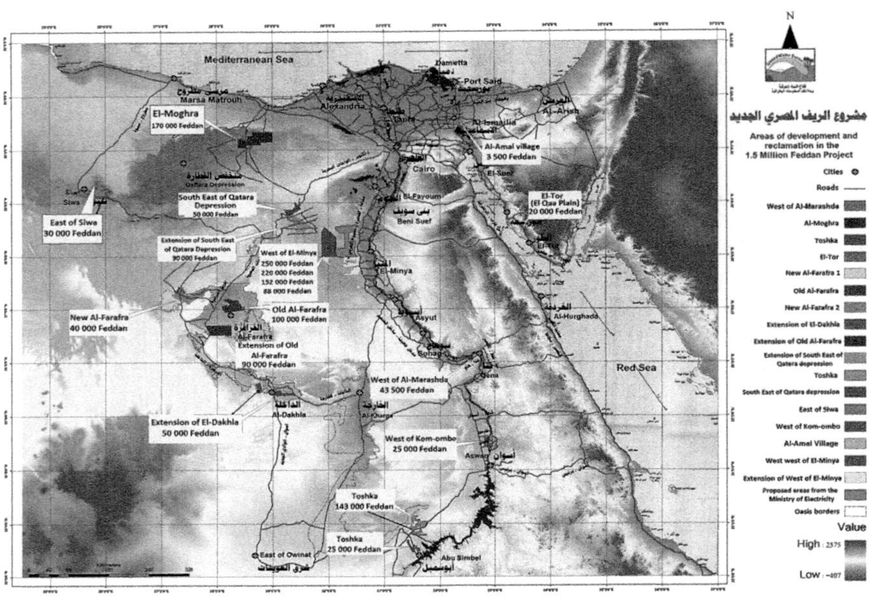

Fig. 3 Locations of the 1.5 million feddan project areas (*source* MWRI [25] and ECDC [26])

The three governmental shareholders are:

- Ministry of Finance
- Ministry of Agriculture and Land Reclamation (General Authority for Rehabilitation and Agricultural Development)
- Ministry of Housing, utilities and urban community (New Urban Cities Authority).

ECDC has a number of clear goals [26]:

- To implement the reclamation of Egypt's desert land according to an overarching master plan and hence prevent ad hoc expansion and settlements into Egypt's desert land.
- To facilitate and encourage entrepreneurship and enterprise creation and also the settlement of youth in the new regions.
- To create decent job opportunities for Egypt's youth along the value chains of agricultural production; logistics, marketing, manufacturing, services, etc.
- Observing the highest standards of environmental practices and the social dimension in the new communities.
- Providing training and technical assistance to investors and small farmers.
- Cooperation and working with any other internationally recognized research institutes for the purpose of improving and supporting the planning and implementation of the project.

The company has created two different bidding mechanisms for large investors and youth & small farmers [26]. In addition, the company aids the already existing Bedouin communities in some areas who occupy the land under the customary system of "Wadaa El-Yad" to register their lands [34]. In the "Summary of Investors' Pre-Qualification Booklet" [24] the following general terms were stated:

1. The applicant should be an Egyptian joint stock company whose activities include desert lands reclamation, cultivation and development.
2. The applicant should be interested in obtaining land areas not less than 2000 feddans.
3. The applicant should not have been subjected to bankruptcy, insolvency or liquidation of any activity inside or outside Egypt or has been convicted of public money offenses, corruption or tax evasion.
4. The applicant should present a credit worthiness certificate from I-Score.
5. An alliance between a group of companies could be carried out on the same Eligibility Booklet where alliance proportions are determined.

The booklet also categories beneficiaries into three categories [24]:

1st category: The applicant is granted the right to apply for obtaining land areas: of a minimum of 2000 feddans and a maximum of 6000 feddans.

2nd category: of a minimum of 2000 feddans and a maximum of 12,000 feddans.

3rd category: of a minimum of 2000 feddans and a maximum of more than 12,000 feddans.

It's an integration approach that allows benefits to both large investors and small farmers. The majority of jobs are expected to be created along the value chains of agricultural production in all types of economic activities; pre and post-harvesting, logistics, services, manufacturing, etc. ECDC also encourages green jobs in fields like recycling, biogas and solar energy.

Obligations of Large Investors:

- To reclaim 2000+ feddans, according to their bids.
- Will be responsible for infrastructural work based on guidelines by ECDC.
- Should create employment opportunities for the youth in agriculture as well as along the value chains of agricultural production.
- Will be encouraged to engage with the youth and small farmers through contract farming.
- Their investments should bring along positive externalities for the youth.

Support for youth and small farmers:

- Will be allocated lands in selected regions close to inhabited areas and equipped with water wells
- Will be helped with housing and living facilities
- Agreements with agricultural research centers concluded to support youth with agricultural research.
- Will be provided with training and technical assistance.

MWRI'S Proposed Strategy to Ensure Sustainability of the Project

The following eight pillars of the MWRI strategy are directly translated from the official website [14]:

First: Aquifer management

- MWRI is solely responsible for designing and planning of well fields required for the project. (Number of wells–Distance between wells–Depth of wells–daily discharge rate–depth of pumps–source of power used to run the wells…etc..(
- The ownership of wells may go to the beneficiaries but the MWRI has full authority to monitor, assess and determine the discharge rate of wells based on changing conditions.
- The MWRI should be the sole authority responsible for the running of wells. **Remote automatic control systems** should be preinstalled and programmed based on the groundwater potentiality of the area and the types of crops cultivated there and with no violation to the water use related laws and regulations. These systems will allow the full monitoring and control of wells to prevent any attempt of violation to the safe yield of well on time.
- The wells should run intermittently to give enough time for the recovery of water in the well through the horizontal flow of groundwater in the aquifer (transmissivity).
- If electrical pumps are licensed, the wells should not be permitted to run continuously and, mostly, no more than 8 h per day.
- Monitoring systems must be installed to follow up any change in water level and to the quality of water.
- Assigning water quotas based on the actual groundwater assessment studies and not based on the political decisions taken by the Ministries of Agriculture, Housing, Industry, tourism…etc. Such decisions used to be taken in the past just to satisfy the beneficiaries regardless of the sustainability of the aquifer in the long term.
- Cultivated crops should be selected based on the safe yield of the wells in the area taking into consideration the concepts of food security and the economic unit value of the water.
- In case of any unexpected deterioration of groundwater quality or quantity, the laws and regulations of water use should be modified to relieve the problem and severe penalties should be imposed in case of any violation.
- Participation of water users in monitoring, assessment and planning measures.
- Training and qualifying the human resources in the areas of development regarding all fields related to monitoring and assessment of groundwater in addition to water use ethics.

Second: Land management, water usage and irrigation

- Prohibiting the use of environmentally unfriendly types of pesticides and cements and encouraging organic agriculture to preserve the water quality of aquifers and also of Lake Nasr.

- Directing users to reuse the discharged agricultural wastewater after mixing it with water from the wells.
- Implementing water conservative methods of irrigation like sprinkle and drip irrigation and completely banning flood irrigation.
- Prohibiting the cultivation of water-loving crops like rice, banana, alfalfa and any other crop not permitted by the Ministry of agriculture.
- Cultivating at least 50% of the land with fruit trees and crops which require intensive labor but produce a profit in relation to the economic value of water.

Third: Energy sources for running wells (using Solar Energy)

One of the very ambitious goals of the project is to prohibit the use of traditional energy sources (like fossil fuels and electricity) for the running of wells and substitute these sources with solar energy. The use of solar energy is regarded as one of the sustainability pillars of this project.

Solar energy provides the following:

- A clean and renewable source of energy with zero negative impact on the environment.
- Using solar energy is economically efficient in the long term as it will save the government from increasing their imports of oil to supply the project and also will secure the project against the instability of oil prices in the long term.
- Solar energy is a naturally driven system which obliges users to abide by the assigned working hours for each well.
- Solar energy is more stable than other renewable sources of energy such as wind energy.
- The economic feasibility of using solar energy compared to diesel power is very acceptable both on the average and for the long term (see Fig. 4).

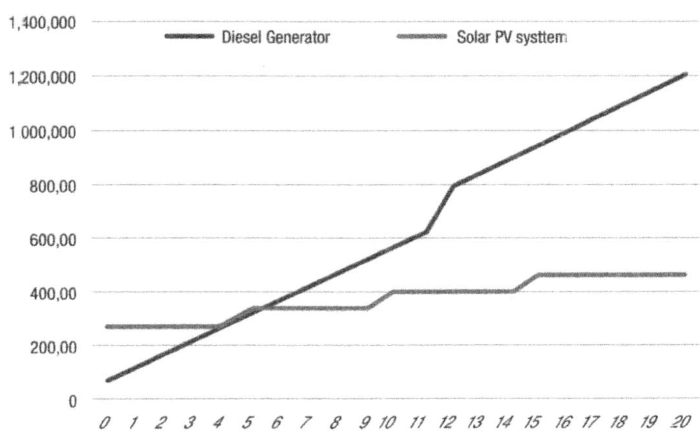

Fig. 4 Accumulated expenses of two different systems (diesel generator and solar PV system) [3]

Fourth: Human resources

Experience of similar kinds of projects has proved that choosing the right beneficiaries and personnel is a very important factor for the long-term success of the project. The project's goal is not only limited to increasing the economic value of groundwater resources and land in those new areas, but also to develop new communities in order to relieve the overcrowding in the Nile valley and Delta. Based on this, the ECDC should take full responsibility for the following:

- Specifying the number and categories of the human personnel required for the project to provide both public and private services (schools–Hospitals–Engineering services–agricultural services…etc.)
- Advertisements for vacancies in the project should include all the advantages and challenges of the project and should also include the desired specifications for the new inhabitants of those new areas.
- Applications for the project should be studied carefully to choose the most suitable personnel. Medical and social history should be examined.
- Comprehensive study centers should be established in the new areas to provide essential services regarding all related fields (like water management, agriculture, power supply, Environment…etc.).

Fifth: Regulation of energy sources used

- Running wells should be done using solar energy taking into consideration that each well should have its power source.
- Fossil fuels could only be allowed on a limited range during nightfall to run small pumps used to discharge water from storage tanks to the irrigation system.
- Fossil fuels should be sold using smart cards while taking all environmental precautions to protect the aquifer and the cultivated land from contamination.
- Using Hybrid systems to provide power supplies needed for houses, schools and factories throughout the day.
- Preparing plans to connect the solar power system with the electrical system in the project areas.

Sixth: Transportations

- Prohibiting the use of traditional vehicles (which run on fossil energy) inside of the villages and encourage the use of electrical cars and bikes instead.
- Providing public vehicles to encourage people to abandon their private vehicles.
- Using smart cards to sell fossil fuels and taking into consideration the gradual cutting of subsidy on fossil fuels.

Seventh: Education

- Raising people's awareness of water scarcity with all its challenges and developing study modules for elementary and secondary schools to identify the best ways to deal with it.

- Training and qualifying a new generation of graduates specialized in the hydrology of groundwater and all related fields.
- Improving human resources to provide highly qualified technicians to work on electrical and mechanical fields required for the running of well using solar energy and also for the new irrigation systems.

Eighth: Agriculture

- Specifying the suitable crops for the developed areas based on climatic conditions, water quality and soil type of each area.
- Specifying the irrigation system suitable for each area also based on climatic condition, water quality and soil type of the area.
- Proposing recommendations regarding the improvement of soil quality and water-holding capacity, decreasing of infiltration and evaporation (and hence decreasing the water need of plants) and raising the efficiency of the irrigation system.
- Providing economic feasibility studies for each crop taking into consideration the economic value of water and climatic conditions.
- Maximizing the economic value per water unit through the use of alternating irrigation, developing irrigation systems, implementing hydroponics agriculture and greenhouses, using water storage tanks… etc.)

Implementation of the Project

Due to the urgency of the project, implementation has proceeded simultaneously with development planning. The government has already dug the wells in all of the proposed areas of the project before assigning the land to the beneficiaries or selling it to investors. Since feasibility studies could not be conducted in advance, the project will involve a process of learning by doing, and continuous iterative review. Management of this process will be critical to its success.

The new lands of the project are allocated in the form of shares in joint stock companies. The price per share is determined inclusive of the value of basic infrastructure, utilities and facilities implemented by the government. The number of installments and duration of payments are determined also by the government depending on the form of land allocation [3].

The implementation of the project involves several preparations steps [27] all of which are organized by the ECDC in integration with the implementing concerned Ministry:

- Providing water sources (Wells digging)
- Providing power source (electricity and solar energy)
- Providing infrastructure and services areas
- Providing roads and transportation

- Providing telecommunications and mobile phones services.
- Providing desalination stations (only for drinking in the areas of high salinity water).

To avoid fragmentations of the agriculture lands, the assigned locations are divided into blocks of 238 feddans as Shareholding Company, each block is supported by a water well and its ownership could be shared by a minimum of 10 and maximum of 23 owners [3].

The implementation of the project is still going on at the time of the writing of this chapter. According to plan, the project should contribute to increasing the populated area in Egypt from 6 to 10%, in addition to increasing the agricultural farmlands from 8 million feddans to 9.5 million feddans, by an increase of 20% [18]. This section briefly presents a few major events along the development of the project:

Cultivating Ten Thousand Feddans at Farafra (Sahl Baraka): On December 30, 2015, President Abdel-Fattah El-Sisi launched the first phase of the 1.5-million-feddan project, at Farafra City, New Valley [18, 28]. The ten thousand feddans was reclaimed by the Armed Forces Engineering Authority along with the establishment of two agricultural villages and an integrated service. The Head of the Armed Forces Engineering Authority said that the first phase that has been completed includes the agricultural lands with an area of 10 thousand feddans, 7500 feddans of which are irrigated by pivot irrigation system and 2500 drip irrigation system. He added that 40 water wells with a depth of 800 m were drilled, 15 oxidation basins were established each with a capacity of 22,000 m^3 to treat the water extracted from the wells and remove the minerals from them at a rate of a basin per 500 feddans and 15 pumping rooms were built to draw water from the oxidation tanks and pump it into irrigation networks [28].

The first stage of the project included the establishment of three villages: two agricultural and one service-oriented to serve workers in the area. Every agricultural village is set up on a 400-feddan area, whereas the service village is built on 800-feddan area and is planned to accommodate a population density of 10–15 thousand people. The villages include 2000 farmhouses; each of an area of 200 m^2[18].

In order to provide required housing for staff and personnel working in administrative and service facilities, 40 housing buildings, incorporating a total of 480 housing units were built. Additionally, other utilities, such as; schools, health units, ambulance, police station, civil protection and veterinary units, markets, agricultural company, and others, were established [18]. On May 5, 2016, President Abdel Fattah al-Sisi launched the wheat harvest at Sahl Baraka.

In the following years after the inauguration of the project at old Farafra the implementation of the project continued in the following areas:

- Old Farafra oasis (locally known as Sahl Baraka) located north of the new valley governorate almost 630 km from Cairo city with total area of 82,000 feddans. All the area was leveled except for limited areas that are undulated. boreholes with a salinity level between (250–500) ppm [29].

- Moghra region is in the desert hinterland of Matrouh governorate northeast of El Qattara depression almost 250 km from Cairo city with a total area of 2,450,000 feddan. All the area is leveled except for limited areas that are undulated. Boreholes with a salinity level between (3000–7000) ppm [29].
- West of El-Minya Extension region is the desert hinterland of El-Minya governorate in the westwards opposite side, which extend from central El-Minya city northwards and central Maloy city southwards with total area of about 157,000 feddan extended in the desert areas bordering Asyut Western desert road. Boreholes with a salinity level between (600–1000) ppm [29].
- West of West El-Minya region is in the desert hinterland of El-Minya governorate in the westwards opposite side, which extend from Bani Mazar northwards to El-Minya city southwards with total area of about 185,000 feddan extended in the desert areas hinterland about 50 km from Asyut Western desert road. All the area is leveled except for limited areas that are undulated. Boreholes with a salinity level between (1200–2000) ppm [29].

Up to December 2018, the total number of wells already established for the project was (949) wells with an approximate total cost of 1.1 billion EGP [30]. No more wells are being dug by the government in the current developed areas of the project. Solar systems were already installed in (50) wells in El-Farafra with an approximate total cost of 68.46 million EGP [30] In Toskha, another 50 wells have been prepared in cooperation with the National Organization for Military Production (NOMP), applying integrated systems including: solar PV panel, smart pumps, surface storage tanks (capacity 2000 m^3) and cameras (for monitoring) [3]. More solar systems are being installed to other wells [30].

On the 10th of December, 2018 Eng. Atter Hanora, the former head of the ECDC has announced on "Hona El Asema" TV show, that almost 500 thousand feddan has already been contracted to both large Investors and youth & small Farmers and that they are currently bidding for the next phase of 300 thousand feddan [31]. Later on, more feddans were contracted to beneficiaries.

On March 2020, an unusually powerful storm named "Dragon Storm" has slammed into Egypt and other nations in the region, unleashing torrential rain, destructive winds and towering walls of dust [32]. Later the association of Moghra investors and farmers announced that 200 million Egyptian pounds worth of damage had been incurred due to the Dragon storm and previous floods that also happened in November 2020. They pleaded for the Egyptian government to compensate them for their loss [33].

Later on June 2020, the ECDC announced their readiness to compensate all the affected farmers and investors in Moghra and West of Al-Minya for their great financial loss due to the storm [34].

Also on 3 June 2020, the ECDC awarded the private company "Enara" a utility scale contract for producing and distributing renewable energy to service the agricultural land currently developed in the 1.5 million feddans project. Under this contract Enara Energy will generate and distribute nearly 400 MW of renewable energies to over 2000 land tenants and agricultural investors in areas mainly located in El Moghra and west of Minya. The cost of providing electricity to the land developed has been estimated to approximately 6.5 billion Egyptian pounds [35].

On Sunday 15 November 2020, during a cabinet meeting, President Abdel Fattah El Sisi directed the government to speed up the completion of the infrastructure works including roads, agricultural draining systems and service areas, in coordination with the authorities concerned. He also ordered officials to prepare a comprehensive marketing study for the project and to facilitate access to finance for small and medium-sized enterprises [36].

According to the Central Agency for Public Mobilization and Statistics (CAPMAS), the total reclaimed area reached 313,323 feddans in the fiscal year of 2017/2018 [37]. Under the title of "Statement of the project one and a half million feddan", the agency published the following table showing the distribution of the reclaimed area according to Governorates (see Table 4).

To date, no public monitoring system has yet been unveiled. A geographic information system is yet under preparation. No information is yet available concerning stakeholder engagement events for sustainability and disaster planning.

Table 4 The land area of the project is one and a half million feddan distributed according to governorates in 2017/2018 (CAPMAS, 2019) [37]

Area feddan المساحة:بالفدان						
المحافظة	المشروع	المساحة المزروعة Area planted	المساحة المستصلحة بالفدان Reclaimed area	المساحة قابلة للاستصلاح Area can be reclaimed	Project	Governorate
الإجمالى العام		115,670	313,323	515,331	Gross total	
المنيا	غرب المنيا	70,000	116,999	259,685	West Menia	Menia
مطروح	المغرة	35,000	133,679	159,313	Al-Maghrah	Matruh
مطروح	شرق سيوة	0	0	30,000	East Siwa	Matruh
جنوب سيناء	الطور	670	0	25,000	Tur	South of Sinai
الوادى الجديد	الفرافرة القديمة	10,000	62,645	41,333	Old Farafra	Elwady Elgded

Discussion and Debates Surrounding the Project

Since much skepticism and criticism arose both socially and politically regarding the possibility and feasibility of such a project, this section illustrates the main points of both the proponents and opponents. This debate has never stopped. In light of this, there is widespread support amongst the population for the implementation of more effective environmental monitoring systems. Success in this will allow the government to avoid the problems of past development on groundwater.

Main points of the proponents

- The project is a must due to the continuous increase in population which imposes the need for more food, houses, job opportunities…etc.
- Mega projects are a very good tool to stimulate the economy and allure foreign investments and this is one of the biggest concerns of all the consecutive governments following the 2011 revolution due to the significant deterioration of the economic status of the country that followed the social and political turbulence and unrest.
- The country should aim at expanding the use of its groundwater reserve to overcome the challenges imposed by a possible decrease of water availability from the Nile due to dry seasons or the construction of storage dams on the Nile's tributaries. The drainage water and renewable groundwater reserve in Egypt are not fully exploited which makes development projects based on groundwater resources possible.
- The government has the scientific capacities and expertise needed to manage the groundwater resources and conserve the strategic national reserves of non-renewable fossil water for uses and ensure national security for the long- and short-term.
- Expanding arable land by desert reclamation has long been used as a method to extend agricultural production in Egypt. and has achieved some successes
- The project aims at developing new communities and villages away from the Nile valley and Delta which has always been a goal for the government to solve the increasing problems caused by the overcrowding around the River Nile.
- The government officials are keen in their planning and vision of the project to avoid all the past mistakes with similar projects and also to guarantee the sustainability of the project which is one of the main elements of assessing the success of the project.

Main points of the opponents

- The project threatens the groundwater reserve of the country because in almost all of the reclamation areas, groundwater is nonrenewable and hence the sustainability of the project is called into question.
- The success of these kinds of projects depends on three main elements: The availability of water, the economic availability of energy and the suitability of

land aimed for cultivation. All of these three elements are not assured for the sustainability and hence the success of the project in the proposed areas.

- Groundwater is nonrenewable and in most areas very high in salinity.
- The great depth of wells in the proposed areas requires a lot of energy and the country is already struggling to provide energy for its already existing developed areas.
- The land is only suitable for certain types of crops which tolerate dry weather and high salinity as most of the proposed areas are desert areas.
- The project lacks any comprehensive study for the economic feasibility and the net beneficial outcome. This situation led to investing too much money in the project before even assessing the economic return to the country.
- From the economic point of view, it could have been better to let investors dig their own wells based on their own investment plans. However, the government has already dug the wells in all of the proposed areas of the project before assigning the land to the beneficiaries or selling it to investors. From the government's point of view, this increases the monetary value of the land and moreover encourages small farmers and investors to participate in the project as the government also provides the infrastructure for all the needed services to the new inhabitants of the developing areas.
- The project has been too rushed. It would have been better to start with a pilot area and based on the full assessment of that area to then progress with other areas.
- The project faces a great challenge because of the harsh and remote nature of the areas assigned for it. The registration of lands owned by Bedouin communities in some areas under the customary system of "Wadaa El-Yad" is another challenge because many of those Bedouin didn't apply or didn't continue the registration process for their lands [34].

Conclusions

We can state that the 4 million feddan reclamation project is indeed a mega project with very high hopes. The country needs such projects to stimulate the economy. The government has a very clear vision of the project and its goals, however, more comprehensive feasibility studies are needed to guarantee success for the long term. As yet, no such studies have been conducted (-to our knowledge) and none is publicly available, nor published in any peer-reviewed international scientific journal.

The government has already invested a lot of money and efforts into the project. However, the challenges are significant and require continuous monitoring, assessment and revision of the implementation steps in addition to the cooperation of all of the factions participating in the project. Some of the main challenges of the project include:

(1) **The harsh nature of the developing areas**. Most of the developing areas are desert areas that are remote from urban areas and lacking roads to reach them, no power supply, no communication coverage…etc. This is a real challenge for the well-digging companies and workers of the project. This condition also maximizes the cost investment required to provide the necessary infrastructure for the future inhabitants of those areas.

(2) **Installation of solar systems and remote automatic control systems**. This kind of technology has never been used in Egypt on such a scale before. Moreover, taking into consideration that the government lacks the experience and qualified personnel to run those systems, there will be further challenges for the government and hence the continuity of the project unless the capacities of the water users are engaged to create solutions.

(3) **Preventing violation of water regulation rules and land in the developing areas**. This challenge might be overcome using the new remote automatic control systems. However, the successful establishment of such systems and the ability to maintain and upgrade them is a must for the continuity and hence the success of the project in the long term. This is likely to require engagement of the water users and intended beneficiaries of the project.

(4) **Building agricultural communities in the newly reclaimed lands**. Taking into consideration the past experience with such projects, this will not be an easy task. The government used its past experience to develop a comprehensive plan regarding the newly formed communities in the developing areas. However, the challenge lies here in converting this beautiful vision into reality and in attracting youth and small investors to participate in this new sustainable model of economic development.

Although the objectives of the project appear relatively similar to previous land reclamation projects in Egypt, the current project has the benefit of all of the lessons and experiences of the past. There are also strengthened groundwater management institutions in place and improved scientific and social media-based methods available for monitoring groundwater conditions to enhance its sustainable management. In addition to this, there are additional capacities available in the Egyptian environmental and economic development sectors. Effective use of the available knowledge, capacities and institutions should enable the necessary studies and adaptive management processes to be supported and applied.

The best available scientific capacities and expertise must be applied to enable the management of the groundwater resources, conserve the strategic national reserves of non-renewable fossil water and ensure national security for the long- and short-term. The establishment of the environmental monitoring system should be a high priority both for the groundwater resource users and for the government. This should be integrated with participatory planning for long-term sustainability, environmental risk assessment and disaster preparedness. At present, we cannot give a full assessment of the project as the implementation is still going on. But we can say that the few coming years will determine the fate of the project.

References

1. Othman E, Ahmed A (2013) Challenges of mega construction projects in developing countries. Org Technol Manag Constr Int J 730–746. https://doi.org/10.5592/otmcj.2013.1.10. https://hrcak.srce.hr/file/154726. Accessed 11 Apr 2019
2. Climate Risk Profile: Egypt (2020) The World Bank Group. https://climateknowledgeportal.worldbank.org/sites/default/files/2021-01/15723-WB_Egypt%2520Country%2520Profile-WEB.pdf&ved=2ahUKEwiOz-DjvsPuAhVjqnEKHVKACfQQFjABegQIARAB&usg=AOvVaw2PrssXnwpxsk1Hk45CbLd6. Last accessed on 30 Jan 2021
3. Regional Center for Renewable Energy and Energy Effciency, RCREEE, Abou-Khodier I, Mahmoud M (2017) Market assessment study of socio-economic impacts of solar pumping systems in terms of local job and value creation in Egypt. Produced by: RECREEE—Implemented by: German Society for International Cooperation (GIZ). https://www.rcreee.org/content/market-assessment-study-socio-economic-impacts-solar-pumping-systems-terms-local-job-and
4. Worldbank (2007) Making the most of scarcity, Appendix 1: water resources data. s.l.: Worldbank
5. Sims D (2015) Egypt's desert dreams: development or disaster? The American University in Cairo Press, Cairo
6. United States Department of Agriculture, USDA (2016) Global Agriculture Information Network (GAIN) report, "Egyptian Land Reclamation Efforts", Date: 16/5/2016
7. Osama S, Elkholy M, Kansoh RM (2017) Optimization of the cropping pattern in Egypt. Alexandria Eng J 56(4):557–566. Alexandria: Faculty of Engineering, Alexandria University. https://doi.org/10.1016/j.aej.2017.04.015
8. Central Agency for Public Mobilization and Statistics, CAPMAS (2017) Annual bulletin of labor force survey. Cairo, Egypt: Central Agency for Public Mobilization
9. Alary V, Aboul-Naga A, Osman MA, Daoud I, Abdelraheem S, Salah E, Juanes X, Bonnet P (2018) Desert land reclamation programs and family land dynamics in the Western Desert of the Nile Delta (Egypt), 1960–2010. World Dev 104:140–153
10. The World Bank (2018) Egypt enabling private investment and commercial financing in infrastructure. Overview and government's objectives. International Bank for Reconstruction and Development/The World Bank Sector
11. Kassim Y, Mahmoud M, Kurdi S, Breisinger C (2018) An agricultural policy review of Egypt: first steps towards a new strategy. MENA RP Working Paper 11. Washington, DC and Cairo, Egypt: International Food Policy Research Institute (IFPRI). http://ebrary.ifpri.org/cdm/ref/collection/p15738coll2/id/132788
12. Isin T, Panos K (2017) Issue paper: Agricultural policies, trade and sustainable development in Egypt. March 2017
13. Ministry of Agriculture and Land Reclamation, MALR (2014) Sustainable agricultural development strategy towards 2030. Cairo: Arab Republic of Egypt
14. Ministry of Water Resources and Irrigation, MWRI[1] (2021) Sustainability of Aquifers, "إستدامة مخزون الخزان الجوفى.pdf". https://www.mwri.gov.eg/index.php/ministry-2/ministry-25/item/27. Last accessed 30 Jan 2021
15. Saleh WSMA (2009) Environmental management of groundwater resources in the Nile delta region. PhD thesis, Faculty of Engineering, Cairo University, Giza, Egypt
16. Cassing J, Nassar S, Siam G, Mousa H (2009) Arab Republic of Egypt. In: Anderson K, William A (eds) Distortions to agricultural incentives in Asia. Masters, pp 71–98. World Bank. http://elibrary.worldbank.org.offcampus.lib.washington.edu/docserver/download/9780821376621.pdf
17. European Union, EU (2018) Directorate-General for External Policies, Policy Department, Study: A stable Egypt for a stable region: socio-economic challenges and prospects. EP/EXPO/B/AFET/2017/06 January 2018—PE 603.858. https://www.readkong.com/page/a-stable-egypt-for-a-stable-region-socio-economic. Accessed 11 Apr 2019

18. Zarad AAEH (2016) The Mega National Projects… A Locomotive of Development, State information services. http://sis.gov.eg/section/337/4683?lang=. Accessed 11 Apr 2019
19. IFAD (2017) President's report proposed loan and grant to the Arab Republic of Egypt for the Promoting Resilience in Desert Environments (PRIDE) Project. In EB 2017/122/R.23/Rev.1, 22. Rome: IFAD
20. Ministry of Water Resources and Irrigation, MWRI[2] (2021) MWRI's vision, "رؤية وزارة الموارد المائية والرى.pdf" . https://www.mwri.gov.eg/index.php/ministry-2/ministry-25/item/28. Last accessed 30 Jan 2021
21. Attia FAR (1989) Use of groundwater for irrigation in the Nile Valley. Water Resources Development 5:91–98
22. King C, Salem B (2012) A socio-ecological investigation of options to manage groundwater degradation in the western desert Egypt. Ambio 41:490–503
23. Villholth KG, Lopez-Gunn E, Conti K, Garrido A, Gun JVD (2017) Advances in groundwater governance, 1st edn. Boca Raton, London, New York, Leiden: CRC Press, pp 621
24. General Authority for Investment and Free Zones, GAFI (2021) One and a half million feddan project. http://mapadmin.investinegypt.gov.eg/English/pages/reef.aspx. Last accessed 30 Jan 2021
25. Ministry of Water Resources and Irrigation, MWRI[3] (2021) Phases of the 1.5 million feddan project. "مراحل مشروع المليون ونصف فدان.pdf","step2.pdf". https://www.mwri.gov.eg/index.php/ministry-2/ministry-25/item/29. Last accessed 30 Jan 2021
26. The Egyptian Countryside Development Company, ECDC (2021) http://www.elreefelmasry.com/. Last accessed 30 Jan 2021
27. Atter H, (2020) Interview on "Hewar khas حوار خاص" TV show, Nile TV channel. https://www.youtube/dmTyA5AeFaA
28. Egyptian Presidency Official Site (2015) Developing and reclaiming 1.5 million feddans in Farafra. https://www.presidency.eg/en/%D8%A7%D9%84%D9%85%D8%B4%D8%A7%D8%B1%D9%8A%D8%B9-%D8%A7%D9%84%D9%82%D9%88%D9%85%D9%8A%D8%A9/%D8%A5%D8%B7%D9%84%D8%A7%D9%82-%D9%85%D8%B4%D8%B1%D9%88%D8%B9-%D8%AA%D9%86%D9%85%D9%8A%D8%A9-%D9%88%D8%A7%D8%B3%D8%AA%D8%B5%D9%84%D8%A7%D8%AD-1-5-%D9%85%D9%84%D9%8A%D9%88%D9%86-%D9%81%D8%AF%D8%A7%D9%86-%D8%A8%D8%A7%D9%84%D9%81%D8%B1%D8%A7%D9%81%D8%B1%D8%A9/. Last accessed 30 Jan 2021
29. Egyptian Commercial Service, ECRG (2021) Berlin office official site. https://www.ecrg.de/images/Download_pdf/National_Project_for_Reclaiming__Developing.pdf&ved=2ahUKEwjxgY2oxcLuAhXIbsAKHR2PC0AQFjAWegQIDhAB&usg=AOvVaw12XzmhGf-o-_gJa1ca3qKo. Last accessed 30 Jan 2021
30. Ministry of water resources and irrigation, MWRI[4] (2018) Groundwater sector, Giza, Cairo
31. Atter H (2018) Interview on ""Hona El Asema هنا العاصمة" TV show, CBC TV channel. https://www.youtube/dCcuIuNSCfA; https://www.youtube/Hm_e57WL7BQ
32. Samenow J, Cappucci M, Washington Post Electronic Site (2020) Fierce 'dragon' storm hits Egypt with flooding rain and damaging winds. Israel braces for impact. https://www.washingtonpost.com/weather/2020/03/12/fierce-dragon-storm-hits-egypt-with-flooding-rain-damaging-winds-israel-braces-impact/. Accessed 30 Jan 2021
33. Salem A, Albawabh News Arabic Electronic Site (2020) "Farmers of 1.5 millhion Feddan in "ElMongra" pleading for help due to floods crises مزارعو ال 1,5 مليون فدان بـ" المغرة" يتسغيثون بسبب "أزمة السيول" , http://www.albawabhnews.com/4037690. Accessed 30 Jan 2021
34. Hanora A (2020) interview on "AlGed'an الجدعان" TV show, AlQahra We El-Nas TV channel. https://www.youtube/BAPxXKisM9k
35. Enterprise Press (2020) ENARA awarded the provision of renewable energy to a landmark national agricultural development project in Egypt. http://enterprise.press/wp-content/uploads/2020/06/20200603_ECDC_pr_en.pdf&ved=2ahUKEwjZi8_qucFuAhXPTRUIHWIYAjgQFjABegQIAhAB&usg=AOvVaw01LlPh9JsLkNZuKwV00ef7. Accessed 30 Jan 2021

36. State information Service Official Site, SIS (2020) Sisi follows up developments of 1.5-million-feddan reclamation project. https://www.sis.gov.eg/Story/153036/Sisi-follows-up-developments-of-1.5-million-feddan-reclamation-project?lang=en-us. Last accessed 30 Jan 2021
37. Central Agency for Public Mobilization and Statistics, CAPMAS (2019) The annual bulletin of lands reclamation. Ref No—71-22128-2018. https://www.capmas.gov.eg/Pages/Publications.aspx?page_id=5104&YearID=23539. Last accessed May 2019

Optimum Economic Uses of Precious Costly Ground Water in Marginal and Desert Lands; Case Study in Egypt

Nader Noureldeen Mohamed

Abstract This chapter explains the optimum economic using the precious, costly and nonrenewable desert ground water in different sectors. Egypt is a country suffering from water scarcity where the water share per capita/year does not exceed 600 m, and the total water shortage reached 42 billion cubic meter/year in year 2018. Thus any new discovered ground water especially deep or spring ones will need economic scientific thinking and wise decision for its uses. The first logical choice for the new ground water should be to reduce the current water gap; but in case of adaptation with this water scarcity; the second choice will be to deliver it into the high income sectors such as hotels, tourisms, industry, and finally agriculture sectors. The municipal and domestic sectors will be also in the focus to meet the demands of the new settlements for the next generation. The least economically feasible choice of using the valuable ground water is to use it in agriculture sectors with its low income, where the return back of using unit of water in industrial sector reached tenfold than agricultural sector. Sometime, the lack of enough foreign currency needed to imports the needed essential food obliged the country to uses desert ground water in producing food. Ground water in Egyptian desert is mostly nonrenewable, deep and costly. The uses of delta and valley renewable shallow ground water in irrigates alluvial soils are completely different than using the desert deep and costly and nonrenewable ground water in cultivates desert lands. The feasibility of using desert ground water in agriculture and growing of agronomy crops, fruit trees and vegetable crops is a waste of a valuable and costly natural resource. Kingdom of Saudi Arabia and Egypt have had negative results for exploiting the ground water in greening desert by planting wheat, barely, forage, vegetables, and fruit crops. In Egypt, most of ground water wells in Cairo-Alexandria desert road farms are exploited and be salinized. Officially, agriculture sectors consumes 62.5 BCM/yr, and shares in Egyptian GDP by only 11.9%, while the industrial sector, consume a little amount of water as low as 2.4 BCM/yr but shares in GDP by 17.1. According to the deep Egyptian water gap, and the valuable of desert water, the uses of ground water in agriculture extension (if necessarily) should be in the north of Egypt where the temperature is moderate, winter rains, high humidity with low water consumptive use. The different between

N. N. Mohamed (✉)
Department of Soil and Water Sciences, Faculty of Agriculture, Cairo University, Giza, Egypt
e-mail: nader.mohamed@agr.cu.edu.eg

A. Negm and A. Elkhouly (eds.), *Groundwater in Egypt's Deserts*, Springer Water,
https://doi.org/10.1007/978-3-030-77622-0_15

the temperature in the North and South Egypt reached 15 centigrade especially in the summer season. Historically all the successes project of desert reclamation located in North Egypt and no any one single success project locates in warm, dry and low humidity South Egypt. In Egypt's desert, the water consumptive use in the warm and dry Upper Egypt is almost double of those in the temperate humid North of Egypt with limited valid of crop types that can be cultivated. Organic agriculture, cash crops, export crops and green house agriculture should be considered as a good investment in desert agriculture using ground water in irrigation to maximize both of net profit and the return back of unite of water.

Keywords Ground water · Desert · Industry sector · Agriculture · Water shortage · Organic farms · Cash crops · Feasibility · South and north Egypt

Introduction and Desert Soils

From the view of physical **geography; desert** is a land area that is hyper arid because it receives little amounts of precipitation; usually in the form of rain but also as well as snow, mist or fog, thus, it has little vegetation, and a low possibility to contain a good non saline ground water in logical depth [1]. **Desert biome** is a tolerant ecosystem that forms due to the low water level of rainfall received each year. **Deserts** cover about 30% of the earth surface. There are four major types of **desert** according to their **biome**—hyper arid and arid, semiarid, coastal, and cold. These low amounts of biomass are able to inhabit animal life that is able to survive there (Fig. 1).

Deserts may have high temperatures, which during the day may exceed 50 °C, but at night it may fall below 0 °C. This wide range between the day and night

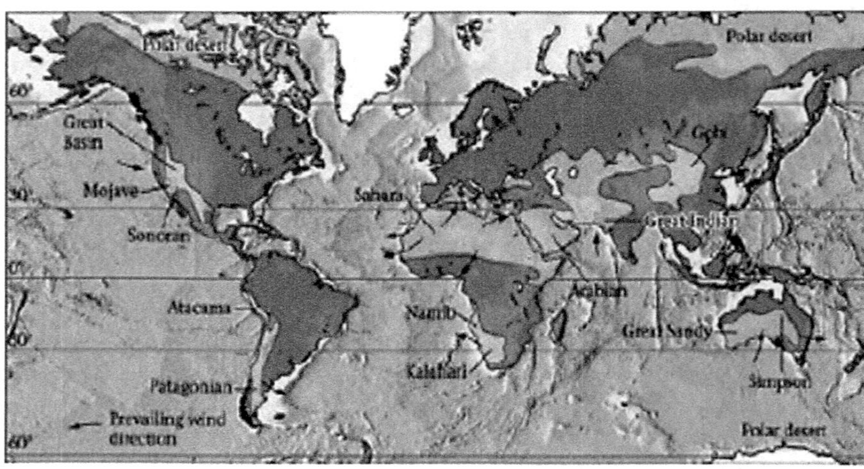

Fig. 1 Desert distribution in the world [1, 2]

temperature may cause a shock to most ordinary plant and only the desert flora can tolerate this wide range of temperatures. Deserts usually received less than 25 mm (10 inch) of rainfall a year, thus the temperature in not important in the definition of desert but the drought, dryness and the lack and water scarcity is the main rule of the desert definition. Some geographical features of the Sahara are **sand dunes**, **dry valleys**, **salt flats and salt crusts**, in addition to several hills and mountain ranges. The Sahara has a number of physical features, such as ergs, regs, hamadas, and oases. Ergs cover 20% of the Sahara [2].

According to [2, 3]**; deserts** are areas that have hardly any rain, are extremely dry and are usually have extreme temperature during day. Animals and plants have adapted to that severe environment and can survive under these extreme conditions of high temperature and little availability of water. The **geographic define the desert as** a "landscape form or region that receives very little precipitation". Generally **deserts** are defined as "areas that receive an average of scatter annual precipitation of less than 25 mm (10 inches)" [3]. Physically deserts are large areas with a lot of bare soil and low vegetation cover. Plants, animals and other organisms that live in deserts have adapted and evolved to survive with these harsh conditions, scarce water and **barren** landscapes. Some desert habitants usually have a short lived cycle to adapt with the season when the rain comes [3].

Desert land surfaces characterize by [1]:

- Lack and deep poor quality ground water.
- Exposed parent rock.
- Clasts accumulation.
- Un-weathered sediment and dominant of primary minerals with almost no secondary minerals.
- Dominance of mechanical weathering only; mostly by temperature and absence of chemical weather.
- High soil salinity and salt crust.
- Mainly wind erosion and little flash flood erosion and sand Mirages.
- Sand dunes and sand shifting.

Marginal Lands

Several definitions have been suggested for the marginal lands but all of them agreed about their low grade soils; soils under unfavorable climatic conditions, has no any feasibility or economic value in in potential production in agriculture which never give over 25% or the regular yield crop and finally it may be contaminated soils (www.miscomar.eu). Any discussion of marginal range lands should consider the impact of marginal soil properties on crop yield; the potential for production to improve soil health; and economically viable end uses of the crop in relation to biomass quality.

Land can be defined as marginal or under-utilized due to a range of factors. These can include:

- Unstable and vulnerable land.
- Economically sub-feasible—due to low returns on their current land use.
- Economically low-optimal—due to physical limitations such as high degradation, low soil fertility, shallow soil and lack of access and good water quality.

Marginal land is "land on which cost-effective food and feed production is not possible under given site conditions and cultivation techniques" [4].

The term marginal land as it is used in the context of the "global land rush' generally refers to land that is arable yet difficult to farm". This is largely determined by biophysical characteristics such as shallow soil profile, temperature, rainfall and topography.

Marginal lands are characterized by extreme climate, poor physical and chemical characteristics to be cultivation. They include areas with limited precipitation, extreme temperatures, low soil quality, and several restriction problems for agriculture [5]. Examples include deserts condition as lack of water area, high mountains, land affected by salinity, waterlogged or marshy land, barren rocky areas, and glacial areas. Evidently, not all of these areas are suitable for agriculture. Various marginal land categories are identified as the following:

- Bare and herbaceous areas (not in use or with only low intensive pastoralism).
- It is not includes lands with intensive and extensive pastoralism.
- Lands with moderate (8–16%) and steep (16–30%) slope.
- Lands with soil problems such as shallow soils (depth < 50 cm)—poorly and weak drained soils—soils with low to moderate fertility—coarse textured—soils with heavy cracking clays—salt-affected soils (Saline–Sodic–Saline sodic and calcareous soils)—soils with gypsic and lime horizons—Acid soils—and peat soils [5].

Degraded Lands

Definitions of Most Degraded Lands and Other Types of Land Associated with Degraded Land

The low quality soils and degraded lands can be concluded in the following types [6]:

Abandoned agricultural land is "land that was previously used for agricultural production or as pasture but that has been abandoned and not converted to forest or urban areas" [6].

Degraded land is "land that has experienced the long-term loss of ecosystem function caused by disturbances from which the system cannot recover unaided" [7].

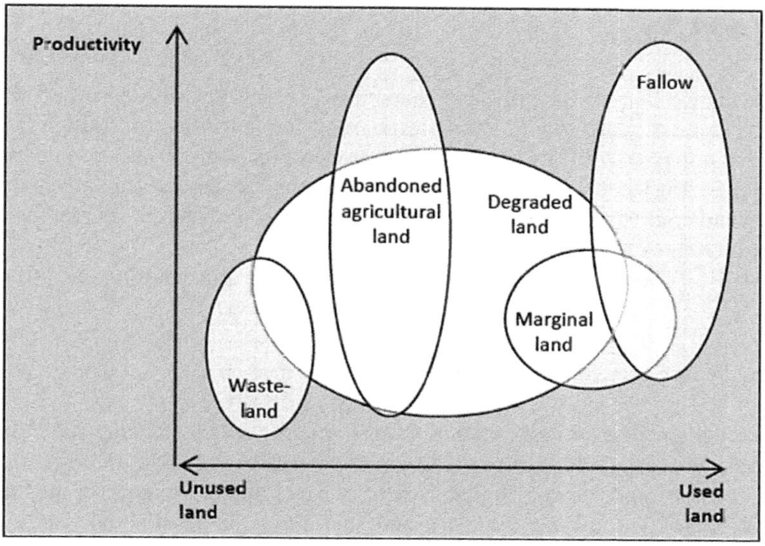

Fig. 2 The different types of lands according to soil quality [6]

Fallow land is "land on which cultivation has been temporarily suspended for one or more vegetation periods to allow recovery of soil fertility".

Low/high-productive land: "spectrum on which land gradually changes from low to high productivity for agriculture and forestry (reclaimed and improved soils)".

Marginal land is "land on which cost-effective food and feed production is not possible under given site conditions and cultivation techniques and never gives more than 25% of the yield crop standard".

Wasteland: is characterized by "natural physical and biological conditions that are unfavorable for land-associated human activities" [8].

The relation between the different soil types is given in the next figure (Fig. 2).

Water Scarcity, Water Shortage and Related Concepts

Before discussing the value of water in a hyper arid country such as Egypt; we will explain the concept of water scarcity to know the value of water as a costly and precious nature resource.

Water Scarcity

"An imbalance between supply and demand of freshwater in a specified domain (country, region, catchment, river basin, etc.) as a result of a high rate of demand compared with available supply, under prevailing institutional arrangements (including price) and infrastructural conditions". Its features are: over demand, tensions and conflicts between users, high competition of different sectors for water, over-extraction of groundwater [9].

In addition, there are more definitions related with water scarcity.

Chronic Water Scarcity

"The level when all freshwater resources available for uses are being used". Beyond this level, water supply for use can only be made available through the use of non-conventional water resources such as agricultural drainage water, treated and non-treated wastewater (industry and sanitation) or desalinated water, or by managing demand and changes agricultural policies. A range between 500 and 1000 m^3/person/year has often been used as a proxy to indicate chronic water scarcity [10].

Absolute Water Scarcity

"Insufficiency of water supply to satisfy aggregate total demand (agriculture, municipal, environment and industry) after all feasible options to enhance supply and manage demand has been implemented". This situation leads to restrictions on water use (such as forbidden of rice, sugar can, forage, and banana cultivation). A threshold of 500 m^3/person/year is often used as a proxy to indicate absolute water scarcity [10].

Water Shortage

"A shortage of water supply of an acceptable quality; low levels of water supply at a given place and a given time relative to design supply levels as a result of insufficient water resources, lack of infrastructure or poorly maintained infrastructure; or low levels of water resources as a result of annual or seasonal differences in precipitation, climate or a range of hydrological or hydro-geological factors" [9, 10].

Water Stress

"The symptoms of water shortage, causing conflict between users, and competition for water, declining standards quality of reliability and service, harvest failures and food insecurity". This term is used to describe a variety of circumstances and causes [9, 11].

Type of Water Scarcity

Three main dimensions of water scarcity can be summarized as follows [9]:

1. "Scarcity in water resources of acceptable quality with respect to total demands", which call "physical water shortage";
2. "Scarcity due to the lack of adequate technical technology and infrastructure, regardless of the level of water resources". This due to financial, technical or other constraints; "technical and economical water scarcity" and
3. Scarcity in access to water services, because of the failure of institutions (including legal rights) in place to ensure reliable, secure and equitable supply of water to users [11].

Some Important Parameters Related to Water Uses in Scarce and Arid Areas

In developing and poor countries especially that located in arid regions, which are usually suffering from various types of water scarcity; there are some parameters that should be considered during their little water managing such as:

Cost of Water

The cost of water relates to the direct expenses to delivering the service of water supply. Full supply cost includes operation, maintenance and replacement costs. The full cost of water to society should include its opportunity cost, and both economic and environmental externalities associated with water supply [12, 13].

Water Pricing

Is the action of establishing a price for a water service. The price can be calculated to cover all or part of the costs of the water service and its delivery, or to induce a change of behavior in water use through less wasteful water use. In irrigation, it can be calculated per area of land, type of crop, or on a volumetric basis. The price assigned to a water service is often called "water tariff".

Therefore in water resource planning, cost-benefit analysis needs to adjust observed prices or estimate prices altogether. These adjusted or estimated prices are commonly referred to as shadow prices [12, 13].

Impacts of Water Pricing

Pricing water will create several problems especially in developing countries and downstream rivers such as:

1. If the downstream river countries sell the river water to their users, the upstream river countries will have a right to obtain a price of their boarder transient water. Then the water will be a good not a service. The downstream country could claims this will be the cost of water delivery not a price of water.
2. The International Rivers such as Nile River will be a private river to upstream countries not a cross boarding natural water resource for all riparian countries; which will create conflicts.
3. The downstream countries may be obliged to pay all its GDP for obtaining their river water share, i.e., if the pricing is similar to the cost of desaline of sea water as the World Bank recommended.
4. No country had chosen its location in the world map, thus no advances to the upstream countries and no any fault or mistakes on downstream countries.
5. This matter will open the door to the upstream countries to sell their water to any other countries out of the river basin or even by shipping or piping for any delivery orders.
6. Water pricing will cause rising of food prices which will make severe problems in the poor's and threaten the UN principal of "Right to food".

Water Productivity

"The quantity or value of output in relation to the volume of water used to produce this output". Crop water productivity is the amount "kg" or "calories" or "values" that product per unit of water supply per cubic meter [12, 13].

Water Situation in Egypt

Officially, according to the ministry of Water Resources and irrigation [14, 15], Egypt is suffering from a deep shortage of water resources where the water shortages reaches 42 billion cubic meter a year (BCM/yr). The total water resources is a sum of 62 BCM/yr (55.5 from the Nile River + 5.5 ground water + 1.3 rainfalls) for 104 citizens which present the total population of Egypt. This population deserves

Table 1 Conventional definitions of levels of water stress [16]

Annual renewable fresh water (m^3/person/yr)	Level of water stress
<500	Absolute water scarcity
500–1000	Chronic water shortage
>1750	Occasional of local water stress

104 BCM/yr: at rate of one southland cubic meter per capita to be above the stress level of water scarcity [16]. Thus the differences between total aggregates demands by the various sectors (domestic, municipal, industry, agriculture) in addition to the water needed for environmental protection, cannot be satisfied by the existing supply of the resources (at a given place and at any moment in time (FAO) [9] reached 42 BCM/yr (Table 1).

Questions About the Water Shortage in Egypt

Under this deep water shortage we need to answer the following questions:

1. When we have a new ground water in desert and marginal lands, what the optimum economic using of this water?! Should we use it to lower the water shortages level?!, or should we use it in some economic activity to increase the GDP and to cover the people demands; such as agriculture extension, industry, hotel, domestic and housing uses to provide new houses with drinking water for the next generations?!
2. What is the priority of using this new ground water in different sectors?! should be agriculture extension, or industry, or municipal and new housing?!. Or it is better to be with economic vision of the highest return back of using unit of water?!
3. Regarding the economic value and the return back from the unit of water, should water shortage countries omit the priority of its agriculture sector because it has the less return back value and change their economic activity to industry and hotel/tourist activity?!, which means losing their food security and rely on importing food with its rapid change and soaring of food price from time to time?!.
4. If the decision maker command in using the new water in agriculture extension in desert lands; where should we apply; warm or cold areas; rainy or dry lands?! And in which crops (cash or regular crops)?!.
5. What agriculture policy should be followed?! Increase the food production and enhance food security or crop for export?! Cash crops of organic farms?! Regular and traditional crops or to translocate some crops suffering from epidemic chronic diseases in the old lands such as potato, onion and tomato to be cultivated in the new pure virgin reclaimed areas?!

Water Use Priority in Egypt

Let us answer the previous questions:

The Priority of Using Ground Water

Groundwater provides drinking water to at least 50% of the global population and accounts for 43% of all water used for irrigation. Groundwater also sustains the base flows of rivers and important aquatic ecosystems [17].

A country suffering from deep water shortage (such as Egypt), where the share water per capita is 600 m and became so narrow from absolute water scarcity, should be thinking seriously on ground water. Moreover; when the desert ground water in this country is limited and unrenewable; this needs more and more thinking about the use of this precious ground water.

The first question may come in the mind of policy makers; is: should water be used to refill or to lower the deep water shortage?! Some researcher believes that; when the people adapt with water scarcity; we should invest this new water in economic sectors to increase the country income and to achieve development progress. Policy maker in arid and poor countries believes that; the person in Africa and Asia doesn't need the amount of 1000 cubic meter/year; as the person who lives in industrial and developed countries in Europe and North America. Thus the people in developing countries may adapt and alive without suffering at a rate of water share of 500 m^3/year. This trend is so theoretical because most of developing countries especially that located in hyper and arid regions rely on the agriculture sector as a main activity on their GDP and this cost them more water consumption than in developed countries. In developing countries the agriculture sector consumes 82% of the total water resources compared with 30% in the industrial developed countries. On vice versa; the industrial sector in developed countries uses 59% of the total water resource compared with 10% in the developing countries [18, 19], as seen in Fig. 3.

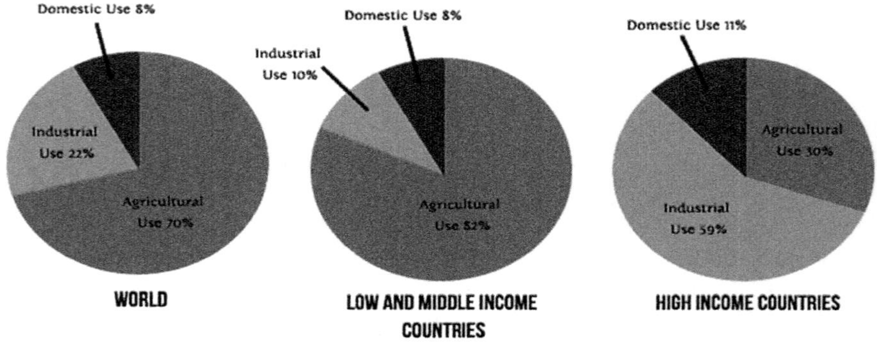

Fig. 3 Competing water uses by sectors in high and low income countries [18, 19]

This data reflects that the return back of the water unite in the industry sector is higher than that of the agriculture sectors. Moreover, some report stated that only 2–6% of the total population employment in developed countries works in agriculture sector and still they are able to feed their people and make sufficiency for export. On the contrary, 33–66% of the population's labors work in the agriculture sector in developing country but they are not able to feed their people and they suffering from food insecurity [20]. For example, Egypt imports 65% of their total essential food which costs Egypt 15 billion dollars/year, with 33% of agriculture employment; and sharing in GDP by only 11.9% [21]. This data reflects the advanced techniques applied in agriculture sector in developed countries compared with those that are primary and out of time ones applied in developing countries and the urgent needs for modernization in this sector [22]. Tables 2 and 3 shows the low rate of using chemical fertilizers and agriculture machines in the Arab Countries compared with world rate and the low production of agriculture labor production in Arab world due to relying on hand work labors not on machine. Figures 4 and 5 explains the employment in Agriculture sectors in developed and developing countries and the low productivity of cereal in Arab countries.

Concluding remarks

According to the Fig. 7 and Table 4, Agriculture sectors in Egypt consume 62.5 BCM/yr, and share in Egyptian GDP by only 11.9%, while the industrial sector,

Table 2 Some agriculture development indicators in the Arab world and world [22]

Item	Unit	Arab world 2013	World 2013
Food standard production	%	118.8	122.9
Rate of using chemical fertilizers	Kg/ha	82.4	141.3
Rate of using machine	Tractor/1000 ha	11	20
Cereal production	Ton/ha	1.74	3.85
Rate of rural/urban	%	41.6	47

Table 3 Productivity of labor in agriculture sector in Arab and developed countries [24]

Country	Agric. labor productivity per dollar (2003–2005)	Cereal production (kg/ha)	Average of chemical fertilizers (kg/ha)
Egypt	497	7545	572
Syria	1196	1786	73
Morocco	719	1243	52
Sudan	371	650	4
Saudi Arabia	5523	4559	99
Holland	23,396	8309	564
United State	23,066	6443	114

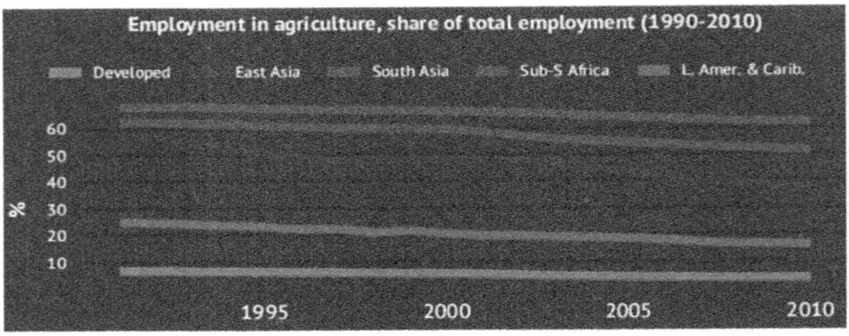

Fig. 4 Employment in agriculture, share of total employment (1990–2010) [20]

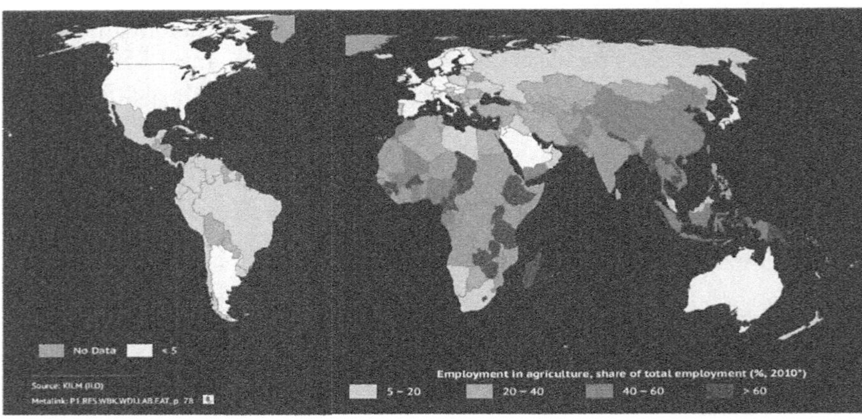

Fig. 5 Employment in agriculture, share of total employment (1990–2010) [20]

consumes a little water as low as 2.4 BCM/yr, but shares in GDP by 17.1. Moreover, we should add the sales of retail trade to the industry sector because all of their products are manufacturing and belongs to the industrial sector, then the total share of industrial sector in GDP will be 31.1%, from only 2.4 BCM/yr of water. In addition and according to Figs. 4, 5, 6 and Tables 3, 4, Agriculture sector in Egypt neither modern nor smart which lead to heavy intensive labor work with low usage of agriculture machines and chemical fertilizers with low production with high cost and low rate of labor production.

In this case industrial sector should have the first priority in using this valuable and costly ground water (in the water shortage and water scarcity countries) to enhance their weak economy and help in eliminate hunger and reduce poverty. This recommendation is also so fitted to Egypt and North Africa countries (Fig. 7).

Table 4 Water demands for different sectors in Egypt at different scenarios [26]

Water demands (BCM/yr)	Current scenario (BCM/yr) (2015)	2025 Normal scenario	2025 Ambitious scenario	2015 Extra scenario
Agriculture sector	62.5	66.6	33.2	33.2
Deep ground water consumes	9	7.4	5.9	5.9
750,000 Feddan reclamation project	–	5.1	4.1	4.1
Domestic sector	9.9	11.2	10	10
Industrial sector	2.4	4	3.3	3.3
Total	80.8	94.2	56.6	56.6

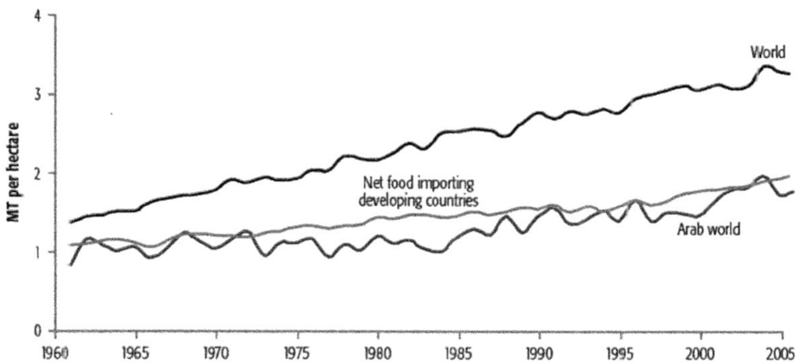

Fig. 6 Low cereal productivity in Arab states compared with world [23]

Using Ground Water in Desert Cultivation

In the scarce water countries especially those in warm and dry weather, the use of desert ground water should be carefully and according to deep scientific background. These countries should address (locate) firstly its needs for increase food shortage under the lack of their foreign currency needed to imports it. In case of adapting people with water scarcity as well as the possibility of attracting new investments to construct industrial factories should be put under consideration. This will be controlled according to its natural resource and the presence of some important raw and crude materials. The national natural resources (blue economy) needed for industry extension such as lime stone for cement production, iron hydro-oxides for steel industry, mud and some clay minerals and silica for ceramic industry, Bauxite for aluminum industry, and enough electricity power for all these factories in addition to the building of new chemical fertilizers factories and so on.

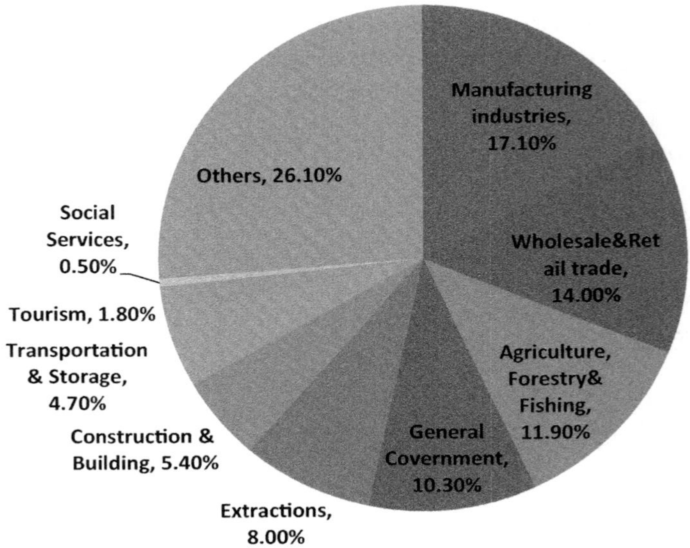

Fig. 7 GDP breakdown in Egypt by economic activity [25]

If one single country is suffering from economic and technical water scarcity (Scarcity due to the lack of adequate technical technology and infrastructure, regardless of the level of water resources [9]), then the use of their water in agriculture will be the easy way because the agricultural is usually the job of the poor's especially in the primitive and out of time agriculture (African countries).

Country with physical water scarcity (Scarcity in water resources of acceptable quality with respect to total demands) should think out of box (not out of scientific logic) and forget completely the revolution emblem or logo to create a fever in their follower people. For example, if one country such as Egypt, is suffering from a deep water gap that reached 42 BCM/yr, reduced it to be 22 BCM/yr by reusing of all types of wastewater. This country should not be planes for reclaiming four million acre (1.7 million hectare) in their warm tropical dry desert lands. The priority of this country and in the current time should be to use this water in industrial or housing and hotel (resorts) sectors. But, if they succeed on raising the efficiency of water distribution (delivery) in the irrigation canals, and also raise the irrigation efficiency inside the fields, that may save 30% of the total water resources (almost of 20 BCM) especially if the country relies on surface open (un-cemented, un-piped) irrigation and drainage canals, then in this case, the water saved may permit the country to use the new ground water in desert cultivations, but In which desert land we should start?!

Specifying the Location of Agricultural Extension

Egypt consists of three main climatic zones as seen in Fig. 8.

Delta and North Egypt (Lower Egypt)

North Egypt includes all Delta districts (8 governorates holding 55% of total population), the north cost of Mediterranean (Alexandria and Matrouh Governorate in the west and North and Sinai in the East), and three Suez Canal Governorates; Port Said, Ismailia and Suez). This area belongs to Mediterranean climate zone which has temperate weather, high humidity especially in the summer time, low evapotranspiration rate, and appreciable winter precipitation, which allow cultivating miscellaneously and large numbers of economic crops (vegetable, fruits, and field crops). This area has a good advantage for low water consumptive uses of all crops and high yield in addition to having a cold rainy winter that helps to save much water. This zone has a good advantage of the narrow differences between night and day temperatures and no frost or sharp decreases in temperature in the early morning

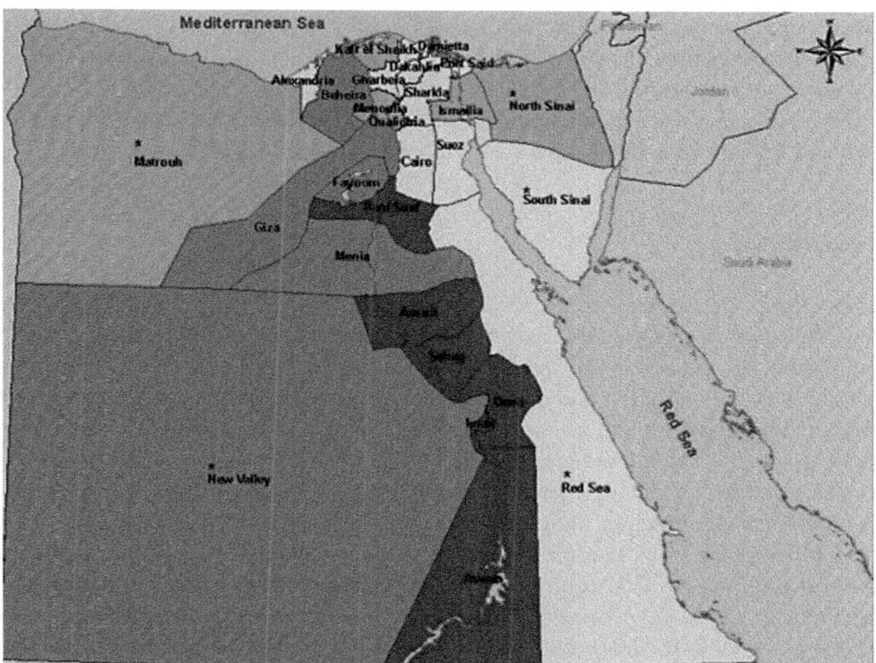

Fig. 8 Maps of Governorates in Egypt [27]

during winter season and never having frost that hurt the plant leafs so much, as it frequently happens in middle Egypt.

Middle Egypt

The middle Egypt is located south to the capital Cairo and includes another 6 Governorate in the upper Nile Valley. These governorates are: Giza, Fayoum, Beni-swef, Minia, Assuit, Sohag in addition to the north part of Red Sea Governorate in the east, South Sinai. Moreover, it includes the north part of the largest desert Governorate area in the western desert which is "New Valley or Wadi Jadid" that occupies 44% of the total Egypt area. In the Middle Egypt zone, the temperature starts to increase by 2° to 5° than the Delta and North Egypt, the humidity is lower and behaves in a reversed way to the Delta where it decreases in the summer and increases in the winter, but never exceeds 50% while it may be saturated as high as 90% in the Delta. The weather in the Middle Egypt almost has no rain, but some of their parts exposed to extreme flushing floods (heavy rains in short time) causing severe hazards even in the Nile Valley or south Sinai and Red Sea Governorates, as seen in Fig. 8. Middle Egypt is characterized by sharp decreases in night temperature especially in the winter months (December to March) which may become as low as below zero degrees with frost formations that may cover and harm the growing plants. Thus, Middle Egypt has higher evapotranspiration and water consumption than north Egypt due to high temperature, low humidity and high dryness. Middle Egypt weather allows for cultivation of all crops that are cultivated in North Egypt such as sugar beet, alfalfa, beans, with the first appearance of sugar cane cultivation in the Mania Governorate.

Upper Egypt

Upper Egypt means the extreme south of Egypt that includes, Qina, Luxor, Aswan, the south part of Wadi Jadid that includes Toshka and Sharq El-Oainate (east of the Oainate mountain) in the west, and the south part of the Red Sea Governorate and the triangle of Halieb and Shalateen (Fig. 9). Upper Egypt records the highest temperature, as seen in Fig. 10 may become 14° more than the North Coast, as seen in the temperature map in the summer, with high aridity, zero rainfall in addition to the lowest humidity percent in Egypt. Moreover, the aridity in Upper Egypt also increases in winter and decreases in summer which leads to high water consumptive uses by plants. Under this condition the evapotranspiration and water consumptive uses are the highest in Egypt. The numbers of different kinds of plants that can tolerate the high temperatures in the Upper Egypt are so limited and need high water demands. The sugar cane is the main field crop in Upper Egypt with some of aromatic and medicine halophyte crops such as hibiscus and some spicery crops. Table 5 shows the comparative water consumptive uses on different crops in north, middle and south Egypt.

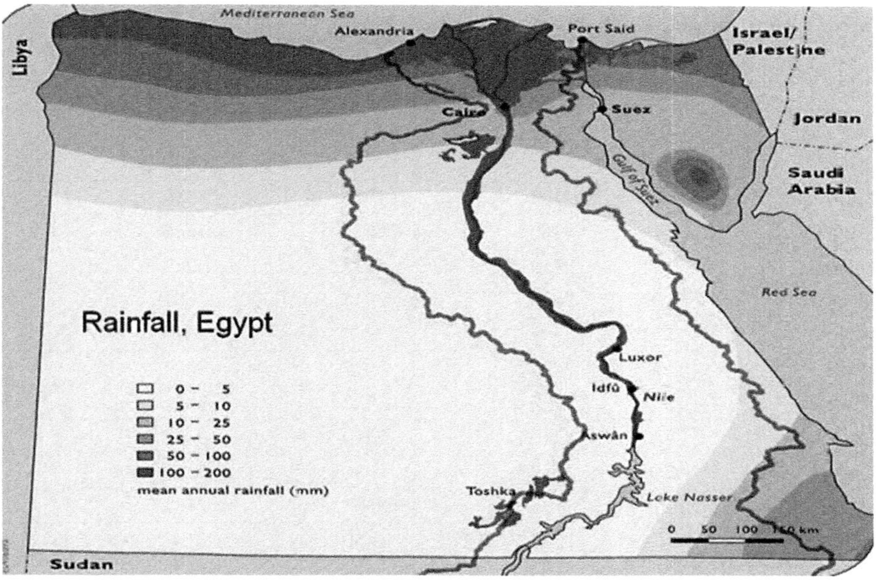

Fig. 9 Rainfall distribution in Egypt [28]

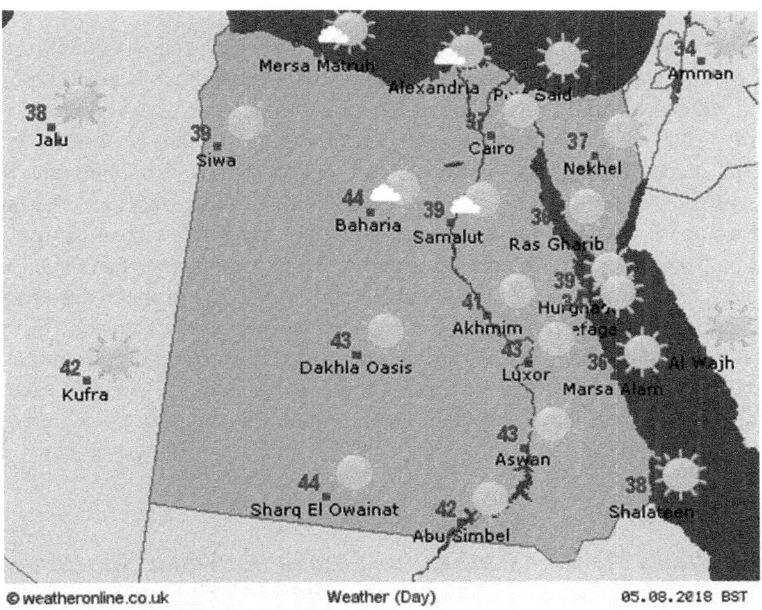

Fig. 10 Example for maximum temperature in upper, middle and lower Egypt (august, 4th, 2018) [29]

Table 5 Water consumption by crop type and region, Egypt [30]

Crop	Water consumptive by region (m^3/acre)			
	Delta	Middle Egypt	Upper Egypt	National average
Sugar cane	–	10,000	12,750	11,375
Rice	6470	–	–	6470
Fruits	5610	5798	7522	6310
Cotton	2645	4960	5434	4346
Peanuts	1162	2752	5738	3217
Maize	2400	2657	3947	3001
Wheat	2152	2795	3072	2673
Malts	1670	2520	3015	2402
Summer veg	1695	2138	3280	2371
Onions	2280	2410	2410	2367
Sesame	1866	2152	2630	2216
Beans	1790	2195	2558	2181
Winter veg	1604	2250	2450	2103
Lentil	1865	2105	2263	2078
Average of all	2555	2672	3409	

From the last discussion, and according to the deep Egyptian water shortage, the uses of ground water in agriculture extension should be in the north of Egypt where the temperature is moderate with low water consumptive use. Even when we follow the history of land reclamation projects and greening the desert projects we find that all the successful projects are located in Lower Egypt (such as: Nubaria, Salhia, Moderiat El-tahrer south Alexandria, Ramses, Wadi El-Natroun, Cairo Alexandria, Cairo Faioum, Cairo Ismailia desert roads, …) no one single successful project is located in Upper Egypt (complete familiar of Toshka project so far). Thus, Egypt should concentrate on having the land reclamation projects in the north desert and delta Frings, and use Upper Egypt for industry projects and extension. Egypt has several raw materials in Upper Egypt that are needed for different types of manufactories such as lime stone for cement, muds and clay minerals for ceramic, electricity and phosphates for chemical fertilizers, high quality fine sand for solar mirrors and lenses and many others.

Egypt Reduces the Rice Cultivation Due to the Water Shortages

The rice cultivation becomes the only alternative choice of the Nile floods which stopped after the construction of Aswan High Dam that ended on 1971. After this period, the salinity building up in the Delta lands continues according to the sea water intrusion to the shallow water table (water logging) and the active capillary rises in the clayey soil of the Delta. The absence river floods after High Dam, and the absence of leaching of the delta soils due to the lack of water results in; that the rice cultivation becomes the only valid and guaranteed way to leach out salts from some of the delta soils every year especially those located close to the Mediterranean coast, the main source of salinity. Delta lands are the capital of agriculture production in Egypt because it contains 72% of the total alluvial soils and produce 62% of agriculture production in addition to it having 55% of the Egyptian population [31]. The soil salinity in the north part of the delta reaches 46% of the total delta area, and in the middle delta 37 and 24% in the south delta with average of 37% for all the total area of the Delta which reaches 4.5 million acres [31]. Thus, the average area of saline soils in the Delta is 1.7 million acre which should be cultivated by rice every year to control the salinity building up and to save the soil's health and fertility. The rice cultivation is the only economic crop and at the same time is the reclamation crop that protects the delta lands from degradation.

As a result of the Ethiopian declaration about the beginning of the first fill of the lake of the Great Ethiopian Renaissance Dam (GERD), on June 2018; The Egyptian Government obliged a new policy for rice cultivation and reduced its regulate area to be only 700,000 instead of 1.5 million acre during the last 20 years.

For wondering, the agriculture policy recommended to increase the investment in the field of red meat production from cattle investment, cotton cultivation (long period crop needs 9 months growth period), forage crops such as alfalfa (Egyptian Beseem with 5 cuts), which all have high water consumption than rice; with no touching of other high water consumptive use crops such as sugar cane, banana and other wide leaf vegetable; just limited the rice cultivation. Moreover, the policy of greening the desert is still as it is and is recommended to reclaim 4 million acre (1.7 million ha) in the desert lands which needs at least 20 billion m^3/yr at a rate of 5000 m per acre!! No talk also about the possibility of getting usefulness from this the new discovered ground water to reduce the gap of water shortage in Egypt by delivering it to the domestic use or to the Nile and its canals.

Egypt Agricultural policy needs urgent reconstruction and to have new policies more suitable for our limited water resources and put good priorities for using every drop of water.

Conclusion and Recommendation

In countries that are suffering from water shortage especially in the arid and warm regions, strategy of priority of using ground water should be planned. The return back of the unit of water in industrial sector is ranged between 10 and 300-fold than the agriculture sector. The industry sector uses a little of water which does not exceed 10% of the total water resources in the developing countries, but share in the GDP of the country 10 times more than the agriculture sectors which consume 85% of the total water resources. The labor income in the industrial sectors is higher as 2–10 times that of the labor income in agriculture sector. Under specific conditions in some developing countries that are obliged to use the new discovered ground water in agriculture to reduce the food insecurity and the lack of foreign currency; scientific and logical policy should be followed. The priority of using ground water in agriculture should be to the low temperature and seasonal rainfall area. Thus the new greening desert and agriculture extension projects in Egypt should be in the North and the Mediterranean coast of North Egypt which have a low temperature than south Egypt by 14 centigrade. The south, warm and low humidity region should have a priority for industry investments to reduce poverty and maximize the return back of water unite.

References

1. CE/SC 10110-2011: Planet earth: earth portrait of a planet, 5th edn. https://www3.nd.edu/~cneal/PlanetEarth/Chapt-21-Marshak.pdf. Accessed on 1st June 2018
2. Walker AS, U.S. Department of the Interior/U.S 1992. Geological survey. Deserts: geology and resource U.S. government printing office: 1992 0-305-515. https://pubs.usgs.gov/gip/7000004/report.pdf. Accessed 1st June 2018
3. Hillman N, Weichman T (2016) Education deserts: the continued significance of "Place" in the Twenty-First Century. Viewpoints: voices from the field. American Council on Education, Washington, DC
4. Cai X, Zhang X, Wang D (2010) Land availability for biofuel production. Environ Sci Technol 2011(45):334–339
5. Energy Working Group August (2009) Assessment of biomass resources from marginal lands in APEC economies. In: Asia-Pacific Economic Cooperation (APEC) Energy Working Group under EWG 11/2008A
6. Wiegmann K, Hennenberg KJ, Fritsche UR, (2008) Degraded land and sustainable feedstock production–Issue paper. In: Joint international workshop on high nature value criteria and potential for sustainable use of degraded lands. Paris, June 30–July 1. Öko-Institute, Darmstadt http://www.unep.fr/energy/activities/mapping/pdf/degraded.pdf. Retrieved 25 June 2009
7. UNDP (2007) The miracle on saline land: "Forest-pulp-paper integration" strategy of Huatai. United Nations Development Programme, New York. http://www.bdsknowledge.org/dyn/bds/docs/735/China_Huatai%20FINAL.pdf. Retrieved 13 Sep 2010
8. Oldeman LR, Hakkeling RTA, Sombroek WG (1991) World map of the status of human-induced soil degradation—an explanatory note—global assessment of soil degradation, GLASOD. International Soil Reference and Information Centre, Wageningen, United Nations Environmental Programme, Nairobi. http://www.isric.org/isric/webdocs/Docs/ExplanNote.pdf. Retrieved 24 Oct 2008
9. FAO water report number 38 (2012) Coping with water scarcity: an action framework for agriculture and food security. ISBN 978-92-5-107304-9
10. Falkenmark M (1989) The massive water scarcity threatening Africa—why isn't it being addressed? Ambio 18(2):112–118
11. International Bank for Reconstruction and Development (2017) UNCHARTED WATERS: the new economics of water scarcity and variability. ISBN (print): 978-1-4648-1179-1; ISBN (electronic): 978-1-4648-1180-7
12. FAO (2004) Water charging in irrigated agriculture. An analysis of international experience. Cornish G, Bosworth B, Perry C, Burke J. FAO Water Reports, no. 28
13. GWP [Global Water partnership] (2000) Integrated water resources management. TAC background paper, No. 4. Technical advisory committee. Stockholm, Global Water Partnership. p 67. http://www.gwp.org/Global/GWP-CACENA_Files/en/pdf/tec04.pdf. Accessed 4 July 2012

14. Ministry of Water Resources and Irrigation, Egypt (2019) Water scarcity in Egypt: the urgent need for regional cooperation among the Nile Basin countries. http://www.mfa.gov.eg/SiteCollectionDocuments/Egypt%20Water%20Resources%20Paper_2014.pdf. Accessed 7 Nov 2016
15. Ministry of Agriculture and Land Reclamation (2020) The reclamation project to reclaim 1.5 million feddan. https://olx.com.eg/ad/130-ID6TZBk.html;https://www.facebook.com/ezrmisr/photos/a.706234316197524.1073741828.466575000163458/706234302864192/?type%BC1%26theater
16. Falkenmark M, Widstrand C (1992) Population and water resources: a delicate balance. Population Bulletin. Population Reference Bureau, Washington, USA
17. FAO (2010) Statistic on groundwater use for agricultural irrigation, FAO: AQUASTAT—FAO's global information system on water and agriculture. FAO, Rome. http://www.fao.org/nr/aquastat
18. UN World Water Development www.unesco.org/water/wwap/wwdr/index.shtml
19. Enviro-supply (2012) Water usage by industries worldwide. https://envirosupply.net/blog/water-usage-by-industries-worldwide-interesting-facts-and-figures/. Accessed 26th July 2018
20. FAO (2015) Agriculture employment. http://www.fao.org/docrep/015/i2490e/i2490e01b.pdf Accessed 28th July 2018
21. The Egyptian Center for Economic Studies (ECES) (2017) Egypt's economic profile and statistics. http://www.eces.org.eg/MediaFiles/Uploaded_Files/e7b0ad3f.pdf. Accessed 28th July 2018
22. Arab Organization for Agriculture development (2014) Annual report for Arab agricultural statistics, vol 34. http://aoad.org
23. World Bank, FAO, IFAD (2009) Improving food security in Arab countries. The World Bank, Washington, DC
24. Arab Organization for Agriculture development report (2010) Arab Food Security outlook. http://aoad.org
25. Bank Audi's Group Research Department, Central Bank of Egypt (2018) Egypt Economic Report. http://www.bankaudigroup.com/GroupWebsite/openAudiFile1.aspx?id=3329. Acceessed 8th March 2018
26. Mohie El Din MO, Ahmed MAM (2016) Water management in Egypt for facing the future challenges. J Adv Res 7:403–412. https://ac.els-cdn.com/S2090123216000229/1-s2.0-S2090123216000229-main.pdf?_tid=4debc6ae-6fc7-472c-860c-2bf6a062f2c4&acdnat=1533037862_a68a011369a6bc5dc9b4ef8c0330ddee. Accessed 31st July 2018
27. Earth Sciences beta (2018) https://earthscience.stackexchange.com/questions/8518/why-is-egypt-so-dry
28. Mohamed HI (2007). Egypt's water resources management under pressure of development. In: First engineering conference for young researchers, engineering between theory and practice. Assiut University. https://www.researchgate.net/publication/269929103_EGYPT'S_WATER_RESOURCES_MANAGEMENT_UNDER_PRESSURE_OF_DEVELOPMENT. Accessed 5th Aug 2018
29. Weather on line (2020) https://www.weatheronline.co.uk/Egypt.htm
30. Ministry of Water Resources (2019) Irrigation, planning sector report
31. Medany M, Attaher S, Abou-Hadid AF (2010) Land-use change and adaptation in the Nile Delta region, regional assessment of the FP6 project of adaptation of agriculture in European regions at environmental risk under climate change (Adagio)

Assessment of Water Resources in Egypt: Current Status and Future Plan

Mohamed Elkholy

Abstract Water resources are nowadays one of the most important issues of common concern among the countries of the world. Studies have confirmed that the world is facing new risks and challenges unless water is used rationally and effectively. The danger lies in the continuous decline in the per capita average of pure water, particularly in Egypt, where it has been affected by the scarcity of water, which threatens its water security, especially with the continuous increase in population, high standard of living and the rapid development of the industrial activities. Water balance deficit has become noticeable. The gap between resources and needs in 2019 for Egypt is estimated at 21 billion cubic meters per year. In addition, the quantities of water needed to provide food security through food imports in the form of virtual water are estimated at 34 billion m^3. Recent studies and results indicate that by 2050, this gap will increase more, prompting the institutions at various levels to use non-conventional water sources on the widest scale. This chapter presents the available water resources and required demands to help the government to take urgent actions to face challenges concerning water deficiency. The bright side is that the government has already started implementing short-, medium- and long-term plans to face water challenges until 2050 as a result of increased demands. Different scenarios for the water budgets for the year 2025 and 2050 are presented from different studies using different assumptions taking into account the effect of population growth, climate change, upper Nile projects, crop water consumptions, irrigation efficiencies, expansion of the industrial activities, and losses in the water networks considering all components of conventional water resources (Nile water share, deep groundwater, rainfall, and desalination) and non-conventional water resources (shallow groundwater, reuse of agriculture water and wastewater).

Keywords Water Budget · Water Demands · Water Resources · Water deficit · Groundwater · Conventional · Non-conventional · Agriculture · Domestic · Reuse · Wastewater

---

M. Elkholy (✉)
Irrigation Engineering and Hydraulics Department, Faculty of Engineering, Alexandria University, Alexandria, Egypt
e-mail: mohamed.elkholy@alexu.edu.eg

A. Negm and A. Elkhouly (eds.), *Groundwater in Egypt's Deserts*, Springer Water,
https://doi.org/10.1007/978-3-030-77622-0_16

Introduction

Water resources management in Egypt is considered a strategic priority. The planning of water resource is not considered as a simple closed negotiation between experts in the government, but various stakeholders should be included in the dialogue process, cultural minorities and user communities that are often excluded from such dialogues.

The water use of the Nile is utilized for irrigation, water supply, hydro-electric power production, and protection of public health. In the upper basin states like Democratic Republic of Congo, Rwanda, Burundi, Ethiopia, Eretria, Kenya, Uganda, Tanzania, and South Sudan, agriculture is considered one of the main economic activity. However, in the lower basin states like Egypt and Sudan who are also relying on agriculture, but in contrast with the upper states, their agriculture is largely irrigation-based. Egypt is unique among the world nations due to its extraordinary dependence upon almost a single source of water, which is the Nile River. Egyptians have been managing the Nile water for irrigation purposes through a series of hydraulic structures like dams and barrages. In southern Egypt, Aswan High Dam is controlling the water passage while a couple of barrages are located to the north as Zefta and Edfina Barrages. Except a few scattered oases in the desert, Egypt is known for its rich agricultural land consisting of 1,200 km strip along the river Nile and ending with the Nile Delta.

Agriculture has been for years the main contributor to economic growth in Egypt, contributing about 20% to Growth Domestic Product (GDP) and contributes close to 40% of the Egyptian total employment [1]. Around 55% of the population are dependent on this sector for their livelihood and agriculture needs. As mentioned in [2], the rural society's population, who represent the large portion of Egyptians depending on this sector, reached 44.88 million in 2010 (57% of the total population in Egypt).

The most critical constraint facing Egypt is the shortage of water resources that grows with a very high rate accompanied by the deterioration in the water quality and the population growth which reached 98.8 million in 2019 compared to 33.9 million in 1970 and is expected to grow by 54.9% from 2010 to 2030 [3]. According to the United Nation [4], Egypt falls in the category of high-water stress countries, since the amount of water withdrawal compared to the available water reaches almost 40%. Moreover, nowadays, this figure is even much worse. This is because of the Nile acting as the only surface source of renewable freshwater in Egypt, other than very small amounts of rainfall on the coast and some flashfloods in the eastern desert and Sinai Peninsula and a little contribution from the groundwater.

Water Resources

Water resources in Egypt may be divided into two main forms:

- Conventional resources: comprising the Nile river water, deep groundwater, rainfall and flash floods, and desalination.
- Non-conventional resources: comprising the shallow groundwater in the Nile valley and delta, reusing of agricultural drainage water and treatment of wastewater since they are not considered as independent water resources and cannot explicitly be added to the share of the fresh water in Egypt.

Conventional Water Resources

a. Nile River Water

The River Nile with its two major distributaries, the Rosetta and Damietta and its five Rayahs (El-Tawfeki, El-Nasseri, El-Behairy, El-Mounofi, and El-Abassi) that divert water to intensive networks of canals through several control structures, supplies about 97% of the annual renewable water resources in Egypt.

Eleven countries are sharing the water of the Nile River, which is considered as the primary source for Egypt. The first Water Agreement for the Nile River between Egypt and Sudan was signed in 1929 where 48 Billion m^3/year (BCM/yr) is allocated to Egypt, and 4 BCM/yr is given for Sudan. The remaining could thereby flow unused to the Mediterranean Sea. However, Ethiopia was not given in this agreement any rights to use the Nile waters and also still binds Kenya, the United Republic of Tanzania and Uganda and even bans them from using the water supply of Lake Victoria. During the cotton peak demand from May to July, these latter countries were not allowed to extract any Nile water in order to secure it for this important Egyptian crop [5]. Then, in the year 1959, the water agreement between Egypt and Sudan assigned to Egypt 55.5 BCM/yr, which is measured at Aswan near the border of Sudan. The agreement was based on the average flow of the Nile during the 1900–1959 period, which was 84 BCM/yr and still in use, minus evaporation from the artificial Lake Nasser. The average annual evaporation and other losses were estimated at 10 BCM/yr from the Aswan High Dam and reservoir (Lake Nasser) leaving thus a net flow of 74 BCM/yr, of which 55.5 BCM/yr was allocated to Egypt and 18.5 BCM/yr to Sudan (Fig. 1).

About, 84% of the Egyptian Nile water is contributed by Ethiopia, and 16% comes from the Equatorial Plateaus of Central Africa (Victoria, Albert, and Kyoga lakes).

b. Deep Groundwater

Deep Groundwater (up to 1500 m in some areas) are mostly located in the Western Desert (WD) and Sinai and are non-renewable which make it hard to utilize since the cost of pumping will be high [1]. These deep storages in the Western Desert are estimated to be 40,000 km^3 with a salinity of 400–700 ppm as illustrated by [13].

Figure 2 shows the amount of abstraction from the deep groundwater (GW)

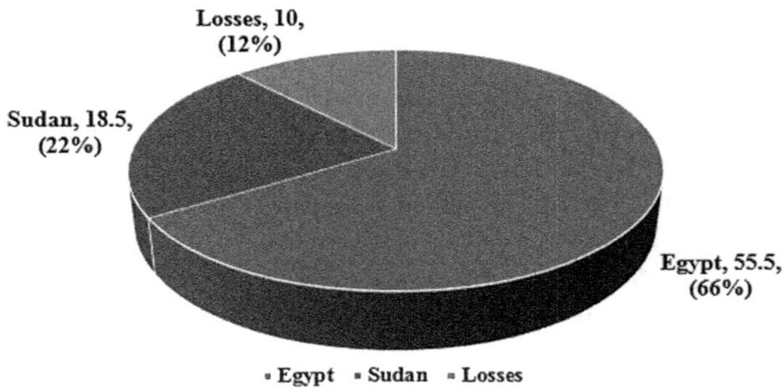

Fig. 1 Nile water share according to the 1959 Nile waters agreement between Egypt and Sudan

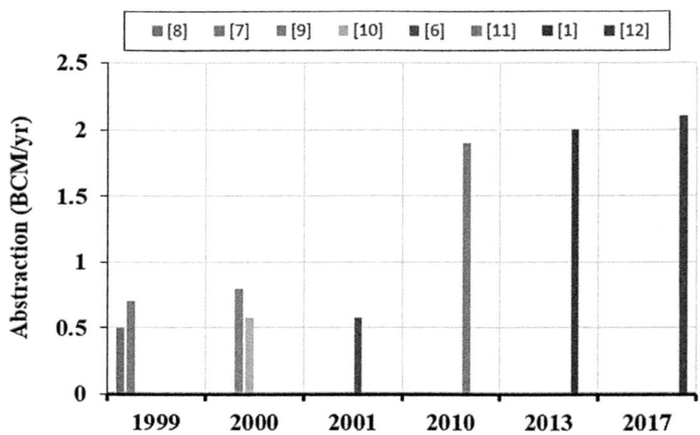

Fig. 2 Total abstractions from deep groundwater (BCM/yr)

according to different studies. From the figure, the abstraction from deep groundwater has increased over the recent years due to irrigation strategies and reached almost 2.1 BCM/yr in 2017 according to [12].

c. Rainfall and Flash Floods

Rainfall is very scarce both temporally and spatially in Egypt and cannot be considered as a dependable source of water, except in a very narrow region along the coastal areas in the northern parts of the country. Scarce flash floods (torrents) also occur in the eastern dessert, Sinai Peninsula, and Upper Egypt [8]. The average range of precipitation in winter season over the northern coasts is around 200 mm (150 mm/yr is estimated in Alexandria and around 75 mm in Port-Said) decreasing gradually to the South to be around 25 mm in Cairo [8]. The water available is merely enough for pastoral purposes. Mostly it is used for cultivating some olives and figs in the

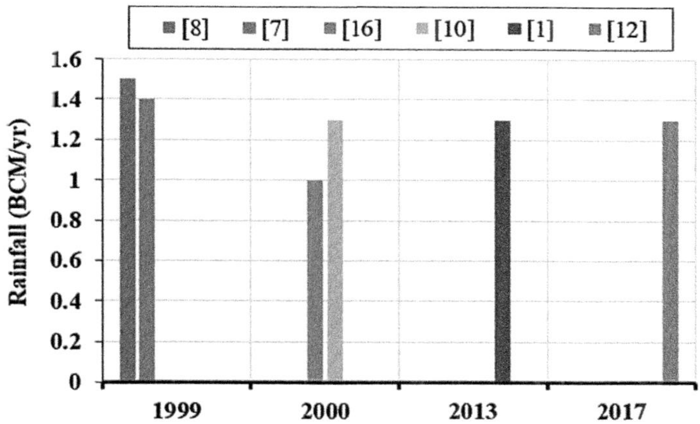

Fig. 3 Total amount of rainfall (BCM/yr)

western zones and peaches in the eastern zones [14]. Torrents (heavy short-term rains or flash floods), may be encountered once every three years [10], have a destructive environmental effect in the Red Sea and South Sinai. However recently, attentions are made for collecting this water and hence recharging the groundwater table. From Fig. 3, the average amount of rainfall is about 1.3 BCM/yr. The Academy of Scientific Research and Technology held the first scientific symposium on the maximum utilization of rainwater in 1994, which showed that there is a possibility of harvesting rainwater and floods estimated at about 2 BCM annually [15].

d. **Desalination**

In some remote areas, along the Mediterranean Sea, the Red Sea coasts and in Sinai Peninsula, desalination has been used to provide domestic water supply for some locations [8]. It was given less priority as the cost of treatment is very high (about 0.5–2 USD/m^3 [13] compared to other resources. The use of desalinated water may be crucial in the future since the water demands are expanding and are expected to exceed all other available water resources. However, the technological advance in this field and its associated cost will determine its future [1]. According to [11], the estimated amount of desalinated water is about 0.2 BCM/yr. The Minister of Housing, Utilities and Urban Communities, Dr. Assem Al-Jazzar, stated in March 2019 that the total capacity of seawater desalination plants is expected to reach 1.7 million m^3/day by 2020, about 35 desalination plants are being developed in the governorates of North Sinai, South Sinai, Matrouh, Port Said and Red Sea. Besides 58 already existing, with a total capacity of 440,000 m^3/day.

Non-Conventional Water Resources

a. Shallow Groundwater

Meanwhile, shallow groundwater in the Nile aquifer cannot be considered a separate source of water. They are recharged only by seepage losses from the irrigation canals and drains, the Nile, and infiltration losses from irrigated lands. Hence, their yield must not be added to Egypt's total water resources. Their storage capacity in the Nile Valley and Delta system (with a salinity of 800 ppm) is estimated to be 200 and 300 km^3, respectively [13]. Although shallow wells with low pumping costs can be used to extract the shallow groundwater in relatively large quantities estimated about (100–300 m^3/hr.), it is extremely susceptible to pollution. Figure 4 shows the abstractions from shallow groundwater. From Fig. 4, there is a slight increase in the abstraction over the years. Table 1 shows the total abstractions from groundwater.

b. Reusing of Agriculture Drainage Water

One of the most economically viable alternatives for irrigation is the reuse of agricultural drainage water if used on an environmentally sustainable and sound basis. Usually, drains are located at the downstream ends of irrigation canals where the cultivated lands that are subject to irrigation water shortages are located. The officially reused drainage water increased from 2.6 BCM/yr in the 1980s to about 4.2 BCM/yr in the early 1990s [8]. About 2.0 BCM/yr of unofficial reuse is taking place in many locations [19]. About 6.6 million feddans were covered by the main surface drainage network. Drainage water quality may be improved by either treating it in small stations on the sub-drains or treating the water of the main drains before being mixed or through the isolation of sewage arid industrial water and reducing

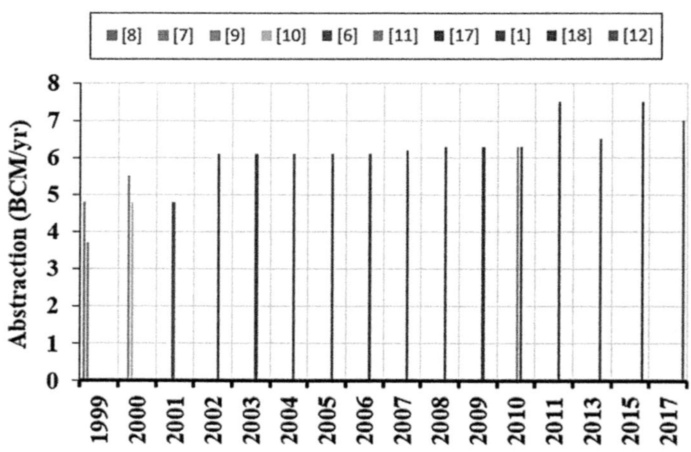

Fig. 4 Total abstraction from shallow groundwater (BCM/yr)

Table 1 Abstractions from deep and shallow groundwater

Reference	year	Deep groundwater (BCM/yr)	Shallow groundwater (BCM/yr)	Total groundwater (BCM/yr)
[8]	1999	0.5	4.8	5.3
[7]	1999	0.7	3.7	4.4
[9]	2000	0.8	5.5	6.3
[10]	2000	0.57	4.8	5.37
[6]	2001	0.57	4.8	5.37
[11]	2010	1.9	6.15	8.05
[1]	2013	2	6.5	8.5
[12]	2017	2.1	7.0	9.1

the utilization of pesticides and fertilizers. The reuses of agricultural drainage water and treated sewage water are like rainfall which cannot be considered as dependent resources. However, they help supplement the fresh water supply in some regions. This recycling process of the previously used Nile water certainly improves the overall efficiency of the irrigation water system [20]. Figure 5 shows the amount of reused agricultural drainage water reported by different researchers. From the figure, the average value of the drainage reused water is about 6.6 BCM/yr in the period 1998–2012.

However, the estimated amount by [21] in the year 2013 is 16 BCM/yr of which 6.33 BCM/yr are reused in agriculture in 2013 [3]. Numbers of reused drainage water projects in the southern part of the Delta and Fayoum governorate use about

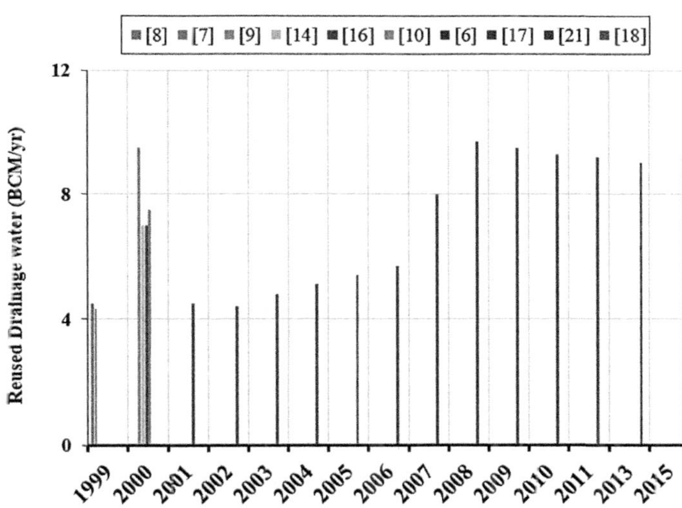

Fig. 5 Reused agriculture drainage water (BCM/yr)

4 BCM/yr. In addition to that, farmers unofficially pump water directly from drains which is estimated to be large quantities of drainage water and is difficult to measure [11] but according to [1] is estimated to be about 2.7 BCM/yr. Besides, there are unofficial reuses practiced along Bahr Hadus, Bahr Baqar, Gharbia, Umoum, Edko drains. There are 89 agricultural drains which directly flow into the river Nile. Hence, the agricultural sector is considered a major dangerous source of pollution affecting water quality. All drainage networks in upper Egypt drain in the Nile river. Also, in the Giza governorate, some drains end in Rosetta branch, like El-Moheet, and drains like Sobol, and Talaa in Mounfia governorate. Most of them collect volumes of wastewater, either municipal or industrial. Also, Moghazy [12] estimated that the use of drainage water in 2017 reached 15 BCM because of the unofficial reuse.

Besides, there is a limit for reusing of the agricultural drainage water which according to [15] is estimated to be from 8 to 9 BCM/year. This is because of the increase in the pollution and salinity of the water which may lead to health and environmental damages and degradation of the soil. So, for example, Rosetta branch receives 0.68 BCM/yr from drainage water with a high degree of pollution and all installed pump stations taking from them have significantly been affected by this low quality of the water. This is because of the low flow rates and the big cross-sectional area that lower the flow velocities.

Adding to this, the effect of the quality of using these waters on the productivity of the crops. Saad-El-Din [22] found that the average net yield of the irrigated feddan of wheat crop using freshwater achieved the highest average net yield, making it the highest cost compared to using mixed water and to the one irrigated by agricultural drainage water only. The main reason for the decline of the productivity of the crops is that the use of mixed water and agricultural drainage water loaded with salts, pesticides, fertilizers, and some other pollutants lead to increased salinity of the soil and increase pollution and spreading of weeds, which significantly affects productivity.

There are three levels for reusing of the agricultural drainage water in the Delta:

(1) Main level: 23 mixing stations are being used and operated by the high-level authorities in the ministry. Many of them are not functioning anymore because of the high level of pollution in the water.
(2) Intermediate level: small pump stations (0.5 to 1.0 m^3/s) operated by the administration of irrigation to fix the water shortage problems at the end of canals.
(3) Single level: And this level is unofficial where the farmer uses directly the drainage water.

In April 2019, the Ministry of Irrigation and Water Resources announced the start of the construction of 92 intermediate lifting stations for the reuse of agricultural wastewater for agricultural purposes in Kafr El Sheikh, Dakahlia, Sharqia and Menoufia governorates which will be completed by 2020.

c. **Municipal Wastewater**

In recent years, the use of treated wastewater has become increasingly significant in the management of water resources for both environmental and economic reasons. It starts early in the 1930s where some practices in areas with sandy soil like Abou Rawash and Al-Gabal Al-Asfar near Cairo use it. In 1980, interest in using treated wastewater, as an alternative for freshwater in irrigation, has accelerated. Egypt has adopted a strategy of wastewater reuse to lessen the pressure imposed by increasing demands on freshwater resources. Egypt produces an estimate of 5.5–6.5 BCM/yr of wastewater in 2010 of which 2.97 BCM/year is treated, but only 0.7 BCM/yr is being used in irrigation. This treated wastewater is used directly in desert areas or indirectly after being mixed with agricultural drainage water [19].

Moreover, according to a report prepared by [21], produced municipal wastewater is estimated at 7 BCM in the year 2012. Around 57% or 4.0 BCM of this amount is treated. Finally, a 1.3 BCM of treated municipal wastewater is directly used. The drainage system that receives the excess irrigation water also collects municipal wastewater, particularly in the upper part of Egypt, which then discharges itself into the Nile or the Northern Lakes or the sea [10]. Different studies have reported the amount of recycled sewage water. From Fig. 6, there is almost a steady annual treated amount of sewage water from 2009 to 2015 estimated at about 1.3 BCM.

Figure 7 shows the quantities of primary treated wastewater in different governorates in 2011. Alexandria and Giza governorates have the highest rate of treating the wastewater. Moreover, in Fig. 8, Cairo governorate is among the most governorates that makes a second treatment to the wastewater.

Overall, the total amount of water resources is estimated at 82.8 BCM/yr in 2017. Figure 9 shows the values of different water resources for the year 2017. The Nile River represents about 67% of the total water resources followed by the reused drainage water with 19% then shallow groundwater with 8% and then comes to the rest of the water resources.

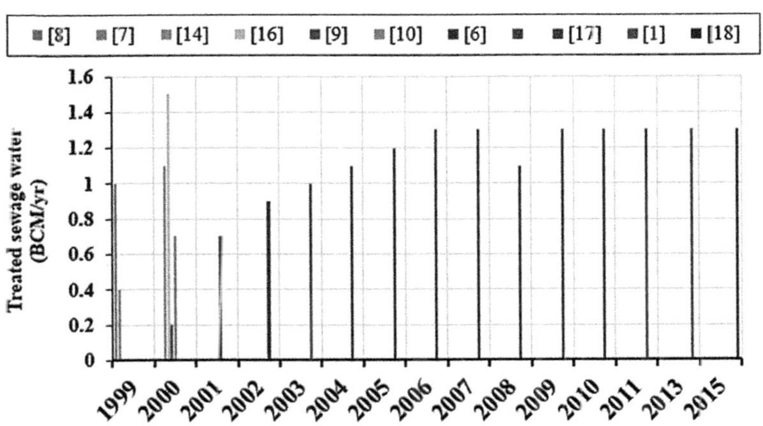

Fig. 6 Treated sewage water (BCM/yr)

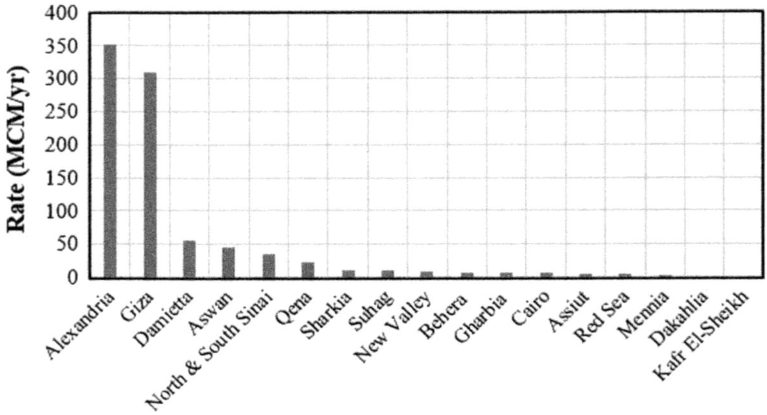

Fig. 7 Primary treated wastewater in 2011 (MCM/yr)

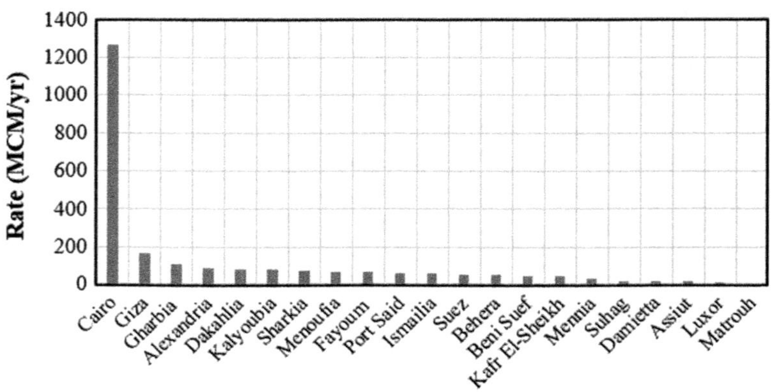

Fig. 8 Secondary treated wastewater in 2011 (MCM/yr)

Water Demands

As a result of population growth followed by an increase in the need for food which depends on agricultural and industrial expansion, the increase in Egypt's water demands has become expected in many sectors as discussed in the next sections.

a. **Agricultural Sector**

This sector is considered the most water consumer in Egypt. Different values have been reported for it: about 86% [10], 83.4% [8, 19], and more than 85% [1] of the available freshwater resources. The agriculture land base consists of the old lands in the Nile Valley and Delta, several oases, rain-fed areas and reclaimed lands in the desert. The total area of irrigated land in 2011 was estimated at 8.8 million feddans

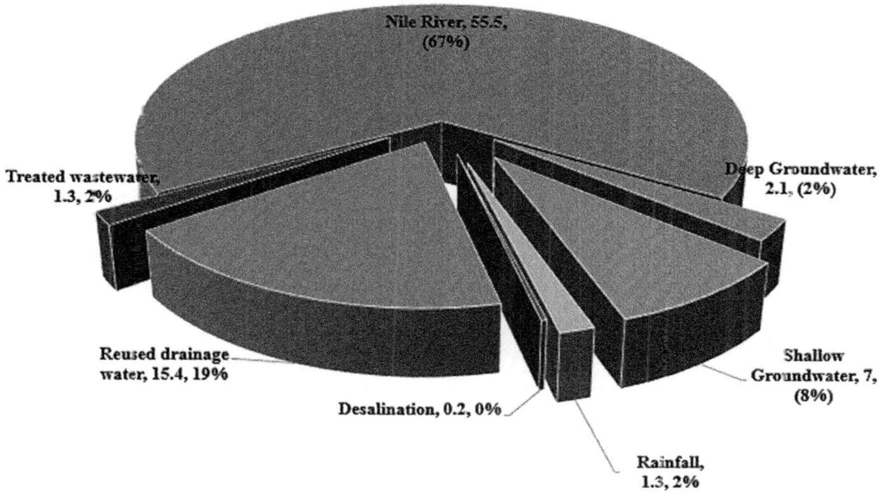

Fig. 9 Total amount of water resources in 2017 in BCM/yr (Moghazy, 2018)

with a total demand of 62.0 BCM/yr [3]. This makes the average water consumption per feddan with an average irrigation efficiency of 65% about 7000 m^3/ feddan/yr.

Agriculture sector contributes by 12% of Egypt's annual GDP [23]. Significant changes have been observed in Egypt crops production and trade during the last 50 years. For example, 322% increase in rice production was observed, accompanied by 590% increase in rice net exports during the period from 1961 to 2015. During the same period, 570% increase in wheat production was observed, accompanied by 1456% increase in wheat net imports [24]. Accordingly, due to the current state of the water crisis in Egypt, a more inclusive perception of the role of water in the trading and consumption in Egypt is needed. This water is used through evaporation and evapotranspiration, which differs according to the cropping pattern and location. Crops with higher water requirements consume more irrigation water like rice and sugarcane that are the two most crops that need irrigation water. Rice uses approximately 15% of the total agricultural water consumption [7]. Also, the water consumptions of wheat, maize, cotton, rice, and winter barseem represent about 60% of the agricultural water demand.

It should be mentioned that many families depend on agriculture as the main source of their incomes, where it contributes nearly by 40% of total employment [1]. In addition, increasing in population and living standards causes an increase in food demands.

Depending on the previous studies and as shown in Fig. 10, a very slight increase in agricultural water demand is observed at the rate of 0.3 BCM/yr for the period 2000–2013. The average value of water demand for the agricultural sector for this period is around 60.3 BCM/yr after excluding Alnaggar's data [16]. Moreover, from 2015 to 2017, its annual estimate is about 61 BCM. However, this increase is not certain

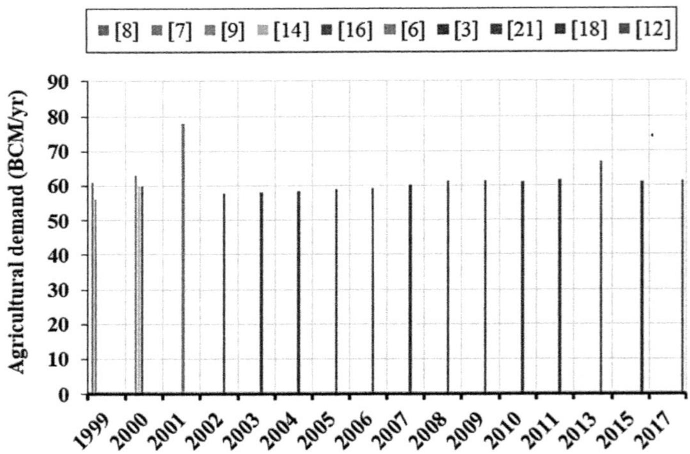

Fig. 10 Water demand for the agricultural sector (BCM/yr)

enough since there have been reports stating that there is a loss in the agricultural land due to other factors like urbanization, degradation, and brick making. The latter was felt significantly in the period 1983–1995. According to MWRI [10], the estimated loss of agricultural land is 45000 feddan/yr. This estimate should be constant by now since the brick making has almost stopped, but the urbanization is still expanding everywhere.

b. **Domestic and Industrial Sectors**

In the year 2000, the domestic and industrial sectors consume 6% and 8% of the total water supply [10], respectively. For the domestic consumption, the daily consumption per capita and the losses from the distribution system, which is calculated from the conveyance efficiency, are considered the main factors affecting it. Distribution losses in the municipal sector are evaluated at 60% in rural areas and at 40% in urban regions [16]. Surface water and groundwater mainly represent the source of municipal water, where they can supply about 83% and 17% of total municipal demands [1], respectively. According to MWRI, the increase in the total domestic and industrial demand is by a factor of 1.4 and 2.0 between 2000 and 2017, respectively. The total municipal water use is estimated to be 8.9 BCM/yr in 2011 [25, 26] and reached to 10.6 BCM/yr in 2017 [12]. A portion of that water is consumed, and the remaining returns back to the water system, either by seeping to the groundwater or to the sewage collection system. In some regions like Suez canal, Alexandria, and in desert areas, this discharge cannot be recovered back.

From Fig. 11, an increase in domestic water demand is significantly observed from 1999 to 2017 with an annual rate of increase of about 0.48 BCM on average.

The demand for drinking water is expected to grow more because of the increase in the water demands as population increases, the high standard of living, besides the

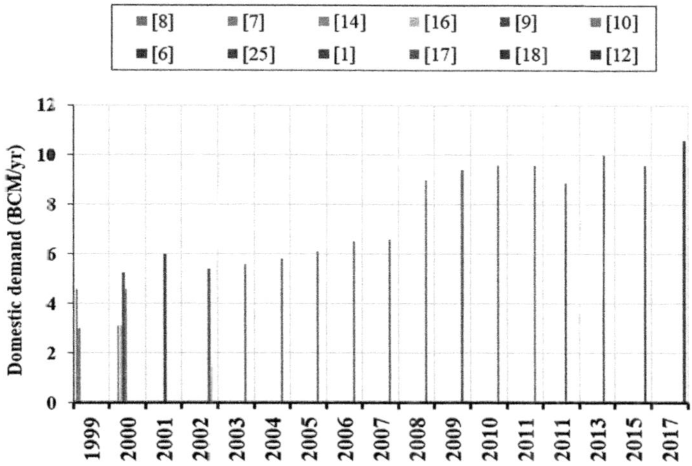

Fig. 11 Water demand for domestic sector (BCM/yr)

governmental policies aiming to reclaim more lands and encouraging the industrial sectors to grow more.

The problem here is the high losses encountered in the networks due to losses in the stage of production, consumption and distribution, low maintenance either in house connections or primary and secondary networks.

However, for the industrial water demand, more industrial expansion, providing employment for a large sector of the population, is expected to play a major role in the socio-economic development of the country. It is estimated that the annual requirement of the industrial sector was 7.5 BCM during the year 2011 (most of which returns back to the system through agriculture drains and sanitary sewer system); making the net water demand for the industrial sector about 1.2 BCM/yr in 2011 [3] and 5.4 BCM/yr in 2017 according to [12]. Figure 12 shows the total water demands consumed by the industrial sector.

c. **Other Sectors**

- Losses from other sectors may include an estimated amount of 2.5 BCM/yr as evaporation losses from canals/drains systems.
- About 0.5 BCM/yr as environmental balance [12].
- An additional amount also is to be accounted for Fishery where about 50% of the fishery comes from the northern lakes that need about 4 BCM/yr from drainage water to keep its environmental balance.
- Also, it is estimated that 8.32 BCM/yr is required as the minimum drainage water to flow into the lakes of Buroulos, Manzala, Edko, and Maryout and the Mediterranean Sea to maintain the salinity levels in these waters within an acceptable range for the fish species. In 2010 according to CAPMAS (2014), there is an amount of about 12 BCM/yr (with an average salinity of 2700 PPM) of drainage water

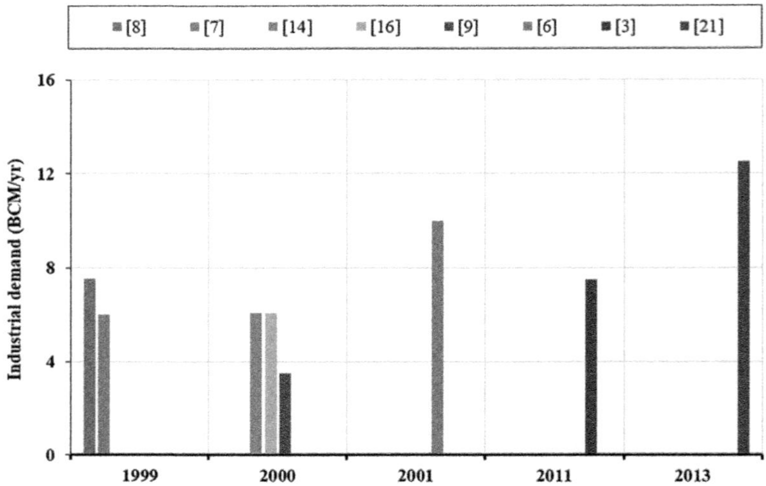

Fig. 12 Water demand for industrial sector (BCM/yr)

discharged to the sea. Further studies need to be done to confirm the minimum amount to be discharged to the sea/lakes that will not affect the fish species and its environmental impact on these areas.

Overall, it can be seen from Fig. 13 in the year 2017 that the agricultural sector has

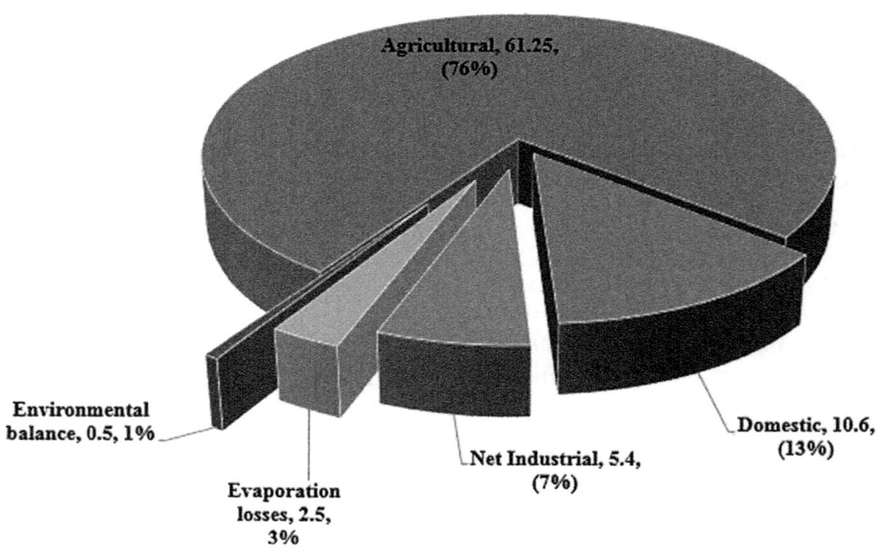

Fig. 13 Total water demands by various sectors for the year 2017 in BCM/yr (Moghazy, 2018)

consumed about 76% of total water demands with an average value of 61.2 BCM/yr followed by the domestic sector, which has consumed about 13% and finally 7% of net total water demand has been used by the industrial sector.

The natural resources of various types and how to maintain them and exploit them safely are the main determinants of sustainable and comprehensive development across generations. The issue of water is considered one of the most important challenges facing the countries of the world today, especially the developing countries.

Future Water Expectations

Some studies focused on adopting new strategies to give the best scenarios for water budgets for 2017, 2020, 2025 and 2050, like [6–8, 15, 16] and the study of water resources and rational use in Egypt reported by [17]. In the next paragraphs, different scenarios by different studies for rational use of water supplies and demands for the year 2025 and 2050 are introduced briefly.

Water scenarios for the year 2025

(a) According to Alnaggar [16], it is expected that the total conventional water resources will increase to 64.3 BCM in the year 2025. This increase is the result of:

1. Reducing of evaporation losses from High Aswan Dam by 2.5 BCM,
2. Seawater desalination by 1.00 BCM,
3. Reusing of agricultural drainage and wastewater that will add 7 and 1.5 BCM, respectively.

On the other side, the total water demand will increase to 81.9 BCM in the year 2025 by 35.6% compared to the year 2000, a shortage of 17.6 BCM. This gap between the water resources and the water demand will be covered by reuse of agricultural drainage water by 7.0 BCM/yr and treated wastewater by 1.5 BCM/yr and there will be still a shortage of about 9.1 BCM in 2025 as shown in Fig. 14 since he did not include the contribution of the shallow groundwater.

(b) Another attempt has been carried out by [27] in which he developed a simplified approach for water resources vulnerability assessment keeping a balance between water resources and consumptions and taking the effect of climate change. He put two scenarios for the Nile flows, and both assume that:

- Proposed dams to be constructed in the Nile Basin countries would reduce our annual share by 5 and 8 BCM in 2025 and 2050, respectively.

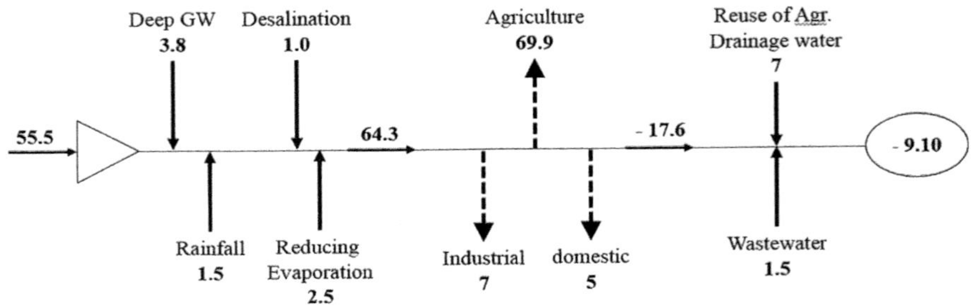

Fig. 14 Water balance scenario for the year 2025 (values are in BCM) (after Alnaggar [16])

- The estimated rise in the mean air temperature would be 1.0° and 1.7 °C in 2025 and 2050, respectively.
- Estimated change in the mean evapotranspiration would be 4 and 8% in 2025 and 2050, respectively.

The first scenario estimates that there will be a drying period that reduces the Nile flows by 6% and 15% in 2025 and 2050, respectively. Thus, our annual Nile resources would be reduced to (55.5–5) $\times$ 0.94 = 47.47 BCM in 2025 and to 40.38 BCM in 2050. Moreover, in the second scenario, there will be a wetting period that increases the Nile flows by 10% and 21% in 2025 and 2050, respectively. Thus, our annual Nile resources would be increased to (55.5–5) $\times$ 1.10 = 55.55 BCM in 2025 and to 57.48 BCM in 2050.

Fig. 15 a and b show the water budgets in 2025 for the two scenarios and it can be seen that in both scenarios, there will be annual shortage by 20.85 BCM on average.

Water scenarios for the year 2050

(a) According to Nour El Din [27], the water budgets in 2050 with both scenarios mentioned previously would have shortage by 20.5 BCM on average. Figure 16a and b show the water budgets in 2050 for both scenarios.

It can be noticed that, for the case of the drying period (Scenario—1), the agricultural sector is the one mostly affected by the shortage of water compared to the other water demands sectors.
According to CAPMAS [17], three different scenarios for water resources and uses in 2050 are presented depending on some assumptions, as shown in Table 2.

Reference values for population and total cultivated area are taken from data in 2010. The discussions and results of these scenarios can be summarized as follows:

(1) ***First Scenario*:**

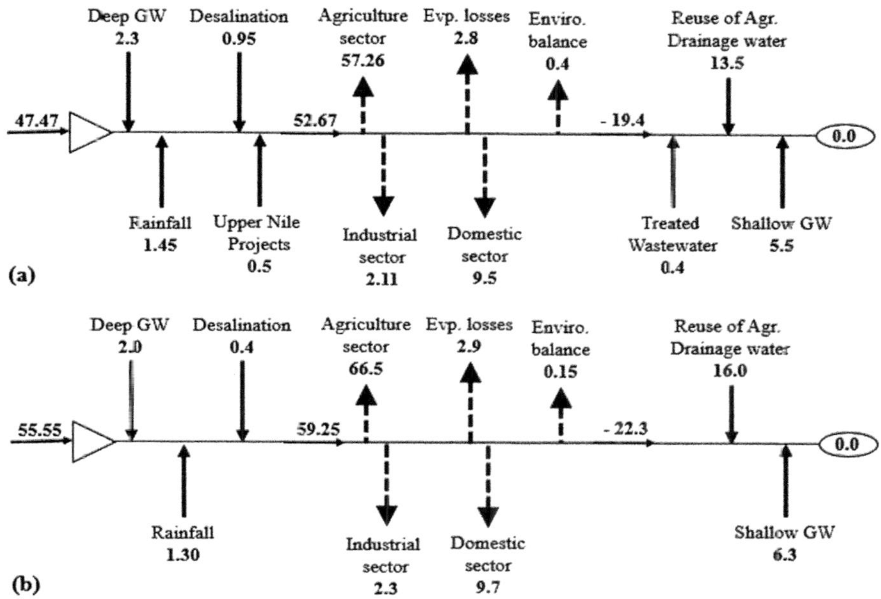

Fig. 15 Water budget in 2025 for a) 1st scenario (drying season) and b) 2nd scenario (wetting season) (values are in BCM) (after Nour El Din [27])

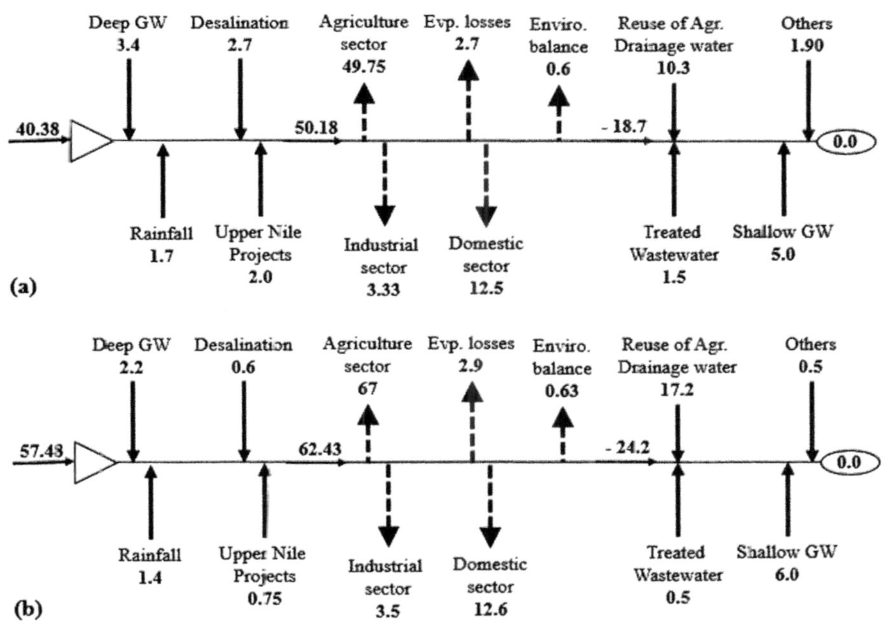

Fig. 16 Water budget in 2050 for a) 1st scenario (drying season) and b) 2nd scenario (wetting season) (values are in BCM) (after Nour El Din [27])

Table 2 Assumptions for scenarios suggested by CAPMAS (2014)

Item	Scenarios		
	1	2	3
	Critical	Balanced	Optimistic
Contribution of upper Nile projects (BCM/yr)	–	2	4
Rate of population growth (%)	2	1.8	1.62
*Estimated population (Million)	172.5	158	146.4
Reclaimed area (thousands feddan/year)	60	80	100
The area lost from urbanization (thousands feddan/year)	25	25	20
**Area that can be cultivated (Million fed.)	10	10.8	11.8
Water consumptive use (m^3/fed/yr)	4500	4400	4300
Water efficiency in agricultural sector (%)	65	70	75
Total consumptive use in agricultural sector (BCM)	45.0	47.5	50.7
Efficiency of water distribution networks (%)	70	85	100
Water use/capita (Lit./cap/day)	150	192	234
Rate of economic growth	low	moderate	high
Rate of increase in water use in industrial sector (%)	0.65	1	1.35

* Population in 2010 is about 77.8 million
** Cultivated area in 2010 is estimated at 8.6 million feddan

This scenario is considered as a *critical scenario*, where there are no abundant water resources. It is assumed that:

- No upper Nile projects are to be constructed and hence Egypt's Nile share of water would remain the same and equals to 55.5 BCM/yr.
- The use of deep groundwater should be expanded to increase from 2 to 4 BCM/year.
- Rainfall and torrents should be consumed in the best manner and thus will contribute by 1.5 BCM/year.
- Desalinated water would have a maximum value of 1 BCM/year.
- So, summing up all the above values would give 62 BCM/year as the total conventional water resources.

- The rate of population growth is assumed = 2% and so, the total estimated population in 2050 shall reach 172.5 million.
- The estimated reclaimed area = 60,000 fed/year and the urbanization effect would reduce it by 25,000 fed. So, the net annual increase in the reclaimed area would be 35,000 fed. Moreover, since the total cultivated area in 2010 is about 8.6 million fed. Then, the total area that can be cultivated in 2050 may reach 10 million fed. Moreover, if it is assumed that the water consumptive use and irrigation efficiency are 4500 m^3/fed/yr and 65%, respectively. So, the total water demand for the agricultural sector would be = $4500 \times 10/0.65/1000 = 69.2$ BCM.

- The efficiency of the water distribution networks equals to 70% and that the water use per capita = 150 lit/cap/day, then the total water demands for the domestic sector would be = 150*365*172.5/0.7 ≅ 13.5 BCM.
- The rate of economic growth is low and the water use in the industrial sector amounts to 0.65%, then the total actual consumption of the water in the industrial sector would be about 1.8 BCM. Assuming the water use efficiency is 35%, then the total water demands for the industrial sector would be equal to 3.1 BCM after subtracting 2 BCM that can be exploited from the water distribution network.
- Evaporation losses from canals and drains = 2.5 BCM.
- As a result, the total water demands would sum up to, 69.2 + 13.5 + 3.1 + 2.5 = 88.3 BCM.
- The shortage in water would be about 62–88.3 = -26.3 BCM (or the water stress index ≅ 142%). This shortage can be covered by the reuse of drainage water from agricultural and industrial sectors and the treatment of sewage water amounting to about 18 BCM and the rest from using the shallow groundwater.
- ***Second Scenario*:**

This scenario is considered as a *balanced scenario*, where it is assumed that:

- The first phase in Jongeli project would be accomplished, thereby increasing the Nile share by 2 BCM/yr and hence Egypt's Nile share of water would be 57.5 BCM/yr.
- The use of deep groundwater should be expanded to increase from 2 to 4 BCM/year.
- Rainfall and torrents should be consumed in the best manner and thus will contribute by 1.5 BCM/year.
- Use of desalinated water would expand in the Mediterranean, Red sea and the tourists' places in order to reduce the dependence on the Nile water. So, the estimated amount that can be obtained = 1.5 BCM/year.
- So, summing up all the above values would give 64.5 BCM/year as the total conventional water resources.
- The rate of population growth is assumed = 1.8% which is the actual growth rate as of 2019 and so, the total estimated population in 2050 shall reach 158 million.
- The estimated reclaimed area = 80,000 fed/year and the urbanization effect would reduce it by 25,000 fed. So, the net annual increase in the reclaimed area would be 55,000 fed. Moreover, since the total cultivated area in the year 2010 is about 8.6 million fed, then, the total area that can be cultivated in 2050 may reach to 10.8 million fed. Moreover, if it is assumed that the water consumptive use and irrigation efficiency are 4400 m^3/fed/yr and 70%, respectively as a result of adopting new irrigation methods in the old lands. So, the total water demand for the agricultural sector would be = 4400 × 10.8/0.7/1000 = 67.9 BCM.
- The efficiency of the water distribution networks equals to 85% and that the water uses per capita = 192 lit/cap/day, then the total water demands for the domestic sector would be = 192*365*158/0.85 ≅ 13 BCM.

- The rate of economic growth is moderate and the water use in the industrial sector amounts to 1.0%, then the total actual consumption of the water in the industrial sector would be about 2.1 BCM (16.7% increase compared to the 1st scenario). Assuming the water use efficiency is 40%, then the total water demands for the industrial sector would be equal to 3.2 BCM after subtracting 2 BCM that can be exploited from the water distribution network.
- Evaporation losses from canals and drains = 2.5 BCM.
- As a result, the total water demands would sum up to, 67.9 + 13.0 + 3.2 + 2.5 = 86.6 BCM.
- The shortage in water would be about 64.5–86.6 = −22.1 BCM (or the water stress index ≅ 134%). This shortage can be covered by the reuse of drainage water from agricultural and industrial sectors and the treatment of sewage water amounting to about 14.6 BCM and the rest from using the shallow groundwater.
- ***Third Scenario*:**

  This scenario is considered as an *optimistic scenario*, where it is assumed that:

- Jongeli project would be completed thereby increasing the Nile share by 4 BCM/yr and hence Egypt's Nile share of water would be 59.5 BCM/yr.
- The use of deep groundwater should be expanded to increase from 2 to 4 BCM/year.
- Rainfall and torrents should be consumed in the best manner and thus will contribute by 1.5 BCM/year.
- Use of desalinated water would expand to its maximum limits and advanced techniques would be applied in the desalination plants. So, the estimated amount that can be obtained = 2 BCM/year.
- So, summing up all the above values would give 67 BCM/year as the total conventional water resources.
- The rate of population growth is assumed = 1.62% which is the less than the average growth rate in the last ten years and so, the total estimated population in 2050 shall reach 146.4 million.
- The estimated reclaimed area = 100,000 fed/year due to the increase in the reclaimed areas and the urbanization effect would reduce it by only 20,000 fed. So, the net annual increase in the reclaimed area would be 80,000 fed. Moreover, since the total cultivated area in 2010 is about 8.6 million fed, then, the total area that can be cultivated in 2050 may reach to 11.8 million fed. Moreover, if it is assumed that the water consumptive use and irrigation efficiency are 4300 m^3/fed/yr and 75%, respectively as a result of adopting new irrigation methods. So, the total water demand for the agricultural sector would be = 4300 × 11.8/0.75/1000 = 67.7 BCM.
- The efficiency of the water distribution networks equals to 100% (i.e., no leakages and better house connections) and that the water use per capita = 234 lit/cap/day, then the total water demands for the domestic sector would be = 234*365*146.4 ≅ 12.5 BCM.
- The rate of economic growth is high and the water use in the industrial sector amounts to 1.35%, then the total actual consumption of the water in the industrial

sector would be about 2.3 BCM (27.8% increase compared to the 1st scenario). Assuming the water use efficiency is 40%, then the total water demands for the industrial sector would be equal to 3.7 BCM after subtracting 2 BCM that can be exploited from the water distribution network.

- Evaporation losses from canals and drains = 2.5 BCM.

- Therefore, water demands would be = 67.7 + 12.5 + 3.7 + 2.5 = 86.4 BCM.
- The shortage in water would be about 67 – 86.4 = -19.4 BCM (or the water stress index ≅ 129%). This shortage can be covered by the reuse of drainage water from agricultural and industrial sectors and the treatment of sewage water amounting to about 11.9 BCM and the rest from using the shallow groundwater.

The summary of the results of these scenarios is outlined in Table 3 and Fig. 17.

(c) According to the current study and all the relevant data, a new scenario is proposed for the water budget in 2050. In this scenario, it is assumed that there are no upper Nile projects (or it is not yet completed) to contribute to the Nile water share. Each water demand sector is analyzed separately and then the whole water balance is presented.

- **Agricultural Sector**

Table 3 Summary of the results of the three scenarios

	1	2	3
	Critical	Balanced	Optimistic
Conventional Water Resources (BCM)			
Nile share	55.5	57.5	59.5
Deep GW	4	4	4
Rainfall	1.5	1.5	1.5
Desalination	1	1.5	2
Total =	**62**	**64.5**	**67**
Water Demands (BCM)			
Total water demands for agricultural sector	69.2	67.9	67.7
Total water demands for domestic sector	13.5	13.0	12.5
Total water demands for industrial sector	3.1	3.2	3.7
Evaporation losses	2.5	2.5	2.5
Total =	**88.3**	**86.6**	**86.4**
Shortage (BCM) =	**-26.3**	**-22.1**	**-19.4**
Water Stress index =	**142**	**134**	**129**
Non-conventional Water Resources (BCM)			
Shallow GW	8.3	7.5	7.5
Reuse of agricultural drainage, industrial and sewage water	18	14.6	11.9
Total =	**26.3**	**22.1**	**19.4**

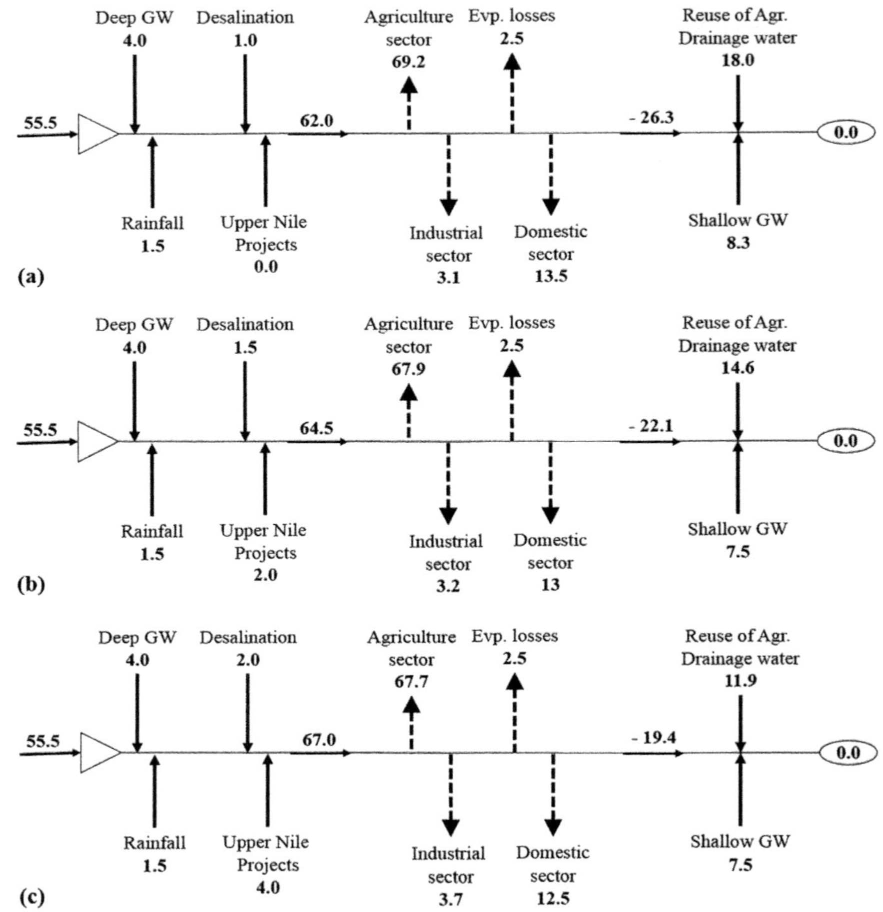

Fig. 17 Sumamry of the different scenarios suggested by CAPMAS [17], **a** 1st critical scenario, **b** 2nd balanced scenario and **c** 3rd optimistic scenario

The Total cultivated area reached 9.10 million feddan in 2015/2016 compared to 9.09 million feddan in 2014/2015, an increase of 1.0%. According to an analysis by [17] done for the period from 2001 to 2007, the rate of increase of the total cultivated area is about 60,000 fed/yr and the rate of urbanization is 25,000 fed/yr. However according to recent studies from 2011 to 2015 as shown in Table 4, the rate of increase has an average of 100,000 fed/yr.

Table 4 Total cultivated area (Million feddan)

Year	2011	2012	2014	2015
TCAr	8.8	8.94	9.09	9.1

The area-weighted average water rations for different crops during the three different seasons (winter, summer and Nili) are illustrated in Fig. 18. From the figure, it is clear that during the winter season, sugar beet has the highest water ration of 4104 m^3/fed and similarly sugarcane during the summer season has 10,797 m^3/fed.

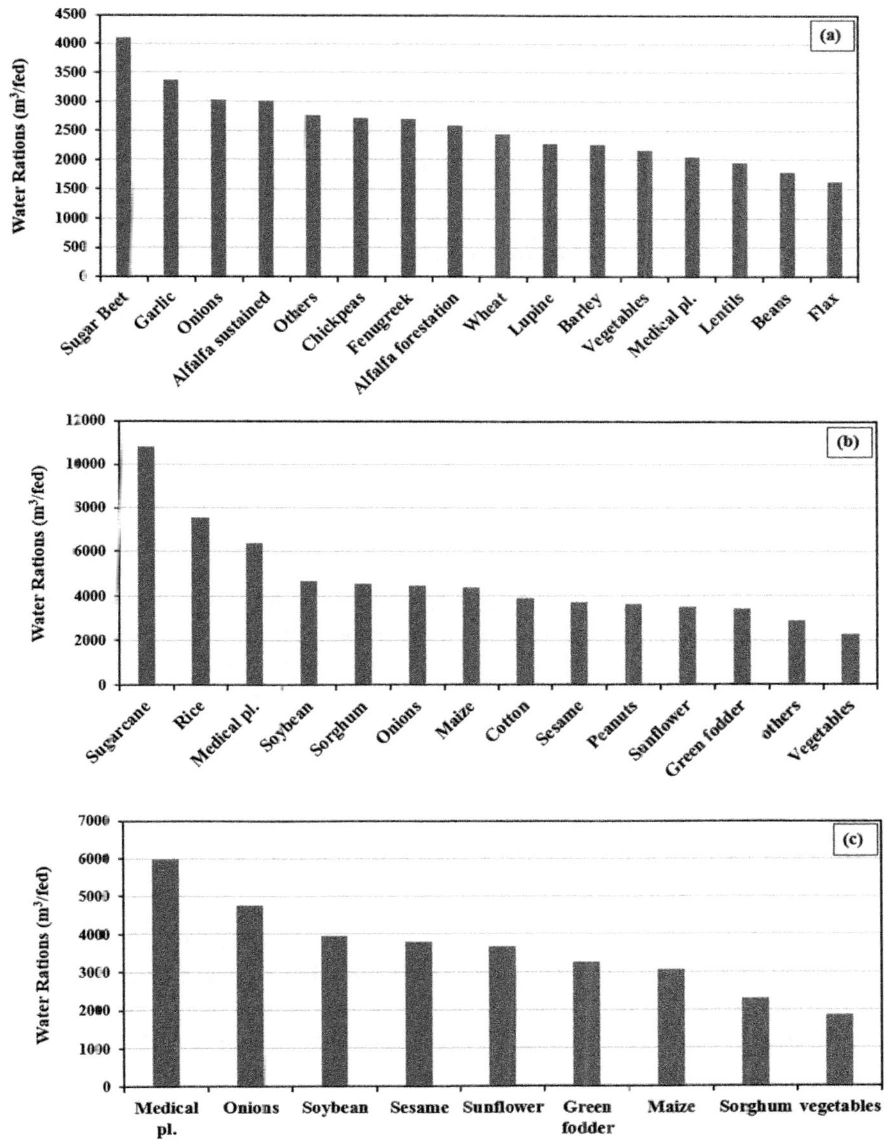

Fig. 18 Water rations for different crops during different seasons in m^3/feddan **a** Winter crops, **b** Summer crops and **c** Nili crops

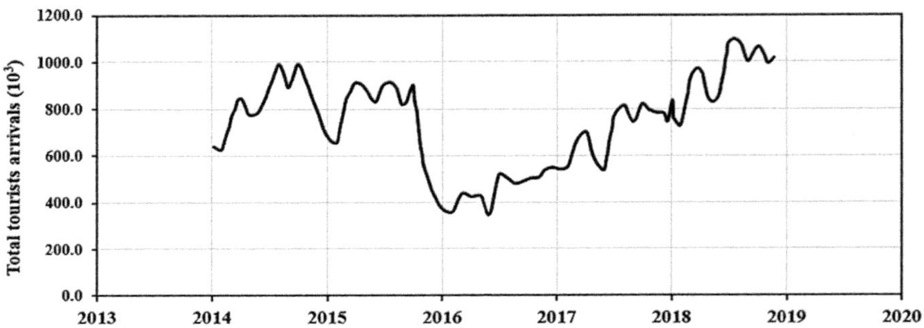

Fig. 19 Total tourists arrivals in Egypt in thousands

The area-weighted average water ration calculated for the year 2015 is equal to 4072 m^3/fed.

Also, from the analysis, the copped area of the wheat is the largest compared to all winter crops and equals to 52% of the total area followed by alfalfa with 28% and then sugar beet and then come to the rest of the winter crops. However, for the summer crops, maize and rice come at the top of the summer crops. And for Nili crops, maize and vegetables at the top of the list regarding their crop areas.

- **Domestic sector**

The domestic water demand in Egypt is estimated using population data of 2017 for a permanent population of each governorate [18], as well as average monthly data for tourists from trading economics indicators [28] as shown in Fig. 19. On average, the total number of tourists' arrivals in Egypt is about 757,000 $\times$ 12 $\approx$ 9.0 million persons. The water demand for public uses is estimated as 10% of the permanent population demands. Moreover, the distribution network leakage is calculated as 30% of the total domestic demands.

The population growth rate is calculated based on the actual population data reported by CAPMAS from 2000 to 2019. Table 5 shows the rate of population growth. From the table, it can be seen that the rate of growth is declining from 2015 to 2019 and can be assumed constant with an average value of 1.93% during the last 20 years.

The introduction of the rural and urban types of population is considered necessary, as there is a difference in their water consumption. In this study, the water consumption for urban, rural and tourists are 170, 100 and 270 lit/cap/day, respectively.

Table 6 shows the population in 2017 for each governorate in Egypt according to

Table 5 Rate of population growth during the last 20 years

Year	2000	2005	2010	2015	2016	2017	2018	2019
% growth	1.87	1.89	1.84	2.20	2.04	1.95	1.87	1.80

Table 6 Domestic water demands calculations for 2050

Governorate	Population in 2017 (million)	% from total	Population in 2019 (million)	Population in 2050 (million)	% Rural	Domestic water demand (MCM)			
						Permanent popluation	Public uses	Leakage	Total
Cairo	9.539	10.06	9.943	17.98	0	3.06	0.31	1.01	4.37
Alexandria	5.163	5.45	5.382	9.73	1.3	1.65	0.16	0.54	2.35
Port-Said	0.749	0.79	0.781	1.41	0	0.24	0.02	0.08	0.34
Suez	0.728	0.77	0.759	1.37	0	0.23	0.02	0.08	0.33
Damietta	1.496	1.58	1.559	2.82	60.6	0.36	0.04	0.12	0.51
Dakahlia	6.492	6.85	6.767	12.24	71.7	1.47	0.15	0.48	2.10
Sharkia	7.163	7.56	7.466	13.50	75.1	1.59	0.16	0.52	2.27
Kalyoubia	5.627	5.94	5.865	10.61	57.3	1.38	0.14	0.45	1.97
Kafr Sheikh	3.362	3.55	3.504	6.34	76.1	0.74	0.07	0.24	1.06
Gharbia	4.999	5.27	5.211	9.42	71.9	1.13	0.11	0.37	1.61
Menoufia	4.301	4.54	4.483	8.11	79.3	0.93	0.09	0.31	1.33
Behera	6.171	6.51	6.432	11.63	81.8	1.31	0.13	0.43	1.88
Ismailia	1.303	1.37	1.358	2.46	55.5	0.32	0.03	0.11	0.46
Giza	8.632	9.11	8.997	16.27	38.9	2.32	0.23	0.77	3.32
Beni-Suef	3.154	3.33	3.288	5.95	79.9	0.68	0.07	0.22	0.97
Fayoum	3.596	3.79	3.748	6.78	77	0.79	0.08	0.26	1.13
Menia	5.497	5.80	5.730	10.36	82	1.17	0.12	0.39	1.67
Asyout	4.383	4.62	4.569	8.26	74.1	0.98	0.10	0.32	1.40
Suhag	4.967	5.24	5.177	9.36	78.8	1.08	0.11	0.35	1.54
Qena	3.164	3.34	3.298	5.96	81.2	0.67	0.07	0.22	0.97
Aswan	1.473	1.55	1.535	2.78	58.9	0.36	0.04	0.12	0.51
Luxor	1.25	1.32	1.303	2.36	59.6	0.30	0.03	0.10	0.43
Red	0.3598	0.38	0.375	0.68	3.5	0.11	0.01	0.04	0.16
El Wadi Elgadid	0.241	0.25	0.251	0.45	54	0.06	0.01	0.02	0.09
Matrouh	0.425	0.45	0.443	0.80	37.3	0.12	0.01	0.04	0.16
North Sinai	0.4503	0.48	0.469	0.85	74.5	0.10	0.01	0.03	0.14
South Sinai	0.102	0.11	0.106	0.19	46.5	0.03	0.00	0.01	0.04
Sum =	**94.8**		**98.8**	**178.7**		**23.15**	**2.32**	**7.64**	**33.11**

[18] and the percentage of rural area in each governorate. Population in 2019 is based on the fact that the total population in 2019 is 98.8 million. Using the average rate of growth, 1.93%, the estimated population in 2050 is calculated for each governorate and then the domestic water demand is calculated based on the above assumptions.

From the table, the estimated population in 2050 is about 178.7 million and the total daily water demands for domestic uses amount to 33.11 million m^3. Adding the tourists' consumption, this will sum up to 12.09 BCM/year.

In Egypt, appropriate data on industrial units and production sizes are not available. Even the water consumed by this sector is not clearly available. However, according to our data in 2013 and previous years, the water demands by the industrial sector is about 2% out of the total water demands.

So, for this scenario, the assumptions are:

- Nile water share would remain the same at 55.5 BCM/yr.
- The use of deep groundwater should be to the maximum limit of the aquifer and amounts to about 4 BCM/year.

- Rainfall and flash floods should be consumed in the best manner and thus will contribute by no less than 2 BCM/year.
- Use of desalinated water would expand to its maximum limits and advanced techniques would be applied in the desalination plants. All coastal cities and touristic places should count on using desalination plants. Research centers and institutions should focus on the most recent technique to reduce the cost of the desalination process. As a result, the estimated amount that can be obtained may reach not less than 2 BCM/year.
- So, summing up all the above values would give 63.5 BCM/year as the total conventional water resources.
- The estimated reclaimed area = 100,000 fed/year due to the increase in the reclaimed areas and the urbanization effect would reduce it by only 20,000 fed. So, the net annual increase in the reclaimed area would be 80,000 fed. And since the total cultivated area in 2015 is about 9.1 million fed then, the total area that can be cultivated in 2050 may reach to 11.9 million fed. And since the average water consumption calculated previously in 2016 is 4072 m^3/fed and assuming that the irrigation efficiency would rise as a result of adopting advanced irrigation techniques and be equal 80%, then the water demand for the agricultural sector would be = 4072 × 11.9/0.80/1000 = 60.57 BCM. Adding to this the new reclaimed areas including Toskha (540,000 fed.), Sharq Owinat (230,000 fed.), El-Salam (485,000 fed.) and One Million and half fed excluding the part for Toshka (1362,000) and using an irrigation efficiency of 90% in the new lands, then the annual water demand for these new areas would be = 12.32 BCM. As a result, the total water demand of the agricultural sector by 2050 won't exceed 72.89 BCM.
- The water demands for the domestic sector is calculated equal to 12.09 BCM.
- The approaches discussed in the previous sections for rationalizing and optimizing water uses should be followed.
- The rate of economic growth is high, the government is able to use brackish water in industries, and if it is assumed that the total water demands for the industrial sector is 3% from the total demands. Then, it would reach 2.69 BCM.
- Evaporation losses from canals and drains are reduced as a result of replacing open ditches with pipelines = 2.0 BCM.
- Water for environmental balance could be used from the available freshwater and then treated and pumped back to the system and hence will not contribute to the water demands sector.
- As a result, the total water demands would sum up to, 72.89 + 12.09 + 2.69 + 2.0 = 89.67 BCM.
- The shortage in water would be about not less than 63.5 – 89.67 = -26.17 BCM (or the water stress index ≅ 141%). This shortage can be covered by the reuse of drainage water from the agricultural sector, the treatment of sewage and from using the shallow groundwater.

The summary of the results of this scenario is outlined in Table 7.

Table 7 New scenario from the current study for 2050

Water supply (BCM/yr)		Demand by sector (BCM/yr)		Shortage/Surplus (BCM/yr)	
Conventional water resources		Drinking	12.09		
NILE (HAD)	55.5	Industry	2.69		
Deep groundwater	4	Agriculture	72.89		
Rainfall & flash floods	2	Evap. Losses	2.0		
Desalination	2	Env. Balance	0.0		
Total conventional	63.50	**Total water usage**	89.67	**Shortage**	−26.17
Unconventional water resources				**Water stress %**	141
Shallow groundwater	8.17				
Reuse of Agr. Drainage water	16				
treated wastewater	2				
Total unconventional	26.16				
Total water available	**89.66**	**Total water usage**	**89.67**		

Estimates of water use and resources may not necessarily be met. These scenarios are best placed under comprehensive rationalization policies and intensive water resource development programs. Physical, technical, political, and social supports may not be available to fulfill these proposed scenarios.

Conclusions

Today, water resources are at the forefront of the world's concerns, and many countries have taken great interest in these resources to develop and maintain them, increase their efficiency and maximize their revenues, and use modern methods to meet the increasing demands for water resources in all sectors. The problem is how to rationalize the use of water in the current situation of increasing demands for water and the misuse of water resources.

The National Water Plan in Egypt includes full strategic projects being implemented on the ground at an accelerated pace in order to overcome problems related to water shortage.

As of 2019, there is a shortage of water by not less than 21 billion cubic meters that can be compensated by the reuse of the agricultural drainage water, shallow groundwater and the use of treated wastewater. However, by the year 2050, the situation will be much worse and the shortage is expected to reach 26 billion cubic meters. Many studies are presented in this work showing the different scenarios for the water balance in 2050. The population is expected to increase to about 178 million and hence the net water share per capita will go to less than 400 m^3/yr. This will put Egypt in a severe water scarcity limit. Meanwhile, the domestic water uses will rise to meet this accelerating population growth and it is estimated that the total amount needed for domestic uses is going to reach to about 12 to 13.5 billion cubic meters according to different assumptions. In addition to that, the share of each capita in the cultivated land will decrease and the amount of water needed to satisfy the agricultural sector will barely be enough.

Recommendations

There are many recommendations for reviewing our water budget and developing our water policies, for example, it is necessary to have a quick stop to re-evaluate what can be provided for the future under the current and expected conditions in terms of financial and water problems, including problems of pollution, deterioration of water networks, the problems of growing rice and sugarcane, the fragmentation of agricultural property and its impact on the efficiency of water use, and the situation with the Nile Basin countries. As well as, to reassess the efforts to rehabilitate water networks for irrigation and drainage, reevaluate the surface irrigation development projects and its long-term effects, develop a tight system to collect data for the actual groundwater levels, more accurate data on the quantity of rainfall especially in Sinai Peninsula and the eastern dessert and precise estimate of losses within the water networks and increase public awareness regarding water problems.

References

1. MWRI, Ministry of Water Resources and Irrigation (Egypt) (2014) Water scarcity in Egypt: the urgent need for regional cooperation among the nile basin countries. Report No. 6 p
2. El Sayed LM (2012) Determining an optimum cropping pattern for Egypt. Doctoral dissertation, The American University in Cairo, p 233
3. CAPMAS E (2012) Central agency for public mobilization and statistics. Study of water resources and rational use in Egypt
4. UN (1997) United nations comprehensive assessment of the freshwater resources of the world, Report of the secretary-general. United nations commission on sustainable development, 5–25 April 1997
5. El Qausy DE, Shideed K, Oweis T, Karrou M (2011) The role of current policies and institutional setups in Egypt in achieving improved and sustainable irrigated agriculture. In: Karrou M, Oweis T, Benli B, Swelam A (2011) Improving water and land productivities in irrigated systems. Community-based optimization of the management of scarce water resources in agriculture in CWANA. Report no. 10. International center for agricultural research in the dry areas (ICARDA), Aleppo, Syria

6. Abdin AE, Gaafar I (2009) Rational water use in Egypt. In: Technological perspectives for rational use of water resources in the Mediterranean region, Series A, vol 88, pp 11–27
7. Mason SA (2004). From conflict to cooperation in the Nile Basin. In: Doctoral dissertation, swiss federal institute of technology Zurich. ETH Zurich, Switzerland
8. El Fellaly SH, Saleh EMA (2004) Egypt's experience with regard to water demand management in agriculture. In: Eighth international water technology conference, IWTC8 2004. Alexandria, Egypt, pp 5–29
9. Allam MN, Allam GI (2007) Water resources in Egypt: future challenges and opportunities. Water Int 32(2):205–218
10. MWRI, Ministry of Water Resources and Irrigation (Egypt) (2005) Water for the future—national water resources plan 2017. Technical report
11. ICARDA, AusAID (2011) Water and agriculture in Egypt. In: International center for agricultural research in the dry areas and Australian government
12. Mohazy H (2018) Egyptian water security. Lectures
13. Hamza W, Mason S (2004) Water availability and food security challenges in Egypt. In: Proceedings international forum on food security under water scarcity in the middle east: problems and solutions. Como, Italy, pp 24–27
14. Abdel-Shafy H, Aly R (2002) Water issue in Egypt: resources, pollution and protection endeavors. Navigation 49(3.1):4–6
15. Allam (2001) Water and agricultural lands in Egypt.
16. Alnaggar D (2003) Water resources management and policies for Egypt. In: Presented at the workshop on policies and strategies options for water management in Islamic countries (Tehran). Ministry of water resources & irrigation, technical documents in hydrology, no 73, pp 55–69
17. CAPMAS E (2014) Central agency for public mobilization and statistics. In: Study of water resources and rational use in Egypt
18. CAPMAS E (2018) Central agency for public mobilization and statistics. Informatic bulletin
19. Wahaab RA, Mohy El-Din O (2011) Wastewater reuse in Egypt: opportunities and challenges, consultation meeting for experts in the arab world on the situation of wastewater management. Dubai, UAE
20. Ahmed M, EL-Kadi A (2013) Personal communication with MEWINA local office personnel and responses to Misr Consult RAR short form questionnaire
21. FAO, AQUASTAT (Food and Agriculture Organization of the United Nations) (2016) Country profile—Egypt. http://www.fao.org/3/i9729en/I9729EN.pdf
22. Saad El-Din A (2008) The economic effects of the use of agricultural drainage water on the productivity of some major crops in Egypt. PhD dissertation, faculty of agriculture, Ain shams university
23. World Bank (2017) Agriculture, value added (% of GDP). https://data.worldbank.org/indicator/NV.AGR.TOTL.ZS?view%C2%BCchart
24. FAO (2017) FAOSTAT [WWW Document]. FAOSTAT database. http://www.fao.org/faostat/en/#data/QC
25. HCWW, Moawad A, Fahmy K, Afifi M, Mahmoud R (2013) Responses to Misr Consult questionnaire
26. EWRA (2011) Egyptian water regulatory agency fourth water and wastewater report.
27. Nour El-Din M (2013) Climate change risk management in Egypt. Report on the proposed climate change adaptation strategy for the ministry of water resources & Irrigation in Egypt
28. Trading Economics (2019) Trading economics indicators. https://tradingeconomics.com/egypt/tourist-arrivals

Conclusions and Recommendations

Update, Conclusions, and Recommendations for "Groundwater in Egypt's Deserts"

Abdelazim M. Negm, Elsayed E. Omran, and Ahmed Elkhouly

Abstract The present situation in Egypt is framed by the scarcity of water that is under serious pressure. Water resources, on the one hand, are at the core of sustainable development and critical to socio-economic growth. On the other hand, water utilized from groundwater is considered an important source of water supply in Egypt after the River Nile. This chapter captures the groundwater in Egypt's deserts (in terms of findings and suggestions) and provides ideas extracted from the volume cases. In addition, some (update) findings from a few recently published research work related to the groundwater covered themes. This chapter provides the present problems faced by the Groundwater in Egypt's Deserts with a set of recommendations to safeguard the water to supply the populations and farmers with water. It covers topics that include groundwater occurrence, ecosystem services of groundwater, water logging, quantity and quality of groundwater in Egypt's deserts and their Management and future use of groundwater in expanding areas.

Keywords Assessment · Sustainability · Groundwater · Deserts · Environment · Egypt · Socio-economic · Scarcity · Water resources · Water logging · Water quality · Ecosystems

A. M. Negm (✉)
Water and Water Structures Engineering Department, Faculty of Engineering, Zagazig University, Zagazig 44519, Egypt
e-mail: amnegm@zu.edu.eg

E. E. Omran
Soil and Water Department, Faculty of Agriculture, Suez Canal University, Ismailia 41522, Egypt

Institute of African Research and Studies and Nile Basin Countries, Aswan University, Aswan, Egypt

A. Elkhouly
Plant Ecology and Range Management Department, Desert Research Center, Cairo, Egypt

A. Negm and A. Elkhouly (eds.), *Groundwater in Egypt's Deserts*, Springer Water,
https://doi.org/10.1007/978-3-030-77622-0_17

Introduction

Over the previous 300,000 years, Egypt's desert areas have hosted many wet phases. Surface water has been channeled through drainage patterns, some of which are now exposed, while others are covered by Aeolian sand. Groundwater is one of Egypt's largest water resources. It ranks after the Nile River as the second source. Furthermore, in the Nile decision-makers area, there are various groundwater aquifers with variable significance for exploitation. They range from shallow local aquifers, recharged by rainfall, to deep aquifers that cannot be replenished. The first consists of the Nile Valley and Delta system groundwater. The second category of aquifers is the non-renewable form in the "Western Desert-Nubian Sandstone Aquifer." It is focused primarily in the Western Oasis of the Desert. Groundwater is found primarily in three distinct water-bearing aquifers in Sinai. This book contains additional data on groundwater issues in Egypt, places, origin, quantity, quality and quantity, contamination, natural aquifer treatment, protection, uses, future overview, and conclusion. Therefore, the intention of the book is to address and improve understanding of the following main three theme.

- Groundwater Occurrence and Ecosystem services
- Groundwater Exploration, Quantity, Quality, and Their Management
- Potential Use of Groundwater and Future Expansion.

The next section presents a brief of the important findings of some of the recent (updated) published studies on the groundwater in Egypt's deserts, then the main conclusions of the book chapters in addition to the main recommendations for researchers and decision-makers. The update, conclusions, and recommendations presented in this chapter come from the data presented in this book.

Update

The following are the major update for the book project based on the main book theme.

Groundwater Occurrence and Ecosystem Services

Three chapters are identified in the book related to groundwater occurrence and ecosystem services. These chapters integrate more than one technique to give an accurate estimation of the groundwater potential. The first integration technique is Groundwater occurrences in West Nile Delta, Egypt. West of the Nile Delta area is one of the most important areas in Egypt for agricultural, industrial and recreation investments due to its high groundwater potentials, accessibility and facilities provided by the government in Wadi El-Natrun, Sadat city and Wadi El-Farigh. The area of Wadi El-Farigh is one of the areas in the western Nile Delta fringes subjected to intensive land reclamation and other agricultural, industrial activities. Such activities are probably not controlled, have seriously affected the potentials of the groundwater resources in the area. This effect is quite noticed in the large farms where, groundwater over-pumping is very high to fulfill the crop requirements leading to serious problems in both the quality and quantity of withdrawing water. The yearly gradual drawdown of water level in these farms and deterioration of the groundwater quality are recorded in aquifers tapping this area. So, the evaluation of the groundwater occurrences needs integration of more than one technique to give an accurate estimation of the groundwater potential. Since, the West Nile Delta suffered from the deterioration in groundwater due to the intensive exploitation which resulted in both groundwater depletion and salinity increase.

The second integration technique is the numerical simulation of groundwater Flow in the Nubian Sandstone Aquifer System (NSAS), Egypt. The well known NSAS has a large areal extent across the Western Desert of Egypt. The NSAS Basin covers the western desert and south-east Libya, north-east Chad, and North Sudan. The huge groundwater reserves contained in its various water-bearing formations, implies serious consideration towards optimizing the utilizations of these most vital natural water resources especially where extreme aridity prevails. The importance of the groundwater management of these aquifers has long been recognized. Its uppermost water-bearing formations have been sporadically used by nomads and settlers centuries. The western desert in Egypt contains 18 depressions covering NSAS. The habitant depressions like El Dakhla's depression will be the target for numerical simulation of the groundwater flow and management. Numerical models have been used to design optimal strategies. Moreover, aquifer management models that combine simulation with optimization help in understanding how social and economic forces interact with the water resources allocation. A simulation model is a tool to understand the physical behavior of an aquifer system. Moreover, Moharram et al. [1] used the combination of MODFLOW and optimization techniques to solve the optimization problem of groundwater management in El-Farafra Oasis, western desert with equality constraints.

The third integration technique is characterizing ecosystem services to human well-being in groundwater-dependent desert environments. In areas affected by drought and long-term dessication trends, groundwater can secure human life, providing a continual source of water for drinking, food production, human settlement and other economic activities [2]. Threats to the continued supply of groundwater for drinking and economic development are causing increasing concern for human security in the drier regions of the world, where dwindling supplies of groundwater are extracted at rates that exceed natural recharge [3]. Since regulatory measures are virtually ineffective to stop over-extraction under conditions of water scarcity [2], there is increasing interest in economic approaches to manage groundwater. Valuation techniques that capture the *in-situ* asset value of groundwater reserves to secure continued human and ecosystem needs are receiving increased attention as a means to quantify and weigh these tradeoffs [4]. Conceptually, ecosystem service assessment would appear an ideal frame through which to capture the full value of services provided by groundwater to secure human water needs with minimal energy and infrastructure for delivery or purification [5]. Management scenarios including different future configurations and valuation of ecosystem services may involve either qualitative or quantitative valuation [6]. This evolving approach to environmental economic decision-making has been the focus of a series of recent conceptual guides for decision-makers and also practical applications for integrated resource management in dryland contexts [7–9].

Groundwater Exploration, Quantity, Quality and Their Management

Eight different techniques are used to assess groundwater resources and their management.

First the geophysical sequential program was applied to determine the depth to the basement rocks and the subsurface structures affecting the groundwater aquifer geometry. The land-magnetic results confirm the results obtained from the DC resistivity survey. Also, it illustrated that the basement rocks have been affected by two predominant sets of faulting systems trending in NE and NW directions. Meanwhile, the correlation of the locations of these faults and the water resources show that the faults having an NE trend are primarily responsible for conduction and recharging of the water aquifer. Also, groundwater potentiality and its Quality in Siwa and Baris Oases was investigated. Hydrogeologically, the groundwater system that underlies Siwa Oasis consists of two productive aquifers. These are the Lower Cretaceous Nubian Sandstone and the Middle Miocene fractured limestone. Besides the Quaternary sand and clay uppermost layer is water-bearing due to water logging. In Baris area the Nubian Sandstone aquifer represents the main aquifer. It is classified into four productive zones and composed of sand and sandstone separated by water confining interbeds of clay and shale. In Siwa, the total salinity groundwater of the Nubian Sandstone aquifer, in terms of TDS, varies from 200 ppm to more than 1,500 ppm, while in Baris Oasis, the total salinity groundwater of the Nubian Sandstone aquifer is fresh and ranging from 353 to 694 mg/l and the chemical eater type is Sodium chloride.

The third is groundwater and characteristics of the tertiary-quaternary aquifer system west of Mallawi, upper Egypt. The development of the Upper Egypt governorates desert areas attracted the attention of the decision-makers and the investors which is achieved by the reclamation of more desert lands and building up new communities. This natural expansion for agricultural, industrial and civil activities in the Western Desert of Egypt necessitates more exploration activities for groundwater resources.

The fourth is the groundwater characterization and quality assessment in Nubian Sandstone Aquifer, Kharga Oasis, Egypt. This chapter evaluates the groundwater quality in Kharga Oasis area using hydrochemistry, GIS, and Factor analysis methods to determine the factors controlling groundwater quality in the study area. The obtained data were interpreted according to the WHO standard and the Egyptian water standards (EHCW). Geographic information system (GIS) is an important tool for plotting. GIS may help as a database system to generate distributed maps of ions concentration and assessment of groundwater quality assessment [10–12]. Statistical factor analysis has been widely used to study groundwater geochemistry [11–13]. Factor analysis is a multivariate statistical method, used to classify the main factors influencing the groundwater quality and water types in the study area. The factor analysis differentiates between dependent and independent variables. Factor analysis, generating the overall relationship between measured factors by rearranging them in a method that better explains the basic structure.

The fifth is the assessment of groundwater potential and quality for agricultural irrigation: case study—Bahariya Oasis, Egypt. Bahariya Oasis is one of the Egyptian oases that mainly rely on the groundwater for all purposes. Recently, Egypt plan is to increase the agricultural land by 20% through a mega project called "One and Half Million Fadden" using mostly groundwater. The main purpose of the present chapter is to re-evaluate the hydrogeological conditions of the groundwater aquifers (Nubian Sandstone Aquifer) in Bahariya Oasis through determination of their potentialities for optimum exploitation, using field measurements, analysis of water samples to investigate the local water exploitation effect on water level and quality. Measurements of the hydrogeological parameters through pumping tests were studied for the groundwater.

The sixth is hydrogeology and water quality of Moghra Aquifer, Western Desert, Egypt. A desert lake called Moghra occurs at the northeastern tip of the Qattara depression. It may represent a remnant of larger paleo lake fed by several channels flowed from the southeast during glacial and interglacial stages of post Miocene period, and Moghra aquifer water table with the ground surface [14]. The Egyptian government tries to achieve the greatest project in this area by planning a new agricultural and industrial society. The target of the project in Moghra is to cultivate 250,000 fadden and digging 1,352 water wells. Assessment of the Moghra aquifer system has great importance as it is the main source of agriculture in the region. The present chapter aims mainly to investigate the hydrogeological characteristics of the Moghra aquifer using geophysical methods and geochemical analysis of the groundwater. The water samples were collected from 140 productive wells to determine the physicochemical characteristics of the groundwater. Also, it is focused on the analysis and interpretation of the well logging data for 48 deep productive wells. Well logging was used to determine the variations in thickness for sedimentary deposits that affect the quality of groundwater.

The seventh approach is related to the transboundary groundwater management with its initiatives have so far focused on the generation of studies with the intent to share data between countries on the hydrological conditions of the aquifer. However, in additional to national coordination, there is growing recognition of the importance of local management institutions to groundwater governance, and the role of ecological as well as hydrogeological knowledge. The presents a discussion of the challenges and way forward highlights the opportunity for increased engagement of local institutions in groundwater management. They are the ones that can conduct or facilitate monitoring of groundwater conditions. They should also use ecological knowledge to implement land and water management practices that conserve the health of the land storing the groundwater reserves and enabling their replenishment. In order to succeed in this task, they must also secure immediate benefits for the dependent communities. This will enable them to ensure the sustainability of the multi-layered aquifer systems locally, and across the system as a whole.

The eighth chapter is devoted to assessing the quality of the groundwater of in the eastern part of the Nile Delta close to Ismailia canal where some of land were deserts and reclaimed. This is important in order to ensure that irrigation water should achieve the crop productivity, keep soil from degradation, and conserve soil properties to maintain the productivity of agricultural lands and sustainable food production as main requirements for eliminating hunger, achieving food security, improving nutrition and health, and ensuring sustainability for a long time to come.

The nineth is the overall assessment of groundwater resources in Egypt's Deserts. Groundwater in the Nile Basin mainly occurs in different compositions of rock systems with confined and unconfined conditions. Egypt needs to re-evaluate the role of groundwater to reallocate them as freshwater sources to meet the projected water shortage in the near future due to the high population growth of 2.05% annually, the increasing demand and the expected negative impacts after the operation of the Ethiopian Renaissance dam. It is necessary to plan to increase water resources and encourage investments aimed at treating and desalinating semi-saline groundwater in non-coastal provinces. Hydrogeological mapping of groundwater resources becomes one of the key tools for the development of groundwater. Integrating remote sensing and GIS have been successfully utilized in groundwater mapping.

Potential Use of Groundwater and Future Expansion

Three potential uses of groundwater are updated. The first issue is groundwater exploitation in Mega Projects: Egypt's 1.5 million Feddan Project. Mega construction projects represent a strategic option for achieving sustainable development objectives in many countries [15]. Government actions to construct infrastructural, industrial, educational, cultural, transportation, medical, and residential projects can aim to provide societies with their needs and fulfill their requirements [16]. The reclamation of 1.5-million-feddan in the desert is a mega project created by the government to boost development following the 25th of January revolution in 2011. This is the first stage of an ambitious plan to reclaim 4 million feddans from the desert. The aims are not only to increase the agricultural farmlands but also to create an integrated society in the new lands and to achieve comprehensive agro-industrial development. Moreover, the project also aims to reduce the need for food imports. Environmental impact assessment, monitoring and periodic evaluation is an important element of project management. This is required to enable the project managers and intended beneficiaries to ensure and enhance the sustainability of the project.

The second issue related to the optimum economic uses of precious costly groundwater in marginal and desert lands in Egypt. Egypt is a country suffering from water scarcity where the water share per capita/year does not exceed 600 m, and the total water shortage reached 42 billion cubic meters/year in the year 2018. Thus any newly discovered groundwater especially deep or spring ones will need economic scientific thinking and wise decision for its uses. The first logical choice for the new groundwater should be to reduce the current water gap; but in the case of adaptation with this water scarcity. The second choice will be to deliver it into the high-income sectors such as hotels, tourism, industry and finally agriculture sectors. The least economically feasible choice of using the valuable groundwater is to use it in agriculture sectors with its low income, where the return back of using a unit of water in the industrial sector reached 10 folds than the agricultural sector. From the view of physical geography; the desert is a land area that is hyper-arid because it receives little amounts of precipitation; usually in the form of rain but also as well as snow, mist or fog. Thus, it has little vegetation, and a low possibility to contain good non-saline groundwater in logical depth [17]. The term "food desert" was first used by Scotland in the 1990s [18] to refer to areas and people that lacked access to healthy, nutritious and affordable food. It became obvious that food deserts means food is distributed randomly and unfair across the landscape. The meaning of the term "Desert food" is unequal access to food and not applying the UN human right item of "right to food".

The third is the assessment of water resources in Egypt: Current status and future plan. Water resources management in Egypt is considered a strategic priority. The water use of the Nile is utilized for irrigation, water supply, hydro-electric power production, and protection of public health. In the upper basin states like the Democratic Republic of Congo, Rwanda, Burundi, Ethiopia, Eretria, Kenya, Uganda, Tanzania, and South Sudan, agriculture is considered one of the main economic activities. However, in the lower basin states like Egypt and Sudan who are also relying on agriculture, but in contrast with the upper states, their agriculture is largely irrigation-based. Agriculture has been for years the main contributor to economic growth in Egypt, contributing about 20% to Growth Domestic Product (GDP) and contributes close to 40% of the Egyptian total employment [19]. Around 55% of the population is dependent on this sector for their livelihood and agriculture needs. As mentioned in [14], the rural society's population, who represent a large portion of Egyptians depending on this sector, reached 44.88 million in 2010 (57% of the total population in Egypt). The most critical constraint facing Egypt is the shortage of water resources that grows with a very high rate accompanied by the deterioration in the water quality and the population growth which reached 98.8 million in 2019 compared to 33.9 million in 1970 and is expected to grow by 54.9% from 2010 to 2030 [20].

Conclusions

Throughout the course of the present book project, several conclusions drawn from this book were reached by the editorial teams. In addition to methodological ideas, the chapter draws important lessons from the book-cases, in specific the promising aspects of Groundwater in Egypt's Deserts. In order to improve sustainable water supply in Egypt, these findings are crucial. Based on the materials described in all chapters of this volume, the following findings could be indicated:

1. The condition in Wadi El-Farigh is a serious indicator of the mismanagement of water resources. Actually, the continuous decrease in the saturated thickness of the aquifer will lead to a tragedic problem in agricultural development in this area. Bearing in mind that the saturated thickness of the Miocene aquifer in Wadi El-Farigh was decreasing with time as clear from the three years 2003, 2012 and 2015, where the thickness is 83.9–175.3 m, 78.5–156.1 m and 73–153 m respectively. It is noticed that, in 2003 the dominant value of the saturated thickness of the Miocene aquifer is more than 120 m while in 2015; the dominant value is less than 120 m reaching in some areas to 73 m. One should give a warning for the applied water exploitation rates. The groundwater depletion ranges from 6 m to more than 30 m during this period of time and the dominant depletion is approximately ranged between 1.3 m/year and 1.7 m/year. Also, the quality of groundwater was deteriorated (salinization problems, low water quality for drinking only 33% are suitable for drinking purposes and high concentrations of some trace elements due to natural or anthropogenic sources). All these problems are considered a very good indicator of mismanagement of the groundwater resources in Wadi El-Farigh area.
2. The groundwater resources management model is proposed based on the combined use of the simulation groundwater flow model. In the proposed management model, MODFLOW is used as the simulation tool to model groundwater flow in NSAS in El-Dakhla depression. The performance of the proposed model is tested on the groundwater management problem (maximization of total pumping rate from an aquifer at steady state). The results show that the simulation model may be used to solve management problems in groundwater modeling. This model makes it feasible to solve the groundwater problems for three dimensions complex aquifer, complicated boundary conditions, steady and transient state. The performance of the proposed model when applied in NSAS in El-Dakhla depression to develop the optimal pumping rate under different scenarios establish the optimal pumping rate of 595,978 m^3/day under the current reclamation activity beside new reclamation of 3000 ha with a total number of 154 deep productive wells. Therefore, this model gives a feasible solution to optimize the number of wells according to the optimal pumping rate.
3. The ecosystem services framework can support decision-making through the valuation of the tradeoffs associated with different management options.

However, a range of challenges for the effective valuation of these services has been identified. Some are associated with the particular qualities of instability and insecurity that are inherent in many dryland contexts, others reflect the perennial difficulties that have previously been observed in relation to other frameworks for groundwater management and economic assessments. There are also limitations that are specific to the ecosystem services concept because it isolates the value of services to human well-being from the natural capital value of unused groundwater. In order not to encourage the over-extraction of groundwater, the case study presented has included characterization of the services and values associated with groundwater storage and buffering against water scarcity. In the selected case study, as most other dryland contexts, limitations on data availability and potentially high costs of additional research needed to fill these gaps for a full assessment of the value of ecosystem services associated with groundwater are inevitable. However, these should not be seen as a barrier to the development of the initial phase of an iterative assessment of groundwater-related ecosystem services.

4. The Moghra groundwater is brackish with TDS ranging from 3090 to 5350 ppm with an average of 4220 ppm, due to saline lakes towards the east, seepage of saltwater from the Mediterranean Sea, low recharge of groundwater and leaching of clay and shale lenses. The contents of major ions and TDS increase to the northwest. The Moghra aquifer is slightly alkaline with pH values range between 7.2 and 8.7 with an average of 8.0 due to the alkaline chemical composition of the aquifer rocks and the effect of the seawater. The Moghra groundwater is unsuitable for drinking and domestic purposes due to the high content of salinity Na, Cl, Ca, Mg, B, Fe, and Mn. It is also unsuitable for livestock and poultry. Under certain conditions of permeable soil, good drainage system, and using agricultural fertilizers, it may be suitable for the irrigation of salt-tolerant and semi-tolerant crops.
5. The groundwater flow in El Bahariya Oasis is from southwest to northeast confirming with the regional general flow, the hydraulic gradient of the groundwater flow is 0.8 m/km. The transmissivity values of the aquifers are 1,134 m^2/d for zone (S1), 471 m^2/d for zone (S2), and 1,023 m^2/d for zone (S3). The deep (S1) and upper (S3) zones are classified as highly potential aquifers (>500 m^2/d), while the middle zone is a moderate potential aquifer (5–50 m^2/d). The hydraulic conductivity ranges from 2.9 m/d for S1 to 5.1 m/d for S3, indicating very fine sand reservoir rock (1–5 m/d).
6. The annual amounts of water extraction were 18.5, 5, and 84.5 Mm^3/y from the deep zone (S1), middle zone (S2) and shallow zone (S3), respectively. The maximum expected drawdown will range between 3 and 26 m for the minimum water allocation after 25 years, which represents less than 5% of the saturated thickness. It will range between 4 and 32 m after 50 years. The groundwater potentiality in Siwa and Baris Oases are excellent. However, the groundwater in the Nubian Sandstone in these oases was recharged during the Pleistocene pluvial periods tens to hundreds of thousands of years ago. The recharge of the aquifer in modern times is negligible due to the sever aridity prevailing

the region. At Siwa Oasis, the fractured carbonate aquifer is recharged mainly from the underlying artesian Nubian Sandstone aquifer by upward leakage that finds its ways through the vertical and sub vertical open fractures. In Siwa, the total groundwater salinity of the Nubian Sandstone aquifer varies from 200 to more than 1,500 mg/l, while in Baris Oasis, the total salinity groundwater of the Nubian Sandstone aquifer is fresh and ranging from 353 to 694 mg/l and the chemical water type is Sodium-chloride.

7. The dominated aquifers in the west Mallawi area fall into two broad categories. The unconsolidated aquifers (granular) represented by Quaternary and Oligocene-Pleistocene. The consolidated fractured aquifer (fractured rock) represented by Eocene fractured limestone which considered as karst aquifer. The Quaternary aquifer in the flood plain is considered as highly productive where the aquifer has transmissivity values ranging between 6592 and 12,700 m^2/day and specific capacity ranging between 11.8 and 46.0 m^3/h/m. The water salinity of this aquifer ranges between 203.5 and 549.4 ppm increasing from east to west. The main source of recharge of this aquifer is the Nile water as indicated by the ion relationships. The Oligocene-Pleistocene aquifer is classified as moderately productive having transmissivity values ranging between 1256 and 6800 m^2/day and hydraulic conductivity ranges between 13.22 and 59.9 m/day with saturated thickness ranges between 91.5 and 113.5 m. This aquifer has specific capacity values between 6.9 and 15.9 m^3/h/m. The water salinity of this aquifer ranges between 719 and 801 ppm. rNa^+/rCl^- ratio, hypothetical salts combinations. The groundwater of this aquifer is in the progress stage of evolution than the other two aquifers (Quaternary and Eocene fractured) where it has sodium-chloride chemical type. For the Eocene fractured limestone aquifer, the aquifer is considered as a highly productive aquifer having transmissivity values range between 1083 and 14,054 m^2/day and attaining specific capacity ranges between 37.0 and 480 m3/h/m. Its water salinity ranges between 462 and 845 ppm. The groundwater of this aquifer is, chemically, in the middle stage of evolution as its groundwater is a mixture from the other two aquifers (Quaternary and the Oligocene-Pleistocene) due to the direct hydraulic connection between the Quaternary aquifer and the Eocene fractured limestone aquifer through faulting. The conclusion of this situation is that, the Eocene fractured limestone aquifer is recharged directly from the Quaternary aquifer and then acts as a rechargeable source for the Oligocene-Pleistocene aquifer.
8. The water samples were analyzed for various physiochemical attributes in Kharga Oasis. Groundwater is somewhat alkaline and largely various in chemical composition; e.g., electrical conductivity (EC) ranges from 276.5 to 2256.3 ppm. The analyzed groundwater samples have TDS values under the allowable limit. The plenty of the major ions is as follows: $Na^+ > Ca^{2+} > Ma^{2+} > K^+$ and $Cl^- > SO_4^{2-} > HCO_3^-$. The study area includes three major hydro chemical facies (Ca–Cl, Na–Cl, and mixed Ca–Mg–Cl). The results of WQI indicate that 88.2% are Excellent for drinking. Based on TDS, TH, Na%, PI%, MH%, SAR, it is clear that the majority of the collected groundwater samples

are suitable for irrigation. It is also observed that the Iron element value exceeds the standard value 0.5 ppm permitting for Drinking (88.2% of samples), and according to the water quality index, there is 4 wells are categorized as poor wells for drinking and 88.2%, 10.4% are Excellent and Good respectively. The hydrogeochemical analysis reveals that the groundwater of the study area is valid for Drinking and irrigation except for some wells which are considered Poor for Water (4 wells) according to the water quality index and PSS. Almost all the groundwater wells of the study exceeded the permissible limits of Fe (81.8%) and K (98.5%). The factor analysis results of the groundwater quality indicate three factors, which described 83.13% of the total variance. Factor 1 describes 56.34% of the total variance, has a strong positive loading on Na^+, Ca^{2+}, Mg^{2+}, Cl^-, SO_4^{2-}, EC, TDS, and TH and a negative loading on pH and HCO_3^-. Factor 2 describes 18.16% of the total variance, has a strong positive loading on HCO_3^- and a moderate positive loading on pH. Factor analysis 3 describes 8.63% of the total variance, has only strong positive loading on Mn. The factor analysis results show that the concentrations of pH and HCO_3^- do not contribute to other chemical parameters.

9. Groundwater is ranked second after the Nile River in the list of water resources in Egypt. There are aquifers that vary from shallow renewable groundwater aquifers to deep groundwater aquifers. Groundwater in Egypt is found in six main water basins, namely: the Nile aquifer, which is located under the Nile Valley and the Delta and is a renewable and vulnerable groundwater reservoir. The deep Nubian sandstone aquifer covering the Western Desert and is also located below the rest of the aquifers in other areas. It is non-renewable and characterized by water that is free of contaminants. Moreover, Fissured carbonate aquifer that spans about 50% of the area of Egypt in the eastern and western deserts and the Sinai Peninsula. Al Moghra and Hardrock aquifers, although water is poor in quantity and quality, it can be used in some saline-tolerant crops. Lastly, the coastal aquifer along the coastal strip of the Red Sea and the Mediterranean Sea. From recent studies, it is certain that Egypt desert is rich in groundwater and is enough to satisfy its demands. However, the government must develop good plans for its exploitation and management and to prepare long- and short-term plans with periodic review to monitor the behavior of the underground reservoir to follow the changes that may occur in terms of quantity and type and the need to use the latest techniques to ensure sustainable future projects that rely on groundwater.
10. In the countries that are suffering from water shortage especially in the arid and warm regions, the strategy of priority for using groundwater should be planned. The return back of the unit of water in the industrial sector is ranged between 10 and 30 folds than the agriculture sector. The industry sector uses a little of water which does not exceed 10% of the total water resources in the developing countries, but share in the GDP of the country 10 times more than the agriculture sectors which consume 85% of the total water resources. The labor income in the industrial sectors is higher as 2–10 times that of the labor income in the agriculture sector. Under specific conditions in some

developing countries that are obliged to use the newly discovered groundwater in agriculture to reduce the food insecurity and the lack of foreign currency; scientific and logical policy should be followed.

11. The government has already invested a lot of money and effort into the project. However, the challenges are significant and require continuous monitoring, assessment and revision of the implementation steps in addition to the cooperation of all of the factions participating in the project. Some of the main challenges of the project include: Megaprojects always need big finance. The harsh nature of the developing areas. Installation of solar systems and remote automatic control systems. Preventing violation of water regulation rules and land in the developing areas. Building agro-based communities in the newly reclaimed lands.
12. It is evident that is no physical or technological barrier nor any other reason why Egypt and its partners should not achieve the sustainable utilization of their aquifer systems. However, this will require a major concerted scientific and institutional effort that is far beyond the existing recommended international norms that have so far been prescribed for transboundary water management.
13. The groundwater in the east of the Nile Delta where the desert land was reclaimed is facing potential hazards of pollution by the irrigation return flow and seepage from drains especially in the desert areas in the eastern part of the study area, salinization, uncontrolled and unmanaged water abstraction and using sewage and drainage water for irrigation without adequate treatment.
14. Today, water resources are at the forefront of the world's concerns, and many countries have taken great interest in these resources to develop and maintain them, increase their efficiency and maximize their revenues, and use modern methods to meet the increasing demands for water resources in all sectors. The problem is how to rationalize the use of water in the current situation of increasing demands for water and the misuse of water resources. The National Water Plan in Egypt includes full strategic projects being implemented on the ground at an accelerated pace in order to overcome problems related to water shortage. As of 2019, there is a shortage of water by not less than 21 billion cubic meters that can be compensated by the reuse of the agricultural drainage water, shallow groundwater and the use of treated wastewater. However, by the year 2050, the situation will be much worse and the shortage is expected to reach 26 billion cubic meters. Many studies are presented in this work showing the different scenarios for the water balance in 2050. The population is expected to increase to about 178 million and hence, the net water share per capita will go to less than 400 m^3/yr. This will put Egypt in a severe water scarcity limit. Meanwhile, the domestic water uses will rise to meet this accelerating population growth and it is estimated that the total amount needed for domestic uses is going to reach about 12 to 13.5 billion cubic meters according to different assumptions. In addition to that, the share of each capita in cultivated land will decrease and the amount of water needed to satisfy the agricultural sector will barely be enough.

Recommendations

The capacity to adapt to future demands is the main element of groundwater sustainability. We contend that to accomplish this objective, sustainable groundwater needs integrated flexibility. The editorial teams observed certain aspects that could be explored for further enhancement throughout the course of this book project. Based on the results and conclusions of the contributors, this chapter provides a number of recommendations that provide suggestions for future scientists to exceed this book's scope.

1. Some recommendations related to groundwater occurrences in West Nile Delta are highlighted. The drilling should penetrate the full thickness of the Miocene aquifer to increase the saturated thickness and, hence, the aquifer transmissivity through the reaching of the drilling to the top surface of the basaltic sheet, which as the base of the Miocene aquifer. The well design should be constructed I a proper way to isolate the salty layers from the productive zone of the aquifer. The spacing between each two successive drilled wells should be not less than 500 m to avoid well interference, and hence, the water quality degradation. The whole area of the west Nile Delta should be having monitoring networks of observation wells. Well drilling in the area, should be strongly controlled and licensed. The modern irrigation systems should be applied. The crop pattern in the study area should be changed to avoid the crops of high consumptive use of water.
2. Inclusive natural resource accounting is an important tool that can help to support the assessment of ecosystem services and the underlying natural capital. Sensitivity analysis and distributional analysis through an iterative, participatory and cross-sectoral assessment approach can help to overcome other challenges in ecosystem service assessment. Although no assessment will ever be exhaustive enough to capture all of the value provided by groundwater to human well-being, there is clearly merit in this framework to begin the pressing task to characterize and value those that most urgently require decision-makers' attention. The iterative approach offers options for groundwater managers in Egypt to coordinate across sectors of government and overcome all assessment challenges in a progressive manner.
3. It is highly recommended to set up a network of monitoring wells in El-Dakhla depression to monitor the groundwater levels required for checking the model results besides applying Genetic Algorithm techniques as an optimization tool.
4. Generally, it is recommended to increase the drilling depth in the Eocene aquifer in order to maximize, as much as possible, the groundwater potentials of this karstic phenomenon. Since the study area is considered as a new development area, it is highly recommended to construct a mathematical model in order to predict the future situation through the expected heavy pumping programs by investors. The management should take into consideration the safe distance between drilled wells as well as the pumping rates. Since the fractured Eocene limestone has good quality and quantity of groundwater, exploitation

of water should be focused on this aquifer which directly will increase the reclamation of new areas. The modern irrigation system should be applied. The coordination between the Ministry of Irrigation & Water Resources and the Ministry of Agriculture and Land Reclamation is very important to make the required management on the groundwater exploitation in the study area.

5. The following recommendations are highlighted for the Moghra groundwater: Monitoring the water quality to discover the changes in water types by continuous uses. Converting two water wells with different depths (125 and 200 m) to be test wells for monitoring the groundwater level. Carrying out more chemical analysis and trace elements to archive the optimum uses of groundwater in the study area. Reviewing the development plan of water uses according to it is chemical properties. Using modern irrigation systems for irrigation of salt-tolerant and semi-tolerant crops.
6. The groundwater exploration in the area of Gabal Kamel needs detailed survey and acquisition. It is recommended that The NE Jabal Kamel area is suitable for drilling water wells for some human activities and drinking purposes for military camps.
7. Some further suggestions and recommendations should be taken into consideration to maintain the groundwater quality in the study area: Controlling the groundwater overpumping to avoid salinity hazards. Continuous monitoring of groundwater quality to avoid degradation. To overcome the problem of iron and manganese in groundwater, aeration method (or using compact unites in case of low concentration for domestic purposes).
8. Most of the aquifer zones in El Bahariya Oasis have freshwater averaging 286 mg/l. The extension of using the groundwater should be directed to industries, truism, and development services more than agriculture to achieve sustainable development. It is strongly recommended to monitor the productive water wells in terms of water level and quality.
9. In Siwa and Baris oases, Recommendations concerning mitigation of water logging problem through improvement of irrigation and drainage. The increase of lifting stations' efficiency is highly recommended since it will decrease the groundwater demand.
10. The priority of using groundwater in agriculture should be to the low temperature and seasonal rainfall area. Thus the new greening desert and agriculture extension projects in Egypt should be in the North and the Mediterranean coast of North Egypt which have a low temperature than south Egypt by 14 centigrade. The south, warm and low humidity region should have a priority for industry investments to reduce poverty and maximize the return back of water unite.
11. It seems that Egypt should further develop, articulate and evaluate its strategies to reshape groundwater management, sustain the balance between groundwater extraction and recharge and conserve land and water quality. This involves building local capacities to manage land quality and recharge patterns in the surface layers, while also studying and conserving the reserves that are stored in the deeper layers.

12. The best available scientific capacities and expertise must be applied to enable the management of the groundwater resources, conserve the strategic national reserves of non-renewable fossil water and ensure national security for the long- and short-term. The establishment of the environmental monitoring system should be a high priority both for the groundwater resource users and for the government. This should be integrated with participatory planning for long-term sustainability, environmental risk assessment, and disaster preparedness.
13. To ensure that the groundwater in suitable for irrigation, basic tests should be done. To improve the quality of the irrigation water in the eastern of the Nile Delta, which originated from the groundwater, it is essential to provide sanitation to rural area to prevent pollution of the groundwater. Therefore, enhancement of the sanitary drainage network in the urban and rural areas. Prohibition and conviction of using sewage and drainage water for irrigation without adequate treatment according to the soil conditions and the types of plants. Periodical monitoring for both surface and groundwater quality. Controlling salinization by artificial recharging.
14. There are many recommendations for reviewing our water budget and developing our water policies. For example, it is necessary to have a quick stop to re-evaluate what can be provided of water for the future under the current and expected conditions in the future in terms of financial and water problems, including problems of pollution, deterioration of water networks, the problems of growing rice and sugarcane, the fragmentation of agricultural property and its impact on the efficiency of water use, and the situation with the Nile Basin countries. As well as, to reassess the efforts to rehabilitate water networks for irrigation and drainage, reevaluate the surface irrigation development projects and its long-term effects, develop a tight system to collect data for the actual groundwater levels, more accurate data on the quantity of rainfall especially in Sinai Peninsula and the eastern desert and precise estimate of losses within the water networks and increase public awareness regarding water problems.

Acknowledgements Abdelazim Negm and Ahmed Elkhouly acknowledge the support of the Science, Technology, and Innovation Authority (STIFA) of Egypt in the framework of the grant no. 30771 for the project titled "a novel standalone solar-driven agriculture greenhouse—desalination system: that grows its energy and irrigation water" via the Newton-Mosharafa funding scheme.

References

1. Moharram SH, Gad MI, Saafan TA, Khalaf S (2011) Optimal groundwater management using genetic algorithm in El-Farafra Oasis, Western Desert, Egypt. Water Res Manage. https://doi.org/10.1007/s11269-011-9865-3
2. WWAP (2014) The United Nations world water development report 2014: water and energy. Paris, UNESCO. United Nations World Water Assessment Programme, UNESCO, Paris, p 230

3. UNDP (2013) Water governance in the Arab region—managing scarcity and securing the future Regional Bureau for Arab States (RBAS), Republic of Lebanon, p 182
4. UN (2014) System of Environmental-Economic Accounting 2012 Central Framework UN, EU, FAO, IMF, OECD, WB, p 378
5. Tuinstra J, Jv W (2014) Ecosystem services in sustainable groundwater management. Sci Total Environ 485–486:798–803
6. Henrichs T, Zurek M, Eickhout B, Kasper Kok, Raudsepp-Hearne C, Ribeiro T, Vuuren Dv, Volkery A (2010) Chapter 5 Scenario development and analysis for forward-looking ecosystem assessments. In: Ash N, Blanco H, Brown C, Garcia K, Henrichs T, Lucas N, Raudsepp-Hearne C, Simpson RD, Scholes R, Tomich TP, Vira B, Zurek M (eds) Ecosystems and human well-being a manual for assessment practitioners, pp 151–220
7. Ahmed MT (2013) Ecosystem services. In: UNEP (ed) Arab millennium ecosystem assessment synthesis report, pp 58–84
8. Favretto N, Stringer LC, Dougill AJ, Perkins JS, Akanyang L, Dallimer M, Atlhopheng JR, Mulale K (2014) Assessing the socio-economic and environmental dimensions of land degradation: a case study of Botswana's Kalahari. Report for the Economics of Land Degradation Initiative ELD Initiative, Leeds, UK, p 28
9. Jabarin A (2013) Environmental economic valuation of the HIMA system: the case of Zarqa River Basin–Jordan. Natural Resource Economic Valuations IUCN, Jordan, p 36
10. El-Rawy M, Ismail E, Abdalla O (2019) Assessment of groundwater quality using GIS, hydrogeochemsitry, and factor statistical analysis in Qena governorate, Egypt. Desalin Water Treat 1–16. https://doi.org/10.5004/dwt.2019.24423
11. Fathi H, El-Rawy M (2018) GIS-based evaluation of water quality index for groundwater resources nearby wastewater treatment plants, Egypt. Poll Res 37:105–116
12. Ismail E, El-Rawy M (2018) Assessment of groundwater quality in West Sohag, Egypt. Desalin Water Treat 123:101–108. https://doi.org/10.5004/dwt.2018.22687
13. Khan SD, Fathy MS, Abdelazeem M (2014) Remote sensing and geophysical investigations of Moghra Lake in the Qattara Depression, Western Desert, Egypt 207 (15):10– 22. https://doi.org/10.1016/j.geomorph.2013.10.023
14. El Sayed LM (2012). Determining an optimum cropping pattern for Egypt. Doctoral dissertation, The American University in Cairo, p 233
15. Kudoda AM, Abdalla OAE (2015) Hydrochemical characterization of the main aquifers in Khartoum, the capital city of Sudan. Environ Earth Sci 74:4771–4786. https://doi.org/10.1007/s12665-015-4464-x
16. Othman A (2013) Challenges of mega construction projects in developing countries. Organisat Technol Manage Construct (Int J 730–746). https://doi.org/10.5592/otmcj.2013.1.10. https://hrcak.srce.hr/file/154726. Web. 11 Apr. 2019
17. CE/SC 10110-2011: Planet earth: earth portrait of a planet, 5th edn. https://www3.nd.edu/~cneal/PlanetEarth/Chapt-21-Marshak.pdf. Accessed 1 June 2018
18. U.S. Department of Agriculture (2014) Food deserts. http://apps.ams.usda.gov/fooddeserts/foodDeserts.aspx
19. MWRI, Ministry of Water Resources and Irrigation (Egypt) (2014) Water scarcity in Egypt: the urgent need for regional cooperation among the Nile Basin Countries. Report no. 6p
20. CAPMAS E (2012) Central Agency for Public Mobilization and Statistics. Study of water resources and rational use in Egypt

Printed by Printforce, the Netherlands